The
Retroviridae
Volume 1

THE VIRUSES

Series Editors
HEINZ FRAENKEL-CONRAT, *University of California*
Berkeley, California

ROBERT R. WAGNER, *University of Virginia School of Medicine*
Charlottesville, Virginia

THE VIRUSES: Catalogue, Characterization, and Classification
Heinz Fraenkel-Conrat

THE ADENOVIRUSES
Edited by Harold S. Ginsberg

THE ARENAVIRIDAE
Edited by Maria S. Salvato

THE BACTERIOPHAGES
Volumes 1 and 2 • Edited by Richard Calendar

THE HERPESVIRUSES
Volumes 1–3 • Edited by Bernard Roizman
Volume 4 • Edited by Bernard Roizman and Carlos Lopez

THE INFLUENZA VIRUSES
Edited by Robert M. Krug

THE PAPOVAVIRIDAE
Volume 1 • Edited by Norman P. Salzman
Volume 2 • Edited by Norman P. Salzman and Peter M. Howley

THE PARAMYXOVIRUSES
Edited by David W. Kingsbury

THE PARVOVIRUSES
Edited by Kenneth I. Berns

THE PLANT VIRUSES
Volume 1 • Edited by R. I. B. Francki
Volume 2 • Edited by M. H. V. Van Regenmortel and Heinz Fraenkel-Conrat
Volume 3 • Edited by Renate Koenig
Volume 4 • Edited by R. G. Milne

THE REOVIRIDAE
Edited by Wolfgang K. Joklik

THE RETROVIRIDAE
Volume 1 • Edited by Jay A. Levy

THE RHABDOVIRUSES
Edited by Robert R. Wagner

THE TOGAVIRIDAE AND FLAVIVIRIDAE
Edited by Sondra Schlesinger and Milton J. Schlesinger

THE VIROIDS
Edited by T. O. Diener

The Retroviridae

Volume 1

Edited by
JAY A. LEVY
University of California
San Francisco, California

PLENUM PRESS • NEW YORK AND LONDON

Library of Congress Cataloging in Publication Data

The Retroviridae / edited by Jay A. Levy.
    p.   cm. – (Viruses)
  Includes bibliographical references and index.
  ISBN 0-306-44074-1 (v. 1)
  1. Retroviruses. I. Levy, Jay A. II. Series.
QR414.5.R48 1992   92-26459
576′.6484 – dc20   CIP

ISBN 0-306-44074-1

© 1992 Plenum Press, New York
A Division of Plenum Publishing Corporation
233 Spring Street, New York, N.Y. 10013

Printed in the United States of America

Contributors

John M. Coffin, Department of Molecular Biology and Microbiology, Tufts University School of Medicine, Boston, Massachusetts 02111

David J. Garfinkel, National Cancer Institute, Frederick Cancer Research and Development Center, ABL-Basic Research Program, Frederick, Maryland 21701-1201

Christine A. Kozak, Laboratory of Molecular Microbiology, National Institute of Allergy and Infectious Diseases, National Institutes of Health, Bethesda, Maryland 20892

Nancy J. Leung, Department of Pathology, School of Medicine, University of California, Davis, California 95616

Paul A. Luciw, Department of Pathology, School of Medicine, University of California, Davis, California 95616

Gerald Myers, Theoretical Biology and Biophysics, Los Alamos National Laboratory, Los Alamos, New Mexico 87545

George N. Pavlakis, Human Retrovirus Section, National Cancer Institute, Frederick Cancer Research and Development Center, ABL-Basic Research Program, Frederick, Maryland 21701-1201

Laurence N. Payne, AFRC Institute for Animal Health, Compton Laboratory, Compton, Newbury, Berkshire, RG16 0NN, England

Sandra Ruscetti, Laboratory of Molecular Oncology, National Cancer Institute, Frederick Cancer Research and Development Center, Frederick, Maryland 21702-1201

Howard M. Temin, McArdle Laboratory, University of Wisconsin, Madison, Wisconsin 53706

Preface

It should be no surprise that the field of retroviruses has generated such tremendous interest in recent years. Not only have infections by these agents in humans created clinical and public health problems worldwide, but their recognition in animal species such as cows, horses, goats, and other farm animals, as well as pets (e.g., cats), has had both commercial and emotional ramifications. This first volume on the Retroviridae introduces a series of volumes that reviews in depth the biologic, molecular, immunologic, and pathologic features of this fascinating virus family.

Retroviruses were among the first viruses identified in nature. Equine infectious anemia virus (discussed in Volume 2 of this series) was discovered in 1904 as a filtrable agent but was only recently recognized as a member of the lentivirus group of retroviruses. Moreover, as early as 1908 and 1911, the retrovirus group of *Oncovirinae* was discovered with the identification of the avian leukosis and sarcoma viruses (Chapter 6). Since the acceptance of these viruses as agents in malignancy (it took nearly 50 years), the research on retroviruses in cancers, particularly in the murine system, has been impressive. Many insights into human cancer have come from studies of the avian and murine retroviral species (Chapters 6 and 7).

The RNA nature of these viruses was determined relatively early with genomic analyses, and in the 1950s and 1960s these agents were generally termed RNA tumor viruses. Then, with the discovery in 1970 of reverse transcriptase (Chapters 1 and 5), tumor viruses became classified as retroviruses (subfamily *Oncovirinae*) along with, surprisingly, other viruses that make up this somewhat diverse family. These included the foamy virus (*Spumavirinae*), and the visna and equine infectious anemia viruses (*Lentivirinae*). Thus, after 1970, retroviruses were noted to be associated with many different diseases.

The biologic properties of these viruses have now been linked to well-defined molecular events in the replicative cycle (Chapter 5). Their existence as infectious agents in many species has been recognized, as well as their presence as retrovirus-like elements in lower organisms (Chapters 2 and 4). Whether the latter agents represent evolutionary progenitors is not known, and several theories exist on their origin (Chapter 1). The ability of retroviruses to evolve, as described in Chapters 1 and 3, does indicate the extent to which this family of viruses can adapt or change over time in hosts from various species.

As these volumes on the Retroviridae will reveal, these viruses share a common enzyme and genomic structure. Nevertheless, great diversity exists among the seven recently defined retroviral groups that were once simply classified as the three subfamilies noted above (Chapter 2). The study of these viruses has enabled us to look at new molecular events that could have relevance to eukaryotic and prokaryotic processes (Chapter 5 of the present volume; also see Volume 2). Transactivation, splicing, and the recent work on RNA-binding proteins are just a few examples of the potential riches to be gained in research on the Retroviridae. Moreover, studies of these viruses have indicated, almost as a paradigm, how a virus and a host interact to influence pathogenesis. Humoral and cellular immune responses form the basis for control or enhancement of disease (Volume 3). And, in some instances, retroviruses may be involved in normal developmental processes (Volume 3).

The ongoing challenges to the understanding of these naturally occurring agents is evident as we examine the variety of examples in this diverse family of viruses. Clearly we have learned and will continue to learn a great deal about viruses in general, and living organisms themselves, through the study of the Retroviridae.

Jay A. Levy

University of California
School of Medicine
San Francisco, CA

Contents

Chapter 3

Evolutionary Potential of Complex Retroviruses

Gerald Myers and George N. Pavlakis

Chapter 4

Retroelements in Microorganisms

David J. Garfinkel

Chapter 5

Mechanisms of Retrovirus Replication

Paul A. Luciw and Nancy J. Leung

Chapter 6

Biology of Avian Retroviruses

Laurence N. Payne

Chapter 7

Retroviruses in Rodents

Christine A. Kozak and Sandra Ruscetti

CHAPTER 1

Origin and General Nature of Retroviruses

HOWARD M. TEMIN

I. SIGNIFICANCE OF RETROVIRUSES

Retroviruses are a family of animal viruses with RNA as their genetic material in virus particles and DNA as their genetic material in cells. Retroviruses have some properties like those of other viruses and also some properties like those of cellular movable genetic elements. When considered as viruses, retroviruses share some properties with DNA viruses and other properties with RNA viruses. The fundamental pattern of replication of retroviruses involving reverse transcription and transcription is also shared by some other animal and plant DNA viruses, for example, hepadnaviruses and caulimoviruses, and also with many cellular movable genetic elements, for example, retrotransposons and retroposons (Temin, 1989a). The retrotransposons, in fact, share with retroviruses details of genomic organization, for example, long terminal repeats, and replication strategy, for example, "jumping" during reverse transcription.

All retroviruses and their relatives are parasites of the usual DNA-based organisms, and they are very successful parasites. For example, over 10% of the human genome is composed of sequences resulting from reverse transcription (these sequences are called retrosequences),

HOWARD M. TEMIN • McArdle Laboratory, University of Wisconsin, Madison, Wisconsin 53706.

The Retroviridae, Volume 1, edited by Jay A. Levy. Plenum Press, New York, 1992.

and the retrovirus HIV (human immunodeficiency virus) is an increasingly important human pathogen (Temin, 1985; Chin *et al.*, 1990).

In addition to their role in human diseases, for example, acquired immune deficiency syndrome (AIDS), adult T-cell leukemia, and tropical spastic paraparesis, retroviruses are important in biotechnology and molecular biology in general. Reverse transcriptase is an important reagent for making cDNAs from mRNAs of important proteins, and it has been used in the cloning of several medically important proteins. Vectors made from modified retroviruses have already been used in human experimentation to mark tumor-infiltrating lymphocytes, and there are protocols proposing the use of modified retroviruses for somatic gene therapy of some human genetic diseases—for example, adenosine deaminase deficiency—and for cancer therapy (Temin, 1989b). Some laboratories are also using retroviruses to make transgenic farm animals.

In molecular biology, the study of retroviruses led to the discovery of new levels of control of gene expression by the retrovirus-encoded regulatory proteins *tat* and *rev/rex*, and the analysis of HIV and AIDS is resulting in new approaches to understanding virus structure and the development of antiviral therapies (see Chapter 5; Wong-Staal, 1990).

Bacterial retrons also code for reverse transcriptases, and eukaryotic telomerases are also reverse transcriptases (Inouye *et al.*, 1989; Yu *et al.*, 1990). Both retrons and telomerases also have important RNA components, the RNA of msDNA of retrons and the template in telomerase. These RNAs may be a relic descendant of the first reverse transcriptase, which probably was a ribozyme, and the ancestor of the associated RNA component (Blackburn, 1990). Later the protein component (reverse transcriptase) may have evolved along with or from other (protein) polymerases. There are some amino acid similarities among all polymerases, suggesting an evolutionary homology among them (Poch *et al.*, 1989).

The existence of a reverse transcriptase as a telomerase demonstrates that some reverse transcriptases have an important normal cellular function. However, so far, in spite of numerous investigations, there is still little evidence that other reverse transcriptases have a role in normal somatic cellular processes (see Brosius, 1991; Gardner *et al.*, 1991; Inouye and Inouye, 1991). The primary role of reverse transcription seems to be as part of the parasitic retroelement life cycle, and its effect on host organisms is primarily at an evolutionary level.

If an RNA world was a precursor to the present DNA-based world, the transfer of information from RNA to DNA must have involved reverse transcriptase activity, perhaps first by ribozymes and then by protein enzymes. The evolution of proteins with more flexible biochemical properties than ribozymes would have resulted in replacement of the reverse transcriptase ribozyme with the reverse transcriptase protein and the evolution of new genetic systems (retrons, retroposons, retro-

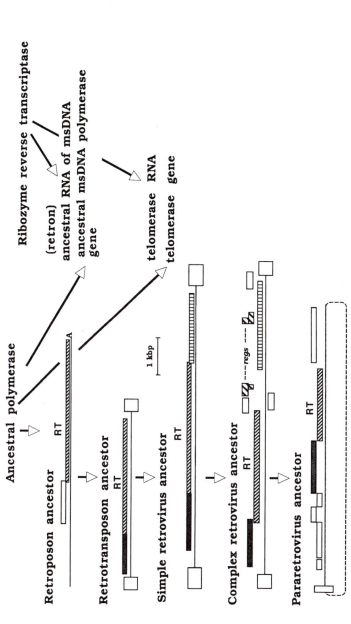

FIGURE 1. Possible evolutionary scheme for different retroelements and reverse transcriptase. Straight lines are element sequences; RT is the *pol* gene coding for reverse transcriptase; open boxes are long terminal repeats; dotted rectangles are *gag*-like genes; lined rectangles are *env*-like genes; open rectangles are other genes; *regs* are genes for regulatory proteins. It is proposed that an ancestral ribozyme reverse transcriptase was ancestral to the RNAs of retrons and telomerase. The gene for an ancestral polymerase, which might have been a reverse transcriptase, was the ancestor for non-LTR retroposons, the ancestors of present-day L1 elements. This ancestor was in turn the ancestor for the ancestor of present-day retrotransposons, like Ty1, which have LTRs, but do not have *env* genes. In turn, the ancestor of simple retroviruses evolved from the retrotransposon ancestor by gaining an *env* gene. Further evolution to more complex retroviruses and then to pararetroviruses occurred by the acquisition of yet other genes.

transposons, and retroviruses) (Fig. 1) (Chapter 4). (I am not aware of a consensus as to whether protein evolved before or after DNA in these evolutionary reconstructions.) Although comparisons of the nucleotide sequences of these retroelements cannot give an order of their appearance in evolution, the simplest hypothesis is that the simpler retroelements appeared first. The alternative hypothesis that the more complex retroelements appeared first faces the great difficulty of explaining the origin of the more complex retroelements. Thus, in agreement with the old protovirus hypothesis, it appears that retroviruses evolved from cellular retroelements (Temin, 1970). The associated RNA of teleromase, the tRNA primer of retroviruses, and also the tRNA 3' end of some plant viruses, may be a relic of their origin in an RNA world (Weiner and Maizels, 1987; Blackburn, 1990).

II. BRIEF OVERVIEW OF RETROVIRUSES

Viruses are nucleic acid molecules that can enter cells, replicate in them, and code for proteins capable of forming protective shells around the viral nucleic acid. Viruses are obligate parasites; that is, they only replicate in living cells. Thus, viruses are not small cells, and they are not plasmids, which are intracellular only.

All viruses have infection cycles with common steps, which include attachment to specific receptors, entrance into cells, synthesis of proteins and nucleic acids, assembly of progeny virus, and release from cells.

There are many different kinds of viruses. They are divided into families and classified according to the presence or absence of an envelope, RNA or DNA genome, single- or double-stranded nucleic acid, and size.

Retroviruses are RNA viruses that replicate through a DNA intermediate. Thus, retroviruses have both RNA and DNA genomes. Retrovirus virions are enveloped and contain a dimer RNA (two identical single-stranded RNA molecules), with each molecule about 10 kilobases (kb).

Retroviruses had previously been classified into three subfamilies: *Oncovirinae*, *Lentivirinae*, and *Spumavirinae*. But newer classifications, based on nucleotide sequence organization and amino acid sequence comparisons, now identify seven genera, among them the avian leukosis virus-like (RSV-like, for Rous sarcoma virus-like), murine leukemia virus-like (MLV-like), human T-cell leukemia virus-1-like (HTLV-like), lentiretroviruses, and spumaretroviruses (Coffin,

Chapter 2; Doolittle *et al.*, 1989; Xiong and Eickbush, 1988). Lentiretroviruses include HIV, the virus which causes AIDS. Spumaretroviruses
include human foamy virus and have been less studied than the other
groups (Flügel, 1991; Volume 2).

Retrovirus virions have a distinctive morphology, which was also
the basis for a previous classification (Chapter 2). A-type particles are
only found inside cells and have a clear center surrounded by a shell.
They are divided into intracytoplasmic and intracisternal A particles.
B-type particles have doughnut-shaped cores (like intracytoplasmic A
particles) at budding and eccentrically located cores within the budded
enveloped particles. C-type particles have crescent-shaped cores at budding and a centrally located core in virions. D-type particles have more
elongate, electron-dense cores in virions (Weiss *et al.*, 1982). Recent
work suggests that some of these differences may involve changes in
only a few amino acids (Rhee and Hunter, 1990).

The genomes of all retroviruses have some common features. The
viral DNA is bounded by long terminal repeats (LTRs) which contain
enhancer, promoter, and 3' RNA processing sequences. Viral proteins
are expressed either from full-length or from spliced mRNAs. Figure 2
illustrates the RNA and DNA genomes of a retrovirus of the RSV- or
MLV-like retrovirus subfamilies. HTLV-1-like retroviruses, lentiretroviruses, and spumaretroviruses have these basic genes, but they also
have additional genes between *pol* and *env*, or *env* and the 3' LTR, or
both. These additional genes seem to code primarily for regulatory
proteins.

The infection cycle of all retroviruses contains certain similar pro-

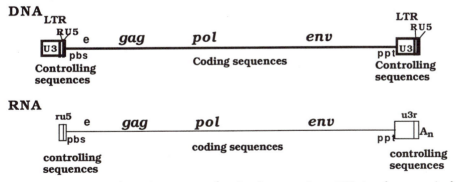

FIGURE 2. RNA and DNA genomes of a simpler retrovirus. LTR is a long terminal
repeat; U3 is a unique 3' region; R is a repeat; U5 is a unique 5' region; *gag, pol,* and *env* are
genes coding for virion proteins; pbs is a primer binding site; e is an encapsidation sequence; ppt is a polypurine track.

cesses, although HTLV-like, lenti-, and spumaretroviruses have additional steps in their replicative pathway (Coffin, 1990; Cann and Chen, 1990; Wong-Staal, 1990). Figure 3 shows the steps common to all retroviruses such as the RSV-like and MLV-like viruses. HTLV-like, lenti-, and spumaretroviruses have more complex infection cycles, with the initial formation of regulatory proteins from multiply spliced mRNAs, preceding the synthesis of genomic RNA and *gag–pol* and *env* mRNAs, which are unspliced or singly spliced (Fig. 4). The later stage (Fig. 4b) is similar to the later stages of replication of RSV-like and MLV-like retroviruses.

Retroviruses that replicate in permissive cells in the absence of any other retrovirus are designated replication-competent. However, many retroviruses, especially most retroviruses containing oncogenes and retrovirus vectors, lack essential protein-coding sequences and can only replicate in the presence of a replication-competent retrovirus, in this context also called helper virus. Such retroviruses are designated replication-defective. (In addition, replication-defective retroviruses can be propagated in specially engineered cells called helper or packaging cells.)

Since retroviruses integrate, and the provirus is replicated along with the cell genes, there are retrovirus proviruses in the germ line. These proviruses are termed endogenous, in contrast to the more usual viruses, which are termed exogenous. Endogenous retroviruses can be replication-competent or replication-defective. In addition, since endogenous retroviruses persist as proviruses, they often lack essential *cis*-acting sequences and so are unable to replicate as a virus even in the presence of a helper virus. Thus, they often are more profoundly defective than the replication-defective retroviruses. Such proviruses without essential *cis*-acting sequences may be referred to as defective proviruses.

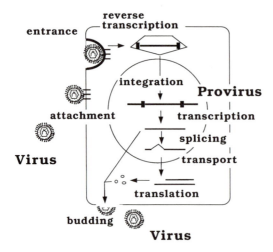

FIGURE 3. Infection cycle of a simpler retrovirus.

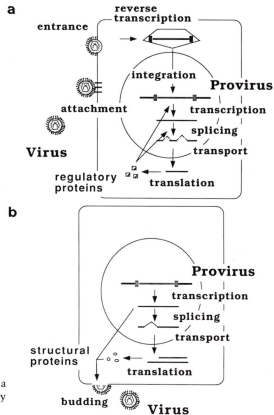

FIGURE 4. Infection cycle of a complex retrovirus. (a) The early stages; (b) the later stages.

III. SOME IMPORTANT DATES IN THE STUDY OF RETROVIRUSES

See Gross (1970) and Witkowski (1990) for additional references.

1904 Equine infectious anemia virus (EIAV) was the first retrovirus described (Vallee and Carre, 1904). EIAV is not oncogenic, and thus, EIAV was not classified with the RNA tumor viruses, as retroviruses were called before 1970.

1908 Avian myeloblastosis and avian erythroblastosis viruses (precursors to present-day avian myeloblastosis and erythroblastosis viruses) were the first oncogenic retroviruses to be described, but the diseases they induced were not accepted as leukemias, and the leukemias were not accepted as real cancers. Chicken leukemias were originally called leukoses.

1911 Rous sarcoma virus was isolated.

1933 Avian lymphomatosis virus was isolated.

1936 Mouse mammary tumor virus was described. Initially, a genetic factor resulting in an increased incidence of mammary tumors in certain mice was discovered. Then an "agent" was described which turned out later to be a retrovirus.

1938 Chorioallantoic membrane assay for RSV was the first focal assay for a retrovirus.

1951 Murine leukemia virus was isolated from a strain of mice selected for high frequency of leukemias. Vertical transmission of retroviruses was first described.

1958 Quantitative focus assay for transformation in cell culture was described based on transformation of cells in culture by RSV.

1960 Provirus hypothesis was first proposed.

Early 1960s Several other murine leukemia and sarcoma viruses were isolated.

Late 1960s Endogenous murine and avian viruses were discovered.

1970 Reverse transcriptase was discovered.

1972 Infectious retrovirus DNA was described.

1978 LTRs were described.

1980 Jumping scheme for reverse transcription was proposed.

1980 Human T-cell leukemia virus-1, the first oncogenic human retrovirus, was isolated.

1983–1984 HIV-1 was isolated and demonstrated to be the cause of AIDS.

1989 First use of retrovirus vectors for genetic engineering in humans (Rosenberg *et al.*, 1990).

IV. RETROVIRUSES AND DISEASE

Retroviruses were first described because of their role in causing diseases, especially cancers. However, most retrovirus infections do not cause disease. (It is generally true that most infections by viruses do not cause disease.) In the past, retroviruses were classified as RNA tumor viruses or oncoviruses if they caused cancer or were related to viruses causing cancer, and they were classified as slow viruses or lentiviruses if they caused diseases other than cancer and had a long latent period for disease development.

The RSV- and MLV-like retroviruses have been separated into three groups: highly oncogenic or acutely transforming retroviruses, weakly oncogenic or slowly transforming retroviruses (also called chronic leukemia viruses), and nononcogenic retroviruses (Chapters 6 and 7). Highly oncogenic or acutely transforming retroviruses cause cancer rapidly and efficiently (that is, within days or weeks), and only a small number of infectious virus particles are necessary to cause the cancer.

Highly oncogenic RSV- or MLV-like retroviruses have been isolated from chickens, a turkey, mice, rats, cats, and a woolly monkey. They cause sarcomas and different types of leukemias. With the exception of some strains of RSV, all highly oncogenic retroviruses are replication-defective, and virus is only produced in the presence of a helper virus. Some examples of highly oncogenic retroviruses are listed in Table I.

Weakly oncogenic or slowly transforming RSV- or MLV-like retroviruses (also called chronic leukemia viruses) also cause cancer (leukemias and mammary carcinomas), but they do so less rapidly and less efficiently than do highly oncogenic RSV- or MLV-like retroviruses. Weakly oncogenic RSV- or MLV-like retroviruses exist in nonlaboratory populations, whereas highly oncogenic RSV- or MLV-like retroviruses exist only as laboratory populations. Some weakly oncogenic MLV-like retroviruses also cause anemia and immune depression; in fact, the syndromes called mouse or feline AIDS are caused by these types of retroviruses (Hoover *et al.*, 1987).

The nucleotide sequences of nononcogenic RSV- or MLV-like retroviruses only differ slightly from the nucleotide sequences of weakly oncogenic retroviruses. These differences relate primarily to the characteristics of the envelope proteins and the enhancer sequences in the viral LTRs. Nononcogenic RSV- or MLV-like retroviruses can become endogenous; that is, part of the germ line. In many cases, these endogenous viruses are replication-defective. However, in other cases they are replication-competent, and infectious virus can be produced from the germ-line provirus. In some cases this activated virus can undergo genetic changes and become a weakly oncogenic retrovirus; for example, endog-

TABLE I. Examples of Highly Oncogenic Retroviruses

Cancer	Species	Virus
Sarcomas	Chicken	Rous sarcoma virus (RSV)
		Fujinami sarcoma virus (FSV)
	Rat	Harvey murine sarcoma virus (Ha-MSV)
		Kirsten murine sarcoma virus (Ki-MSV)
	Mouse	Moloney murine sarcoma virus (Mo-MSV)
		FBJ-murine sarcoma virus (FBJ-MSV)
	Cat	Snyder-Theilen feline sarcoma virus (ST-FeSV)
		McDonough feline sarcoma virus (M-FeSV)
	Woolly monkey	Simian sarcoma virus (SSV)
Leukemias	Chicken	Avian myeloblastosis virus (AMV)
		Avian erythroblastosis virus (AEV)
		Avian myelocytomatosis virus MC29 (MC29 virus)
	Mouse	Abelson murine leukemia virus (A-MLV)
Leukemias and carcinomas	Chicken	Avian myelocytomatosis and carcinoma virus MH2 (MH2 virus)

enous *akv* MLVs can undergo recombination with other endogenous viruses and become leukemogenic (Laigret *et al.*, 1988).

HTLV-1 is associated with adult T-cell leukemia (in over 1% of those infected as children) and tropical spastic paraparesis or HLTV-1-associated myelopathy. There also are some cases of immune depression caused by HTLV-1.

Lentiretroviruses are involved in immune dysfunction (HIV, simian immunodeficiency virus, caprine arthritis encephalitis virus), central nervous system disorders (HIV, visna virus), and anemia (equine infectious anemia virus). Cancer also occurs at a higher frequency in some lentiretrovirus-infected organisms, possibly both from the immune depression and from a direct effect of a viral-replication-controlling protein, tat.

V. GENETIC STRATEGIES OF RETROVIRUSES

As discussed above, organisms using a mode of replication involving a reverse transcription step have been very successful in evolution, as illustrated by their high representation in the eukaryotic genome and the number of different kinds of genetic elements that use reverse transcription. I now discuss some of the genetic strategies that led to this success.

Since all retroviruses are parasitic, I shall first compare retroviruses, parasites using reverse transcription, to parasites using more classical DNA and viral RNA replications.

RNA viruses are very successful parasites. Perhaps two-thirds of all viruses have RNA genomes. RNA viruses also mutate rapidly. They have on the average 0.1–1 base substitution per replication cycle, and they can recombine. As a result of this high error rate and recombination, RNA viruses replicate as a swarm or population, even though they have a very small genome relative to higher organisms (Eigen and Biebricher, 1988). They also spread widely and quickly.

The quasispecies theory of Eigen describes how the replication of small single-stranded RNA viruses can be described as taking place in sequence space (Eigen and Biebricher, 1988). When there is strong conservative selection, the population remains grouped around a master sequence; when there is selection for a new characteristic, such as resistance to a particular antibody, the population can evolve rapidly. For example, a new strain of influenza A virus appears almost each year as a result of genetic drift (Fitch *et al.*, 1991; Gorman *et al.*, 1991); there are many antigenically different strains of human rhinoviruses in circulation in the human population at any one time (Halsey, 1986); and the antigenic type of foot-and-mouth disease virus in a cattle herd changes frequently (Gebauer *et al.*, 1988). However, RNA viruses can conserve a particular sequence when there is strong selection for that sequence. For example, the antigenic types of polio viruses and measles virus and the sequence of the 3D protein of foot-and-mouth disease virus are highly conserved over time. In addition, phenotypic mixing (packaging genomes of mutant virus in proteins of a wild-type virus) displaces the selective consequences of some mutations from the mutant genome. All of these properties apply both to true RNA viruses and to retroviruses.

True RNA viruses, however, cannot pass on their genotype as an unchanged piece, even though certain epitopes can be conserved by selection. The high rate of mutation and the existence of recombination results in a high rate of evolution of RNA viruses. Kimura has shown that $k = f_0 v_T$, where k is the rate of evolution in nucleotide substitutions per generation; f_0 is the fraction of neutral mutations; and v_T is the total mutation rate per generation (Kimura, 1968). Organisms with higher mutation rates evolve faster than organisms with lower mutation rates; that is, RNA-based genetic systems (RNA viruses) evolve faster than DNA-based microorganisms. RNA virus populations contain more variant genomes than do DNA-based microorganisms, and thus, they will more likely be successful under changing selective conditions. Recombination further increases the amount of variation in a population and thus further increases the probability of success under altered selective conditions. Again, this high rate of evolution is a property of both true RNA viruses and retroviruses.

However, true RNA viruses cannot become integrated into DNA-based genetic systems, and they cannot become truly latent; that is, true RNA viruses are not capable of going from a replicating state to a non-replicating state and back to a replicating state again. In contrast, retroviruses form proviruses in which the viral genotype is replicated as one unchanged piece (except for the low rate of cellular mutation), and retroviruses can easily become latent and later be activated. Thus, whereas other RNA viruses can only maintain a consensus sequence by strong conservative selection, retroviruses can maintain a particular sequence as a provirus for a long time even in the absence of selection.

The length of time a retrovirus sequence is maintained unchanged as a provirus will depend only upon the rate of mutation of cellular DNA. Furthermore, since there can be reactivation of latent proviruses and since there also can be recombination or gene conversion between proviruses (or RNA transcribed from them) and exogenous viruses, a proviral sequence remains able to contribute all or part of its genome to a replicating retrovirus population when selective conditions favor this contribution, until the provirus is finally completely disabled by spontaneous germline mutations. However, since the rate of mutation of chromosomal DNA is roughly a factor of one million less than that of retrovirus replication, this inactivation will take a very long time.

DNA viruses, transposons, and plasmids are successful parasites using usual DNA–RNA–protein-based replication. DNA transposons are much more prevalent in prokaryotes than are retrotransposons. In fact, no retrotransposons are known in prokaryotes. Furthermore, DNA transposons can undergo duplicative transposition in which they form a new copy of the transposon while maintaining the original copy. However, they, and plasmids, can only move from one organism to another by conjugation. Bacteriophage are a form of parasitic DNA that is able to move from one prokaryotic organism to another in the absence of conjugation. Temperate bacteriophage share some properties of cellular transposons and some of viruses, as do retroviruses. Temperate bacteriophage differ from retroviruses in having to undergo a lytic cycle, which is lethal to the host cell, to produce progeny phage and in having only a DNA genome.

In eukaryotes there are no viruses capable of setting up a lysogenic infection in a way analogous to that established by temperate phage. There also seem to be relatively fewer DNA transposons in eukaryotes compared to the numbers of retroposons, retrotransposons, and retroviruses. Three possible explanations for the relative dearth of the usual DNA-based parasitic DNA in eukaryotes compared to prokaryotes are (1) the advantages of the high rate of mutation involved in the transcription–reverse transcription steps of retroreplication, (2) the ease of releasing copies of the parasitic retrosequences from the eukaryotic genome, and (3) the larger size and more complex organization of eukaryotic cells. However, the first two explanations also apply to prokaryotes. Therefore, the difference in size or chromatin composition between prokaryotes and eukaryotes might explain the difference in the nature of their genomic parasites. Alternatively, the larger size of eukaryotic cells may make it difficult to maintain a sufficient concentration of repressor molecules to maintain a lysogenic state. Therefore, the retrovirus mode of parasitism has been selected in eukaryotic viruses rather than the lysogenic mode.

Retroviruses are rarely, if ever, spread by the respiratory or fecal-oral routes. They are also rarely spread by arthropod vectors, although EIAV, and perhaps bovine leukemia virus, are spread this way (Teich, 1982). Retroviruses usually spread vertically, in infected blood, by sex, and in saliva. The lack of spread by the respiratory and fecal–oral routes and the rareness of arthropod spread limits the speed of retrovirus spread by horizontal transmission in most host populations. The spread of HIV-1 in some populations of intravenous drug users and prostitutes indicates that under some circumstances retroviruses can spread very rapidly, but this is unusual (Slutkin *et al.*, 1990). Thus, in spite of rapid genetic change and a large pool of variants, passage from organism to organism seems to be too slow to allow sufficient passage generations between organisms to establish a dominant strain except in relatively restricted populations.

In evolution, the retrovirus DNA phase gives stability and allows maintenance of a parental strain. Since this parental strain was able to form a provirus originally, it had to have some selective advantage. The mutations and recombinations occurring among progeny retroviral genomes provide the material for further selection of a fitter variant if and when one appears. However, the parental DNA provirus remains and continues to produce parental virus while selection determines if any of the variant progeny genomes are more fit. Thus, retroviruses have the cake of a stable genome with the provirus and the eating of it with an error-prone viral replication cycle.

VI. RECOMBINATION AND SEX IN RETROVIRUSES

Retroviruses have two molecules of RNA (dimer RNA) in their virions, so they can have recombination (Temin, 1991). The hypothesis of two identical molecules of RNA in retrovirus virions was first proposed by Vogt (1973). All retrotransposons and retroviruses seem to have two identical copies of viral genomic RNA in their virions. Thus, at this stage they appear to be diploid. However, it has been shown definitively that only a single provirus is formed from each virion (Hu and Temin, 1990; Panganiban and Fiore, 1988).

With other viruses, recombination occurs among molecules in the pools of replicating intermediate molecules formed during virus replication (Ramig, 1990). For influenza and other segmented viruses, recombination is primarily by reassortment. For other RNA viruses and certainly for DNA viruses, recombination is by copy-choice and breakage-reunion mechanisms.

However, the retrovirus life cycle, as described above, never involves a pool of replicative intermediates. The input RNA genome is reverse transcribed in a virion-derived particle, which isolates this process from other input genomes, and then the DNA copy is integrated independently of other viral genomes into the host-cell DNA. There is a pool of progeny viral RNA molecules, including full-length mRNAs, but progeny viral RNA does not form a replicative intermediate with the potential for recombination until it is inside a mature virion. Thus, there never is a pool of replicative intermediates during retrovirus replication.

Recent experiments establish that formation of virions containing two different genomic RNAs (heterodimer RNA) is required for recombination (Hu and Temin, 1990). That is, the parental cell has to be producing two different kinds of genomic viral RNA in order for retrovirus recombination to be observed. Thus, the dimer RNA in retrovirus virions appears necessary to allow recombination. Since in these recombination experiments, the two parental viral molecules differed only by 8 bp, there is every reason to think that recombination also takes place even when it cannot be seen because the dimer RNA molecules are identical.

Retroviruses have a very high rate of variation as a result of the high frequency of errors that arise during reverse transcription. Therefore, it is hard to postulate that the additional contribution from recombination to the amount of variation and to the speed of incorporation of beneficial mutations into a retrovirus population is critical, since there could be rapid incorporation of beneficial variants into a retrovirus population by selection. However, recombination during minus-strand DNA synthesis allows the repair of single-stranded breaks, since the DNA growing point can switch from the broken strand to the other parental strand, thus forming a viable progeny genome from two parental genomes with deleterious mutations, that is, single-strand RNA breaks (Fig. 5) (Coffin, 1979). In addition, the cost of the synthesis of a retrovirus virion with

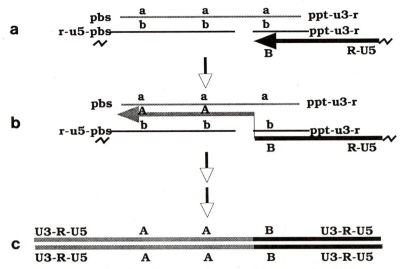

FIGURE 5. Minus-strand or forced copy-choice recombination. (a) and (b) are two molecules of viral RNA. A and B are the molecules of DNA reverse-transcribed from (a) and (b). Step (a) is after the minus-strand DNA has jumped to the 3' end of the RNA molecule during reverse transcription. Step (b) is after strand switching, and step (c) is the recombinant DNA molecule. This recombination results in a functional provirus even though there was a break in viral RNA.

dimer RNA is to the cell, not to the virus. Therefore, it is a free attribute of retrovirus replication.

Usually, sex has an evolutionary cost in that it decreases the contribution of parent to offspring to one-half, while asexual division or parthenogenesis results in the parent having a 100% genetic contribution to the progeny. Since sex is so widely distributed among living organisms, it must have strong countervailing benefit(s). Sex allows recombination. However, although evolutionary biologists agree that recombination is beneficial, they do not agree on the reasons. Proposed reasons include removing damage to the genome such as mutations and insertions, increasing the speed at which beneficial mutations are incorporated into the population, and increasing the variability in a population by increasing the number of combinations of mutations (Michod and Levin, 1988).

It is true that under stable environmental conditions, some groups of organisms frequently evolve to asexual forms. However, in general, the asexual forms are smaller and are not as long-lasting in evolutionary time. Therefore, it seems clear that usually there are evolutionary benefits to sex. However, certain parasitic protozoans may be an exception to this assertion (Tibatrenc et al., 1990).

Since retroviruses replicate without a pool of replicative intermediates, retroviruses would have no sex or recombination if there were not two genomic molecules in one virion. Therefore, since there are benefits to recombination and no apparent costs to the virus, there would have been strong selection for a dimer RNA in retrovirus virions. The primary benefit of retrovirus recombination is probably repair, although the ability to increase variation may also be important (Temin, 1991).

VII. FUTURE OF RETROVIROLOGY

Work with retroviruses started as basic research because of its relationship to animal cancers. Now with the advent of AIDS and human gene therapy, retroviruses are the stuff of front page newspaper articles, and they are of direct clinical relevance.

In the future, AIDS is going to be even more of concern to public health as a cause of illness and death (Chin et al., 1990). In addition, retrovirus-mediated gene therapy appears to have the potential to ameliorate some single-gene hereditary diseases, and it may also be useful in other kinds of diseases. It is also possible that some other chronic human diseases will be found to be caused by a retrovirus.

The Human Genome Project, while intended to turn up new genes responsible for human disease, will certainly provide materials to construct a detailed picture of the retroelements in the human genome and much material for their evolutionary study—a topic of great interest to retrovirologists.

Thus, it is likely that, in the future as at present, retroviruses will be the center of intense biological study, and it is also likely that retroviruses will be the source of further novel and general findings.

VIII. GLOSSARY

Dimer RNA: Two identical molecules of viral RNA in retrovirus virions

Endogenous retroviruses: Retrovirus genomes transmitted in the germ line

Exogenous retroviruses: Retroviruses transmitted from somatic cell to somatic cell

Long terminal repeats (LTRs): Directly repeated sequences at ends of retrovirus DNA genomes

Oncogenes: Dominantly acting genes whose expression results in cancer

Provirus: Integrated DNA form of retrovirus

Retrons: Bacterial retroelements coding for msDNA

Retrosequences: DNA sequences resulting from reverse transcription

Retrotransposons: Cellular movable genetic elements using reverse transcription

Retrovirus vectors: Modified retroviruses containing inserted genetic sequences

Retrovirus-mediated human gene therapy: Use of retroviruses to introduce functioning genes to replace the function of a defective gene

Reverse transcriptase: DNA polymerase capable of copying information from RNA to DNA and from DNA to DNA

Ribozyme: RNA enzyme

Telomerase: Reverse transcriptase responsible for synthesizing chromosome ends

Transposon: Cellular movable genetic element

Virion: Virus particle

ACKNOWLEDGMENTS. I thank G. Pulsinelli for useful comments on this manuscript. The research in my laboratory is supported by Public Health Service grants CA-22443 and CA-07175 from the National Cancer Institute. I am an American Cancer Society Research Professor.

IX. REFERENCES

Blackburn, E. H., 1990, Telomeres and their synthesis, *Science* **249**:489.

Brosins, J., 1991, Retroposons—Seeds of evolution, *Science* **251**:753

Cann, A. J., and Chen, I. S. Y., 1990, Human T-cell leukemia virus types I and II, in: *Virology*, 2nd ed., (B. N. Fields, D. M. Knipe, *et al.*, eds.), Vol. 1, pp. 1501–1527, Raven Press, New York.

Chin, J., Sato, P. A., and Mann, J. M., 1990, Projections of HIV infections and AIDS cases to the year 2000, *WHO Bull.* **68**:1.

Coffin, J. M., 1979, Structure, replication, and recombination of retrovirus genomes: Some unifying hypotheses, *J. Gen. Virol.* **42**:1.

Coffin, J. M., 1990, Retroviridae and their replication, in: *Virology*, 2nd ed. (B. N. Fields, D. M. Knipe, *et al.*, eds.), Vol. 1, pp. 1437–1500, Raven Press, New York.

Doolittle, R. F., Feng, D.-F., Johnson, M. S., and McClure, M. A., 1989, Origins and evolutionary relationships of retroviruses, *Q. Rev. Biol.* **64**:1.

Eigen, M., and Biebricher, C. K., 1988, Sequence space and quaisispecies distribution, in: *RNA Virus Genetics* (E. Domingo, J. J. Holland, and P. Ahlquist, eds.), Vol. III, pp. 211–245, CRC Press, Boca Raton, Florida.

Fitch, W. M., Leiter, J. M. E., Li, X., and Palese, P., 1991, Positive Darwinian evolution in human influenza A viruses, *Proc. Natl. Acad. Sci. USA* **88**:4270.

Flügel, R. M., 1991. Spumaviruses: A group of complex retroviruses, *J. AIDS* **4**:739.

Gardner, M. B., Kozak, C. A., and O'Brien, S. J. 1991, The Lake Casitas wild mouse: Evolving genetic resistance to retroviral disease, *Trends Genet.* **7**:22.

Gebauer, F., De la Torre, J. C., Gomes, I., Mateu, M. G., Barahona, H., Tiraboschi, B., Bergmann, I., Auge de Mello, P., and Domingo, E., 1988, Rapid selection of genetic and antigenic variants of foot-and-mouth disease virus during persistence in cattle, *J. Virol.* **62**:2041.

Gorman, O. T., Bean, W. J., Kawaoka, Y., Donatelli, I., Guo, Y., and Webster, R. G., 1991, Evolution of influenza A virus nucleoprotein genes: Implications for the origins of H1N1 human and classical swine viruses, *J. Virol.* **65**:3704.

Gross, L., 1970, *Oncogenic Viruses*, 2nd ed., Pergamon Press, New York.

Halsey, N. A., 1986, The epidemiology of viral disease, in: *Virology in Medicine*, (H. Rothschild and C. Cohen J. C., ed.), pp. 89–132, Oxford University Press, New York.

Hoover, E. A., Mullins, J. I., Quackenbush, S. L., and Gasper, P. W., 1987, Experimental transmission and pathogenesis of immunodeficiency syndrome in cats, *Blood* **70**:1880.

Hu, W.-S., and Temin, H. M., 1990, Genetic consequences of packaging two RNA genomes in one retroviral particle: Pseudodiploidy and high rate of genetic recombination, *Proc. Natl. Acad. Sci. USA* **87**:1556.

Inouye, M., and Inouye, S., 1991, msDNA and bacterial reverse transcriptase, *Ann. Rev. Microbiol.* **45**:164.

Inouye, S., Hsu, M.-Y., Eagle, S., and Inouye, M., 1989, Reverse transcriptase associated with the biosynthesis of the branched RNA-linked msDNA myxococcus xanthus, *Cell* **56**:709.

Kimura, M., 1968, Evolutionary rate at the molecular level, *Nature* **217**:624.

Laigret, F., Repaske, R., Boulukos, K., Rabson, A. B., and Khan, A. S., 1988, Potential progenitor sequences of mink cell focus-forming (MCF) murine leukemia viruses: Ecotropic, xenotropic, and MCF-related viral RNAs are detected concurrently in thymus tissues of AKR mice, *J. Virol.* **62**:376.

Michod, R. E., and Levin, B. E., 1988, *The Evolution of Sex*, Sinauer Associates, Sunderland, Massachusetts.

Panganiban, A. T., and Fiore, D., 1988, Ordered interstrand and intrastrand DNA transfer during reverse transcription, *Science* **241**:1064.

Poch, O., Sauvaget, I., Delarue, M., and Tordo, N., 1989, Identification of four conserved motifs among the RNA-dependent polymerase encoding elements, *EMBO J.* **8**:3867.

Ramig, R. F., 1990, Principles of animal virus genetics, in: *Virology*, 2nd ed. (B. N. Fields, D. M. Knipe, *et al.*, eds.), Vol. 1, pp. 95–122, Raven Press, New York.

Rhee, S. S., and Hunter, E., 1990. A single amino acid substitution within the matrix protein of a type D retrovirus converts its morphogenesis to that of a type C retrovirus, *Cell* **63**:77.

Rosenberg, S. A., Aebersold, P., Cornetta, K., Kasid, A., Morgan, R. A., Moen, R., Karson, E. M., Lotze, M. T., Yang, J. C., Topalian, S. L., Merino, M. J., Culver, K., Miller, A. D., Blaese, R. M., and Anderson, W. F., 1990, Gene transfer into humans—Immunotherapy of patients with advanced melanoma, using tumor-infiltrating lymphocytes modified by retroviral gene transduction, *N. Engl. J. Med.* **323**:570.

Slutkin, G., Chin, J., Tarantola, D., and Mann, J., 1990, Use of HIV Surveillance Data in National AIDS Control Programmes, a Review of Current Data Use with Recommendations for Strengthening Future Use, WHO Document WHO/GPA/SF1/90.1.

Teich, N., 1982, Taxonomy of retroviruses, in: *RNA Tumor Viruses*, 2nd ed. (R. Weiss, N. Teich, H. Varmus, and J. Coffin, eds.), pp. 25–207, Cold Spring Harbor Laboratory, Cold Spring Harbor, New York.

Temin, H. M., 1970, Malignant transformation of cells by viruses, *Perspect. Biol. Med.* **14**:11.

Temin, H. M., 1985, Reverse transcription in the eukaryotic genome: Retroviruses, pararetroviruses, retrotransposons, and retrotranscripts, *Mol. Biol. Evol.* **2**:455.

Temin, H. M., 1989a, Retrons in bacteria, *Nature* **339**:254.

Temin, H. M., 1989b, Retrovirus vectors: Promise and reality, *Science* **246**:983.

Temin, H. M., 1991, Sex and recombination in retroviruses, *Trends Genet.* **7**:71.

Tibatrenc, M., Kjellberg, F., and Ayala, F. J., 1990, A clonal theory of parasitic protozoa: The population structure of *Entameoeba, Giardia, Leishmania, Naegleria, Plasmodium, Trichomonas,* and *Trypanosoma* and their medical and taxonomical consequences, *Proc. Natl. Acad. Sci. USA* **87**:2414.

Vallee, H., and Carre, H., 1905, Sur infectieuse de l'anemie du cheval, *Comptes Rendus Acad. Sci. Press* **139**:331.

Vogt, P. K., 1973, The genome of avian RNA tumor viruses: A discussion of four models, in: *Possible Episomes in Eukaryotes, Proceedings of the Fourth Lepitit Colloquium, 1972* (L. Silvestri, ed.), pp. 35–41, North-Holland, Amsterdam.

Weiner, A. M., and Maizels, N., 1987, tRNA-like structures tag the 3' ends of genomic RNA molecules for replication: Implications for the origin of protein synthesis, *Proc. Natl. Acad. Sci. USA* **84**:7383.

Weiss, R., Teich, N., Varmus, H., and Coffin, J., eds., 1982, *RNA Tumor Viruses (Molecular Biology of Tumor Viruses)*, 2nd ed., Cold Spring Harbor Laboratory, Cold Spring Harbor, New York.

Witkowski, J. A., 1990, The inherited character of cancer—An historical survey, *Cancer Cells* **2**:229.

Wong-Staal, F., 1990, Human immunodeficiency viruses and their replication, in: *Virology*, 2nd ed. (B. N. Fields, D. M. Knipe, *et al.*, eds.), Vol. 2, pp. 1529–1543, Raven Press, New York.

Xiong, Y., and Eickbush, T. H., 1988, Similarity of reverse transcriptase-like sequences of viruses, transposable elements, and mitochondrial introns, *Mol. Biol. Evol.* **5**:675.

Yu, G.-L., Bradley, J. D., Attardi, L. D., and Blackburn, E. H., 1990, *In vivo* alteration of telomere sequences and senescence caused by mutated *Tetrahymena* telomerase RNAs, *Nature* **344**:126.

Structure and Classification of Retroviruses

JOHN M. COFFIN

I. INTRODUCTION

The retroviruses encompass a large family of infectious agents (Retroviridae) unified by a common virion structure and mode of replication. Retroviruses have been isolated from most vertebrate species in which they have been sought, and have been found to display a remarkable diversity in their association with the host. Table I gives a list of some of the more commonly encountered viruses. At the one end of the diversity, infections with some retroviruses can lead to uniformly fatal conditions, such as AIDS, a variety of malignancies, neurologic diseases, and other clinical conditions. At the other end, some retroviruses induce only a benign viremia with no outward adverse effects, and can even become established as DNA in the germ line and passed as "endogenous" viruses from generation to generation. Indeed, the line between endogenous viruses and the retrotransposable elements found in large numbers in the genome of all eukaryotes is very fine (Chapters 1 and 4). This chapter will be concerned with a discussion of general properties of the retroviruses, the structure of their virions, and their classification. Its scope will be limited to those elements which are demonstrably viruses.

Retroviruses are unified into a family, the Retroviridae, by impor-

JOHN M. COFFIN • Department of Molecular Biology and Microbiology, Tufts University School of Medicine, Boston, Massachusetts 02111.

The Retroviridae, Volume 1, edited by Jay A. Levy. Plenum Press, New York, 1992.

TABLE I. Some Common Retroviruses

Abbreviation	Full name
ALSV	Avian leukosis-sarcoma viruses
ALV	Avian leukosis (leukemia) virus
BIV	Bovine immunodeficiency virus
BLV	Bovine leukemia virus
CAEV	Caprine arthritis-encephalitis virus
CSRV	Corn snake retrovirus
EIAV	Equine infectious anemia virus
FeLV	Feline leukemia virus
FIV	Feline immunodeficiency virus
GALV	Gibbon ape leukemia virus
HIV	Human immunodeficiency virus
HSRV	Human spuma retrovirus
HTLV	Human T-cell leukemia (lymphotropic) virus
MLV (MuLV)	Murine (mouse) leukemia virus
MMTV	Mouse mammary tumor virus
MPMV	Mason-Pfizer monkey virus
REV	Reticuloendotheliosis virus
RSV	Rous sarcoma virus
SFV	Simian foamy virus
SIV	Simian immunodeficiency virus
SMRV	Squirrel monkey retrovirus
SNV	Spleen necrosis virus
VRV	Viper retrovirus

tant features of virion structure and replication cycle (Chapter 5). Virions of retroviruses are enveloped particles about 100 nm in diameter with an internal spherical or conical core and a dimeric genome of polyadenylated RNA 7–10 kilobases (kb) in length. The virion contains several enzymatic activities—reverse transcriptase, ribonuclease H, and integrase—necessary for early events in replication, as well as protease used in processing the virion proteins. Shortly after entry of the virus into the cell, the genome is copied into a double-stranded DNA molecule by the combined action of reverse transcriptase and RNase H. Subsequently, this DNA is covalently joined to the genomic DNA of the host cell to form the integrated provirus. Once integrated, the provirus is stable and serves as template for viral mRNA and protein synthesis. Assembly of the capsid and release of the virions are accomplished by association of unprocessed precursor polyproteins containing the capsid protein (coded for by *gag*) or capsid plus enzymatic proteins (*gag–pro–pol*) with the genome RNA, the cell membrane, and one another. This process leads to budding of the virion from the cell, and is followed by cleavage of the *gag* and *gag–pol* protein precursors to separate the various domains from one another. Cleavage is accompanied by obvious morphological changes, particularly condensation of the core. During budding, the separately synthesized *env* glycoproteins are incorporated into the virion envelope.

Infection of a cell with a retrovirus does not inevitably lead to death of the cell, although it sometimes does so. In many cases, the provirus remains stably integrated without directly affecting cell viability. This central feature of retrovirus biology enables the diversity of host–virus interaction associated with this group. Indeed, many of the features associated with retrovirus infection are due to rare side effects: the occasional acquisition of cellular sequences as oncogenes, and the disruption or deregulation of cellular genes by rare integrations in their vicinity.

II. VIRION STRUCTURE

A. Morphology

A schematic illustration of the current perception of the structure of the retrovirus virion is shown in Fig. 1, depicting the inferred location of the various gene products. Within this structural framework there is some amount of morphological variation as illustrated by the electron micrographs in Fig. 2. To a certain extent, virion morphology is taxonomically useful: closely related viruses (i.e., those within the same genus) are usually identical in appearance in electron micrographs. However, similarity in structure itself does not imply close relationship. Like much of retroviral nomenclature, description of virion structure is not

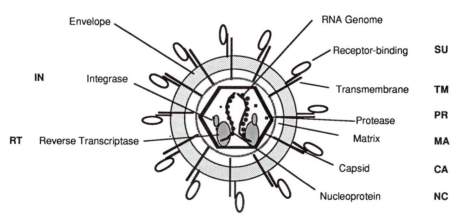

FIGURE 1. The retrovirus virion. The figure is schematic, depicting the current ideas of the locations and relationships of the various components. The shapes of the structures as depicted are fanciful.

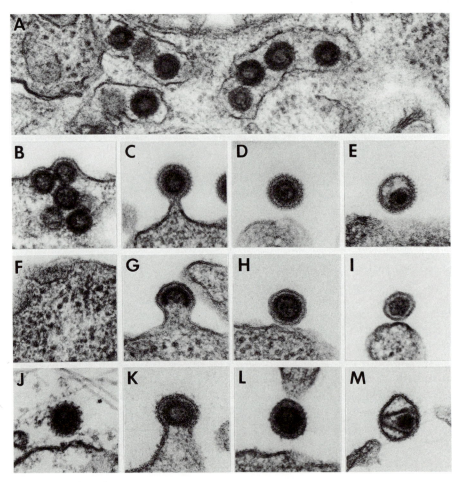

FIGURE 2. Morphology and morphogenesis of retroviruses. The photographs from left to right in each row show intracellular (if present), budding, immature, and mature forms of each virus. All magnifications are 75,600×. [All photographs courtesy of K. Nagashima and M. Gonda. Reproduced in part from Gonda *et al.* (1989).] A. Intracisternal A particles (IAPs). B–E. Type B virus (mouse mammary tumor virus). F–I. Type C virus (murine leukemia virus). J–M. Type D virus (Mason-Pfizer monkey virus).

consistent. The original, but slightly extended, classification of Bernhard (1960; Fine and Schochetman, 1978) into A, B, C, and D particles is still used to describe those viruses (sometimes called oncoviruses) well known in the 1960s and 1970s. For the more recently described genera (lentiviruses and spumaviruses), no specific appellations have been assigned.

In general, the different virion structures are classified according to mode of assembly, location and shape of core (nucleocapsid) in the mature virion, and appearance of the surface glycoproteins.

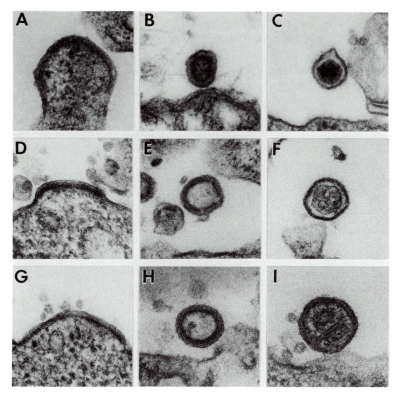

FIGURE 2. (*Continued*) HTLV–BLV group viruses. A–C. Bovine leukemia virus. D–F. HTLV-1. G–I. HTLV-2.

A particles are not virions, but are strictly intracellular structures composed of nucleocapsids (genome plus *gag* and *gag–pol* proteins) containing unprocessed precursors. Intracytoplasmic A particles are intermediates in assembly of type B and D virions, which subsequently associate with the cell membrane to initiate budding (Fig. 2A, parts B–E and J–M). All other retroviruses bud by a mechanism involving simultaneous assembly and budding in which preformed cores are never visible (Fig. 2A, parts F–I).

A large number of endogenous provirus-related elements in rodents (Coffin, 1982b; Stoye and Coffin, 1985) and other mammals—including humans (Ono, 1986)—also give rise to A particles. These are usually viewed budding into intracellular membrane vesicles and are often referred to as intracisternal A particles (IAPs) (Fig. 2A, part A), although strictly cytoplasmic forms are also seen. Proviruses of A particles are particularly abundant—the laboratory mouse has been estimated to contain more than 1000 copies (Lueders and Kuff, 1977; Kuff *et al.*,

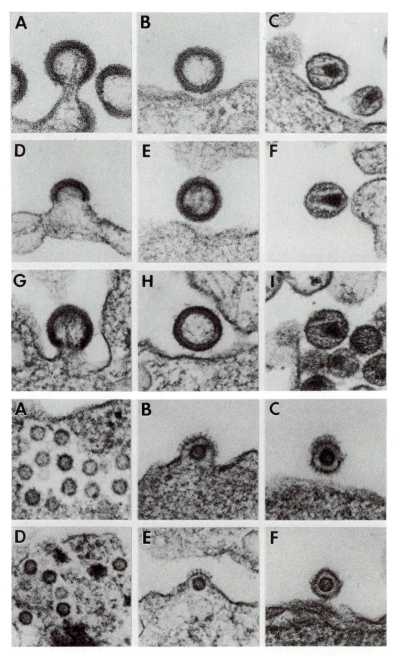

FIGURE 2. (*Continued*) Top. Lentiviruses. A–C. HIV-1. D–F. HIV-2. G–I. Bovine immunodeficiency virus. Bottom. Spumaviruses. Note that the immature and mature forms are
similar in appearance. A–C. Chimpanzee syncytial (foamy) virus. D–F. Bovine syncytial
(foamy) virus.

1968; Kuff and Lueders, 1988)—and they are abundantly expressed in certain cell types, such as mature B cells and tumors derived from them (Dalton, et al., 1961) as well as embryos. Their high prevalence and tissue specificity of expression have caused them to be frequently rediscovered in the course of cloning experiments designed to find other types of genes (for example, Ono, et al., 1980). Although obviously related to retroviruses (Kuff and Lueders, 1988; Ono et al., 1980; Mietz et al., 1987; Ono et al., 1985), and capable of at least occasional reintegration (Heidmann and Heidmann, 1991; Heberlein et al., 1990; Horowitz et al., 1984; Hawley et al., 1984), IAPs are not known or believed to encode infectious virus, but rather are thought to represent retrotransposons—elements which strongly resemble retroviruses in organization and mechanism of replication, but lack *env* genes and do not have an extracellular phase in their life cycle (see Chapter 1). Although retrotransposons are often viewed as ancestral to retroviruses, the survival of a highly degraded remnant of the *env* gene in IAPs and related elements (Kuff and Lueders, 1988; Ono et al., 1980, 1985) implies that this group is relatively recently descended from retroviruses, not vice versa.

B particles are represented by the mature virions of mouse mammary tumor virus (MMTV) (Fig. 2A, parts B–E). B particles are assembled via budding of A particles into an immature form resembling an A particle with an envelope, and mature into a virion characterized by a tightly condensed acentric core and prominent surface projections (the *env* protein complex).

C particles are the virions of two retrovirus groups—the mammalian and avian type C viruses (Fig. 2A, parts F–I). These groups were formerly classified together (as type C oncoviruses), but are now recognized as having only a very distant relationship (see Section III). C particles are assembled directly at the cell membrane with no visible cytoplasmic intermediate, and have an immature form with a large, open spherical core, which matures into a centrally located condensed form, with visible but less than prominent surface projections.

D particles are characteristic of most of the retroviruses which have been found in primates, including the Mason-Pfizer virus and "simian AIDS" (SAIDS) viruses (Fine and Schochetman, 1978; Thayer et al., 1987; Heidecker et al., 1987; Marx et al., 1985) (not to be confused with the simian immunodeficiency virus—SIV—a lentivirus). They resemble virus B particles in assembly, maturation, and morphology (Fig. 2A, parts J—M), but have less prominent surface projections and a characteristic cylindrical core.

Lentivirus virions have a distinctive, but as yet unnamed, morphology (Fig. 2C). In the electron microscope, their assembly and budding resemble type C particles, leading to immature virions with hollow cylindrical cores. Mature particles have a distinctive core in the shape of a truncated cone.

Spumavirus virions also resemble type C viruses in morphology, but have much more prominent surface projections and a less condensed core (Fig. 2D).

B. Virion Components

Despite differences in morphology (and biology), all retroviruses are assembled from a very similar set of virion components.

1. Envelope

The envelope of retroviruses is composed of a lipid bilayer derived by budding from the plasma membrane. The similarity of the pattern of lipids and other membrane components of the virus and cell implies that there is no virus-specified process that actively modifies the small molecules of the membrane. The process of budding does exclude the bulk of cell surface proteins, although traces of normal cell proteins, such as histocompatibility antigens, can be found in the purified virus preparations (Henderson *et al.*, 1987). (Whether these are really embedded in the virion envelope remains questionable, since even highly purified virus is never completely free of contamination with cellular vesicles.)

2. RNA

The retrovirus genome consists of a dimer of identical single-stranded RNA molecules, each 7–10 kb in length. Viruses with genomes greater than about 8 kb are those that have genes in addition to *gag, pol,* and *env*. The genome RNA is modified in ways reflecting its synthesis and processing by cellular machinery: It is capped at the 5′ end, polyadenylated at the 3′ end, and contains some internal 6-methyl modifications of A residues, which are added at specific sites, but whose function is uncertain (Kane and Beemon, 1985; Csepany *et al.*, 1990). The genome is of plus sense—it is capable of serving as a messenger for capsid proteins and enzymes, even though it does not do so during the early stages of infection.

In addition to the genome, retrovirus virions contain a variety of small RNA molecules (and even some DNA molecules), most of which have no obvious role in replication and are believed to be accidental cellular components included in the virion (Coffin, 1982a). The one small molecule with a clearly defined role is the single molecule of tRNA specifically associated with the genome by base-pairing to the primer-binding (PB) site near the 5' end of the genome. A variety of different tRNAs are used by genomes of different retroviruses (see below, Section III). The tRNA molecule serves the crucial role of providing the primer for initiation of DNA synthesis shortly after infection (see Chapter 5).

The genome itself can be divided into terminal noncoding regions, necessary in *cis* for its replication, and internal regions, encoding the virion proteins and (in some viral groups) proteins involved in control of expression. The terminal regions are arranged in the same order in all retroviruses, but their sizes and certain other internal features vary from group to group as detailed in the last section of this chapter. Details of the function of these regions are presented in Chapter 5. Briefly, they include the following at the 5' end (Fig. 3):

R: A short sequence directly repeated at each end of the genome used during reverse transcription to ensure correct end-to-end transfer of the growing chain.

U5: A unique sequence (usually fairly short) near the 5' end of the genome between R and PB.

PB: The sequence of 18 bases complementary to the 3' end of the specific tRNA primer.

L: An untranslated leader region between PB and *gag* which includes a specific signal for packaging of genome RNA [usually called ψ (Mann *et al.*, 1983) or E (Watanabe and Temin, 1982)] [for review see Linial and Miller (1990)]. The leader region usually contains the splice donor site for generation of subgenomic mRNAs for expression of *env* (and other genes if present).

Important *cis*-acting sequences near the 3' end include:

PP (for polypurine): An AG-rich sequence which becomes the primer for plus-strand DNA synthesis during reverse transcription.

U3: A unique sequence between PP and R which contains signals used in the provirus to specify and regulate its transcription and the processing (e.g., polyadenylation) of the transcripts.

R: The 3' copy of the terminal repeat.

The process of reverse transcription (Chapter 5) causes the terminal sequences to be fused into the structure U3–R–U5, found at each end of the DNA and universally referred to as the "long terminal repeat" or LTR. Enhancer, promoter, and regulatory sequences in U3 (and sometimes R and U5 as well) specify and regulate the initiation of transcription (and thus determine the 5' end of genomes and all mRNAs) at the beginning of R. Signals specifying cleavage and poly(A) addition determine the 3' end of all species at the end of R. The ends of the LTR (i.e., the 5' end of U3 and the 3' end of U5), which are present in the RNA adjacent to the primer sites, contain sequences recognized by the integration system for joining the viral to cellular DNA (Chapter 5). Reflect-

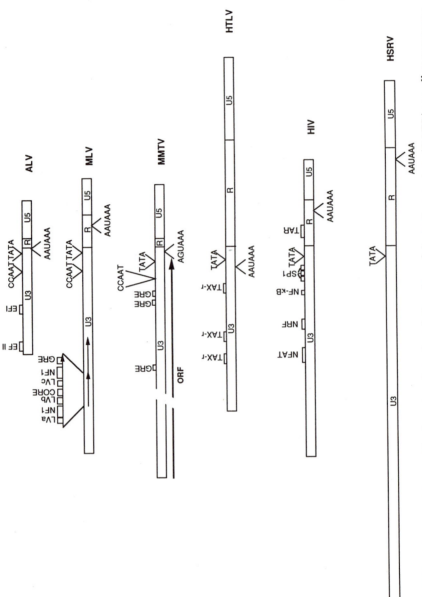

FIGURE 3. Retroviral LTRs. The diagrams show the relative sizes of the various regions (5' LTR) as well as consensus signals for transcription initiation (CCAAT and TATA boxes) and 3' processing of transcripts (AAUAAA) as well as binding sites for specific transcription factors and other features. See Table I for virus abbreviations.

ing their similar function, these short (ca. 12–15 bases) signals often have a similar sequence, arranged in inverted order relative to one another.

The disposition of coding regions in the genome of various viruses is shown in Fig. 4. All retroviruses include the *gag*, protease (*pro*), *pol*, and *env* genes in this order, with *gag* and *gag–pro–pol* proteins synthesized by translation of mRNAs identical to the genomes, and *env* protein synthesized from a spliced subgenomic mRNA. The relationship between the *gag*, *gag–pro*, and *gag–pro–pol* protein precursors is determined by partial translational readthrough of a translational terminator (Yoshinaka *et al.*, 1985) or frameshift site at one or both of the gene boundaries (Jacks, 1990; Jacks and Varmus, 1985). The relative translational organization of these genes is characteristic of each group of viruses as indicated in the figure.

The *env* gene usually overlaps the 3' end of the *pol* reading frame (except in lentiviruses and spumaviruses) and the splice acceptor for *env* usually lies within the *pol* coding region. The extent to which this splice acceptor is used determines the ratio of genome to *env* mRNA, and its usage is specified in a demonstrable, but complex and poorly understood way by sequences near the splice sites and within the *gag* and *pol* genes (Arrigo *et al.*, 1987; Arrigo and Beemon, 1988; Katz *et al.*, 1988; Katz and Skalka, 1990; Miller and Temin, 1986).

Two types of genes not coding for virion proteins are also found in some groups of retroviruses. The first type includes genes encoding proteins which are not virion structural proteins, but which seem to play important roles in virus replication—particularly in the regulation of expression. These genes range in number from one in mammary tumor virus to at least six in lentiviruses. They are expressed via additional spliced mRNAs, often with rather complex splicing patterns (Schwartz *et al.*, 1990). The presence and pattern of such genes is characteristic of each individual retrovirus genus (see Chapters 3 and 5).

The other virus genes not encoding virion proteins are copies of cellular genes recently acquired by the virus genome, usually as oncogenes which confer upon the virus the ability to transform normal into malignant cells (see Chapters 6 and 7) and to cause rapidly arising tumors in the host animal. Some two dozen different cellular sequences have been identified as retroviral oncogenes, and these have provided the most incisive entree yet available into molecular mechanisms of carcinogenesis (Bishop and Varmus, 1982; Bishop, 1991). In most cases, oncogenes are cDNA (i.e., intronless) copies of normal cellular genes inserted at various places in the viral genome. They usually replace essential viral genes, and therefore leave the virus dependent on coinfection of a cell with a related intact virus (usually called a helper virus) to provide proteins for its replication. Oncogenes can be expressed either

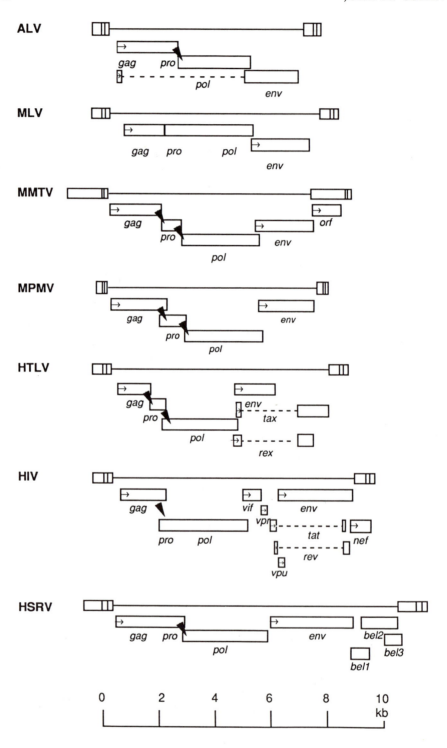

as a fusion with one of the viral proteins or via a separate spliced mRNA. Oncogenes are very recent additions to virus genomes. In most cases, they probably arose within the animal from which the transforming virus was isolated. Oncogenes have been found in only two groups of viruses, the avian leukosis viruses and the mammalian type C viruses.

3. Proteins

Until recently, retrovirus proteins were named by appending their apparent molecular weight to a prefix (p—protein; pp—phosphoprotein; gp—glycoprotein; Pr—precursor). As more viruses were discovered, and the functions of the viral proteins became more understood, this scheme became too cumbersome and confusing. Thus, a more rational nomenclature was developed (Leis *et al.*, 1988). Proteins are now assigned a two-letter mnemonic based on function. Current usage is a hybrid between these systems, with proteins from the most studied viruses (such as HIV) more often called by old, rather than new, nomenclature: p24 instead of CA, gp120 instead of SU, etc. Although this system is often easier in presentation, it is generally desirable in published works to provide the reader (at least once) with the corresponding "official" nomenclature for all proteins mentioned. The correspondence of the names and the sizes of the virion proteins are shown in Table II. Individual proteins are discussed below in the order in which they are encoded in the genome.

a. gag *Proteins*

These constitute the major structural elements of the capsid. They are present in (at least roughly) equal numbers—2000–4000 copies per virion. Retroviruses of all groups encode three *gag* proteins of fairly well-defined function, and some groups encode a fourth whose function is poorly understood. It is important to remember that *gag* proteins fill two roles: as components of the major precursor which participates in and directs assembly and budding, and as individual proteins to support the structure of the virion and play important roles (DNA synthesis, integration) early in infection. This dual role is consistent with the order

---

FIGURE 4. Coding regions of retrovirus genomes. The top lines depict the proviral DNA with the LTRs boxed. Under them are the coding regions, with each box corresponding to a separate reading frame. Horizontal arrows indicate points of translational initiation; vertical lines indicate terminators which may be partially suppressed during translation; diagonal arrows indicate frameshift sites. Dashed lines show reading frames joined by splicing events. See Table I for virus abbreviations.

TABLE II. Proteins of Retrovirus Virions[a]

Current name	ALSV	Mammalian C type	MMTV	D Type	HTLV–BLV	Lentivirus
MA	p19	p15	p10	p10	p19/15	p15–17
?	p10	p12	p21	p18	NP[b]	NP[b]
CA	p27	p30	p27	p27	p24	p24–26
NC	p12	p10	p14	p14	p12	p7–11
PR	p15	p14	p13	—	p14	p17
RT	p68	p80	—	—	—	p66
IN	p32	p46	—	—	—	p32
SU	gp85	gp70	gp52	gp70	gp60	gp95–120
TM	gp37	p15E	gp36	gp22	gp30	gp41

[a] The order of 5' to 3' from top to bottom.
[b] Not present.

of domains in the precursor, corresponding to location of the proteins (outside to inside) in the mature virion. The domain structure of the *gag* protein precursor seems to make sense in terms of functions required for assembly, since simultaneous associations of protein–membrane to cause budding (mediate by MA), protein–protein to form the core (mediated by CA), and protein–RNA to incorporate the genome (mediated by NC) can be readily accomplished (Bolognesi *et al.*, 1978, Wills and Craven, 1991).

MA (for matrix) proteins line the inner face of the virion envelope. The association of the MA domain of the *gag* protein precursor with the membrane is necessary for budding and its structure determines whether assembly proceeds via intracytoplasmic (A) particles or directly at the membrane (Rhee *et al.*, 1990; Rhee and Hunter, 1990). Consistent with its role in membrane association, MA proteins of most retroviruses are modified by the addition of a myristic acid group to their NH_2 terminal (Henderson *et al.*, 1983). This modification is necessary to promote their association with membranes (Copeland *et al.*, 1988; Jorgensen *et al.*, 1988; Rein *et al.*, 1986; Rhee and Hunter, 1987; Bryant and Ratner, 1990).

pX. Some groups of viruses encode this "extra" *gag* protein of unknown function which is located between the MA and CA proteins in the virion—a location corresponding to its position in the *gag* protein precursor (Dickson *et al.*, 1985).

CA (capsid). The CA protein forms the core shell of the virion (Dickson *et al.*, 1982, 1985). With many retroviruses, the CA protein constitutes the most readily detectable antigen and is often the basis for retroviral immunoassays.

NC (nucleocapsid) is the protein in intimate association with the genomic RNA. Consistent with this role, it is always a basic protein, with one or more characteristic Cys–His arrays resembling "zinc finger" structures found in a number of nucleic acid-binding proteins (Katz and Jentoff, 1989). The issue of whether these structures act by binding zinc is controversial. *In vitro*, NC protein behaves as a nonspecific nucleic acid-binding protein (Fu *et al.*, 1985, 1988; Meric *et al.*, 1986; Meric and Spahr, 1986). However, it must also interact specifically with genomic RNA to specify correct packaging, but how this process is accomplished remains a mystery. NC has also been reported to promote RNA–RNA duplex formation (Prats *et al.*, 1988, 1990), presumably necessary to accomplish genome–primer association and genome dimerization.

b. Protease (PR)

The region of the genome encoding the protease (referred to here as *pro*) always lies between *gag* and *pol*, but is expressed differently in different virus groups, depending on the disposition of sites of translational termination and readthrough (Jacks, 1990; Coffin, 1990b). PR is an aspartic protease, whose functional form is a dimer (Oroszlan and Luftig, 1990; Leis *et al.*, 1990; Wlodawer *et al.*, 1989). It is responsible for all the proteolytic cleavages generating the mature *gag* and *pol* proteins during virion maturation. It is the only virion protein whose three-dimensional structure has been determined (Miller, *et al.*, 1989; Weber *et al.*, 1989; Navia *et al.*, 1989). The timing of assembly, budding, and maturation events implies that PR is largely inactive in the precursor form but can be activated by events (including dimerization) occurring after budding. The suggestion that PR might also function early in infection (Roberts and Oroszlan, 1989) awaits confirmation.

c. pol Proteins

The *gag–pro–pol* protein precursor is cleaved to yield two major proteins from the *pol* domain. In some viruses additional products, either uncleaved (Dickson *et al.*, 1982) or additionally cleaved (Di Marzo Veronese *et al.*, 1986), are found as part of a heterodimeric structure with the "correct length" molecule, but are of unknown significance. The role of the *pol* proteins is the synthesis of viral DNA and its integration into host DNA soon after infection. It is important to keep in mind that, although these proteins are capable of carrying out the necessary enzymatic functions *in vitro*, the actual reactions *in vivo* take place in the context of a more complex structure, derived from the viral capsid (Bowerman *et al.*, 1989), which may be necessary to ensure fidelity and specificity.

RT. The reverse transcriptase protein contains all the activities known to be necessary for synthesis of viral DNA: RNA and DNA-directed DNA polymerase activities and the RNA endonucleolytic activities (e.g., RNase H) necessary for degradation of template and primers during synthesis (Varmus and Swanstrom, 1985; Varmus, 1987, 1988; Coffin, 1990b) (Chapter 5). The synthetic and nucleolytic activities lie in distinct, genetically separable domains in the order RT–RNase H (Prasad and Goff, 1989; Tanese and Goff, 1988).

IN. The integrase protein provides most of the enzymatic activities necessary for integration of the viral DNA into the cellular DNA target. These include nucleolytic activities to trim each 3′ end of the viral DNA and to cleave the DNA target, as well as a ligation function to join the viral to the cellular DNA (Brown, 1990; Bushman and Craigie, 1991, Craigie et al., 1990; Bushman et al., 1990; Katzman et al., 1989) (see Chapter 4).

d. env Proteins

The env gene is translated on membrane-bound polysomes to yield a precursor, which is subsequently modified by glycosylation and cleavage into two proteins which remain associated with one another by disulfide and noncovalent interactions. In contrast to the proteins of gag and pol, the env proteins are processed by cellular systems. The env heterodimer is itself organized into a higher-order structure, variously reported as dimer, trimer, or tetramer; depending on the method of analysis and perhaps the virus group (Schwaller et al., 1989; Earl et al., 1990; Yang et al., 1990). The only role of the env proteins is to mediate association of the virion with the host cell and entry into it. They are dispensable for both virion assembly and postpenetration events (Dickson et al., 1982; Coffin, 1990b).

SU. The surface protein is always the larger of the two env proteins and is invariably glycosylated, although to widely varying extents in different virus groups. The extent of glycosylation is a major determinant of the large variation in apparent molecular weight of this protein from group to group (Table II). The SU protein contains the site for interaction of the virion with the host-cell receptor. Retroviruses are unique among the virus families in the extent to which receptor utilization can vary among otherwise closely related viruses (see Chapter 5). In some groups, variation in receptor utilization from one isolate (type or subgroup) to another has permitted localization of the regions of SU involved in receptor interaction (Dorner and Coffin, 1986; Dorner et al., 1985; Vogt et al., 1986; Bova et al., 1988). SU seems to be the virion protein most exposed to the external environment and, not surprisingly, contains the major determinants for recognition by neutralizing antibodies.

The host ranges encoded within SU are useful for identifying and classifying otherwise closely related retroviruses into subgroups within certain species (Weiss, 1982; Teich, 1982; Chapters 6 and 7). Thus, distinct ALV strains have been isolated and designated as subgroups A–F based on their recognition of distinct (but sometimes allelic) cell receptors. Similarly, four receptor recognition specificities of murine leukemia viruses have been identified. In this case, they are named for the distribution of usable receptors among mice and other species. Thus, the receptor for ecotropic virus is found only on mouse cells (as well as some other rodents); that for xenotropic virus is found on cells of most species except mice (the subgroup E ALV receptor has an analogous distribution); and ampho and polytropic virus receptors are found on cells of both murine and nonmurine species. A similar classification can be made within the feline leukemia viruses.

TM. The transmembrane protein is the C-terminal cleavage product of the *env* protein precursor [excepting a small terminal peptide removed during maturation in some virus groups (Dickson *et al.*, 1982)]. It anchors the *env* protein complex to the virus envelope and mediates fusion of the envelope with the host-cell membrane. TM proteins contain several identifiable regions: an amino-terminal hydrophobic region necessary for membrane fusion (Gallaher, 1987; Crane *et al.*, 1988; Bosch *et al.*, 1989; Freed *et al.*, 1990), a hydrophobic membrane-spanning domain, and a cytoplasmic region, which, in at least one retrovirus group, seems dispensable for synthesis of infectious virions (Perez *et al.*, 1987).

III. RETROVIRAL TAXONOMY

A. Principles

The goal of taxonomy is to impose some sort of order on the diversity of viral agents for the purposes of simplifying communication among virologists, of aiding progress in understanding new isolates by comparison to properties of related known isolates, and of illuminating evolutionary relationships among isolates (Chapter 3). Regarding the latter point, it is important to remember that taxonomic assignments must always be made on the basis of incomplete knowledge of true relationships. Thus, they will usually not precisely reflect evolutionary relationships. At the least, however, taxonomy should not violate such relationships as they are understood at the time. For this reason, taxonomic organization is subject to adjustment from time to time as new knowledge of relationships is uncovered. Until recently, classification of retroviruses emphasized virion structure and pathogenicity (Murphy and Kingsbury, 1991; Teich, 1982, 1985; Brown, 1989). With the explosion of nucleotide sequence information (Weiss *et al.*, 1985; Myers *et al.*, 1990), it was apparent that these characteristics did not accurately reflect more fundamental relationships, and the classification scheme presented below was developed to rectify the situation.

Classification of viruses is developed and reviewed by study groups consisting of experts in specific viral fields, appointed by the International Committee on Taxonomy of Viruses (ICTV) (Murphy and Kingsbury, 1991; Brown, 1989). The ICTV meetings are held every 3 years (most recently in 1990) to review the status of viral taxonomy and to act on proposals for changes and additions from the various study groups. In addition to its role in developing taxonomy, the retrovirus study group has also occasionally become involved in the separate subject of nomenclature, taking up issues including the naming of oncogenes (Coffin *et al.*, 1981) and the immunodeficiency viruses (Coffin *et al.*, 1986).

All retroviruses are assigned to the family Retroviridae, which is further divided into seven genera, and then into subgenera and species. In general, there is virtually no detectable nucleotide sequence similarity between genera, but a careful analysis reveals relationships at the level of amino acid sequence (Doolittle *et al.*, 1989, 1990) (Chapter 3). The highest conservation of amino acid sequence between genera is found in *pol* and can be used to derive relationship "trees" like the one in Fig. 4. Similar relationships can be discerned in other viral genes and in general yield the same tree, implying coevolution of all regions of the genome. One exception to the similarity of branch order is in *env* genes. The *env* genes of the type D viruses more closely resemble those of the MLV-like group of viruses, while the rest of the type D genome is more like that of the mammalian type B viruses. This observation implies the involvement of a recombination event, which had the effect of moving an envelope gene from a virus resembling murine leukemia virus into one resembling mammary tumor virus some time in the history of the type D group (Sonigo *et al.*, 1986; Doolittle *et al.*, 1989).

The relationships revealed by the sort of computerized comparison shown in Fig. 5 are very useful for comparative and taxonomic purposes and for exposing broad patterns of retroviral evolution to further development. Great caution should be used in their interpretation, however, particularly in attempts to interpret rates of change and times of divergence. For reasons beyond the scope of this chapter [see Coffin (1990a) and Chapter 3 for a fuller argument], it is fallacious to assume proportionality between amount of divergence (branch lengths) and real time. To attempt to do so—as some workers have done (Smith, *et al.*, 1988)—is to risk errors of many orders of magnitude.

The sequence relationships are useful for establishing the broad taxonomic outlines, but do not themselves set the criteria for creation of

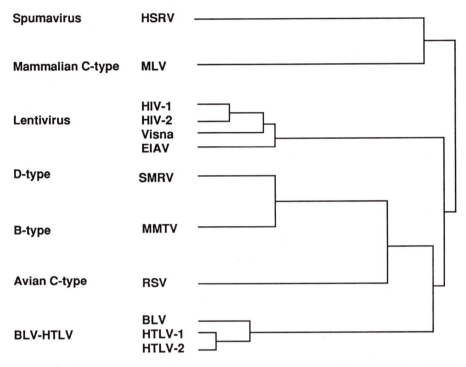

FIGURE 5. Sequence relationships of representative retrovirus RT proteins. See Table I for virus abbreviations. [Adapted from Doolittle *et al.* (1990).]

taxonomic groups or assignment of viruses to them. Instead, these goals are accomplished by analysis of a variety of properties, the most important of which is the overall genome organization. The hierarchy of characteristics and their use to assign virus isolates to distinct taxa is shown in outline form in Fig. 6.

B. Specific Viral Groups

The most recently promulgated classification of the family Retroviridae into genera, subgenera, and species is summarized in Table III. A brief description of each of the major groups follows. It should be noted that the names currently assigned to these groups are descriptive and provisional and do not necessarily adhere to taxonomically correct guidelines. Further elaboration of the biological properties of the individual groups will be found in subsequent chapters.

The *mammalian type B viruses* are a small group of viruses comprising the mammary tumor viruses of mice. Several isolates of exogenous

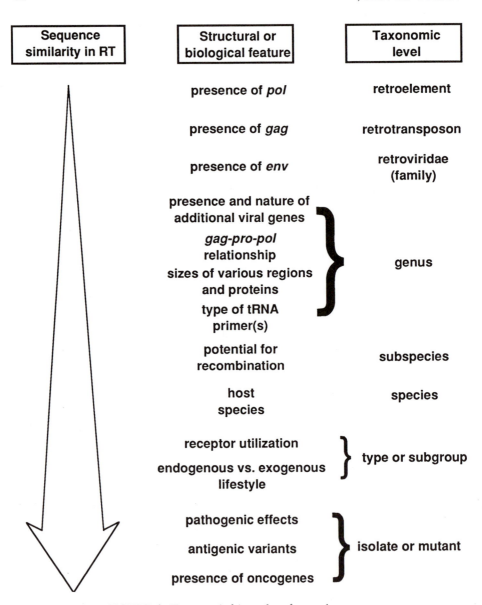

Sequence similarity in RT	Structural or biological feature	Taxonomic level
	presence of *pol*	retroelement
	presence of *gag*	retrotransposon
	presence of *env*	retroviridae (family)
	presence and nature of additional viral genes *gag-pro-pol* relationship sizes of various regions and proteins type of tRNA primer(s)	genus
	potential for recombination	subspecies
	host species	species
	receptor utilization endogenous vs. exogenous lifestyle	type or subgroup
	pathogenic effects antigenic variants presence of oncogenes	isolate or mutant

FIGURE 6. Taxonomic hierarchy of retroelements.

viruses transmitted via milk from mother to offspring have been described (Coffin, 1982b; Teich, 1982), and a number of closely related
endogenous proviruses are found in mice and some other species. Both
endogenous and exogenous viruses have been implicated in the induction of mammary carcinoma (Nusse *et al.*, 1985) and T-lymphoma (Yana-

gawa *et al.*, 1990; Dudley, 1988; Ball and Dekaban, 1987) in mice. Distantly related sequences can be found as endogenous elements in humans and other mammals (Horn *et al.*, 1986; Mariani-Costantini *et al.*, 1989), but the significance of these is uncertain.

The *MLV-related viruses* comprise a very large group of viruses found as exogenous infectious, vertically and horizontally transmitted agents in many mammalian species, including some primates, as well as some birds and reptiles. [Some of these have also been classified by their host range (Levy, 1978 Chapter 7).] Infectious human isolates are not known. A number of oncogene-containing members of the MLV-related viruses have been isolated. Closely related endogenous proviruses are also found in many mammals, and distantly related sequences are found in virtually all mammals, including humans (Cohen *et al.*, 1985; Hehlmann *et al.*, 1988; Harada *et al.*, 1987; Mager and Greeman, 1987). Virions have a type C morphology. These viruses are associated with a variety of pathogenic effects, including a variety of malignancies, immunodeficiencies, neurological degeneration, and other effects. They include some important model systems for oncogenesis, gene transfer, and other studies.

The *type D viruses* include a relatively small number of isolates, almost all from primates. Endogenous members are also found in some primate species. No oncogene-containing members are known. Most primate isolates are associated with immunodeficiencies [e.g., Mason Pfizer, sometimes called simian retrovirus (SRV) AIDS]. The one known nonprimate virus is associated with an infectious pulmonary carcinoma (known as jaagsiekte) of sheep.

The *ALV-related viruses* comprise a number of isolates, mostly exogenous, vertically and horizontally transmitted agents in chickens and a few other avian species (see Chapter 6). Closely related endogenous proviruses are also found in chickens and some pheasants. Many oncogene-containing isolates are known, and these have provided invaluable models for studies of oncogenesis. This group has many biological similarities to the MLV-related viruses, with which it was once grouped, but major differences in genome organization and sequence justify its separation into a distinct genus.

Spumaviruses (or foamy viruses) include many isolates from primates (Teich, 1982) and some other mammals. Related endogenous proviruses are not known to exist. These viruses cause a characteristic "foamy" cytopathology in cell culture. They are not known to cause any disease. Despite their prevalence, they represent the least-studied group of retroviruses.

The *HTLV–BLV group* includes a relatively small number of members of horizontally transmitted agents associated with T-lym-

TABLE III. Characteristics of Retrovirus Groups

Genus Subgenus	Examples	Type	Genome size[a] (kb)	"Extra" genes	tRNA primer	LTR size[a] (bp)	Endogenous members?	Members with oncogenes?	Pathogenesis
Mammalian B type	Mouse mammary tumor virus (MMTV)	B	10	1 (orf)	Lys-3	1300	Yes	No	Mammary carcinoma; T-cell lymphoma
MLV-related Mammalian C type	Murine leukemia virus	C	8.3	None	Pro (Glu)	600	Yes	Yes	Various: malignancies, immunodeficiency and neurological diseases
Reticuloendotheliosis	Avian reticuloendotheliosis virus								
Reptilian C type	Viper retrovirus								
D type Primate D type	Mason-Pfizer monkey virus, simian retrovirus (SRV)	D	8.0	None	Lys-1,2	350	Yes	No	Immunodeficiencies
Ovine D type[b]	Jaagsiekte retrovirus								(Malignancy)

ALV-related	Avian leukosis and sarcoma viruses	C	7.2	None	Trp	350	Yes	Yes	Malignancies, osteopetrosis, immunological disease
Spumavirus	Human, simian, feline foamy virus	S	11	Several	Lys-1,2	1150	No	No	None known
HTLV–BLV	Human T-lymphotrophic virus, bovine leukemia virus	C	8.3	2 (*tax, rex*)	Pro	550–750	No	No	T- or B-cell lymphoma, neurological disease
Lentivirus									
Primate immunodeficiency viruses (human, simian)	HIV, SIV	L	9.2	~6 (*tat, rev, nef, vif, vpu, vpr*)	Lys-1,2 or Lys-3	600	No	No	Immunodeficiencies, autoimmune degenerative, neurological, and other disease
Ovine-caprine lentiviruses	Visna/Maedi virus								
Equine lentiviruses	Equine infectious anemia virus (EIAV)								
Feline lentiviruses	Feline immunodeficiency virus (FIV)								
Bovine lentiviruses	Bovine immunodeficiency virus (BIV)								

[a] Approximate.
[b] Tentative.

phoma in humans and B-lymphoma in cattle. Virion morphology and assembly are similar to type C viruses. One of the human viruses is associated with neurological disorders (Rudge, 1989; Sarin *et al.*, 1989). Oncogene-containing or endogenous members are not known.

Lentiviruses include many intensively studied isolates of primates and other mammals. Members of this genus are exogenous, horizontally and vertically transmitted agents of immunodeficiencies and neurological and other diseases, including AIDS. None is known to contain oncogenes, and related endogenous viruses are not found. Lentiviruses have the most complex genome structure and expression strategy known. On the basis of relatively small differences in genome structure as well as sequence similarity (Myers *et al.*, 1990), known lentiviruses are further subdivided into subgenera. The primate lentiviruses include the human and simian immunodeficiency viruses (HIV and SIV); the ovine/caprine members include visna and the closely related caprine arthritis-encephalitis virus; the equine lentiviruses are represented by equine infectious anemia virus; and the bovine and feline lentiviruses by the corresponding immunodeficiency viruses. More subgenera are likely to be identified with time.

IV. CONCLUSIONS

To the outsider, retroviruses appear as a daunting array of different biological phenomena confounded by an equally bewildering nomenclature. A goal of taxonomy is to impose some order on this apparent confusion. While the formalism of the taxonomic exercise is disdained (or, at best, ignored) by the majority of virologists, it is nonetheless essential for several purposes. For one, it enhances understanding and communication among workers. Second, it has considerable predictive value: knowledge of a few characteristics of a virus provides readily testable hypotheses regarding other properties. Finally, it helps to illuminate relationships among viruses and the relationship between genome and biology of individual groups. To accomplish these goals, a taxonomy must be capable of change to accommodate new insights, even at the price of some additional confusion. Table IV compares the former taxonomic organization of the Retroviridae with that most recently recommended by the ICTV. The newer classification, which replaces reliance on biological features such as host range and pathogenicity with more fundamental molecular characteristics, should prove to be more accurate and useful. Like its predecessor, however, it is not written in stone and will likely change as new viruses, new relationships, and new ways of looking at them are discovered.

TABLE IV. Comparison of "Old" and "New" Taxonomy of Retroviruses[a]

Taxon	"Old taxonomy"	"New taxonomy"
Family	Retroviridae	Retroviridae
Subfamily	*Oncovirinae*	[b]
Genus	Type C oncovirus	MLV-related viruses
Subgenus	Mammalian type C oncoviruses	Mammalian type C viruses
Species	MLV, FeLV, GALV, etc.	MLV, FeLV, GALV, etc.
Species	HTLV, BLV	[c]
Subgenus	Reptilian type C oncoviruses	Reptilian type C viruses
Species	CSRV, VRV	CSRV, VRV
Subgenus	Avian type C oncoviruses	Reticuloendotheliosis viruses
Species	SNV, REV	SNV, REV
Species	ALV, RSV	[c]
Genus	Type B oncovirus	Mammalian type B viruses
Species	MMTV	MMTV
Genus	Type D oncovirus	Type D viruses
Species	MPMV, SMRV	MPMV, SMRV
Genus	[c]	ALV-related
Species	[c]	ALV, RSV
Genus	[c]	HTLV–BLV
Species	[c]	HTLV-I, II, BLV
Subfamily	*Lentivirinae*	[b]
Genus	*Lentivirus*	*Lentivirus*
Subgenus	—	Ovine/caprine lentiviruses
Species	Visna, CAEV	Visna, CAEV
Subgenus	—	Equine lentiviruses
Species	EIAV	EIAV
Subgenus	—	Primate lentiviruses
Species	HIV, SIV	HIV, SIV
Subgenus	—	Feline lentiviruses
Species	FIV	FIV
Subgenus	—	Bovine lentiviruses
Species	BIV	BIV
Subfamily	*Spumavirinae*	[b]
Genus	*Spumavirus*	*Spumavirus*
Species	HSRV, SFV	HSRV, SFV

[a] Abbreviations as in Table I.
[b] The subfamily level of classification is no longer used.
[c] These groups are no longer considered to be sufficiently similar to the mammalian type C viruses to be classified with them.

ACKNOWLEDGMENTS. I thank Dr. Matthew Gonda for providing the electron micrographs shown in Fig. 2. Work in my laboratory is supported by a grant from the National Cancer Institute.

V. REFERENCES

Arrigo, S., and Beemon, K., 1988, Regulation of Rous sarcoma virus RNA splicing and stability, *Mol. Cell. Biol.* **18**:4858.

Arrigo, S., Yun, M., and Beemon, K. 1987, Cis-acting regulatory elements within gag genes of avian retroviruses, *Mol. Cell. Biol.* **7**:388.

Ball, J. K., and Dekaban, G. A., 1987, Characterization of early molecular biological events associated with thymic lymphoma induction following infection by a thymotropic type-B retrovirus, *Virology* **161**:357.

Bernhard, W., 1960, The detection and study of tumor viruses with the electron microscope, *Cancer Res.* **20**:712.

Bishop, J. M., 1991, Molecular themes in oncogenesis, *Cell* **64**:235.

Bishop, J. M., and Varmus, H. E., 1982, Functions and origins of retroviral transforming genes, in: RNA Tumor Viruses (R. Weiss, N. Teich, H. Varmus, and J. Coffin, eds.), pp. 999–1108, Cold Spring Harbor Laboratory, Cold Spring Harbor, New York.

Bolognesi, D. P., Montelaro, R. C., Frank, H., and Schafer, W., 1978, Assembly of type C oncornaviruses: A model, *Science* **199**:183.

Bosch, M. L., Earl, P. L., Fargnoli, K., Picciafuocco, S., Giobini, F., Wong-Staal, F., and Franchini, G., 1989, Identification of the fusion peptide of primate immunodeficiency viruses, *Science* **244**:694.

Bova, C. A., Olsen, J. C., and Swanstrom, R., 1988, The avian retrovirus env gene family: Molecular analysis of host range and antigenic variants, *J. Virol.* **62**:75.

Bowerman, B., Brown, P. O., Bishop, J. M., and Varmus, H. E., 1989, A nucleoprotein complex mediates the integration of retroviral DNA, *Genes Dev.* **3**:469.

Brown, F., 1989, The classification and nomenclature of viruses: Summary of results of meetings of the International Committee on Taxonomy of Viruses in Edmonton, Canada 1987, *Intervirology* **30**:181.

Brown, P. O., 1990, Integration of retroviral DNA, in: *Retroviruses. Strategies of Replication* (R. Swanstrom and P. K. Vogt, eds.), pp. 19–48, Springer-Verlag, New York.

Bryant, M., and Ratner, L., 1990, Myristoylation-dependent replication and assembly of human immunodeficiency virus 1, *Proc. Natl. Acad. Sci. USA* **87**:523.

Bushman, F. D., and Craigie, R., 1991, Activities of human immunodeficiency virus (HIV) integration protein *in vitro*: Specific cleavage and integration of HIV DNA, *Proc. Natl. Acad. Sci. USA* **88**:1339.

Bushman, F. D., Fujiwara, T., and Craigie, R., 1990, Retroviral DNA integration directed by HIV integration protein *in vitro*, *Science* **249**:1555.

Coffin, J. M., 1982a, Structure of the retroviral genome, in: *RNA Tumor Viruses* (R. Weiss, N. Teich, H. Varmus, and J. Coffin, eds.), pp. 261–369, Cold Spring Harbor Laboratory, Cold Spring Harbor, New York.

Coffin, J. M., 1982b, Endogenous viruses, in: *RNA Tumor Viruses* (R. Weiss, N. Teich, H. Varmus, and J. Coffin, eds.), pp. 1109–1204, Cold Spring Harbor Laboratory, Cold Spring Harbor, New York.

Coffin, J. M., 1990a, Genetic variation in retroviruses, in: *Applied Virology Research*, Vol. 2 (E. Karstak, R. G. Marusyk, F. A. Murphy, and M. H. V. Van Regenmortel, eds.), pp. 11–13, Plenum Press, New York.

Coffin, J. M., 1990b, Retroviridae and their replication, in: *Virology*, 2nd ed. (B. Fields, D. Knipe, and R. Chanock, eds.), pp. 1437–1500, Raven Press, New York.

Coffin, J. M., Varmus, H. E., Bishop, J. M., Essex, M., Hardy, W. D., Martin, G. S., Rosenberg, N. E., Scolnick, E. M., Weinberg, R. A., and Vogt, P. K., 1981, A proposal for naming host cell-derived inserts in retrovirus genomes. *J. Virol.* **40**:953.

Coffin, J., Haase, A., Levy, J. A., Montagnier, L., Oroszlan, S., Teich, N., Temin, H., Toyoshima, K., Varmus, H., Vogt, P., and Weiss, R., 1986, Human immunodeficiency viruses, *Science* **232**:697.

Cohen, M., Powers, M., O'Connell, C., and Kato, N., 1985, The nucleotide sequence of the env gene from the human provirus ERV3 and isolation and characterization of an ERV3-specific cDNA, *Virology* **147**:449.

Copeland, N. G., Jenkins, N. A., Nexo, B., Schultz, A. M., Rein, A., Middelsen, T., and Jorgensen, P., 1988, Poorly expressed endogenous ecotropic provirus of DBA/2 mice encodes a mutant Pr65gag protein that is not myristylated, *J. Virol.* **62**:479.

Craigie, R., Fujiwara, T., and Bushman, F., 1990, The IN protein of Moloney murine leukemia virus processes the viral DNA ends and accomplishes their integration *in vitro*, *Cell* **62**:829.

Crane, S. E., Clements, J. E., and Narayan, O., 1988, Separate epitopes in the envelope of Visna virus are responsible for fusion and neutralization: Biological implications for anti-fusion antibodies in limiting virus replication, *J. Virol.* **62**:2680.

Csepany, T., Lin, A., Baldick, C. J., and Beemon, K., 1990, Sequence specificity of N^6-adenosine methyltransferase, *J. Biol. Chem.* **265**:20117.

Dalton, A. J., Potter, M., and Merwin, R. M., 1961, Some ultrastructural characteristics of a series of primary and secondary plasma-cell tumors of the mouse, *J. Natl. Cancer Inst.* **26**:1221.

Di Marzo Veronese, F., Copeland, T. D., De Vico, A. L., Rahman, R., Oroszlan, S., Gallo, R. C., and Sarngadharan, M. G., 1986, Characterization of highly immunogenic p66/p51 as the reverse transcriptase of HTLV-III/LAV, *Science* **231**:1289.

Dickson, C., Eisenmann, R., Fan, H., Hunter, E., and Teich, N., 1982, Protein biosynthesis and assembly, in: *RNA Tumor Viruses* (R. A. Weiss, N. Teich, H. E. Varmus, and J. M. Coffin, eds.), pp. 513–648, Cold Spring Harbor Laboratory, Cold Spring Harbor, New York.

Dickson, C., Eisenmann, R., and Fan, H., 1985, Protein synthesis and assembly, in: *RNA Tumor Viruses* (R. Weiss, N. Teich, H. Varmus, and J. Coffin, eds.), pp. 135–146, Cold Spring Harbor Laboratory, Cold Spring Harbor, New York.

Doolittle, R. F., Feng, D.-F., Johnson, M. S., and McClure, M. A., 1989, Origins and evolutionary relationships of retroviruses, *Q. Rev. Biol.* **64**:1.

Doolittle, R. F., Feng, D. F., McClure, M. A., and Johnson, M. S., 1990, Retrovirus phylogeny and evolution, in: *Retroviruses. Strategies of Replication* (R. Swanstrom and P. K. Vogt, eds.), pp. 1–18, Springer-Verlag, New York.

Dorner, A. J., and Coffin, J. M., 1986, Determinants for receptor interaction and cell killing on the avian retrovirus glycoprotein gp85, *Cell* **45**:365.

Dorner, A. J., Stoye, J. P., and Coffin, J. M., 1985, Molecular basis of host range variation in retroviruses, *J. Virol.* **53**:32.

Dudley, J. P., 1988, Mouse mammary tumor proviruses from a T-cell lymphoma are associated with the retroposon L1Md, *J. Virol.* **62**:472.

Earl, P. L., Doms, R. W., and Moss, B., 1990, Oligomeric structure of the human immunodeficiency virus type 1 envelope glycoprotein, *Proc. Natl. Acad. Sci. USA* **87**:648.

Fine, D., and Schochetman, G., 1978, Type D primate retroviruses. A review, *Cancer Res.* **38**:3123.

Freed, E. O., Myers, D. J., and Risser, R., 1990, Characterization of the fusion domain of the human immunodeficiency virus type 1 envelope glycoprotein gp41, *Proc. Natl. Acad. Sci. USA* **87**:4650.

Fu, S., Phillips, N., Jentoft, J., Tuazon, P. T., Traugh, J. A., and Leis, J., 1985, Site-specific phosphorylation of avian retrovirus nucleocapsid protein pp12 regulates binding to RNA, *J. Biol. Chem.* **260**:9941.

Fu, X., Katz, R. A., Skalka, A. M., and Leis, J., 1988, Site-directed mutagenesis of the avian retrovirus nucleocapsid protein pp12: Mutation which affects RNA binding *in vitro* blocks viral replication, *J. Biol. Chem.* **263**:2134.

Gallaher, W. R., 1987, Detection of a fusion peptide sequence in the transmembrane protein of human immunodeficiency virus, *Cell* **50**:327.

Gonda, M., Boyd, A. L., Nagashima, K., and Gilden, R. V., 1989, Pathobiology, molecular organization, and ultrastructure of HIV, *Arch. AIDS Res.* **3**:1.

Harada, F., Tsukada, N., and Kato, N., 1987, Isolation of three kinds of human endogenous retrovirus-like sequences using tRNA^{Pro} as a probe, *Nucleic Acids Res.* **15**:9153.

Hawley, R. G., Shulman, M. J., and Hozumi, N., 1984, Transposition of two different intracisternal A particle elements into an immunoglobulin kappa-chain gene, *Mol. Cell. Biol.* **4**:2565.

Heberlein, C., Kawai, M., Franz, M.-J., Beck-Engeser, G., Daniel, C. P., Ostertag, W., and Stocking, C., 1990, Retrotransposons as mutagens in the induction of growth autonomy in hematopoietic cells, *Oncogene* **5**:1799.

Hehlmann, R., Brack-Werner, R., and Leib-Mosch, C., 1988, Human endogenous retroviruses, *Leukemia* **2**:167S.

Heidecker, G., Lerche, M. W., Lowenstine, L. J., Lackner, A. A., Osborn, K. G., Gardner, M. B., and Marx, P. A., 1987, Induction of simian acquired immune deficiency syndrome (SAIDS) with a molecular clone of a type D SAIDS retrovirus, *J. Virol.* **61**:3066.

Heidmann, O., and Heidmann, T., 1991, Retrotransposition of a mouse IAP sequence tagged with an indicator gene, *Cell* **64**:159.

Henderson, L. E., Krutzsch, H. C., and Oroszlan, S., 1983, Myristyl amino-terminal acylation of murine retrovirus proteins: An unusual post-translational protein modification, *Proc. Natl. Acad. Sci USA* **80**:339.

Henderson, L. E., Sowder, R., Copeland, T. D., Oroszlan, S., Arthur, L. O., Robey, W. G., and Fischinger, P. J., 1987, Direct identification of class II histocompatibility DR proteins in preparations of human T-cell lymphotropic virus type III, *J. Virol.* **61**:629.

Horn, T. M., Huebner, K., Croce, C., and Callahan, R., 1986, Chromosomal locations of members of a family of novel endogenous human retroviral genomes, *J. Virol.* **58**:955.

Horowitz, M., Luria, S., Rechavi, G., and Givol, D., 1984, Mechanism of activation of the mouse c-*mos* oncogene by the LTR of an instracisternal A-particle gene, *EMBO J.* **3**:2937.

Jacks, T., 1990, Translational suppression in gene expression in retroviruses and retrotransposons, in: *Retroviruses. Strategies of Replication* (R. Swanstrom and P. K. Vogt, eds.), pp. 93–124, Springer-Verlag, New York.

Jacks, T., and Varmus, H. E., 1985, Expression of the Rous sarcoma virus *pol* gene by ribosomal frameshifting, *Science* **230**:1237.

Jorgensen, E. C., Kjeldgaard, N. O., Pedersen, F. S., and Jorgensen, P., 1988, A nucleotide substitution in the gag N terminus of the endogenous ecotropic DBA/2 virus prevents Pr65gag myristylation and virus replication, *J. Virol.* **62**:3217.

Kane, S. E., and Beemon, K., 1985, Precise localization of m6A in Rous sarcoma virus RNA reveals clustering of methylation sites: Implications for RNA processing, *Mol. Cell. Biol.* **5**:2298.

Katz, R. A., and Jentoff, J. E., 1989, What is the role of the Cys–His motif in retroviral nucleocapsid (NC) proteins? *Bioessays* **11**:176.

Katz, R. A., and Skalka, A. M., 1990, Control of retroviral RNA splicing through maintenance of suboptimal processing signals, *Mol. Cell. Biol.* **10**:696.

Katz, R. A., Kotler, M., and Skalka, A. M., 1988, *cis*-Acting intron mutations that affect the efficiency of avian retroviral RNA splicing: Implications for mechanisms of control, *J. Virol.* **62**:2686.

Katzman, M., Katz, R. A., Skalka, A. M., and Leis, J., 1989, The avian retroviral integration protein cleaves the terminal sequences of linear viral DNA at the *in vivo* sites of integration, *J. Virol.* **63**:5319.

Kuff, E. L., and Lueders, K. K., 1988, The intracisternal A-particle gene family: Structure and functional aspects, *Adv. Cancer Res.* **51**:183.

Kuff, E. L., Wivel, N. A., and Lueders, K. K., 1968, The extraction of intracisternal A particles from a mouse plasma-cell tumor, *Cancer Res.* **28**:2137.

Leis, J., Baltimore, D., Bishop, J. M., Coffin, J., Fleissner, E., Goff, S. P., Oroszlan, S., Robinson, H., Skalka, A. M., Temin, H. M., and Vogt, V., 1988, Standardized and simplified nomenclature for proteins common to all retroviruses, *J. Virol.* **62**:1808.

Leis, J., Weber, I., Wlodawer, A., and Skalka, A. M., 1990, Structure–function analysis of the Rous sarcoma virus-specific proteinase, *ASM News* **56**:77.

Levy, J. A., 1978, Xenotropic type C viruses, in: *Current Topics in Microbiology and Immunology,* Vol. 79, pp. 111–213, Springer-Verlag, Heidelberg.

Linial, M. L., and Miller, A. D., 1990, Retroviral RNA packaging: Sequence requirements and implications, in: *Retroviruses. Strategies of Replication* (R. Swanstrom and P. K. Vogt, eds.), pp. 125–152, Springer-Verlag, New York.

Lueders, K. K., and Kuff, E. L., 1977, Sequences associated with intracisternal A particles are repeated in the mouse genome, *Cell* **12**:963.

Mager, D. X., and Greeman, J. D., 1987, Human endogenous retroviruslike genome with type C pol sequences and gag sequences related to human T-cell lymphotropic viruses, *J. Virol.* **61**:4060.

Mann, R. S., Mulligan, R. C., and Baltimore, D., 1983, Construction of a retrovirus packaging mutant and its use to produce helper-free defective retrovirus, *Cell* **32**:871.

Mariani-Costantini, R., Horn, T. M., and Callahan, R., 1989, Ancestry of a human endogenous retrovirus family, *J. Virol.* **63**:4982.

Marx, P. A., Bryant, M. L., Osborn, K. G., Maul, D. H., Lerche, N. W., Lowenstine, L. J., Kluge, J. D., Saiss, C. P., Henrickson, R. V., Shiigi, S. M., Wilson, B. J., Malley, A., Olson, L. C., McNulty, W. P., Arthur, L. O., Gilden, R. V., Barker, C. S., Hunter, E., Munn, R. J., Heidecker, G., and Gardner, M. B., 1985, Isolation of a new serotype of simian acquired immune deficiency syndrome type D retrovirus from Celebes black macaques (*Macaca nigra*) with immune deficiency and retroperitoneal fibromatosis, *J. Virol.* **56**:571.

Meric, C., and Spahr, P.-F., 1986, Rous sarcoma virus nucleic acid binding protein p12 is necessary for viral 70S RNA dimer formation and packaging, *J. Virol.* **60**:450.

Meric, C., Darlix, J. L., and Spahr, P.-F., 1986, It is Rous sarcoma virus p12 and not p19 that binds tightly to Rous sarcoma virus RNA, *J. Mol. Biol.* **173**:531.

Mietz, J. A., Grossman, Z., Lueders, K. K., and Kuff, E. L., 1987, Nucleotide sequence of a complete mouse intracisternal A-particle genome: Relationship to known aspects of particle assembly and function, *J. Virol.* **61**:3020.

Miller, C. K., and Temin, H. M., 1986, Insertion of several different DNAs in reticuloendotheliosis virus strain T suppresses transformation by reducing the amount of subgenomic DNA, *J. Virol.* **58**:75.

Miller, M., Jaskolski, M., Mohana Rao, J. K., Leis, J., and Wlodawer, A., 1989, Crystal structure of a retroviral protease proves relationship to aspartic protease family, *Nature* **337**:576.

Murphy, F. A., and Kingsbury, D. A., 1991, Virus taxonomy, in: *Fundamental Virology* (B. Fields, D. Knipe, and R. Chanock, eds.), pp. 9–36, Raven Press, New York.

Myers, G., Rabson, A. B., Josephs, S. F., Smith, T. F., Berzofsky, J. A., and Wong-Staal, F., 1990, Human retroviruses and AIDS, 1990, Los Alamos National Laboratory, Los Alamos, N.M.

Navia, M. A., Fitzgerald, P. M. D., McKeever, B. M., Leu, C.-T., Heimbach, J. C., Herber, W. K., Sigal, I. S., Darke, P. L., and Springer, J. P., 1989, Three-dimensional structure of aspartyl protease from human immunodeficiency virus HIV-1, *Nature* **337**:615.

Nusse, R., van Ooyen, A., Rijsewijk, F., van Lohuizen, M., Schuuring, E., and van't Veer, L., 1985, Retroviral insertional mutagenesis in murine mammary cancer, *Proc. R. Soc. Lond.* **226**:3.

Ono, M., 1986, Molecular cloning and long terminal repeat sequences of human endogenous retrovirus genes related to types A and B retrovirus genes, *J. Virol.* **58**:937.

Ono, M., Cole, M. D., White, A. T., and Huang, R. C. C., 1980, Sequence organization of cloned intracisternal A particle genes, *Cell* **21**:465.

Ono, M., Toh, H., Miyata, T., and Awaaya, T., 1985, Nucleotide sequence of the Syrian hamster intracisternal A-particle gene: Close evolutionary relationship of type A particle gene to types B and D oncovirus genes, *J. Virol.* **5**:387.

Oroszlan, S., and Luftig, R. B., 1990, Retroviral proteinases, in: *Retroviruses. Strategies of Replication* (R. Swanstrom and P. K. Vogt, eds.), pp. 153–186, Springer-Verlag, New York.

Perez, L. G., Davis, G. L., and Hunter, E., 1987, Mutants of the Rous sarcoma virus envelope glycoprotein that lack the transmembrane anchor and cytoplasmic domains: Analysis of intracellular transport and assembly into virions, *J. Virol.* **61**:2981.

Prasad, V. R., and Goff, S. P., 1989, Linker insertion mutagenesis of the human immunodeficiency virus reverse transcriptase expressed in bacteria: Definition of the minimal polymerase domain, *Proc. Natl. Acad. Sci. USA* **86**:3104.

Prats, A. C., Sarih, L., Gabus, C., Litvak, S., Keith, G., and Darlix, J.-L., 1988, Small finger protein of avian and murine retroviruses has nucleic acid annealing activity and positions the replication primer tRNA onto genomic RNA, *EMBO J.* **7**:1136.

Prats, A.-C., Roy, C., Wang, P., Erard, M., Housset, V., Gabus, C., Paoletti, C., and Darlix, J.-L., 1990, *cis* elements and *trans*-acting factors involved in dimer formation of murine leukemia virus RNA, *J. Virol.* **64**:774.

Rein, A., McClure, M. R., Rice, N. R., Luftig, R. B., and Schultz, A. M., 1986, Myristylation site in Pr65gag is essential for virus particle formation by Moloney murine leukemia virus, *Proc. Natl. Acad. Sci. USA* **83**:7246.

Rhee, S. S., and Hunter, E., 1987, Myristylation is required for intracellular transport but not for assembly of D-type retrovirus capsids, *J. Virol.* **61**:1045.

Rhee, S. S., and Hunter, E., 1990, A single amino acid substitution within the matrix protein of a type D retrovirus converts its morphogenesis to that of a type C retrovirus, *Cell* **63**:77.

Rhee, S. S., Hui, H., and Hunter, E., 1990, Preassembled capsids of type D retroviruses contain a signal sufficient for targeting specifically to the plasma membrane, *J. Virol.* **64**:3844.

Roberts, M. M., and Oroszlan, S., 1989, The preparation and biochemical characterization of intact capsids of equine infectious anemia virus, *Biochem. Biophys. Res. Commun.* **160**:486.

Rudge, P., 1989, HTLV-1 and neurological disease, *Curr. Opin. Neurol. Neurosurg.* **2**:195.

Sarin, P. S., Rodgers-Johnson, P., Sun, D. K., Thornton, A. H., Morgan, O. S. C., Gibbs, W. N., Mora, C., McKhann, G. I., Gajdusek, D. C., and Gibbs, C. J. J., 1989, Comparison of a human T-cell lymphotropic virus type I strain from cerebrospinal fluid of a Jamaican patient with tropical spastic paraparesis with a prototype human T-cell lymphotropic virus type I, *Proc. Natl. Acad. Sci. USA* **86**:2021.

Schwaller, M., Smith, G. E., Skehel, J. J., and Wiley, D. C., 1989, Studies with crosslinking reagents on the oligomeric structure of the *env* glycoprotein of HIV, *Virology* **172**:367.

Schwartz, S., Felber, B. K., Fenyo, E.-M., and Pavlakis, G. N., 1990, Env and vpu proteins of human immunodeficiency virus type 1 are produced from multiple bicistronic mRNAs, *J. Virol.* **64**:5448.

Smith, T. F., Srinivasan, A., Schochetman, G., Marcus, M., and Myers, G., 1988, The phylogenetic history of immunodeficiency viruses, *Nature* **333**:573.

Sonigo, P., Barker, C., Hunter, E., and Wain-Hobson, S., 1986, Nucleotide sequence of Mason-Pfizer monkey virus: An immunosuppressive D-type retrovirus, *Cell* **45**:375.

Stoye, J. P., and Coffin, J. M., 1985, Endogenous viruses, in: *RNA Tumor Viruses* (R. Weiss, N. Teich, H. Varmus, and J. Coffin, eds.), pp. 357–404, Cold Spring Harbor Laboratory, Cold Spring Harbor, New York.

Tanese, N., and Goff, S. P., 1988, Domain structure of the Moloney murine leukemia virus reverse transcriptase: Mutational analysis and separate expression of the DNA polymerase and RNase H activities, *Proc. Natl. Acad. Sci. USA* **85**:1777.

Teich, N., 1982, Taxonomy of retroviruses, in: *RNA Tumor Viruses* (R. Weiss, N. Teich, H. Varmus, and J. Coffin, eds.), pp. 25–208, Cold Spring Harbor Laboratory, Cold Spring Harbor, New York.

Teich, N., 1985, Taxonomy of retroviruses, in: *RNA Tumor Viruses* (R. Weiss, N. Teich, H. Varmus, and J. Coffin, eds.), pp. 1–16, Cold Spring Harbor Laboratory, Cold Spring Harbor, New York.

Thayer, R. M., Power, M. D., Bryant, M. L., Gardner, M. B., Barr, P. J., and Luciw, P. A., 1987, Sequence relationships of type D retroviruses which cause simian acquired immunodeficiency syndrome, *Virology* **157**:317.

Varmus, H. E., 1987, Reverse transcription, *Sci. Am.* **257**:56.

Varmus, H., 1988, Retroviruses, *Science* **240**:1427.

Varmus, H. E., and Swanstrom, R., 1985, Replication of retroviruses, in: *RNA Tumor Viruses* (R. Weiss, N. Teich, H. Varmus, and J. Coffin, eds.), pp. 74–134, Cold Spring Harbor Laboratory, Cold Spring Harbor, New York.

Vogt, M., Haggblom, C., Swift, S., and Haas, M., 1986, Specific sequences of the *env* gene determine the host range of two HC-negative viruses of the Rauscher virus complex, *Virology* **154**:420.

Watanabe, S., and Temin, H. M., 1982, Encapsidation sequences for spleen necrosis virus, an avian retrovirus, are between the 5' long terminal repeat and the start of the *gag* gene, *Proc. Natl. Acad. Sci. USA* **79**:5986.

Weber, I. T., Miller, M., Jaskolski, M., Leis, J., Skalka, A. M., and Wlodawer, A., 1989, Molecular modeling of the HIV-1 protease and its substrate binding site, *Science* **243**:928.

Weiss, R. A., 1982, Experimental biology and assay of RNA tumor viruses, in: *RNA Tumor Viruses* (R. Weiss, N. Teich, H. Varmus, and J. Coffin, eds.), pp. 209–260, Cold Spring Harbor Laboratory, Cold Spring Harbor, New York.

Weiss, R., Teich, N., Varmus, H., and Coffin, J., 1985, RNA Tumor Viruses, Cold Spring Harbor Laboratory, Cold Spring Harbor, New York.

Willis, J. W., and Craven, R. C., 1991, Form, function and use of retroviral gag proteins, *AIDS* **5**:639.

Wlodawer, A., Miller, M., Jaskolski, M., Sathyanarayana, B. K., Baldwin, E., Weber, I. T., Selk, L. M., Clawson, L., Schneider, J., and Kent, S. B. H., 1989, Conserved folding in retroviral proteases: Crystal structure of a synthetic HIV-1 protease, *Science* **245**:616.

Yanagawa, S.-I, Muakami, A., and Tanaka, H., 1990, Extra mouse mammary tumor proviruses in DBA/2 mouse lymphomas acquire a selective advantage in lymphocytes by alteration in the U3 region of the long terminal repeat, *Virol.* **64**:2472.

Yang, Y., Tojo, A., Watanabe, N., and Amanuma, H., 1990, Oligomerization of Friend spleen focus-forming virus (SFFV) env glycoproteins, *Virology* **177**:312–316.

Yoshinaka, Y., Katoh, I., Copeland, T. D., and Oroszlan, S. J., 1985, Murine leukemia virus protease is encoded by the gag–pol gene and is synthesized through suppression of an amber termination codon, *Proc. Natl. Acad. Sci. USA* **82**:1618.

CHAPTER 3

Evolutionary Potential of Complex Retroviruses

Gerald Myers and George N. Pavlakis

I. INTRODUCTION

Before the recent clinical identification of acquired immune deficiency syndrome (AIDS) and the discovery of the human immunodeficiency viruses (HIV), our understanding of retroviral evolution was based largely on studies of the murine and avian oncoviruses. Recent comparisons of oncoviruses and these newly discovered retroviruses suggest that the AIDS viruses and, more generally, the lentiviruses of which they form a subgroup constitute a special case of viral and molecular evolution. These viruses differ from the oncoviruses in several important ways. The most significant of these is that they are able to regulate the expression of their own genes. We will therefore characterize them as "complex retroviruses" (Cullen, 1991; Chapter 2). Other newly discovered retroviruses, not classified with the lentiviruses, that also have this characteristic are the human T-cell leukemia viruses (HTLV) and the human spumaretroviruses (HSRV or HSP). Although there is still much to be learned about these other complex retroviruses, we can explore some interesting similarities and differences between them and the lentiviruses. For instance, HIV is highly variable, whereas HTLV is not. As

GERALD MYERS • Theoretical Biology and Biophysics, Los Alamos National Laboratory, Los Alamos, New Mexico 87545.   GEORGE N. PAVLAKIS • Human Retrovirus Section, National Cancer Institute, Frederick Cancer Research and Development Center, ABL-Basic Research Program, Frederick, Maryland 21701-1201.

The Retroviridae, Volume 1, edited by Jay A. Levy. Plenum Press, New York, 1992.

we shall see, the lentiviruses as a group are among the oxymorons of natural history, degenerate and imprecise in their complexity. When they are examined against a backdrop of other, more stable retroviruses, they provide us with a fuller sense of the retroviral evolutionary potential.

In light of the fact that the genetic variability and evolutionary plasticity of oncoviral genomes has been richly described and analyzed in major reviews, among which are those of Chiu *et al.* (1984), Temin and Engels (1984), Temin (1985, 1988a,b, 1989), Weiss *et al.* (1985), Lowy (1986), Doolittle *et al.* (1989, 1990), Katz and Skalka (1990), Coffin (1990), and Chapters 1 and 2 of this volume, we will focus our attention in what follows upon the "special case" of the evolution of the lentiviruses and other complex retroviruses. This will require us to give particular attention to the molecular structure and variation of the AIDS viruses (HIV), about which numerous recent studies have provided valuable information. HIV is especially interesting for our purposes because it appears to have made a recent ecological and evolutionary breakthrough out of a still to be discovered retroviral reservoir (Myers *et al.*, 1992a).

There is lack of general agreement about the distal evolutionary history of retroviruses; they may be old or they may be new (Doolittle *et al.*, 1989). The oncoviruses, which served as prototypes for retroviral evolution up to the discovery of HIV and its relatives, are thought to be sufficiently old that coevolution with their hosts is common. Thus, oncoviruses from a given host have tended to be decidedly homologous (Lowy, 1986), and divergence, rather than convergence, has accounted for much of what we know about them. The newly discovered human retroviruses—HIV, HTLV, and HSRV—are a curious group of variably complex retroviruses that are only remotely homologous, or "paraphyletically related" (Gould *et al.*, 1987), as we shall see. They therefore invite speculation about convergent retroviral evolution. The lentiviruses, and HIV in particular, do not patently exemplify coevolution with their host, nor do they appear to approximate steady-state evolution (Gould *et al.*, 1987; Myers *et al.*, 1992a) at this time: some HIVs have greater protein and nucleotide sequence homology with simian immunodeficiency viruses (SIV) than with other HIVs. Accordingly, our focus is upon proximal facets of retroviral evolution as manifested in certain emerging retroviruses. The viewpoint is as prospective as it is retrospective.

Lentiviruses are named for their comparatively slow pathogenic effects. So far as is known, they, unlike oncoviruses, have adapted themselves to nondividing host cells (Narayan and Clements, 1990; McCune, 1991). To the extent that this is universally true, it goes a long way toward explaining the complexity and resourcefulness of lentiviruses; in particular, it accounts for their possession of multiple RNA transcripts and diverse accessory proteins. But the lentiviruses are also exceptionally variable, and they have an unusual base composition quite unlike that of the vast majority of host cell genes or, for that matter, almost any other genes. Do these characteristics help us understand their adaptation to relatively stable, differentiated host cells? How can "slow viruses" adapted to slowly dividing cells participate in "fast-forward" evolution? With respect to the emergence of an acutely lethal variant of the simian immunodeficiency virus, Martin (1990) invites us to wonder how fast-acting the "slow viruses" might become. These are the specific questions considered in our review.

In the discussion that follows we will use data provided by molecular sequence analysis to explore the evolutionary potential, that is, the molecular constitutions and processes, the constraints, and, so to speak, boundary conditions, of HIV and the other lentiviruses. We will investigate, on the one hand, the terms of the exceptional complexity of these viruses as compared to other retroviruses, and on the other hand, the apparent excesses of their genetic instability as compared to other known viruses.

This chapter will have two parts. The first, entitled "Kinematics of Lentiviral Evolution," summarizes the phenomena that define, to the best of our knowledge, the molecular evolution of the AIDS virus and its relatives. The second part, entitled "Dynamics of Lentiviral Evolution," considers what is known or theorized about the mechanisms and effects that confer complexity upon these remarkable retroviruses. In short, what are (as Lucretius would put it) the "outer forms and inner workings" of the lentiviral engine?

II. KINEMATICS OF LENTIVIRAL EVOLUTION

A. Phylogenetic Analysis

1. *pol* Gene Homologies

Given a homologous set of retroviral nucleotide or amino acid sequences, it is possible to infer a genealogy of the molecules, that is, a phylogenetic history in the form of a tree. With the advent of rapid DNA sequencing, tree analysis has become both a dependable and a heuristic procedure for extracting the rich information present in nucleotide and amino acid sequences (Nei, 1987; Li and Graur, 1991). There are pitfalls in this approach [excellently reviewed by Felsenstein (1988) and Sidow and Wilson (1990)]: in particular, statistically robust tree analysis requires sufficient information without excessive noise due to multiple hits at a given site, so-called "character conflict," and backmutation. As we shall show, lentiviral nucleotide and amino acid sequence sets, with some noteworthy exceptions, generally satisfy the optimal condition for cluster analysis, which is that there be enough information for consistency without statistically intrusive noise. In the discussion that follows, tree analysis of *pol* gene sequences from complex retroviruses, most particularly the lentiviruses, will be compared to a previously published, similar analysis of oncoviral *pol* gene sequences.

Figure 1 is a phylogenetic tree based upon homologous *pol* gene nucleotide sequences from:

>   I. Primate immunodeficiency viruses (HIV and SIV, human and simian immunodeficiency viruses, respectively).

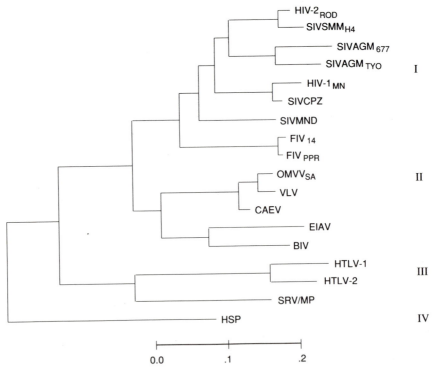

FIGURE 1. Phylogenetic tree based upon *pol* gene nucleotide sequences of (I) primate immunodeficiency viruses; (II) nonprimate lentiviruses; (III) HTLV-1, HTLV-2, and SRV$_{MP}$ (Mason-Pfizer type D retrovirus); and (IV) a human spumaretrovirus (HSP). For the sake of clarity, only a few representative sequences of each group are included in the tree. This minimum-length tree was generated by the PAUP algorithm, as described in T. F. Smith *et al.* (1988b) and Myers *et al.* (1989). The branching order of the primate immuno-deficiency viruses is uncertain for reasons discussed in the text. Evolutionary distances, expressed in terms of percentage nucleotide differences, can be read from the horizontal branch lengths using the scale bar at the bottom of the tree.

II. Other lentiviruses (BIV, bovine immunodeficiency virus; FIV, feline immunodeficiency virus; EIAV, equine infectious anemia virus; CAEV, caprine arthritis-encephalitis virus; VLV, visna lentivirus; $OMVV_{SA}$, the closely related ovine visna virus from South Africa).

III. Human T-cell lymphotropic viruses type 1 and 2 (HTLV-1 and HTLV-2, sometimes abbreviated herein as simply HTLV-1/2) and a simian type D retrovirus (e.g. $SRV_{Mason-Pfizer}$).

IV. A human spumaretrovirus (HSRV or HSP).

This categorization by Roman numerals is merely intended as a brief aid to this discussion; it includes the retroviral types we will be evaluating in terms of variability and complexity.

The study of oncoviral *pol* gene relationships by Chiu *et al.* (1984) provides a comparative backdrop against which the retroviral diversification depicted in Fig. 1 can be quantitatively judged: three distantly related types of exogenous retroviruses—type D virus, and avian and murine oncoviruses—were shown in that study to be homologous by virtue of being 45–65% similar in their *pol* gene nucleotide sequences. The then newly determined HTLV-1 *pol* gene sequence was found to be 40–45% similar to these other oncoviral sequences. Virtually the same degree of homology is observed here for HTLV-1 in relation to the primate immunodeficiency viruses (I, an average of 41.5% similarity), the nonprimate immunodeficiency viruses (II, an average of 41.7%), and the type D SRV_{MP} (43.8%). From these broadly resolved connections, we can conclude that no known retroviral *pol* sequence stands out as unlinked. This strongly suggests that all exogenous retroviruses have a common origin; that is to say, they are monophyletically related "in the broad sense" (Gould *et al.*, 1987). Whether that implied moment of evolutionary radiation was recent or not is a question still to be answered (Doolittle *et al.*, 1989).

Although the various exogenous retroviral types evaluated in this manner are about equidistant from one another, different rates of evolution may underly the lineages. It can be seen by inspection of Fig. 1 that the evolutionary linkages among lentiviruses, measured by *pol* gene homology, are not quantitatively different from what Chiu and co-workers observed among the major types of oncoviruses. For example, the bovine and ovine lentiviral *pol* genes, which one might expect to be highly similar, are only as nearly related (55%) as are the *pol* genes of the squirrel monkey and the mouse mammary tumor retroviruses.

While nonprimate lentiviral sequences have not been given the close inspection now being given in the midst of the AIDS pandemic to primate immunodeficiency viruses, the variation rate of the three visna-related viruses shown in Fig. 1 proves to be as great (see Section II.B) as what is being measured for their primate lentiviral relatives. This finding suggests that diverse forms of feline and bovine immunodeficiency viruses, for example, will probably be discovered among both domestic and feral animals in the near future. A worldwide search for murine lentiviruses is now underway.

Doolittle *et al.* (1989) argue that the proteases of the primate and ungulate lentiviruses are no more similar than are fungal and human proteases. Despite this weak sequence similarity, HIVs were recognized early on to be evolutionarily related in several ways to the ungulate lentiviruses (Chiu *et al.*, 1985; Gonda *et al.*, 1985; Sonigo *et al.*, 1985). This observation is supported by the phylogenetic relationships presented in Fig. 1. Some recent reviews of the lentiviruses are those of Haase (1986), Narayan and Clements (1990), and Gonda (1990).

For the sake of clarity, only a very small number of the known HIV and SIV sequences are included in Fig. 1. For example, ANT-70, a Cameroonian HIV-1 isolate not shown in Fig. 1 (DeLeys *et al.*, 1990), is as distant from a prototypical HIV-1, MN in Fig. 1, as is a newly discovered chimpanzee virus, SIV_{CPZ}. Comparison with another recent chimpanzee isolate (Peeters *et al.*, 1992) is in progress. Furthermore, a Ghanian HIV-2 isolate, D205 (Dietrich *et al.*, 1989), is more distant from the prototypical HIV-2_{ROD} than are the macaque and sooty mangabey viral isolates included in Fig. 1. The African green monkey immunodeficiency viruses are astonishingly more diverse than Fig. 1 depicts (Daniel *et al.*, 1988; Li *et al.*, 1989; Johnson *et al.*, 1990; Fomsgaard *et al.*, 1991). Furthermore, a unique subtype of SIV_{AGM}, isolated from wild-caught West African green monkeys of the *sabaeus* species, has recently been identified (Allan *et al.*, 1991). A fuller examination of HIV and SIV variability will be taken up in Section II.C.

By way of contrast, sequence comparisons for the human T-cell leukemia viruses do not indicate anything like the extent of variation seen in the immunodeficiency viruses: HTLV-1 isolates from diverse geographical regions, and among different patients from a given region, are remarkably similar (Malik *et al.*, 1988; Reddy *et al.*, 1989; Gray *et al.*, 1990). A simian counterpart to HTLV-1, designated STLV-1, has not been sequenced over the *pol* gene and therefore has not been included in Fig. 1; but from homologous long terminal repeat (LTR) and *env* gene sequences, the human and simian viruses are found to be 90% similar (Watanabe *et al.*, 1985, 1986), providing further evidence of HTLV stability. In this exceptional case, retroviral complexity does not coexist with extreme variability.

Simian type D retroviruses can differ by up to 30% at the nucleotide level (Thayer *et al.*, 1987). Comparison of the only human and simian spumaretroviruses characterized to date reveals homologous relationships in their *pol* and *env* gene products of about 20% and 40%, respectively (Mergia *et al.*, 1990a). Thus, the preliminary indications are that both type D retroviruses and spumaretroviruses are highly variable—like lentiviruses and unlike primate T-cell leukemia viruses or prototypical oncoviruses.

2. Possible Simian Origin of HIV

At present, four known types of primate immunodeficiency viruses can be distinguished on the basis of slightly differing genomic organizations: (1) HIV type 1, including a somewhat closely related chimpanzee virus; (2) HIV type 2 and its close relatives found thus far in captive macaques and in both captive and wild-caught mangabeys; (3) diverse viruses isolated from asymptomatic African green monkeys; and (4) a distinct retrovirus isolated from mandrills. Representatives of each type, many of which are found in wild-caught animals (Fukasawa *et al.*, 1988; Tsujimoto *et al.*, 1989; Huet *et al.*, 1990; Marx *et al.*, 1991; Allan *et al.*, 1991), are shown in group I of Fig. 1. Several recent articles have reviewed the rapidly accumulating evidence for a simian origin of the AIDS viruses (Daniel *et al.*, 1988; Hirsch *et al.*, 1989; Doolittle *et al.*, 1990; Huet *et al.*, 1990; Desrosiers, 1990; Karpas, 1990; Grmek, 1990; Johnson *et al.*, 1991b; Gilks, 1991; Contag *et al.*, 1991; Myers *et al.*, 1992a). In this regard, it is astonishing to learn about an "HIV-2" sample taken from an asymptomatic Liberian individual that is indistinguishable from some SIVs (Hahn, 1990). In general, it is not possible to tell from a primate lentiviral sequence whether it was taken from a human or a monkey.

The likelihood that 30% or more of African green monkeys could be harboring HIV-related retroviruses (Kanki *et al.*, 1985) and the anticipated characterization of still more SIV variants (Fomsgaard *et al.*, 1991; Contag *et al.*, 1991; Johnson *et al.*, 1991b; Khan *et al.*, 1991; Novembre *et al.*, 1992) are reasons to expect eventual blending of the sublineages that appear as distinct types in the phylogenetic tree in Fig. 1. The newly discovered "fifth type" of primate immunodeficiency virus taken from wild-caught West African green monkeys (Allan *et al.*, 1991) and the HIV-1-related chimpanzee virus from Gabon (Huet *et al.*, 1990) provide the kind of evidence that should finally clarify the relationships among the primate lentiviruses and the origin of HIV. This issue is of particular interest to the present review insofar as "adaptive radiation," or "bottom-heavy cladism" (Gould *et al.*, 1987), may be the immediate evolutionary consequence of the hypothesized cross-species transmission.

3. Lentiviral Radiation

The retroviral *pol* gene nucleotide distances apparent in Fig. 1 are consistent not only with the comparative study of Chiu *et al.* (1984), but with a now extensive body of molecular analyses grounded upon a variety of computer approaches and derived from protein as well as nucleotide sequences. Nevertheless, the exact relationship of these sequences to one another, in particular, the branching order of the mandrill and green monkey immunodeficiency viruses, stands in doubt. Some studies have placed the AGM sequences, for instance, with the HIV-1s (Doolittle *et al.*, 1990; Querat *et al.*, 1990; Gojobori *et al.*, 1990a; Johnson *et al.*, 1990; Garvey *et al.*, 1990). Other analyses propose that the AGMs are more closely linked to HIV-2s (Fig. 1) (Talbott *et al.*, 1989; Olmstead *et al.*, 1989; Myers *et al.*, 1992a), or that the AGMs are outside both the HIV-1 and HIV-2 groups (Tsujimoto *et al.*, 1989; Gojobori *et al.*, 1990a). This apparently irreducible uncertainty stems from the occurrence of multiple base substitutions, which creates a condition of "noise accumulation," or "mutational saturation" (Kirchoff *et al.*, 1990; Sidow and Wilson, 1990; Myers *et al.*, 1992b), that is present even among the most conserved lentiviral coding sequences.

In an attempt to avoid molecular sequence noise in the phylogenetic tree analysis of Fig. 1, only second base positions in *pol* sequence codons were included. While short of ideal, this particular tree, one of a few equally parsimonious trees, was slightly favored by the "bootstrapping" method of verification (Felsenstein, 1988). Finally, we turned to other relevant information in an attempt to reach a judgment about the phylogenetic history: the fact that the AGM retroviruses have a genomic organization and envelope protein structure more like that of HIV-2 and its simian relatives might be taken as corroborative support for the branching relationships embodied in Fig. 1 (Myers *et al.*, 1992a).

Of greater interest, perhaps, than the exact branching order for the primate immunodeficiency viruses is the intriguing possibility of a common point of radiation: taken together, the four known HIV and SIV lineages appear to have arisen from a single common ancestor (Sharp and Li, 1988; Tsujimoto *et al.*, 1989; Kirchhoff *et al.*, 1990; Johnson *et al.*, 1991b; Myers *et al.*, 1992b). Alternatively, this hypothetical point of radiation may be artifactual, an epiphenomenon resembling the familiar effect of peering down a set of railway tracks to a point at which they seem to intersect, but having as its cause mutational noise. This problem may be resolved before long through application of maximum likelihood methods (Sidow and Wilson, 1990).

B. Time Estimates and Nucleotide Substitution Rates

1. Tree Calibration

Although there is no universal agreement that a time scale can be meaningfully affixed to retroviral phylogenetic trees (Coffin, 1990), a number of independent attempts to calibrate lentiviral evolution have produced a consistent phylogenetic scenario. For, while the mutational noise in lentiviral sequences is a complicating condition of analysis, there is a concomitant abundance of positive information. One strategy for calibrating a phylogenetic tree, such as the one shown in Fig. 1, is to assume a representative or characteristic base substitution rate. Adopting this approach, Yokoyama et al. (1988) inferred that HIV-1 and HIV-2 must have diverged from one another at least 280 years ago. However, to attain this estimate, they assumed an average nonsynonymous substitution rate deduced from murine retroviral gag genes (Gojobori and Yokoyama, 1985).

By assuming a representative nucleotide substitution rate for SIV gag coding sequences, Khan et al. (1991) calculated a divergence date of 1963 for sequence samples taken from macaques thought to have been infected in the 1960s when they were brought to the California Regional Primate Research Center. This calculation of what is called a "lookback" time, albeit based upon some questionable assumptions, turns out to be in good agreement with a similarly derived estimate made by Hirsch et al. (1989). Moreover, these calculations are in accord with other historical evidence, which includes the curious possibility that captive macaques were infected in the 1950s as part of malaria experiments (Gilks, 1991).

If there is reason to think that a particular rate might not apply to the genomic molecules in question, then the alternative strategy is to calibrate a tree based upon two empirical or deduced time points, and to derive from those a likely base substitution rate. Thus, T. F. Smith et al. (1988a,b) utilized the earliest known molecular specimen in the AIDS epidemic, an HIV-1 viral isolate from a 1976 Zairean serum sample (Srinivasan et al., 1989), to calibrate a tree representing HIV-1 and HIV-2 samples collected up to 1986. Their estimate for the minimum time from divergence of HIV-1 and HIV-2 was approximately 40 years. The correlative base substitution rate was about 1% per virus per year based mostly upon largely synonymous changes in the env gene coding sequence; this was about 20-fold higher than the murine retroviral rate assumed in the Yokoyama study.

Using a similar approach, Sharp and Li (1988) drew upon their analytical conclusion that a particular Zairean HIV-1 isolated in 1983 must have diverged from North American HIV-1 forms in about 1969 (Li et *al.*, 1988) to reach an intermediate estimate for the HIV-1/HIV-2 divergence time of 150 years. In this latter study, the deduced *pol* gene substitution rate was nearly 0.1% per year, twice the previously established retroviral rate. These results, and other variously derived estimates which we shall have reason to mention, are summarized in Table I.

2. Equal Evolutionary Rates?

To the extent that these analytical inferences rest upon an assumption of equal rates of evolution for the lineages being compared, they are justifiably regarded with skepticism. However, support for the uniformity assumption and for the timeframe required by the tree analyses of HIV was recently obtained through comparative sequence analysis of visna virus and the ovine Maedi-visna virus from South Africa (VLV and OMVV$_{SA}$ in Fig. 1). Recognizing from the historical record that these viruses had to have diverged from one another 42 years back, Querat *et*

TABLE I. Base Substitution Rates and Divergence Time Estimates[a]

Substitution Rate (per site per year)	Basis	HIV-1/2 Divergence (years)	Reference
0.0005	Retroviral *gag* genes	—	Gojobori and Yokoyama (1985)
	Assumes above rate	280	Yokoyama *et al.* (1988)
0.003–0.016	HIV-1 *env* gene	—	Hahn *et al.* (1986)
0.0116	HIV-1 *env* gene (synonymous)	40	T. F. Smith *et al.* (1988b)
0.01	HIV-1 genomic average (synonymous)	—	Li *et al.* (1988)
0.00096	HIV-1 *pol* gene	150	Sharp and Li (1988)
0.013	HIV-1 *gag* gene (synonymous)	—	Gojobori *et al.* (1990b)
0.005	HIV-1 *env* gene	—	Balfe *et al.* (1990)
0.0095	HIV-1 *env* gene (synonymous)	—	Wolfs *et al.* (1990)
0.0085	SIV-mac *env* gene	—	Burns and Desrosiers (1991)
0.0017	Visna virus genomic average	—	Braun *et al.* (1987)
0.0008	Ovine lentiviral *pol* genes	203	Querat *et al.* (1990)

[a] The first column gives nucleotide substitution rates per site per year and the second column the basis; e.g., HIV-1 *env* gene synonymous substitutions. The third column gives the minimal time elapsed since divergence of HIV-1 and HIV-2 from a common ancestor.

al. (1990) deduced minimal lookback times of 203 years for the HIVs and 430 years for the primate and ungulate lentiviruses. The *pol* gene nucleotide substitution rate for these ungulate lentiviruses was calculated to be 0.08% per year, a rate very close to the one derived by Sharp and Li for conserved HIV-1 *pol* gene coding sequences. An earlier study of two quite closely related visna genomic sequences yielded an average nucleotide substitution rate of 0.17% per year (Braun *et al.*, 1987), and rates at least this great are implied by the EIAV *env* gene variation observed in a single horse (Payne *et al.*, 1988).

The tenfold range of lentiviral nucleotide substitution rates summarized in Table I appears quite plausible in light of the larger examination of relative rates of change reported for retroviruses by Doolittle and co-workers (McClure *et al.*, 1988; Doolittle *et al.*, 1989): retroviral coat proteins (*env*) vary on the average at 2.6 times the rate of retroviral polymerases (*pol*). It remains the case that all rate estimates in Table I for the relatively conserved *pol* gene of lentiviruses prove to be greater than the rate calculated for the murine retroviral *gag* sequences. Moreover, for reasons that become clear below, some of the rate estimates in Table I are minimal estimates insofar as they are derived from culture-selected HIV genomic molecules. (Uncultured HIV typically displays three times more variation than what is seen with cultured viral stocks; see Section III.B.)

In spite of this overall consistency, there remain interesting and legitimate doubts about the divergence time estimates for lentiviral evolution summarized in Table I. We have already raised in a related context the problem of mutational noise and how it clouds the distant history of the lentiviruses. Closely attached to that uncertainty is the questionable assumption of uniform mutation rates and a molecular clock: in particular, newly evolved sublineages of HIV and SIV sequences manifest ratios of synonymous to nonsynonymous substitutions that are very different from ratios seen across "older" sequence lineages (Myers *et al.*, 1992a). Purifying selection (Li and Graur, 1991) is more apparent across greater evolutionary distances (see Table IV below). Furthermore, it becomes difficult to imagine how current prevalence levels of retroviruses descended from a single ancestor could have been globally established in so many different hosts in so short a time as 400 years.

Allan *et al.* (1991), in light of their recent discovery of a unique SIV in West African green monkeys, propose that the green monkey viruses as a group may be on a different evolutionary clock from that of the lentiviruses in humans and domesticated animals: the four species of African green monkeys are thought to have been geographically and ecologically isolated from one another for several thousand years, and each of these species harbors a form of SIV without manifestation of disease. While these AGM retroviruses display as much overall heterogeneity as that seen in HIV and visna virus, the pattern of genomic variation is different: in the latter lentiviruses, the *env* coding sequence is more variable than are the *gag* and *pol* coding sequences, whereas in the SIV_{AGM} strains studied to date, *gag*, *pol*, and *env* sequences do not show this differential, suggesting that diversification in these wild-type lentiviruses may not be immune-driven (Myers *et al.*, 1992a) (see Section III, Table IV). If this molecular indicator of relative protein stability continues to hold up as more of these AGM isolates are characterized, the interpretation of their evolutionary timeframe, as well as the historical account of all the ancestral lentiviruses, might have to be recast by at least an order of magnitude over what is being deduced from the quantities in Table I.

Given how much is still to be discovered about simian, ungulate, and feline lentiviruses, it is prudent to regard the quantities reported in Table I as having analytical validity only with respect to the last 50 years of lentiviral variation (Myers *et al.*, 1992a).

C. Macroscopic and Microscopic Swarming

1. Molecular Epidemiology of AIDS

Many of the rates summarized in Table I are derived from *pol* and *env* gene sequence data taken from closely related human immunodeficiency viruses. Time calibration of HIV phylogenetic trees over the past 50 or so years can, accordingly, be verified to an appreciable extent through correlation with AIDS epidemiologic data. This makes possible new avenues of molecular epidemiology which are simply not available with less rapidly varying pathogenic retroviruses such as HTLV-1.

Retrospective serosurveys of hepatitis B cohort members in Los Angeles, San Francisco, and New York provide consistent snapshots of the temporal onset of the HIV-1 epidemic in the United States: the preponderance of early HIV infections can be traced back to 1977–1978 (Selik *et al.*, 1984; Jaffe *et al.*, 1985; Stephens *et al.*, 1986). Thus, Fig. 2, a phylogenetic tree based upon North American HIV-1 sequences, is a fair representation of HIV-1 *env* gene diversification between 1977 and approximately 1985. For reasons that become clear below, it is no longer possible to think that these sequences derive from stable viral variants, as was earlier proposed by Coffin (1986).

Since most of the viral samples represented by molecular sequences in Fig. 2 were from viruses isolated between 1983 and 1985, a very tentative time calibration based upon a 7-year interval is suggested: the average interpatient nucleotide distance for the data set, constituted of

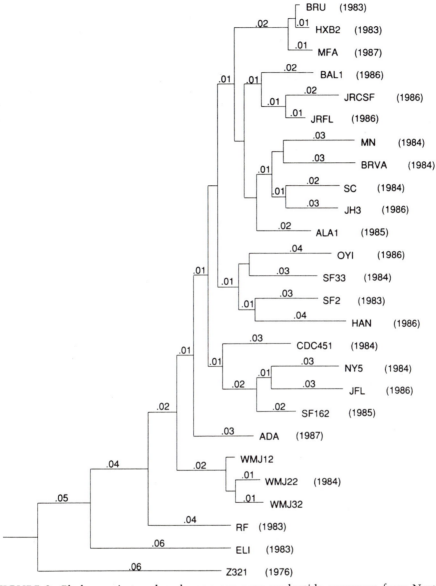

FIGURE 2. Phylogenetic tree based upon *env* gene nucleotide sequences from North American HIV-1s isolated in the years indicated. The average interpatient nucleotide difference is approximately 7% in 1984–1985. For purposes of comparison, two Zairean HIV-1 *env* sequences—ELI and z321—are included. The HAN virus was isolated in Germany and the OYI virus was isolated in Gabon. The computer-determined branch lengths, as in Fig. 1, express percentage nucleotide differences (values less than 0.01 are not shown) based upon approximately 1200 sites in the *env* gp120 coding region; thus, a distance of 7% represents approximately 85 base substitutions or differences.

samples from randomly selected AIDS patients throughout the United States, is 7% (range of 5.5–9.4%), which suggests that the average inter-patient *env* gene nucleotide distance is increasing at about 1% per year. This rough prediction for the base substitution rate is not disturbingly far off from the HIV-1 *env* gene estimates summarized in Table I: in particular, Balfe *et al.* (1990) deduced an average *env* gene substitution rate of 0.55% per year per virus from sequence data similar to but not included in Fig. 2. By this reckoning, and assuming that migration will not make a significant contribution to the variation until later in the epidemic, the average interpatient *env* gene distance for U.S. samples should have reached 13% in 1990. This projection has been recently confirmed, to a first approximation, in a study of more than 40 AIDS patient viral samples from Florida (Ou *et al.*, 1992). By the same reason-ing, Zairean *env* gene sequences, differing by an average of 16% in, say, 1984, will differ by at least 22% in 1992.

Migration of diverse forms of HIV-1 will inevitably play a major role in the molecular epidemiology of HIV-1 and HIV-2. A minimum of five distinct subtypes of HIV-1 is now encountered in the centers of the AIDS pandemic; it is conceivable that these distinct forms (in both *gag* and *env* coding sequences) may have diverged from a single common ancestor as recently as 1960 (Myers *et al.*, 1992a). Cocirculating lineages of HIV-1 are being tracked in Thailand, Brazil, and Gabon. There is no indication that branch lengths in the North American tree are increasing as a function of time; nevertheless, it is conceivable that HIV-1 variation might be epidemic driven. That is to say, the phenomenological rate could vary as some function of the growth rate of the epidemic. (For example, the implied time scale for the tree in Fig. 2 could be a qua-dratic.) In light of these various manifestations of extreme viral variabil-ity, it becomes clear that no two AIDS patients could carry identical forms of HIV.

2. Intrapatient Variation

From the earliest work describing intrapatient HIV variation (Hahn *et al.*, 1986; Saag *et al.*, 1988), it seemed likely not only that the rate of base substitution in HIV was exceptionally high, but there were few discernible selective constraints upon, or within, the intrapatient swarms. No one could say just how many closely related but distinguish-able genotypes were present in a single individual. Subsequent investi-gation revealed that *ex vivo* culture, as a step toward viral isolation, significantly reduced the genetic complexity of the swarms, or quasispe-cies (Goodenow *et al.*, 1989; Meyerhans *et al.*, 1989; Delassus *et al.*, 1991; Vartanian *et al.*, 1991; Kusumi *et al.*, 1992). Thus, HIV *env* gene sequences differing by 1–2% in culture can be routinely selected from an *in vivo* intrapatient pool of sibling viruses having mean nucleotide distances of 4–5% (Balfe *et al.*, 1990). A similar picture of microscopic swarming and the selective effects of tissue culture is seen with SIV (Hirsch *et al.*, 1990; Burns and Desrosiers, 1991; Overbaugh *et al.*, 1991; Johnson *et al.*, 1991a).

Tree analyses performed in conjunction with tracking efforts (molecular epidemiology) have included culture-selected samples with little risk of misrepresentation because quasispecies sequences ("sibling" sequences) are quantitatively very similar: each branch in Figs. 1 and 2 is itself a small tree. No AIDS patient has been found to be multiply infected by distinctly different forms of HIV-1 (dual infection by HIV-1 and HIV-2 has been documented), hence tree analyses based upon cultured samples are probably not misleading in that respect. Intrapatient tracking (Nowak *et al.*, 1991), in contrast, is acutely sensitive to the experimental conditions and approaches involved in the surveillance. In addition to the perturbations of viral swarms brought about by cell culture, researchers must keep in mind the direct (Larder and Kemp, 1989) and indirect (Atwood *et al.*, 1951; Buonagurio *et al.*, 1986) selective effects caused by antiviral drug therapy.

Utmost caution should be taken in the interpretation of the pathogenesis owing to an intrapatient swarm or population: as the above-mentioned studies reveal, culture-selected viral forms can represent minor species, characterized by sequences subtly differing from those of the major species *in vivo* (Wain-Hobson, 1992). There are many indications of HIV selection *in vivo*; however, it would be premature to try to state the precise terms of the selective forces. The focal points of investigation have been pressures from the host immune system (which are taken up in Section III), cytopathogenicity (Cheng-Mayer *et al.*, 1988; Meyerhans *et al.*, 1989; Tersmette *et al.*, 1989), and tropism (Cheng-Mayer *et al.*, 1990, 1991; Hwang *et al.*, 1991; Westervelt *et al.*, 1991). With respect to all three of these considerations, most of the attention thus far has been given to the principal neutralizing determinant, a fascinating peptide (of typically 35 amino acids) found in HIV-1 envelopes (Section III.E.5).

Both noncytopathic (Huet *et al.*, 1989) and highly cytopathic (Spire *et al.*, 1989) variants of HIV-1 have been encountered in the pandemic. HIV-2 heterogeneity is also being examined with relation to infectivity and replicative capacity (Dietrich *et al.*, 1989; Kumar *et al.*, 1990; Castro *et al.*, 1990). In the most dramatic case reported to date, an acutely lethal variant of what was originally a benign form of a sooty mangabey isolate emerged after a single passage in an experimentally infected macaque (Fultz *et al.*, 1989; Martin, 1990). In none of these instances of abnormal course of disease has it been possible to pinpoint the genetic basis for the excess pathogenicity or its absence.

3. Epidemiologically Linked Sequences

HIV sequence information about viruses taken from epidemiologically linked patients is consistent with this general picture of macroscopic and microscopic swarming. For example, genetic heterogeneity has been documented for donor–recipient pairs in transfusion-associated AIDS (Srinivasan et al., 1987). More recently, Burger et al. (1991) reported an env gene nucleotide distance of 3.7% for the major species of viruses taken from sexual partners. Comparable differences (i.e., distances) have been observed between viral sequences taken from mothers and their perinatally-infected year-old infants (Wolinsky et al., 1992). However, in a case of a mother–daughter pair studied 10 years after perinatal transmission had occurred, env gene sequences were found to differ by 8.5% (Burger et al., 1991).

Among six hemophilia patients infected after exposure to a single common batch of factor VIII, the mean env gene interpatient nucleotide distance was 8.3%, about half of which accrued in the infected donor, with the remainder accruing in the 5 years since infection (Balfe et al., 1990). From this latter fact, the authors derived the base substitution frequency, for uncultured viral genomes, of 0.55% per year that was mentioned earlier; a twofold difference in rate was observed from patient to patient. Surprisingly few inactivating substitutions were encountered. Viral env sequences taken from a dentist and five of his patients believed to have been infected through him during invasive dental procedures showed average interpatient sequence differences of about 4% (after 3 years?), while control viral samples from the same geographical area differed on the average by approximately 12% (Ou et al., 1992).

In a study of macaques experimentally infected with molecularly cloned virus from a sooty mangabey, Johnson et al. (1991a) observed extents and rates of variation surprisingly similar to what is seen with HIV-1 in individuals. However, in contrast to what was observed in the above-mentioned hemophilia cohort study, a large number of inactivating substitutions were encountered in two of the four macaques: specifically, the vast majority of tryptophan codons in the env genes of these "hypermutated" animals were changed to stop codons. In two other studies involving experimentally infected animals, nucleotide substitutions in the env gene accumulated at average rates of 0.56% and 0.85% per year, in very close agreement with the estimates summarized in Table I (Overbaugh et al., 1991; Burns and Desrosiers, 1991). What is so remarkable from one of these studies (Burns and Desrosiers, 1991) is that 98% of the substitutions, the vast majority of which were not inactivating mutations, resulted in amino acid changes. This latter finding might have been predicted from the sequence analysis of HIV-1 published by Leigh-Brown and Monaghan (1988). In that study, the rate of change of *env* amino acid sequences relative to the overall rates of change in nucleotide sequences was found to be higher than for any known protein. This evidence of extraordinarily weak purifying selection, not seen in sequences from the feline and murine leukemia retroviruses, will be discussed in further detail in the next section and in Section III.E.

Irrespective of whether inactivating substitutions are encountered, or whether viruses are isolated in culture, most of these molecular studies have uncovered a curious preponderance of G → A substitutions (Goodenow *et al.*, 1989; Balfe *et al.*, 1990; Vartanian *et al.*, 1991; Delassus *et al.*, 1991; Overbaugh *et al.*, 1991; Johnson *et al.*, 1991a; Wain-Hobson, 1992). Discussion of the possible mechanism underlying this mutational tendency, and of hypermutation in general, is left to Section III.B.

4. HIV and Flu Virus

The phenomenon of extreme genetic heterogeneity within a virally infected individual is not unique to AIDS. The quasispecies concept has been invoked, for example, in relation to foot-and-mouth disease virus (Mateu *et al.*, 1989; see also Steinhauer and Holland, 1987, and Wain-Hobson, 1992). The microscopic and macroscopic swarming witnessed in the lentiviruses, however, is not simply the consequence of a high mutation rate, which is a property common to most if not all RNA viruses. For example, the hemagluttinin gene of human type A influenza virus has a base substitution rate of about 0.6% per year, a rate virtually identical to what is seen for the HIV *env* gene (Fitch and Palese, 1991). But the pattern of evolution of flu-A differs strikingly from what is implied in Figs. 1 and 2: the structure of the phylogenetic tree for the flu-A (derived from NS gene sequences) is slender, or "cactuslike" (Buonagurio *et al.*, 1986), whereas the trees derived from HIV and SIV sequences are invariably bushy, as schematized in Fig. 3. While there remain uncertainties about the time calibration of HIVs in Figs. 1 and 2 (uncertainties which do not exist in the case of flu-A, for which sequence samples go back 50 years), there is no main trunk in HIV trees, and we can be certain that the average age of the branches in these trees exceeds that of the side branches of the flu-A tree, which is merely 4.2 years.

Thus, lineages of influenza virus are not at all comparable to HIV swarms. Fitch and Palese reason that influenza virus is undergoing positive selection through immune surveillance at primarily the population level. There remains fierce intraspecific competition among flu variants. While HIV also undergoes positive selection (Leigh-Brown and Monaghan, 1988; Simmonds *et al.*, 1990a; Myers *et al.*, 1992a), it is not at all clear from the phenomenological description of lentiviral variation

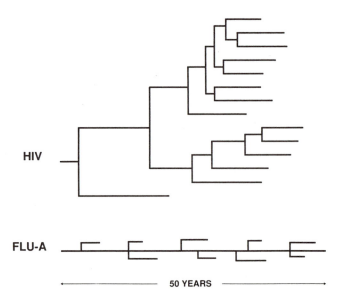

FIGURE 3. Schematic trees contrasting flu-A viral speciation with HIV speciation, as discussed in the text. Sequence samples of influenza A virus go back 50 years. While an equivalent time calibration for the HIV tree is not possible at this time, the two RNA viruses, having similar nucleotide substitution rates, appear to represent very different evolutionary patterns.

that HIV evolutionary dynamics is equivalent to what is seen with influenza virus: with HIV and SIV, the intraspecific competition is not apparent at this time (hence the "bushiness" of phylogenetic trees) and the selection appears to be more at the level of the individual (Nowak *et al.*, 1991).

The remainder of this chapter will be an exploration of the complex molecular and mechanistic properties that any hypothesis about the emergence of these viruses must include.

III. DYNAMICS OF LENTIVIRAL EVOLUTION

A. Rates and Mechanisms

1. Biochemical Studies

The lentiviruses appear to be among the most rapidly evolving genomic molecules. While comparison with other RNA viruses has limited interpretive value, there is a suggestive difference between flu-A, for example, which exhibits a slender phylogenetic path (Buonagurio *et al.*,

1986; Fitch and Palese, 1991), and the immunodeficiency viruses, which exhibit bushy phylogenetic paths (Fig. 3). Later in this review, we shall discuss this lentiviral propensity for a broad front of speciation, which seems to reflect weak intraspecific competition as much as a brisk mutation rate (Steinhauer and Holland, 1987; Leigh-Brown and Monaghan, 1988; Temin, 1989; Fitch and Palese, 1991; Myers *et al.*, 1992a). First, we shall review briefly what has been discovered about HIV mutation rates and the error-specificities of its reverse transcriptase *in vitro*.

Retroviral reverse transcriptases are known to have negligible proof-reading exonuclease activity (Battula and Loeb, 1974; Roberts *et al.*, 1988; Katz and Skalka, 1990). Purified HIV-1 reverse transcriptase (RT) has been recently investigated and found to be approximately tenfold less accurate than the avian myeloblastosis retroviral RT, implying that its error-prone characteristics go beyond the mere absence of a sequence-correcting exonuclease. By extrapolation, Roberts *et al.* (1988, 1989), Preston *et al.* (1988), and Bebenek *et al.* (1989) have estimated the *in vivo* error rate to be 5–10 misincorporations per HIV-1 genome per replicative cycle. In yet another study, Takeuchi *et al.* (1988) conclude that the HIV-1 RT is significantly more error-prone, *in vitro*, than other retroviral RTs. All nucleotide polymerases are known to incur mismatches; this work has led to the finding that extension of mismatched 3' termini is the distinguishing feature of HIV-1 RT infidelity (Perrino *et al.*, 1989; Bebenek *et al.*, 1989; Roberts *et al.*, 1989; Ricchetti and Buc, 1990).

2. Mutational Analyses

These *in vitro* findings are consistent with the relatively higher nucleotide substitution rates inferred for lentiviruses in Table I, and they account for the ease with which resistance to zidovudine (AZT) is achieved by the AIDS viral RT (Larder and Kemp, 1989). It remains to be seen whether this high degree of infidelity is inordinate among retroviruses. Pathak and Temin (1990a,b) have closely examined the "low processivity" of the RT from spleen necrosis virus (SNV) after a single replicative cycle *in vivo*; in addition to substitutions, the occurrence of frameshifts, deletions, and deletions with insertions was assessed. Hypermutation, defined as "the process by which individual proviruses acquire several mutations in a single replication cycle," gave a base-pair substitution rate of 0.02 per cycle in a particular stretch of the SNV genome. This rate was 10,000-fold higher than the average rate measured for SNV and was notably associated with the G → A substitutions commonly observed in HIV and SIV (Section II.C.3). If this *in vivo* rate can be justifiably compared to the above-mentioned *in vitro* rate, the tentative conclusion is that the HIV-1 RT must be in a persistent state of greater or lesser hypermutation. Future investigations of retroviral RTs (RNA polymerase II is not ruled out as the agent of hypermutation in the *in vivo* studies) should reveal to what extent this is the situation with all lentiviruses and whether the effect is genetic or epigenetic. Variant polymerases are proposed by Pathak and Temin (1990a), who reason that the replicative infidelity acts upon itself up to some mutationally critical level.

Insertions and deletions, usually involving multiples of three bases, have been observed in virtually every study of HIV and SIV sequences. The recent work of Balfe *et al.* (1990) beautifully documents the extent to which conspicuous length polymorphism is possible without vitiating coding sequences or generating defective viruses.

HIV-1 intragenic recombination is revealed most clearly in the study by Vartanian *et al.* (1991). Intergenic recombination has not been so obvious: in only one instance, to our knowledge, has an HIV-1 genome been judged hybrid (Li *et al.*, 1988). No one has yet uncovered evidence of multiple infection by a single type of HIV (a necessary condition for intergenic recombination); only dual infection by HIV-1 and HIV-2 has been observed in some African patients (Rayfield *et al.*, 1988; Evans *et al.*, 1988). HIV superinfection may be blocked (Hart and Cloyd, 1990), or simply made difficult to detect by virtue of low copy numbers (Balfe *et al.*, 1990). Be that as it may, heterodimeric viral particles are encountered and have been studied. Hu and Temin (1990) have measured a recombination rate for spleen necrosis virus of approximately 2% per kilobase per replication cycle, from which they conclude that recombination occurs at a frequency of one in two or three wild-type viruses. The polymerase chain reaction, which is being increasingly utilized to characterize heterogeneous HIV populations, can itself mediate recombination, and this must be carefully distinguished from naturally occurring events (Meyerhans *et al.*, 1990; Vartanian *et al.*, 1991).

Gene duplication is the likeliest explanation for the *vpX–vpR* sequence similarity found in HIV-2 and related SIVs (Tristem *et al.*, 1990; Myers *et al.*, 1992a). Gene duplication has been previously described for the lentiviruses: a "protease-like" element, PrL, has been found in all nonprimate lentiviruses as well as in the type D retroviruses, the mouse mammary tumor virus, and even in vaccinia virus, which has a curious similarity not only to other aspartate proteinases (the enzyme family to which all lentiviral proteases belong), but to deoxyuridine triphosphatases (McGeoch, 1990). It is not clear whether the *vpR* and *vpX* coding sequences are responsible for different functions.

Heterologous insertion, or copy-choice recombination, must be invoked to explain the curious sequence similarity between an immunoglobulin heavy-chain coding sequence and a unique noncoding, albeit functional, stretch found in all HIV-2/SIV LTRs (Myers *et al.*, 1989).

B. Hypermutation as a Possible Driving Force

More than one study of the replication of HIV or SIV, in culture or in animals, has encountered extensive and sometimes monotonous G → A hypermutation (Goodenow *et al.*, 1989; Meyerhans *et al.*, 1989; Johnson *et al.*, 1991a; Delassus *et al.*, 1991; Wain-Hobson, 1992). This phenomenon may or may not have been associated with inactivating mutations and the generation of defective proviruses: Balfe *et al.* (1990) observed only two inactivating nucleotide substitutions in a total of 42 kilobases (kb) of *gag* and *env* gene sequences amplified from cells of nine HIV-1-infected patients. In partial contrast, two of the experimentally infected animals studied by Johnson *et al.* (1991) displayed conspicuous G → A inactivating substitutions; yet two other animals infected with the same viral clone did not. In the two cases of hypermutation, it was uncanny how the process left routinely conserved sequences, such as the *env* gene cysteine codons, untouched. What biological sense can be made of retroviral, and specifically lentiviral, hypermutation?

Following up on some earlier studies of intrapatient HIV-1 variation, Vartanian *et al.* (1991) focused upon a situation in which G → A transitions accounted for more than 90% of all substitutions in the various members of a cell-grown viral population. A strong predilection for this hypermutation is observed within GpA dinucleotides (the order of preference is GpA ≫ GpG ≫ GpT > GpC). They proposed as a mechanism "dislocation mutagenesis" (Kunkel and Alexander, 1986), by which slippage or dislocation following mismatch during either (+)-strand or (−)-strand synthesis would explain the statistically significant bias. To the degree that hypermutation yields inactivating substitutions, frequently poly-A runs, it may prove to be merely transient and evolutionarily insignificant. However, the high rate of recombination observed in this study, which is consistent with the occurrence of recombination in SNV (Hu and Temin, 1990), creates a potential for fixing hypermutated stretches.

Whatever the biological importance of G → A hypermutation coupled with recombination, it cannot account completely for the lentiviral instability. If it did, the retroviral genome would be quickly driven toward an extreme base composition, and the GpA dinucleotide frequency would become depleted in short time. In fact, all lentiviruses possess a high adenine content: perhaps through cumulative hypermutation, the *env* gene coding sequence for the principal neutralization determinant,

about which we will have more to say below, has reached 40% A. But as
we shall see in the next section, that unusual molecular condition comes
about at the expense of C, not of G. This points to one or more mutational and selective constraints not yet identified. We shall return to
other facets of retroviral mutation and fitness after some consideration
of lentiviral base compositions and RNAs.

C. Base Composition and Its Possible Implications

1. Lentiviruses Have an Unusual Base Composition

A search of nucleotide sequence databases for homologies with HIV
does not tend to yield more than numerous marginal and submarginal
matches with pseudogenes, introns, and satellite DNA. If we take this
result at face value, it is as if HIV were made from the junk of the cell.
Any attempt to comprehend the dynamics of lentiviral evolution may
have to assimilate this quirky fact, which is best appreciated through
inspection of genomic base composition.

Adaptation of a retrovirus to its host can involve at some fundamental level the base compositions of the interrelated genomic molecules. Thus it was no surprise to find that the codon frequencies in the
complete genome sequence of HTLV-1 match those of humans for the
most part; the genomes are thus said to have compatible "coding strategies" (Grantham and Perrin, 1986). Because preferential codon usage,
and other molecular properties pertinent to this discussion, will be dic-
tated to a large degree by base composition, a compilation of some relevant genomic makeups is given in Table II.

It was previously noted that the lentiviruses possess an unusually
high A content, approximately 35%, compared to the average A content,
25%, found in nearly everything else (Table II). Fewer than 5% of the
entries in the mammalian categories of the GenBank nucleotide sequence database have A contents in excess of 34%. Many of these sequence entries—importantly not all—are intronic DNA, pseudogenes,
and other genetic "debris," which explains the earlier mentioned result
of a routine nucleotide homology search when the HIV genomic molecule is the query sequence.

A larger fraction of the total viral entries, nearly 18%, has a base
composition similar to that of lentiviruses, due to the large number of
influenza virus sequences that also manifest high A contents. Among
known retroviral sequences, the lentiviruses constitute nearly 95% of
the nucleotide sequence entries having this peculiar property. Curiously, what is observed for spumaretroviruses in Table II suggests that
they share this compositional feature with influenza and all lentiviruses

TABLE II. Base Compositions of DNAs and RNAs[a]

Category	Number of bases × 10⁻⁶	Average composition (%)			
		A	C	G	T/U
Primates	7.7	25.4	25.0	25.3	25.0
Rodents	6.4	25.7	24.7	24.7	24.7
Other mammals	1.7	24.6	25.9	25.7	25.9
RNA	0.4	24.5	22.5	28.9	22.5
Viruses	5.8	28.1	23.3	24.2	23.3
Retroviruses	1.0	29.0	23.2	24.0	23.2

Retrovirus	Number of bases × 10⁻³	Composition (%)			
		A	C	G	T/U
HIV-1 Bru	9.2	35.6	17.9	24.2	22.2
HIV-2 Rod	9.7	34.3	20.3	24.8	20.5
SIVAGM155	9.8	33.9	19.5	25.0	21.6
SIVMND	9.2	36.1	16.0	23.8	24.1
FIV14	9.5	37.7	14.5	22.4	25.5
BIV106	8.4	31.9	21.2	23.8	23.0
EIAV	8.3	35.8	16.3	22.1	25.8
CAEV	9.2	38.1	16.0	25.1	20.8
Visna virus	9.2	37.1	15.4	26.1	21.4
HTLV-1	9.1	23.1	34.9	19.0	23.1
HTLV-2	9.0	24.2	35.6	18.2	21.9
Simian spumaretrovirus	11.8	32.5	19.1	20.0	28.3

[a] Sequences represented are taken from GenBank and from Myers *et al.* (1990). The simian spumaretrovirus sequence was kindly provided prior to publication by K. Shaw, P. Luciw, and co-workers, University of California, Davis, California. For the viruses and retroviruses, only sequences of 300 nucleotides or more were included.

reported to date. HTLV-1 and HTLV-2 are strikingly idiosyncratic by virtue of their high C and low G contents.

What meaning can be attached to these quantities and phenomena? Returning briefly to the earlier discussion of hypermutation, it appears that the G → A tendency seen under certain circumstances in both HIV and spleen necrosis virus (albeit more frequently in the former) does not of itself decide the equilibrium base composition: G content is surprisingly stable throughout all of the categories and sequences given in Table II, except for the HTLVs and the type D retrovirus SIV$_{MP}$. If high A content begins with the mutational dynamics described in Section III.B, some additional constraint, played out over a longer time period, is still to be recognized. The total process responsible for the equilibrium implied in Table II invites comprehension through a model along the lines of the mutation-driven scheme published by Sved and Bird (1990). This equilibrium model has the added virtue of tracking CpG dinucleotide frequencies, which enter the picture presently.

2. Codon Usage

Immediately following the earliest publications of HIV-1 genomic sequences, Grantham and Perrin (1986) reflected upon the virtual absence of shared codon preferences between the AIDS virus and its host. Again, they observed this not to be the case with the other known human retrovirus, HTLV-1, nor for that matter with other human pathogens such as adenovirus and Epstein–Barr virus. The latter example turns out to be more complicated than an all-or-none coding strategy might suggest: Karlin et al. (1990) find that codon usage in latent genes of Epstein–Barr virus is statistically different from that in productive genes.

Kypr (1987) claimed that of 1638 genes examined, none is as divergent from typical codon usage as HIV. In a subsequent paper Kypr and Mrazek (1989), characterized these codon choices as destabilizing: point mutations will more often than not lead to changes in the hydrophobicity of the encoded amino acids. This implies that HIV codon usage has specifically evolved to achieve volatility, and escape from immune surveillance, at the expense of protein stability.

In light of the greater similarity of HIV codon choices to those of influenza and cauliflower mosaic viruses than to that of another retrovirus (Moloney murine leukemia virus), Sharp (1986) takes issue with the principle of invoking shared codon preferences as a metric for evolutionary relatedness. Although this criticism sidesteps the similar base compositions of influenza and HIV and ignores the curious relationship of the cauliflower mosaic virus to retroviruses, the point seems to be well taken. Parallel differentiation on the basis of codon usage, by this argument, may be merely an instance of evolutionary convergence. Both Sharp (1986) and Grantham and Perrin (1986) take keen interest in the codon preferences shared by HIV and certain T-cell receptor genes, since a protein product of the latter is known to be a principal target for the AIDS virus.

T-cell receptor coding sequences are a striking example of the small fraction of mammalian coding sequences having high A content. A few other immune system genes have this rare distinction, for example, some immunoglobulin-coding sequences. Cytokine-related coding sequences can also possess the low C and high A content characteristic of lentiviruses: a significant number of interferon- and interleukin-coding sequences are among the more intriguing examples. The equally unusual base composition of the HTLVs—high C, low G (Table II)—lends itself to analogous exploration.

3. Base Composition Similarities

No one has, to our knowledge, systematically assessed the extent to which this base compositional similarity derives from homology rather than convergence, and how the similarity, whatever its origin, shapes the pathogenesis and evolution of the lentiviruses. This is not to say that speculations are not plentiful: Vega *et al.* (1990), for example, recently proposed what could be an important similarity between the HIV-1 *nef* protein and the beta chain of HLA class II histocompatibility antigen; they go on to postulate autoimmune consequences deriving from the similarity. As noted above (Section III.A.2), the only indisputable sequence similarity of this kind is between a stretch of the HIV-2/SIV LTR and a coding region for an immunoglobulin heavy-chain gene (Myers *et al.*, 1989). In this somewhat puzzling example, consisting of a 70% match over nearly 100 nucleotides for a noncoding sequence and a coding sequence, as well as with the *nef*–HLA example, the base composition happens not to be high in A content. One of us (G.M.) regularly conducts similarity searches with lentiviral sequences and the high A sequences of the GenBank nucleotide sequence database. Matches reflecting potential homologies are extremely rare and demand tedious verification. We remind ourselves of the fact that the *pol* gene nucleotide homologies described in Section II.A involved sequences with totally different base compositions, for example, those of HIV and HTLV.

Yet, sequence similarities attained by convergence are probably not so rare. For example, a segment of the HIV-1 *gag* p17 nucleotide sequence is 64% identical over a 76-nucleotide A-rich stretch to a coding region for the human heavy neurofilament subunit. We have no reason to think that the gene sequences involved are homologous (see Section III.E); this degree of base compositional similarity may be pointing to analogous RNA states or shared RNA compartments that have resulted from convergent evolution.

4. Regulatory Implications

Thus far, codon usage, codon volatility as an evolutionary strategy, and sequence homology and similarity have been considered in relation to lentiviral base composition. Base composition may also affect viral integration into the host chromosome. Yet another possible evolutionary significance of the unusual lentiviral base composition revolves around methylation of nucleic acids and the implied regulatory ramifications. Proviral methylation has been reported to correlate negatively with viral transcription; for example, see Bednarik *et al.* (1987) pertaining to HIV. CpG doublets, known to be targets in DNA methylation, are found in HIV at only one-fourth the frequency they are found in HTLV-

1 (Kypr *et al.*, 1989). As a result of mutational bias, these dinucleotides typically occur in any large stretch of DNA, say in vertebrate genomes, at only 20% or so of the *a priori* expected frequency (Ohno, 1988; Sved and Bird, 1990). Shpaer and Mullins (1990) argue that the very low frequency of CpG in lentiviruses, in contrast to CpG in HTLV-1/2, must result from selection relating to viral expression. Further inquiry into this possible consequence of a skewed base composition should examine RNA methylation as well as DNA methylation: in the studies mentioned, CpG enrichment (relatively speaking) is found in lentiviral LTRs, in the untranslated RNA leader, and also in the *rev* coding sequence. As we shall have reason to note in the next section, the *rev*-responsive element (RRE) and the *rev* coding sequence, both embedded in the *env* genes of HIV and SIV, possess more normal A base compositions and higher frequencies of CpG, suggesting that lentiviruses are heterogeneous entities with respect to their RNAs. This mosaicism can be clearly seen in the HIV-1 genomic molecule through the H-curve analysis of Hamori and Varga (1988).

D. RNAs of Complex Retroviruses

1. Multiple mRNAs in Lentiviruses

Retroviruses produce in general two types of transcripts or mRNAs: (1) unspliced mRNA producing the *gag* and *gag–pol* polyproteins (this RNA is also encapsidated as the viral genomic molecule); and (2) spliced mRNA, producing the envelope protein. Lentiviruses have an exceptionally complex array of transcripts; for instance, HIV produces more than 20 mRNAs (Schwartz *et al.*, 1990a,b). Alternative splicing permits HIV to utilize eight different initiator codons (indicated by asterisks in Fig. 6) to generate at least ten distinct proteins; in contrast, many retroviruses require just two start codons. Furthermore, the relative levels of lentiviral mRNAs are observed to be regulated over time, introducing the potential for adjusting viral gene expression in host cells and providing opportunities for regulation of that expression independent of, or in concert with, cellular factors.

The genomic organizations and RNA transcripts for three representative retroviruses, RSV, HTLV-1, and HIV-1, are illustrated in Figs. 4–6. Prototypical oncoviruses such as Rous sarcoma virus (RSV) and Moloney murine leukemia virus, which were shown in Section II.A to be distantly homologous with HIV, are known to produce one additional spliced mRNA, but regulation of this transcript over the course of infection is not observed. HTLV-1/2, sometimes classified with RSV and other oncoviruses, has more of the complexity observed in lentiviruses,

Mo-MLV

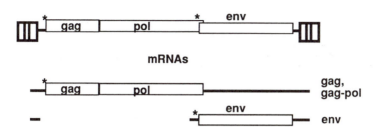

RSV

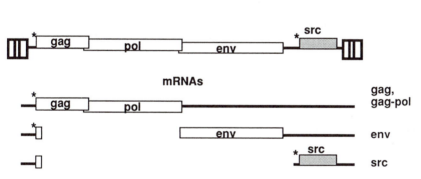

FIGURE 4. Genome structure and splicing pattern of the Moloney murine leukemia virus (Mo-MLV) and Rous sarcoma virus (RSV). Boxes indicate the genomic regions encoding proteins. Lines indicate the RNA sequences found in the known spliced and unspliced viral RNAs. Asterisks indicate the initiator AUG codons known to be used efficiently for translation. While Mo-MLV uses two different initiator AUGs for the production of *gag* and *env* proteins, RSV uses the *gag* AUG for *env* expression.

though not quite the total number of mRNAs. SIV transcripts have a splicing pattern remarkably similar to that of HIV (Viglianti *et al.*, 1990), and visna lentivirus also exhibits similar mRNA complexity (Davis *et al.*, 1987; Mazarin *et al.*, 1988). Complex splicing patterns and RNA structures have also been discovered in the spumaretroviruses (Muranyi and Flugel, 1991).

Inspection of the splicing pattern of HIV-1 shown in Fig. 6 reveals an interesting architecture: almost every expressed coding sequence has a splice acceptor upstream of its initiator codon, creating an array of

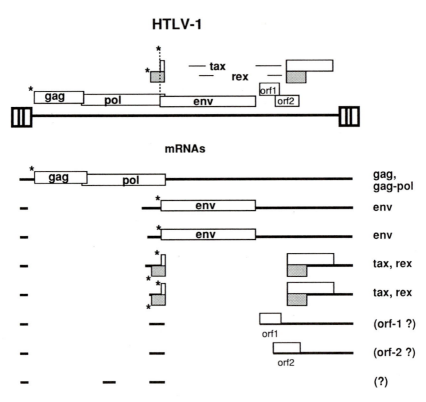

FIGURE 5. Genome structure and splicing pattern of HTLV-1. As in Fig. 4, the asterisks indicate initiator codons. HTLV-1 is a complex retrovirus that may produce some additional mRNAs which have not yet been characterized. The *tax* and *rex* reading frames are seen to be overlapping.

mRNAs sharing the same untranslated leader. The *gag* and *gag–pol* polyproteins are expressed by the unspliced transcript, as in other retroviruses. The protein denoted *vif* is translated from the *vif* mRNA, while the *vpr* protein is translated from the *vpr* mRNA, as shown in Fig. 6. (Our purpose here is to describe the RNAs; for some discussion of the proteins themselves, see Section III.E below.) The complexity of the produced mRNAs is greater in the cases of the other HIV proteins. The virus may produce three different mRNAs encoding one of the forms of the *tat* protein, *tat*-1, by alternative splicing; these mRNAs differ by the presence or absence of the two small exons denoted 2 and 3 in Fig. 6. Nine different mRNAs could generate the *vpu* and *env* proteins, depending upon the utilization of different splice sites (Schwartz et al., 1990b).

A plethora of multiply-spliced transcripts also results from the use of additional splice sites: thus, further splicing generates three mRNAs

HIV-1

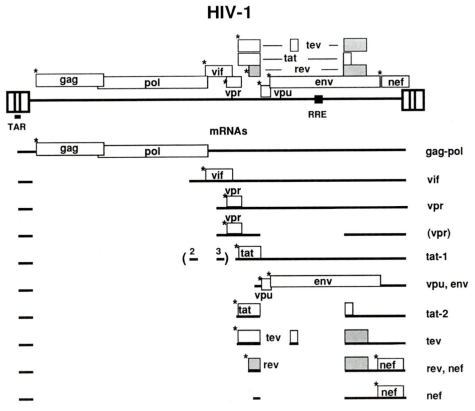

FIGURE 6. Genome structure and splicing pattern of HIV-1. As in Figs. 4 and 5, asterisks indicate initiator codons. The small exons denoted 2 and 3 (indicated in parentheses) are found in many HIV-1 transcripts; they increase the complexity of the produced mRNAs (Schwartz *et al.*, 1990a,b). The alternative, doubly spliced, *vif* and *vpr* mRNAs, also shown in parentheses, are thought to be minor mRNA species. Several of the open reading frames are overlapping, most noticeably those for *tat*, *rev*, and *env*.

producing a second form of the *tat* protein, *tat*-2, as well as three mRNAs producing the *nef* protein. Analogously, six different multiply-spliced species are produced for another HIV-1 protein, *rev*. This degeneracy of transcripts is significant in the case of the *rev* protein because it acts as a negative feedback agent by lowering the levels of proteins produced by the multiply-spliced mRNAs (including its own level) and elevating the levels of structural proteins, which are produced from the unspliced and singly-spliced mRNAs (Sodroski *et al.*, 1985; Felber *et al.*, 1988, 1989b, 1990).

Additional splice donor sites are found downstream of acceptor sites in HIV. Upon consideration of all the possible splicing combinations, many of which have been detected experimentally, a minimum of

20 viral transcripts is predicted. Why is it necessary for the virus to develop such a complex splicing system? Certainly, novel proteins are created: two examples are *tat-2* and a fusion protein that is named *tev*, or *tnv*. This additional protein, *tev*, has been identified in certain HIV-1-infected cells, but is absent in many HIV-1 genomes (Benko *et al.*, 1990; Salfeld *et al.*, 1990). The *tev* protein has the regulatory properties of both *tat* and *rev*, and it can replace both functions in complementarity experiments (Benko *et al.*, 1990; Solomin *et al.*, 1990). The presence of this protein in only some infected cells suggests that part of the biological variability of HIV-1 may result from facile changes in splicing. It seems reasonable to conclude, therefore, that in addition to the advantages of producing extra proteins, the acquisition of new regulatory capabilities is an important factor in the complex design of mRNA splicing.

From time to time, there has been speculation about RNA expression deriving from the "comp" or "plus" strand of retroviral DNA. Miller (1988) has called attention to a potential open reading frame situated on the comp strand of HIV-1 DNA; and Bukrinsky and Etkin (1990) have reported what may be plus-strand RNA products early in HIV-1 infection. Many laboratories have looked for such RNAs and proteins without success.

In a prospective study of SIV variation and pathogenesis, Johnson *et al.* (1991a) report the discovery of interesting forms of proviral DNA that must have been generated from spliced subgenomic RNAs; the significance of these DNA molecules is still to be learned.

2. *Cis*-Acting RNA Elements

The regulatory (*tat* and *rev*) proteins of HIV-1 act by binding directly to specific targets that have been defined on the viral RNA. With respect to the *rev* protein, the specific binding site has been mapped to the *env* transcript and has been named RRE (Felber *et al.*, 1989b; Hadzopoulou-Cladaras *et al.*, 1989) or CAR (Dayton *et al.*, 1988). The minimal continuous RRE sequence for *in vivo* function is a 204-nucleotide stretch within the *env* coding sequence that can be folded into a functional secondary structure (Malim *et al.*, 1989b; Solomin *et al.*, 1990). RRE-like elements have been tentatively identified in other lentiviruses, namely CAEV (Saltarelli *et al.*, 1990) and visna virus (Tiley *et al.*, 1990). All known lentiviral genomes appear to have a reading frame corresponding to the *rev* reading frame of HIV-1; thus, it seems very likely that this group of complex retroviruses shares the property of transcript regulation facilitated through *rev*–RRE interaction (Felber *et al.*, 1989a; Malim *et al.*, 1989a; Stephens *et al.*, 1990; Tiley *et al.*, 1990). The *rex* protein of HTLV-1 is also a posttranscriptional regulator of mRNA transport, stabilization, and utilization. However, the *rex* responsive element of HTLV-1, named by analogy RXRE, is found not in the *env* coding region but in the LTR (Derse, 1988; Felber *et al.*, 1989a; Hanly *et al.*, 1989; Solomin *et al.*, 1990; Toyoshima *et al.*, 1990), and it displays a complex secondary structure that is different from RRE in HIV-1. In addition to being necessary for *rex* function, formation of this RNA structure is necessary for the correct juxtaposition of the polyadenylation signal AAUAAA with the cleavage-poly-A addition site (Seiki *et al.*, 1983). Curiously, the *rex* factor of HTLV-1 can interact with both the bovine leukemia virus RXRE and the HIV RRE (Rimsky *et al.*, 1988; Felber *et al.*, 1989a; Lewis *et al.*, 1990), in addition to transregulating HTLV-2 gene expression (Kim *et al.*, 1991). While the HIV-1 *rev* does not reciprocally interact with the HTLV-1 RXRE, it can bind to, and regulate, the HIV-2 RRE (Malim *et al.*, 1989a). To the extent that heterologous RNA processing has been observed among these complex retroviruses, it has required direct binding of the protein factors to the RNA (Daly *et al.*, 1989; Zapp and Green, 1989; Daefler *et al.*, 1990).

The HIV TAR element is another small, *cis*-acting, DNA and RNA structure of evolutionary importance (Varmus, 1988; Wong-Staal, 1990). As was noted in Section III.A.2, it appears to have been captured by just the primate lentiviruses from a cellular immunoglobulin heavy-chain coding sequence (Myers *et al.*, 1989). In concert with the *tat* regulatory protein, this element serves as a downstream promoter element. The *tat* protein probably also has an effect on the upstream promoter elements, since an analogous protein is found in all lentiviruses and also in HTLV-1/2, where it is named *tax*. The simian foamy virus also encodes a transactivator protein which acts at the level of the LTR (Mergia *et al.*, 1990b).

In the type 2 HIVs and related SIVs, the TAR core element has been duplicated, leading to some divergence within the constraints of a conserved secondary structure. Heterologous transactivation of viral gene expression by means of diverse *tat* proteins and TAR structures, not unlike the heterologous transregulatory effects observed with *rev* and RRE, have been observed (Emerman *et al.*, 1987; Wong-Staal, 1990). Furthermore, many cellular proteins interact with the TAR DNA and RNA sequences (Garcia *et al.*, 1987; Jones *et al.*, 1988; Gaynor *et al.*, 1989; Hart *et al.*, 1991; Tadmori *et al.*, 1991).

3. Base Compositions of Lentiviral RNAs

The unusual base composition of lentiviruses—the high A and low C contents in comparison to those of most mammalian genes and retroviral molecules—was discussed in Sections II.C and III.B. Given the multiplicity of lentiviral mRNAs and *cis*-acting elements, it is noteworthy that many of the HIV-1 mRNAs and *cis*-acting elements having regulatory functions also have more normal base compositions (Table III). In particular, the *tat-2, tev, rev,* and *nef* transcripts shown in Fig. 6, all of which encode regulatory proteins, have nearly normal A and C contents. The alternate *tat* messages are especially intriguing in that they manifest both extreme and nearly normal base compositions. Moreover, the RRE and TAR elements are islands in the genomic molecule that do not possess the unusually high A content seen in the structural protein RNA transcripts.

The higher G–C contents of the HIV *cis*-elements may have been selected to support the strong secondary RNA structures that they adopt. But to begin to understand the differences between some of the regulatory protein transcripts and the structural protein transcripts of lentiviruses, we must return to the possibilities and considerations mentioned in Section III.C—codon usage, codon volatility, methylation, and recombination. With exception of the HIV and SIV TAR element, which appears to have its origin in an immunoglobulin gene sequence, there is no reason to believe that these RNAs possessing more normal nucleotide compositions are the result of divergent rather than convergent evolution. If they are shown to represent "captured" elements,

TABLE III. Base Compositions of Some Retroviral Structural
and Regulatory RNAs

mRNA	Number of bases $\times 10^{-3}$	Average composition (%)			
		A	C	G	U
HIV-1 *gag–pol*	9.2	35.6	17.9	24.2	22.2
HIV-1 *env*	3.9	32.7	18.7	25.3	23.2
HIV-1 *tat-1*	4.1	32.7	18.8	25.3	23.3
HIV-1 *tat-2*	1.8	27.9	22.4	28.3	21.4
HIV-1 *rev, nef*	1.6	27.4	22.8	28.8	21.1
HIV-1 leader (untranslated)	0.3	23.3	24.3	30.2	22.2
CAEV *gag–pol*	9.2	38.1	16.0	25.1	20.8
HTLV-1 *gag–pol*	9.1	23.1	34.9	19.0	23.1

Regulatory RNA (*cis*-element)	Number of bases $\times 10^{-3}$	Composition (%)			
		A	C	G	U
HIV-1 RRE	0.2	26.0	23.0	29.9	21.1
CAEV RRE	0.2	23.9	22.9	28.4	24.9
HTLV-1 RXRE	0.2	14.9	35.7	27.8	21.6

[a] The HIV-1 TAR element constitutes the first 50 bases, approximately, of the mRNA leader. The various HIV-1 transcripts, including this leader, are shown in Fig. 6. The CAEV sequences are reported in Saltarelli *et al.* (1990). The HXB2 sequence is reported for HIV-1s, but the results are representative of HIVs and SIVs in general (Table II).

acquired parts of genes, we must try in the years to come to comprehend why HIV and HTLV, retroviruses with clearly different lineages of descent, have adopted equivalent forms of complexity.

E. Proteins of Complex Retroviruses

1. Accessory Proteins

The exceptional array of virally-encoded proteins that distinguishes the lentiviruses (and, to a lesser degree, HTLV-1/2) from other retroviruses derives in part from larger genomic molecules and mRNA complexity. In the preceding discussion, splicing options were shown to be a major source of protein diversity. One instance of alternate splicing was shown to give rise, in certain strains, to a natural fusion protein (*tev*), and other variants of the *tat* regulatory protein were seen to arise from subtly different transcripts. The *src* oncogene (Fig. 4) and the accessory genes of complex retroviruses (Figs. 5 and 6) typically have their own transcripts, which suggests that they may have been acquired by RT-mediated recombination. What is so astonishing about the emergence of retroviral accessory proteins is the extent of condensation in the reading frames. For example, the *tax* and *rex* reading frames of HTLV-1 (Fig. 5) and the *tat*, *rev*, and *env* reading frames of HIV-1 (Fig. 6) are overlapping. This degree of protein coding density, in combination with the variability of these gene sequences in HIV-1, implies that at some level the evolution of these protein sequences must be coordinated.

Two of the molecular processes that probably contribute to the fine tuning of lentiviral protein evolution are gene duplication and proteolysis, for example, in the derivation of the HIV/SIV *vpx* protein from the *vpr* protein by gene duplication (Tristem *et al.*, 1990; Myers *et al.*, 1992a), and the proteolytic processing of several small proteins from the ends of the HIV *gag* polyprotein (Henderson *et al.*, 1988; Myers *et al.*, 1990). Proteolytically-derived nonstructural proteins are of considerable importance in relation to molecular mimicry (discussed in Section III.E.3).

Ribosomal frameshifting associated with the *gag–pol* reading frames has been well characterized for retroviruses (Coffin, 1990). It is perhaps not surprising, then, that an additional frameshift event in the central region of the HIV-1 genome may result in a 17-kilodalton (kDa) fusion protein (Cohen *et al.*, 1990). While it is not known whether this protein, denoted *vpt*, is made in natural HIV-1 infections, the open reading frame and the frameshift signal are conserved. It must be kept in mind that many of the lentiviral accessory proteins are difficult to detect because they are made in very low amounts. There is some evidence to suggest that the use of alternate initiator and stop codons is also a factor in protein diversification.

2. Homologies, Similarities, and Chance Relationships

We began this review with a phylogenetic tree (Fig. 1) that depicted the homologous relationships of complex retroviruses. In fact, all exogenous retroviruses are found to be homologous in their *pol* gene nucleotide sequences. But it was also noted that retroviral species which one might expect to be rather closely related, namely the bovine and ovine lentiviruses, were merely 55% identical. These observations were based upon the most highly conserved coding sequence for most retroviruses, namely the *pol* gene, which encodes essential proteins having enzymatic functions. Now it is informative to ask about the degree of similarity of the regulatory and other accessory proteins of complex retroviruses, some of which are essential, others of which are not. For instance, are the *rev* and *rex* proteins of lentiviruses and HTLV-1/2 homologous? Or are they merely similar, the results of convergent evolution? (The reader may recall that the base compositions of HIV-1 and HTLV-1 are very different.) Can we at least surmise that the *rev* proteins of two lentiviruses, say HIV-1 and visna virus, are homologous? If these proteins and their coding sequences are not homologous, are they each homologous to something else that can be identified? Are they, like retroviral oncogenes, the evolutionary progeny of captured genes or transcripts?

Answers to these questions are not easy. First, because our investigations of these retroviruses and their proteins are still in an early stage. Second, the general challenges of protein pattern analysis and the identification of protein relationships are heightened in the case of the highly mutable lentiviruses. As a rule of thumb, amino acid sequences (depending on their length) can be 25% similar by chance alone, although, in the extreme, proteins that are less than 25% similar may still be homologous. In the complex retroviruses, proteins with this small degree of similarity can still have the same function. Following Doolittle (1987, pp. 10–15), we shall refer to this level of uncertainty as the "twilight zone." These are not insuperable barriers to inquiry at this time. Nevertheless, in this review we will limit ourselves to sketching the protein relationship problem with the lentiviruses and their distant relatives, and to reporting some interesting findings and possibilities that enlarge our view of retroviral evolution.

3. Lentiviral Protein Relationships

Current strategies for studying protein families and patterns take into account, among other things, conservative amino acid replacements (for example, the substitution of leucine for valine). This effort significantly improves the chances of identifying similarities, and it is essential to the ensuing critique of the similarities. Are the proteins homologous or merely similar? Are they products of divergent or convergent evolution? The approach we have favored is a "regular expression" algorithm developed by R. F. Smith and Smith (1990): an amino acid hierarchical classification scheme, based upon side-chain properties, becomes the basis for quantifying conservative substitutions. The approach is grounded in information theory, but it retains the comfortable feature that quantities can be expressed in terms of amino acid equivalents. Thus, invariant residues in a family of homologous protein sequences are given the information content of 1.0 amino acid equivalents, while completely variable positions have the information content of 0. All other replacements, more or less conservative, take on values between 1.0 and 0 as prescribed by the theory and the predefined, and appropriately controlled, amino acid classes. In general, the information density of a set of homologous proteins will exceed the percentage of invariant residues by about 35%; the additional information is a measure of the variation involving conservative replacements.

Some representative information densities for lentiviral and other proteins are shown in Table IV. The HIV-1 *gag* polyprotein represented

TABLE IV. Protein Information Densities Expressed as Amino Acid Equivalents[a]

Virus	*gag*	*pol*	*rev*	*env*	*nef*
HIV-1/CPZ (8)	0.76	0.85	0.57	0.59	0.64
HIV-2/SIVs (7)	0.84	0.87	0.69	0.72	0.55
HIV-1/HIV-2/SIVs (15)	0.56	0.62	0.30	0.35	0.34
SIVAGMs (3)	0.78	0.76	—	0.75	0.76
Lentiviruses (nonprimate) (5)	0.19	0.31	—	<0.10	—

Protein families: Information densities reported in Smith and Smith (1990)

α-Hemoglobins (107)	0.20
IG heavy-chain precursors (42)	0.21
α-Inferons (32)	0.33
Insulins (31)	0.30

[a] Following alignment of amino acid sequences, the average information densities (IDs) were determined using the "calc-IC" program of R. F. Smith and Smith (1990). The information density expresses the average invariant and conserved amino acid residues at each position of the molecule; for example, there is 0.76 amino acid equivalent of information at each position, on the average, in the HIV-1 *gag* protein set under study. Comparisons within each row involve the same set of genomic sequences: 8 HIV-1/SIVCPZ, 7 HIV-2 and related SIV, 15 HIV-1/HIV-2/SIV, 3 SIV-AGM, and 5 nonprimate lentiviral sequences (FIV, BIV, EIAV, visna virus, CAEV). Obviously these values will decrease to some level as more sequences are compared. Some representative IDs for large protein families (number of sequences in parentheses) are shown for comparison.

by eight isolate sequences (including the HIV-1-related SIV_{CPZ}) is shown to have 0.76 amino acid equivalents, on the average, at each position. The fraction of invariant sites in this *gag* consensus protein is closer to 0.5; thus, more of the information present in the sample of eight *gag* sequences has been captured by the amino acid class covering analysis. Revising the amino acid classification scheme has surprisingly little effect upon the general result (though it could make a critical difference in the prediction of amino acid replacements at particular positions). A far greater effect is seen through inclusion of more sequences: the 15 homologous HIV/SIV sequences, which may have diverged from a common ancestor as recently as 50–200 years ago (Section II.B.1), represent consensus proteins with marked reductions in conservation.

When viewed together, the results for the HIV-1 *gag, pol,* and *env* proteins are what we would expect on the basis of HIV-1 nucleotide sequence variation—*pol* (0.85) is most conserved, *env* (0.59) is least conserved, *gag* is intermediate (0.76). The same correlation holds for the HIV-2s and their SIV relatives and for five nonprimate lentiviruses that were examined in this way. However, the *gag, pol,* and *env* ratios for the three SIV_{AGM} samples do not display this differential, which argues (as we discussed in Section II.A.2) that they are not subject to the same gradient of selection.

The *rev* regulatory protein, which has been a focal point of our inquiry into complexity, is seen in Table IV to be as variable as the *env* protein; in fact, all of the HIV and SIV accessory proteins are as variable as the *env* protein (Myers *et al.,* 1990). It should be kept in mind that several of the reading frames encoding these proteins are overlapping. Envelope variation is thought by some to be driven by immune pressure (as we discuss below). Be that as it may, many of the amino acid replacements in *env* and the other variable proteins are considered to be selectively neutral (Leigh-Brown and Monaghan, 1987; Simmonds *et al.,* 1990a; Gojobori *et al.,* 1990a; Burns and Desrosiers, 1991). The high degree of variability of lentiviral accessory proteins relative to the *pol* polyprotein may also result from indirect selection (Atwood *et al.,* 1951). Finally, balanced polymorphism and complementation among sibling members of a swarm (Eigen and Biebricher, 1988) provide further possibilities for variation, especially if it can be shown that the accessory proteins are rudimentary.

The conspicuous variation—low information densities—of lentiviral regulatory proteins does not necessarily imply that they have rudimentary functions. Some well-characterized protein families were found by R. F. Smith and Smith (1990) to have information densities as low as 0.2–0.3 (Table IV). On the other hand, the replacement of the HIV-1 *rev* function by the *rex* protein from the very distantly related HTLV-1 (Rimsky *et al.,* 1988; Lewis *et al.,* 1990) raises the question of whether these proteins are basically support molecules for cellular proteins. It is noteworthy with regard to this particular example that the target sequence for HTLV-1 *rex* and the HIV-1 target sequence, RRE, are quite dissimilar (Section III.D.2).

The *gag* and *env* information densities in Table IV that are derived from five nonprimate lentiviral protein sequence sets (and which offer the fullest assessment of conservative amino acid replacements) are in or below the twilight zone. If, on the basis of the nucleotide sequence similarities over the most highly conserved sites of the *pol* gene (Section II.A), the *pol* gene products are judged to be monophyletically related, perhaps the envelope and accessory proteins of lentiviruses should be deemed "paraphyletically" related (Gould *et al.*, 1987). These proteins, by this argument, have radically diverged from one another; and the similarities in overall genomic organizations and base compositions strongly argue for divergent evolution. Thus, the putatively homologous visna virus *rev* protein, in concert with its own target sequence, has the ability to rescue an HIV-1 strain devoid of both the *rev* function and RRE element (Tiley *et al.*, 1990). A reasonable case can be made for the homology of the S *orf* protein of BIV and the *tat* protein of HIVs (Garvey *et al.*, 1990). However, the evolutionary relationships among the other lentiviral S *orf* and *tat* gene products remain thoroughly uncertain, beyond the fact that they are involved with regulation by transactivation (Derse *et al.*, 1991; Myers *et al.*, 1992a).

Cross-transactivation with lentiviruses might be taken as evidence of protein relatedness between highly divergent viruses ("paraphyly") were it not for the fact that many different viruses, including DNA viruses, are able to transactivate HIV (Gendelman *et al.*, 1986; Mosca *et al.*, 1987; Siekevitz *et al.*, 1987; Rando *et al.*, 1990). Furthermore, HIV regulation influences, and is affected by, cellular processes (Greene *et al.*, 1986; Siekevitz *et al.*, 1987; M. R. Smith and Greene, 1989; Tadmori *et al.*, 1991). Many of these heterologous interactions with viral and cellular factors are mediated through ubiquitous promoter and enhancer elements such as Sp1 (Jones *et al.*, 1986) and NFkB (Nabel and Baltimore, 1987). Given these equally compelling signs of divergence on the one hand and convergence on the other hand, inquiry into the evolutionary connections among lentiviruses, and between them and their hosts, is currently focusing not so much upon complete proteins with low information contents as upon regions of proteins with high information contents.

4. Molecular Mimicry

To date, not one of the genes of complex retroviruses has been identified as a homologue of a cellular gene. For instance, none of the HTLV-

1/2 accessory genes can be identified as an oncogene (Cann and Chen, 1990). The HIV/SIV *nef* gene might represent an acquired gene: it is found only at the 3' end of primate lentiviral genomic molecules and it has a base composition closer to that of most nuclear genes. By one report, it has been classified with the phosphorylated GTP-binding proteins (cytoplasmic G proteins) that are also produced by the *ras* oncogene (Guy *et al.*, 1987). A more recent study has called into question the GTP or GTPase activity of *nef* (Kaminchik *et al.*, 1990). Both the function and origin of *nef* are controversial at this time (Hammes *et al.*, 1989).

Strong evidence in support of the cellular origin of lentiviral accessory proteins is made unlikely by the rapid rate of lentiviral change. In the absence of such evidence, attention has been turned toward short viral peptides, shared protein motifs that may have arisen by either a divergent or a convergent process. For example, application of the regular expression or covering pattern analysis developed by R. F. Smith and Smith (1990) has identified a common structural pattern in HIV-1 *nef* and HLA class II histocompatibility antigens that would not have been detected in routine protein similarity searches (Vega *et al.*, 1990). An autoimmune complication is postulated to be the result of the molecular mimicry.

Interest in this field of fine-structure protein sequence analysis is stimulated by the recognition that many viral proteins, especially epitopes, mimic host-cell proteins (Oldstone, 1987). Among the types of cellular proteins believed to be mimicked by HIV oligopeptides are hormones (Sarin *et al.*, 1986; Glass, 1991), neuropeptides (Pert *et al.*, 1986; Yamada *et al.*, 1991), cytokines (Reiher *et al.*, 1986), immunoglobulins (Maddon *et al.*, 1986), histocompatibility antigens (Golding *et al.*, 1988; Vega *et al.*, 1990), and peptides interacting with potassium ion channels (Werner *et al.*, 1991). Similarities between HIV proteins and proteins of other viruses suggest instances of convergent mimicry (Argos, 1989; Blomberg and Medstrand, 1990).

Conserved and invariant patterns embedded in the most highly variable regions of HIV proteins, such as those of the envelope, *nef*, and *gag* p17 proteins, may be implicated in these acts of mimicry. Three regions of the HIV *gag* p17 core protein [which displays a high rate of amino acid replacement (Leigh-Brown and Monaghan, 1988)] have been examined in this respect. At the N-termini of HIV and SIV *gag* polyproteins, a highly conserved sequence appears to be uncannily similar to the human gonadoliberin molecule, even to the extent of mimicking a proteolysis site that could generate an amidated C-terminus for the 23-residue peptide (Glass, 1991). A sequence in the carboxyl-terminus of HIV-1 *gag* p17 sequences is found to be highly similar to sequences in envelope or membrane-associated proteins of 12 other RNA viruses (Blomberg and

Medstrand, 1990). While no similarity between this 19-residue peptide and cellular proteins has been identified, individuals who are unlikely to be infected with HIV-1 show cross-reactive antibodies (Blomberg and Medstrand, 1990). Finally, about 25 residues upstream of this site in *gag* is a conserved epitope that is similar to a protein motif in alpha-thymosin (Sarin *et al.*, 1986) and also in human neurofilament protein (G. Myers, unpublished results). Thus, nearly half of the *gag* p17 residues (in HIV-1) have been implicated in molecular mimicry. Because the flanking regions of these sites are not similar to the flanking regions of the protein sequences they are thought to mimic, and the base compositions of the coding sequences are sometimes quite different, convergent evolution becomes one plausible explanation for the similarities.

5. Envelope Variation

Reflecting upon some of the earliest evidence of extreme variability of HIV-1, Coffin (1986) made the interesting observation that the envelope glycoprotein of HIV was about 35% larger than the coat glycoprotein of an oncovirus, ALV. The early studies of HIV envelope diversity by Hahn *et al.* (1985) and Modrow *et al.* (1987) lent support to Coffin's notion that it was mainly the pattern of *env* variability that distinguished HIV from more stable retroviruses. We now know that all of the HIV coding sequences, with the exception of *pol* and *gag* p24, participate in hypermutation (Delassus *et al.*, 1991; Vartanian *et al.*, 1991; Johnson *et al.*, 1991a) and have exceptional rates of amino acid replacement compared to synonymous nucleotide substitution rates (Leigh-Brown and Monaghan, 1988; Simmonds *et al.*, 1990a; Burns and Desrosiers, 1991) (Table IV). Nevertheless, Coffin's (1986) idea that variable regions of the HIV envelope could be serving as "umbrellas" for conserved regions to be found in every retroviral envelope is suggestive.

The well-known antigenic variation of visna virus and EIAV (Clements *et al.*, 1980; Braun *et al.*, 1987; Carpenter *et al.*, 1991; Payne *et al.*, 1987) might also be mentioned in relation to envelope protein size. All of the complex retroviruses, with the exception of HTLV-1/2, have extremely large coat proteins, and those of the primate lentiviruses are excessively glycosylated (Leonard *et al.*, 1990; Mizuochi *et al.*, 1990; Myers *et al.*, 1991, 1992a). It is perhaps no surprise, then, to find HTLV-1/2 to be relatively stable (Section II.A). Nowak (1990) has followed up the length-dependence notion by showing how the mathematical theory which is developed with the size of the entire genomic molecule in mind can be appropriately modified to take into account neutral sites and the length of an immunodominant loop that will navigate an escape from immune surveillance. On this basis, he argues that HIV-1 and other lentiviruses have optimized their mutation rates in order to escape immune surveillance. This theoretical argument is grounded upon the general understanding of lentiviral variation (Narayan and Clements, 1990) and upon the recent findings pertaining to the principal neutralization determinant (PND) of HIV-1 (LaRosa et al., 1990; Simmonds et al., 1990a; Wolfs et al., 1990).

The PND of HIV-1 is located in the so-called V3 loop of the envelope glycoprotein (Meloen et al., 1989; LaRosa et al., 1990). This highly variable structure of about 35 amino acid residues is quite unusual: the A content of the V3 loop coding sequence, for example, is typically 40%. In a study of viral V3 sequences representing six children infected by the same blood donor, the average nucleotide substitution rate was found to be about 0.01 per site per year (Wolfs et al., 1990). This is in close agreement with the env gene values summarized in Section II.B, Table I. While the rate ratio of nonsynonymous (amino acid changing) to synonymous nucleotide substitutions appears to be higher for this protein stretch than any previously reported rate ratio (Simmonds et al., 1990a), it is remarkable that not all of the amino acid replacements are functionally neutral. Some of the changes in the V3 loop affect the ability of the env glycoprotein to induce cell fusion (Freed et al., 1991) and thereby affect cellular tropism (Westervelt et al., 1991; Hwang et al., 1991). Nevertheless, single amino acid changes in the HIV-1 PND can profoundly affect type-specific antibody recognition (Looney et al., 1988; Goudsmit et al., 1989; Wolfs et al., 1990) and cytotoxic T-cell lymphocyte (CTL) specificity (Takahashi et al., 1989).

Further modeling of HIV-1 antigenic variation by Nowak and co-workers (Nowak et al., 1990, 1991) provides a possible account for what appears to be a paradoxical program of lentiviral variability coupled with complexity. The cornerstone for their mathematical model is the hypothesis that AIDS develops after the population of antigenic variants in an HIV-infected patient has exceeded some critical threshold, beyond which the immune system fails. A long latent period, during which, first, slowly replicating viral forms and, later, rapidly replicating forms emerge in opposition to the immune system, is required by the theory to attain the diversity threshold. In its preliminary version, the model emphasizes antigenic drift; thus, it does not explicitly depend upon viral gene modulation during the course of infection. It will be interesting to see how the roles of the highly variable accessory proteins become incorporated into future models of AIDS and other slow diseases caused by complex retroviruses.

IV. CONCLUDING REMARKS

In this review, we have ranged over many facets of the molecular genetics of complex retroviruses. Approximately half of the information included in the review was published in the last 2 years. Although this large body of information is too new to warrant a bold revision of our notions of retroviral evolution, it is intended to direct some of our attention to the awesome evolutionary potential represented by the AIDS viruses (Table V).

In light of all that is now known about the AIDS viruses and their relatives, what might we expect in the future? Will HIV become an endogenous retrovirus? Will the route of transmission change? Can a successful vaccine be made? These are some of the frequently asked questions.

We have examined the origin and evolutionary potential of the complex retroviruses largely with the hope of identifying homologies and similarities that would suggest possible answers to these questions. Thus, in the course of this inquiry, we have specifically looked for prece-

TABLE V. Summary of Exceptional Molecular Properties of HIV and SIV

1. Variability
   Nucleotide substitution rates measured both *in vivo* (Table I, Section III.A.2) and *in vitro* (Section A.1) prove to be high compared to those of other retroviruses
   Hypermutation (Section III.B) and localized, or intragenic, recombination (Section III.A.2) are conspicuous; inactivating mutations need not be the result (Section III.B)
   The ratio of nonsynonymous (amino acid-replacing) to synonymous substitution frequencies is exceptionally high in the *env* gene (Sections II.C.3, III.E.5); variation of the accessory and regulatory proteins is comparable to that of the *env* protein (Table IV, Section III.E.3)
   Microscopic and macroscopic swarms (quasispecies) are the rule (Section II.C)

2. Base composition
   The high A content of *gag*, *pol*, and *env* coding sequences is unusual, although not unique (Section III.C.1); the CpG dinucleotide frequency is the lowest encountered to date (Section III.C.4)
   Codon usage is atypical and volatile: single base changes tend to promote nonconservative amino acid replacements (Section III.C.2)
   The genomic molecule is a mosaic: certain RNA transcripts and *cis*-acting regulatory elements have nearly normal base compositions (Section III.D.3)

3. Complexity
   In comparison to other retroviruses, there are many virally encoded accessory and regulatory proteins (Section III.E.1); the coding density—extent of overlapping reading frames—is remarkably great (Section III.E.1)
   A multiplicity of transcripts is achieved through alternate splicing (Fig. 6, Section III.D.1)
   Unspliced and singly-spliced RNAs for structural proteins are processed differently than the multiply-spliced RNAs for regulatory proteins (Section III.D.1); both RNA processing and transcription are responsive to cofactors and heterologous agents (Section III.D.2)
   Both envelope variation and viral gene regulation are key processes of disease induction (Section III.E.5)

dents for the molecular biology of the AIDS viruses. We have noted that most retroviruses mutate rapidly, though they do not have the ability to regulate the expression of their own genes (what we have called "complexity"). The human T-cell leukemia viruses, on the other hand, are complex in this sense, but mutate slowly. The novelty of the lentiviruses is that they are both variable and complex. Unfortunately, as a result of the combination of complexity and variability there are many ways to be a successful immunodeficiency retrovirus.

Still, a high degree of variability has a fortunate result: because these complex retroviruses change so rapidly, they quickly reveal their inner structure and evolutionary dynamics. After only 5 years of intensive investigation on the part of many researchers, we now have a reasonably good picture of the invariant and conserved protein structures of HIV-1. This new information, along with a rapidly growing understanding of the other complex retroviruses, provides some ground for scientific and technological optimism.

ACKNOWLEDGMENTS. The HIV Sequence Database and Analysis Project that has been the springboard for much of this inquiry is funded by the Vaccine Branch of the AIDS Division of the National Institute of Allergy and Infectious Diseases through an Interagency Agreement with the U.S. Department of Energy (LAUR-91-660). The research of G.N.P. was sponsored in part by the National Cancer Institute, DHHS, under contract number N01-CO-74101 with ABL. We are indebted to Kersti MacInnes, George Nelson and Charles Calef of Los Alamos National Laboratory for their excellent technical assistance and to Howard Temin of the University of Wisconsin for his careful reading and criticism of this manuscript.

V. REFERENCES

Allan, J. S., Short, M., Taylor, M. E., Su, S., Hirsch, V. M., Johnson, P. R., Shaw, G. M., and Hahn, B. H., 1991, Natural infection of West African green monkeys with a unique subtype of SIV-agm, J. Virol. **65**:2816.

Argos, P., 1989, A possible homology between the human immunodeficiency virus core protein and picornaviral vp2 coat protein: Prediction of HIV p24 antigenic sites, EMBO J. **8**:779.

Atwood, K. C., Schneider, L. K., and Ryan, F. J., 1951, Periodic selection in Escherichia coli, Proc. Natl. Acad. Sci. USA **37**:146.

Balfe, P., Simmonds, P., Ludlam, C. A., Bishop, J. O., and Leigh-Brown, A. J., 1990, Concurrent evolution of human immunodeficiency virus type 1 in patients infected from the same source: Rate of sequence change and low frequency of inactivating mutations, J. Virol. **64**:6221.

Battula, N., and Loeb, L. A., 1974, The infidelity of avian myeloblastosis virus deoxyribonucleic acid polymerase in polynucleotide replication, *J. Biol. Chem.* **294**:4086.

Bebenek, K., Abbotts, J., Roberts, J. D., Wilson, S. H., and Kunkel, T. A., 1989, Specificity and mechanism of error-prone replication by human immunodeficiency virus-1 reverse transcriptase, *J. Biol. Chem.* **264**:16948.

Bednarik, D. P., Mosca, J. D., and Raj, N. B., 1987, Methylation as a modulator of expression of human immunodeficiency virus, *J. Virol.* **61**:1253.

Benko, D. M., Schwartz, S., Pavlakis, G. N., and Felber, B. K., 1990, A novel human immunodeficiency virus type 1 protein, *tev*, shares sequences with *tat*, *env*, and *rev* proteins, *J. Virol.* **64**:2505.

Blomberg, J., and Medstrand, P., 1990, A sequence in the carboxyl terminus of the HIV-1 matrix protein is highly similar to sequences in membrane-associated proteins of other RNA viruses: Possible functional implications, *New Biol.* **2**:1044.

Braun, M. J., Clements, J. E., and Gonda, M. A., 1987, The visna virus genome: Evidence for a hypervariable site in the *env* gene and sequence homology among lentivirus envelope proteins, *J. Virol.* **61**:4046.

Bukrinsky, M. I., and Etkin, A. F., 1990, Plus strand of the HIV provirus DNA is expressed at early stages of infection, *AIDS Res. Hum. Retroviruses* **6**:425.

Buonagurio, D. A., Nakada, S., Parvin, J. D., Krystal, M., Palese, P., and Fitch, W. M., 1986, Evolution of human influenza A viruses over 50 years: Rapid uniform rate of change in NS gene, *Science* **232**:980.

Burger, H., Weiser, B., Flaherty, K., Gulla, J., Nguyen, P.-N., and Gibbs, R. A., 1991, Evolution of human immunodeficiency virus type 1 nucleotide sequence diversity among close contacts, *Proc. Natl. Acad. Sci. USA.* **88**:11236.

Burns, D. P. W., and Desrosiers, R. C., 1991, Selection of genetic variants of SIV in persistently infected Rhesus monkeys, *J. Virol.* **65**:1843.

Cann, A. J., and Chen, I. S. Y., 1990, Human T-cell leukemia virus types I and II, in: *Virology*, 2nd ed. (B. N. Fields and D. M. Knipe, eds.), pp. 1501–1528, Raven Press, New York.

Carpenter, S., Alexandersen, S., Long, M. J., Perryman, S., and Cheseboro, B., 1991, Identification of a hypervariable region in the long terminal repeat of equine infectious anemia virus, *J. Virol.* **65**:1605.

Castro, B. A., Barnett, S. W., Evans, L. A., Moreau, J., Odehouri, K., and Levy, J. A., 1990, Biologic heterogeneity of human immunodeficiency virus type 2 (HIV-2) strains, *Virology* **178**:527.

Cheng-Mayer, C., Seto, D., Tateno, M., and Levy, J. A., 1988, Biologic features of HIV-1 that correlate with virulence in the host, *Science* **240**:80.

Cheng-Mayer, C., Quiroga, M., Tung, J. W., Dina, D., and Levy, J. A., 1990, Viral determinants of human immunodeficiency virus type 1 T-cell or macrophage tropism, cytopathogenicity, and CD4 antigen modulation, *J. Virol.* **64**:4390.

Cheng-Mayer, C., Shioda, T., and Levy, J. A., 1991, Host range, replicative and cytopathic properties of human immunodeficiency virus type 1 are determined by very few amino acid changes in *tat* and gp120, *J. Virol.* **65**:6931.

Chiu, I.-M., Callahan, R., Tronick, S. R., Schlom, J., and Aaronson, S. A., 1984, Major *pol* gene progenitors in the evolution of oncoviruses, *Science* **223**:364.

Chiu, I.-M., Yaniv, A., Dahlberg, J. E., Gazit, A., Skuntz, S. F., Tronick, S. R., and Aaronson, S. A., 1985, Nucleotide sequence evidence for relationship of AIDS retrovirus to lentiviruses, *Nature* **317**:366.

Clements, J. E., Pedersen, F. S., Narayan, O., and Haseltine, W. A., 1980, Genomic changes associated with antigenic variation of visna virus during persistent infection, *Proc. Natl. Acad. Sci. USA* **77**:4454.

Coffin, J. M., 1986, Genetic variation in AIDS viruses, *Cell* **46**:1.

Coffin, J. M., 1990, Retroviridae and their replication, in: *Virology*, 2nd ed. (B. N. Fields and D. M. Knipe, eds.), pp. 1437–1489, Raven Press, New York.

Cohen, E. A., Lu, Y., Gottlinger, H., Dehni, G., Jalinoos, Y., Sodroski, J. G., and Haseltine, W. A., 1990, The T open reading frame of human immunodeficiency virus type 1, *J. Acquired Immune Defic. Syndr.* **3**:601.

Contag, C. H., Dewhurst, S., Viglianti, G. A., and Mullins, J. I., 1991, Simian immunodeficiency virus (SIV) from Old World monkeys, in: *The Human Retroviruses* (R. C. Gallo and G. Jay, eds.), pp. 245–276, Academic Press, San Diego, California.

Cullen, B. R., 1991, Human immunodeficiency virus as a prototypic complex retrovirus, *J. Virol.* **65**:1053.

Daefler, S., Klotman, M. E., and Wong-Staal, F., 1990, *Trans*-activating *rev* protein of the human immunodeficiency virus 1 interacts directly and specifically with its target RNA, *Proc. Natl. Acad. Sci. USA* **87**:4571.

Daly, T., Cook, K., Gray, G., Maione, T., and Rusche, J., 1989, Specific binding of HIV-1 recombinant *rev* protein to the *rev*-responsive element *in vitro*, *Nature* **342**:816.

Daniel, M. D., Li, Y., Naidu, Y. M., Durda, P. J., Schmidt, D. K., Troup, C. D., Silva, D. P., MacKey, J. J., Kestler III, H. W., Sehgal, P. K., King, N. W., Ohta, Y., Hayami, M., and Desrosiers, R. C., 1988, Simian immunodeficiency virus from African green monkeys, *J. Virol.* **62**:4123.

Davis, J. L., Molineaux, S., and Clements, J. E., 1987, Visna virus exhibits a complex transcriptional pattern: One aspect of gene expression shared with the acquired immunodeficiency syndrome retrovirus, *J. Virol.* **61**:1325.

Dayton, A. I., Terwilliger, E. F., Potz, J., Kowalski, M., Sodroski, J. G., and Haseltine, W. A., 1988, *Cis*-acting sequences responsive to the *rev* gene product of the human immunodeficiency virus, *J. Acquired Immune Defic. Syndr.* **1**:441.

Delassus, S., Cheynier, R., and Wain-Hobson, S., 1991, Evolution of the HIV-1 *nef* and LTR sequences over a four year period *in vivo* and *in vitro*, *J. Virol.* **65**:225.

DeLeys, R., Vanderborght, B., Haesevelde, M. V., Heyndrickx, L., van Geel, A., Wauters, C., Bernaerts, R., Saman, E., Nijs, P., Willems, B., Taelman, H., van der Groen, G., Piot, P., Tersmette, T., Huisman, J. G., and van Heuverswyn, H., 1990, Isolation and partial characterization of an unusual human immunodeficiency retrovirus from two persons of West-Central African origin, *J. Virol.* **64**:1207.

Derse, D., 1988, *Trans*-acting regulation of bovine leukemia virus mRNA processing, *J. Virol.* **62**:1115.

Derse, D., Carvalho, M., Carroll, R., and Peterlin, B. M., 1991, A minimal lentiviral tat, *J. Virol.* **65**:3877.

Desrosiers, R. C., 1988, Simian immunodeficiency viruses, *Annu. Rev. Microbiol.* **42**:607.

Dietrich, U., Adamski, M., Kreutz, R., Seipp, A., Kuhnel, H., and Rubsamen-Waigmann, H., 1989, A highly divergent HIV-2-related isolate, *Nature* **342**:948.

Doolittle, R. F., 1987, *Of Urfs and Orfs: A Primer on How to Analyze Derived Amino Acid Sequences*, University Science Books, Mill Valley, California.

Doolittle, R. F., Feng, D.-F., Johnson, M. S., and McClure, M. A., 1989, Origins and evolutionary relationships of retroviruses, *Q. Rev. Biol.* **64**:1.

Doolittle, R. F., Feng, D.-F., McClure, M. A., and Johnson, M. S., 1990, Retrovirus phylogeny and evolution, in: *Current Topics in Microbiology and Immunology*, Vol. 157 (P. Vogt and R. Swanstrom, eds.), pp. 1–18, Springer-Verlag, Berlin.

Eigen, M., and Biebricher, C. K., 1988, Sequence space and quasispecies distribution, in: *RNA Virus Genetics* (E. Domingo, J. J. Holland, and P. Ahlquist, eds.), Vol. III, pp. 211–245, CRC Press, Boca Raton, Florida.

Emerman, M., Guyader, M., Montagnier, L., Baltimore, D., and Muesing, M. A., 1987,

The specificity of the human immunodeficiency virus type 2 transactivator is different from that of human immunodeficiency virus type 1, *EMBO J.* **6**:3755.

Evans, L. A., Odehouri, K., Thomson-Honnebier, G., Barboza, A., Moreau, J., Seto, D., Legg, H., Cheng-Mayer, C., and Levy, J. A., 1988, Simultaneous isolation of HIV-1 and HIV-2 from an AIDS patient, *Lancet* ii(1988):1389.

Felber, B. K., Cladaras, M., Cladaras, C., Wright, C. M., Tse, A., and Pavlakis, G. N., 1988, Regulation of HIV-1 by viral factors, in: *The Control of Human Retrovirus Gene Expression* (B. R. Franza et al., eds.), pp. 71–77, Cold Spring Harbor Laboratory, Cold Spring Harbor, New York.

Felber, B. K., Derse, D., Athanassopoulos, A., Campbell, M., and Pavlakis, G. N., 1989a, Cross-activation of the *rex* proteins of HTLV-1 and BLV and of the *rev* protein of HIV-1 and nonreciprocal interactions with their RNA responsive element, *New Biol.* **1**:318.

Felber, B. K., Hadzopoulou-Cladaras, M., Cladaras, C., Copeland, T., and Pavlakis, G. N., 1989b, Rev protein of human immunodeficiency virus type 1 affects the stability and transport of the viral mRNA, *Proc. Natl. Acad. Sci. USA* **86**:1495.

Felber, B. K., Drysdale, C. M., and Pavlakis, G. N., 1990, Feedback regulation of human immunodeficiency virus type 1 expression by the *rev* protein, *J. Virol.* **64**:3734.

Felsenstein, J., 1988, Phylogenies from molecular sequences: Inference and reliability, *Annu. Rev. Genet.* **22**:521.

Fitch, W. M., and Palese, P., 1991, Demonstration of positive Darwinian evolution in human influenza A viruses, *Proc. Natl. Acad. Sci. USA* **88**:4270.

Fomsgaard, A., Hirsch, V. M., Allan, J. S., and Johnson, P. R., 1991, A highly divergent proviral DNA clone of SIV from a distinct species of African green monkey, *Virology* **182**:397.

Freed, E. O., Myers, D. J., and Risser, R., 1991, Identification of the principal neutralizing determinant of human immunodeficiency virus type 1 as a fusion domain, *J. Virol.* **65**:190.

Fukasawa, M., Miura, T., Hasegawa, A., Morikawa, S., Tsujimoto, H., Miki, K., Kitamura, T., and Hayami, M., 1988, Sequence of simian immunodeficiency virus from African green monkey, a new member of the HIV–SIV group, *Nature* **333**:457.

Fultz, P. N., McClure, H. M., Anderson, D. C., and Switzer, W. M., 1989, Identification and biologic characterization of an acutely lethal variant of simian immunodeficiency virus from sooty mangabeys (SIV/SMM), *AIDS Res. Hum. Retroviruses* **5**:397.

Garcia, J. A., Wu, F. K., Mitsuyasu, R., and Gaynor, R. B., 1987, Interactions of cellular proteins involved in the transcriptional regulation of the human immunodeficiency virus, *EMBO J.* **6**:3761.

Garvey, K. J., Oberste, M. S., Elser, J. E., Braun, M. J., and Gonda, M. A., 1990, Nucleotide sequence and genome organization of biologically active proviruses of the bovine immunodeficiency-like virus, *Virology* **175**:391.

Gaynor, R., Soultanakis, E., Kuwabara, M., Garcia, J., and Sigman, D. S., 1989, Specific binding of a HeLa cell nuclear protein to RNA sequences in the human immunodeficiency virus transactivating region, *Proc. Natl. Acad. Sci. USA* **86**:4858.

Gendelman, H. E., Phelps, W., Feigenbaum, L., Ostrove, J. M., Adachi, A., Howley, P. M., Khoury, G., Ginsberg, H. S., and Martin, M. A., 1986, *Trans*-activation of the human immunodeficiency virus long terminal repeat sequence by DNA viruses, *Proc. Natl. Acad. Sci. USA* **83**:9759.

Gilks, C., 1991, AIDS, monkeys and malaria, *Nature* **354**:262.

Glass, J. D., 1991, A sequence related to the human gonadoliberin precursor near the N-termini of HIV and SIV *gag* polyproteins, *J. Theor. Biol.* **150**:489.

Gojobori, T., and Yokoyama, S., 1985, Rates of evolution for the retroviral oncogene of

Moloney murine sarcoma virus and of its cellular homologues, *Proc. Natl. Acad. Sci. USA* **82**:4198.

Gojobori, T., Moriyama, E. N., Ina, Y., Ikeo, K., Miura, T., Tsujimoto, H., Hayami, M., and Yokoyama, S., 1990a, Evolutionary origin of human and simian immunodeficiency viruses, *Proc. Natl. Acad. Sci. USA* **87**:4108.

Gojobori, T., Moriyama, E. N., and Kimura, M., 1990b, Molecular clock of viral evolution, and the neutral theory, *Proc. Natl. Acad. Sci. USA* **87**:10015.

Golding, H., Robey, A., Gates III, F. T., Linder, W., Beining, P. R., Hoffman, T., and Golding, B., 1988, Identification of homologous regions in human immunodeficiency virus I gp41 and human MHC class II beta 1 domain, *J. Exp. Med.* **167**:914.

Gonda, M. A., 1990, Visna virus genome: Variability and relationship to other lentiviruses, in: *Applied Virology Research*, Vol. 2 (R. G. Marusyk *et al.*, eds.), pp. 75–98, Plenum Press, New York.

Gonda, M. A., Wong-Staal, F., Gallo, R. C., Clements, J. E., Narayan, O., and Gilden, R. V., 1985, Sequence homology and morphologic similarity of HTLV-III and visna virus, a pathogenic lentivirus, *Science* **227**:173.

Goodenow, M., Huet, T., Saurin, W., Kwok, S., Sninsky, J., and Wain-Hobson, S., 1989, HIV-1 isolates are rapidly evolving quasispecies: Evidence for viral mixtures and preferred nucleotide substitutions, *J. Acquired Immune Defic. Syndr.* **2**:344.

Goudsmit, J., Zwart, G., Bakker, M., Smit, L., Back, N., Epstein, L., Kuiken, C., d'Amaro, J., and de Wolf, F., 1989, Antibody recognition of amino acid divergence within an HIV-1 neutralization epitope, *Res. Virol.* **140**:419.

Gould, S. J., Gilinsky, N. L., and German, R. Z., 1987, Asymmetry of lineages and the direction of evolutionary time, *Science* **236**:1437.

Grantham, R., and Perrin, P., 1986, AIDS virus and HTLV-1 differ in codon choices, *Nature* **319**:727.

Gray, G. S., White, M., Bartman, T., and Mann, D., 1990, Envelope gene sequence of HTLV-1 isolate MT-2 and its comparison with other HTLV-1 isolates, *Virology* **177**:391.

Greene, W. C., Leonard, W. J., Wano, Y., Svetlik, P. B., Peffer, N. J., Sodroski, J. G., Rosen, C. A., Goh, W. C., and Haseltine, W. A., 1986, *Trans*-activator gene of HTLV-II induces IL-2 receptor and IL-2 cellular gene expression, *Science* **232**:877.

Grmek, M. D., 1990, *History of AIDS: Emergence and Origin of a Modern Pandemic* Princeton University Press, Princeton, New Jersey.

Guy, B., Kieny, M. P., Riviere, Y., Le Peuch, C., Dott, K., Girard, M., Montagnier, L., and Lecoco, J. P., 1987, HIV F/3' *orf* encodes a phosphorylated GTP-binding protein resembling an oncogene product, *Nature* **330**:266.

Haase, A. T., 1986, Pathogenesis of lentivirus infections, *Nature* **322**:130.

Hadzopoulou-Cladaras, M., Felber, B. K., Cladaras, C., Athanassopoulos, A., Tse, A., and Pavlakis, G. N., 1989, The *rev* (*trs/art*) protein of human immunodeficiency virus type 1 affects viral mRNA and protein expression via a *cis*-acting sequence in the *env* region, *J. Virol.* **63**:1265.

Hahn, B., 1990, Biologically-unique, SIV-like HIV-2 variants in healthy west African individuals, in: Fifth Cent Gardes Colloqium on Retroviruses of Human A.I.D.S. and Related Animal Diseases (M. Girard and L. Valette, eds.), pp. 31–38, Pasteur Merieux, Lyon, France.

Hahn, B. H., Gonda, M. A., Shaw, G. M., Popovic, M., Hoxie, J. A., Gallo, R. C., and Wong-Staal, F., 1985, Genomic diversity of the acquired immune deficiency syndrome virus HTLV-III: Different viruses exhibit greatest divergence in their envelope genes, *Proc. Natl. Acad. Sci. USA* **82**:4813.

Hahn, B. H., Shaw, G. M., Taylor, M. E., Redfield, R. R., Markham, P. D., Salahuddin, S. Z., Wong-Staal, F., Gallo, R. C., Parks, E. S., and Parks, W. P., 1986, Genetic variation in HTLV-III/LAV over time in patients with AIDS or at risk for AIDS, *Science* **232:**1548.

Hammes, S. R., Dixon, E., Malim, M. H., Cullen, B. R., and Greene, W. C., 1989, *Nef* protein of human immunodeficiency virus type 1: Evidence against its role as a transcriptional inhibitor, *Proc. Natl. Acad. Sci. USA* **86:**9549.

Hamori, E., and Varga, G., 1988, DNA sequence (H) curves of the human immunodeficiency virus 1 and some related viral genomes, *DNA* **7:**371.

Hanly, S. M., Rimsky, L. T., Malim, M. H., Kim, J. H., Hauber, J., Duc Dodon, M., Le, S.-Y., Maizel, J. V., Cullen, B. R., and Greene, W. C., 1989, Comparative analysis of the HTLV-1 *rex* and HIV-1 *rev trans*-regulatory proteins and their RNA response element, *Genes Dev.* **3:**1534.

Hart, A. R., and Cloyd, M. W., 1990, Interference patterns of human immunodeficiency viruses HIV-1 and HIV-2, *Virology* **177:**1.

Hart, C. E., Westhafer, M. A., Galphin, J. C., Ou, C.-Y., Bacheler, L. T., Petteway, S. R., Wasmuth, J. J., Chen, I. S. Y., and Schochetman, G., 1991, Human chromosome-dependent and -independent pathways for HIV-2 *trans*-activation, *AIDS Res. Hum. Retroviruses* **7:**877.

Henderson, L. E., Benveniste, R. E., Sowder, R., Copeland, T. D., Schultz, A. M., and Oroszlan, S., 1988, Molecular characterization of *gag* proteins from simian immunodeficiency virus (SIV-mne), *J. Virol.* **62:**2587.

Hirsch, V. M., Olmsted, R. A., Murphey-Corb, M., Purcell, R. H., and Johnson, P. R., 1989, An African primate lentivirus (SIV) closely related to HIV-2, *Nature* **339:**389.

Hirsch, V. M., Zack, P. M., and Johnson, P. R., 1990, SIV-infected macaques harbor multiple proviral genotypes: Selection of a predominant genotype in tissue culture, in: *Vaccines 1990: Modern Approaches to New Vaccines* (F. Brown *et al.*, eds.), pp. 379–382, Cold Spring Harbor Laboratory, Cold Spring Harbor, New York.

Hu, W. S., and Temin, H. M., 1990, Retroviral recombination and reverse transcription, *Science* **250:**1227.

Huet, T., Dazza, M.-C., Brun-Vezinet, F., Roelants, G. E., and Wain-Hobson, S., 1989, A highly defective HIV-1 strain isolated from a healthy Gabonese individual presenting an atypical Western blot, *AIDS* **3:**707.

Huet, T., Cheynier, R., Meyerhans, A., Roelants, G., and Wain-Hobson, S., 1990, Genetic organization of a chimpanzee lentivirus related to HIV-1, *Nature* **345:**356.

Hwang, S. S., Boyle, T. J., Lyerly, H. K., and Cullen, B. R., 1991, Identification of the envelope V3 loop as the primary determinant of cell tropism in HIV-1, *Science* **253:**71.

Jaffe, H. W., Darrow, W. W., Echenberg, D. F., O'Malley, P. M., Getchell, J. P., Kalyanaraman, V. S., Byers, R. H., Drennan, D. P., Braff, E. H., Curran, J. W., and Francis, D. P., 1985, The acquired immunodeficiency syndrome in a cohort of homosexual men, *Ann. Int. Med.* **103:**210.

Johnson, P. R., Fomsgaard, A., Allan, J., Gravell, M., London, W. T., Olmsted, R. A., and Hirsch, V. M., 1990, Simian immunodeficiency viruses from African green monkeys display unusual genetic diversity, *J. Virol.* **64:**1086.

Johnson, P. R., Hamm, T. E., Goldstein, S., Kitov, S., and Hirsch, V. M., 1991a, The genetic fate of molecularly cloned SIV in experimentally infected macaques, *Virology* **185:**217.

Johnson, P., Myers, G., and Hirsch, V. M., 1991b, Genetic diversity and phylogeny of non-human primate lentiviruses, in: *Annual Review of AIDS Research*, Vol. I (W. Koff *et al.*, eds.), pp. 47–62, Marcel Dekker, New York.

Jones, K. A., Kadonaga, J. T., Luciw, P. A., and Tijan, R., 1986, Activation of the AIDS retrovirus promoter by the cellular transcription factor, Spl, *Science* **232:**755.

Jones, K. A., Luciw, P. A., and Duchange, N., 1988, Structural arrangements of transcription control domains within the 5'-untranslated leader regions of the HIV-1 and HIV-2 promoters, *Genes Dev.* **2**:1101.

Kaminchik, J., Bashan, N., Pinchasi, D., Amit, B., Sarver, N., Johnston, M. I., Fischer, M., Yavin, Z., Gorecki, M., and Panet, A., 1990, Expression and biochemical characterization of human immunodeficiency virus type 1 *nef* gene product, *J. Virol.* **64**:3447.

Kanki, P. J., Alroy, J., and Essex, M., 1985, Isolation of T-lymphotropic retrovirus related to HTLV-III/LAV from wild-caught African green monkeys, *Science* **230**:951.

Karlin, S., Blaisdell, B. E., and Schachtel, G. A., 1990, Contrasts in codon usage of latent versus productive genes of Epstein–Barr virus: Data and hypotheses, *J. Virol.* **64**:4264.

Karpas, A., 1990, Origin and spread of AIDS, *Nature* **348**:578.

Katz, R. A., and Skalka, A. M., 1990, Generation of diversity in retroviruses, *Annu. Rev. Genet.* **24**:409.

Khan, A. S., Galvin, T. A., Lowenstein, L. J., Jennings, M. B., Gardner, M. B., and Buckler, C. E., 1991, A highly divergent SIV recovered from stored stump-tailed macaque tissues (SIV-stm), *J. Virol.* **65**:7061.

Kim, J. H., Kaufman, P. A., Hanly, S. M., Rimsky, L. T., and Greene, W. C., 1991, *Rex* transregulation of human T-cell leukemia virus type II gene expression, *J. Virol.* **65**:405.

Kirchhoff, F., Dieter Jentsch, K., Bachmann, B., Stuke, A., Laloux, C., Luke, W., Stahl-Hennig, C., Schneider, J., Nieselt, K., Eigen, M., and Hunsmann, G., 1990, A novel proviral clone of HIV-2: Biological and phylogenetic relationship to other primate immunodeficiency viruses, *Virology* **177**:305.

Kumar, S. K., Hui, H., Kappes, J. C., Haggarty, B. S., Hoxie, J. A., Arya, S. K., Shaw, G. M., and Hahn, B., 1990, Molecular characterization of an attenuated human immunodeficiency virus type 2 isolate, *J. Virol.* **64**:890.

Kunkel, T. A., and Alexander, P. S., 1986, The base substitution fidelity of eucaryotic DNA polymerases: Mispairing frequencies, site preferences, insertion preferences and base substitution by dislocation, *J. Biol. Chem.* **261**:160.

Kusumi, K., Conway, B., Cunningham, S., Berson, A., Evans, C., Iversen, A. K. N., Colvin, D., Gallo, M. V., Coutre, S., Shpaer, E. G., Faulkner, D. V., DeRonde, A., Volkman, S., Williams, C., Hirsch, M. S., and Mullins, J. I., 1992, Human immunodeficiency virus type 1 envelope gene structure and diversity *in vivo* and after cultivation *in vitro*, *J. Virol.* **66**:875.

Kypr, J., 1987, Unusual codon usage of HIV, *Nature* **327**:20.

Kypr, J., and Mrazek, J., 1989, Reading frames of HIV genes, *J. Theor. Biol.* **141**:423.

Kypr, J., Mrazek, J., and Reich, J., 1989, Nucleotide composition bias and CpG dinucleotide content in the genomes of HIV and HTLV 1/2, *Biochim. Biophys. Acta* **1009**:280.

Larder, B. A., and Kemp, S. D., 1989, Multiple mutations in HIV-1 reverse transcriptase confer high-level resistance to zidovudine (AZT), *Science* **246**:1155.

LaRosa, G. J., Davide, J. P., Weinhold, K., Waterbury, J. A., Profy, A. T., Lewis, J. A., Langlois, A. J., Dreesman, G. R., Boswell, R. N., Shadduck, P., Holley, L. H., Karplus, M., Bolognesi, D. P., Matthews, T. J., Emini, E. A., and Putney, S. D., 1990, Conserved sequence and structural elements in the HIV-1 principal neutralizing determinant, *Science* **249**:932.

Leigh-Brown, A., and Monaghan, P., 1988, Evolution of the structural proteins of human immunodeficiency virus: Selective constraints on nucleotide substitution, *AIDS Res. Hum. Retroviruses* **4**:399.

Leonard, C. K., Spellman, M. W., Riddle, L., Harris, R. J., Thomas, J. N., and Gregory, T. J., 1990, Assignment of intrachain disulfide bonds and characterization of potential glycosylation sites of the type 1 recombinant human immunodeficiency virus envelope glycoprotein (gp120) expressed in Chinese hamster ovary cells, *J. Biol. Chem.* **265**:10373.

Lewis, N., Williams, J., Rekosh, D., and Hammarskjold, M.-L., 1990, Identification of a cis-acting element in human immunodeficiency virus type 2 (HIV-2) that is responsive to the HIV-1 rev and human T-cell leukemia virus types I and II rex proteins, J. Virol. 64:1690.

Li, W.-H., and Graur, D., 1991, Fundamentals of Molecular Evolution, Sinauer Associates, Sunderland, Massachusetts.

Li, W.-H., Tanimura, M., and Sharp, P. M., 1988, Rates and dates of divergence between AIDS virus nucleotide sequences, Mol. Biol. Evol. 5:313.

Li, Y., Naidu, Y. M., Daniel, M. D., and Desrosiers, R. C., 1989, Extensive genetic variability of simian immunodeficiency virus from African green monkeys, J. Virol. 63:1800.

Looney, D. J., Fisher, A. G., Putney, S. D., Rusche, J. R., Redfield, R. R., Burke, D. S., Gallo, R. C., and Wong-Staal, F., 1988, Type-restricted neutralization of molecular clones of human immunodeficiency virus, Science 241:357.

Lowy, D. R., 1986, Transformation and oncogenesis: Retroviruses, in: Fundamental Virology (B. N. Fields and D. M. Knipe, eds.), pp. 235–263, Raven Press, New York.

Maddon, P. J., Dalgleish, A. G., McDougal, J. S., Clapham, P. R., Weiss, R. A., and Axel, R., 1986, The T4 gene encodes the AIDS virus receptor and is expressed in the immune system and the brain, Cell 47:333.

Malik, K. T. A., Even, J., and Karpas, A., 1988, Molecular cloning and complete nucleotide sequence of an adult T cell leukaemia virus/human T cell leukaemia virus type I (ATLV/HTLV-I) isolate of Caribbean origin: Relationship to other members of the ATLV/HTLV-I subgroup, J. Gen. Virol. 69:1695.

Malim, M., Bohnlein, S., Fenrick, R., Le, S. Y., Maizel, J. V., and Cullen, B. R., 1989a, Functional comparison of the rev trans-activators encoded by different primate immunodeficiency virus species, Proc. Natl. Acad. Sci. USA 86:8222.

Malim, M. H., Hauber, J., Le, S., Maizel, J. V., and Cullen, B. R., 1989b, The HIV-1 rev transactivator acts through a structured target sequence to activate nuclear export of unspliced viral mRNA, Nature 338:254.

Martin, M. A., 1990, SIV pathogenicity. Fast-acting slow viruses, Nature 345:572.

Marx, P. A., Li, Y., Lerche, N. W., Sutjipto, S., Gettie, A., Yee, J. A., Brotman, B. H., Prince, A. M., Hanson, A., Webster, R. G., and Desrosiers, R. C., 1991, Isolation of a simian immunodeficiency virus related to human immunodeficiency virus type 2 from a West African pet sooty mangabey, J. Virol. 65:4480.

Mateu, M. G., Martinez, M. A., Rocha, E., Andreu, D., Parejo, J., Giralt, E., Sobrino, F., and Domingo, E., 1989, Implications of a quasispecies genome structure: Effect of frequent, naturally occurring amino acid substitutions on the antigenicity of foot-and-mouth disease virus, Proc. Natl. Acad. Sci. USA 86:5883.

Mazarin, V., Gourdou, I., Querat, G., Sauze, N., and Vigne, R., 1988, Genetic structure and function of an early transcript of visna virus, J. Virol. 62:4813.

McClure, M. A., Johnson, M. S., Feng, D.-F., and Doolittle, R. F., 1988, Sequence comparisons of retroviral proteins: Relative rates of change and general phylogeny, Proc. Natl. Acad. Sci. USA 85:2469.

McCune, J. M., 1991, HIV-1: The infective process in vivo, Cell 64:351.

McGeoch, D. J., 1990, Protein sequence comparisons show that the "pseudoproteases" encoded by poxviruses and certain retroviruses belong to the deoxyuridine triphosphate family, Nucleic Acid Res. 18:4105.

Meloen, R. H., Liskamp, R. M., and Goudsmit, J., 1989, Specificity and function of the individual amino acids of an important determinant of human immunodeficiency virus type 1 that induces neutralizing activity, J. Gen. Virol. 70:1505.

Mergia, A., Shaw, K. E. S., Lackner, J. E., and Luciw, P. A., 1990a, Relationship of the env genes and the endonuclease domain of the pol genes of simian foamy virus type 1 and human foamy virus, J. Virol. 64:406.

Mergia, A., Shaw, K. E. S., Pratt-Lowe, E., Barry, P. A., and Luciw, P. A., 1990b, Simian foamy virus type 1 is a retrovirus which encodes a transcriptional transactivator, *J. Virol.* **64**:3598.

Meyerhans, A., Cheynier, R., Albert, J., Seth, M., Kwok, S., Sninsky, J., Moreldt-Manson, L., Asjo, B., and Wain-Hobson, S., 1989, Temporal fluctuations in HIV quasispecies *in vivo* are not reflected by sequential HIV isolations, *Cell* **58**:901.

Meyerhans, A., Vartanian, J.-P., and Wain-Hobson, S., 1990, DNA recombination during PCR, *Nucleic Acid Res.* **18**:1687.

Miller, R. H., 1988, Human immunodeficiency virus may encode a novel protein in the genomic DNA plus strand, *Science* **239**:1420.

Mizuochi, T., Matthews, T. J., Kato, M., Hamako, J., Titani, K., Soloman, J., and Feizi, T., 1990, Diversity of oligosaccharide structures on the envelope glycoprotein gp120 of human immunodeficiency virus 1 from the lymphoblastoid cell line H9. Presence of complex-type oligosaccharides with bisecting N-acetylglucosamine residues, *J. Biol. Chem.* **265**:8519.

Modrow, S., Hahn, B. H., Shaw, G. M., Gallo, R. C., Wong-Staal, F., and Wolf, H., 1987, Computer-assisted analysis of envelope protein sequences of seven human immunodeficiency virus isolates: Prediction of antigenic epitopes in conserved and variable regions, *J. Virol.* **61**:570.

Mosca, J. D., Bednarik, D. P., Raj, N. B. K., Rosen, C. A., Sodroski, J. G., Haseltine, W. A., and Pitha, P. M., 1987, Herpes simplex virus type-1 can reactivate transcription of latent human immunodeficiency virus, *Nature* **325**:67.

Muranyi, W., and Flugel, R. M., 1991, Analysis of splicing patterns of human spumaretrovirus by polymerase chain reaction reveals complex RNA structures, *J. Virol.* **65**:727.

Myers, G., Rabson, A. B., Josephs, S. F., Smith, T. F., Berzofsky, J. A., and Wong-Staal, F., 1989, in: *Human Retroviruses and AIDS 1989*, p. III-33, Theoretical Biology and Biophysics, Los Alamos National Laboratory, Los Alamos, New Mexico.

Myers, G., Rabson, A. B., Smith, T. F., Berzofsky, J. A., and Wong-Staal, F., 1990, in: *Human Retroviruses and AIDS 1990*, pp. III-5–III-6, Theoretical Biology and Biophysics, Los Alamos National Laboratory, Los Alamos, New Mexico.

Myers, G., Korber, B., Berzofsky, J. A., Smith, R. F., and Pavlakis, G. N., 1991, in: *Human Retroviruses and AIDS 1991*, pp. III-28–III-43, Theoretical Biology and Biophysics, Los Alamos National Laboratory, Los Alamos, New Mexico.

Myers, G., MacInnes, K., and Korber, B., 1992a, The emergence of simian/human immunodeficiency viruses, *AIDS Res. Hum. Retroviruses* **8**:373.

Myers, G., MacInnes, K., and Myers, L., 1992b, Phylogenetic moments in the AIDS epidemic, in: *Emerging Viruses* (S. S. Morse, ed.), Chapter 12, Oxford University Press, Oxford.

Nabel, G., and Baltimore, D., 1987, An inducible transcription factor activates expression of human immunodeficiency virus in T cells, *Nature* **326**:711.

Narayan, O., and Clements, J. E., 1990, Lentiviruses, in: *Virology*, 2nd ed. (B. N. Fields and D. M. Knipe, eds.), pp. 1571–1585, Raven Press, New York.

Nei, M., 1987, *Molecular Evolutionary Genetics*, Columbia University Press, New York.

Novembre, F. J., Hirsch, V. M., McClure, H. M., Fultz, P. N., and Johnson, P. R., 1992, SIV from stump-tailed macaques: Molecular characterization of a highly transmissable primate lentivirus, *Virology* **186**:783.

Nowak, M. A., 1990, HIV mutation rate, *Nature* **347**:522.

Nowak, M. A., May, R. M., and Anderson, R. M., 1990, The evolutionary dynamics of HIV-1 quasispecies and the development of immunodeficiency disease, *AIDS* **4**:1095.

Nowak, M. A., Anderson, R. M., McLean, A. R., Wolfs, T. F. W., Goudsmit, J., and May, R. M., 1991, Antigenic diversity thresholds and the development of AIDS, *Science* **254**:963.

Ohno, S., 1988, Universal rule for coding sequence construction: TA/CG deficiency-TG/CT excess, *Proc. Natl. Acad. Sci. USA* **85**:9630.

Oldstone, M. B. A., 1987, Molecular mimicry and autoimmune disease, *Cell* **50**:819.

Olmsted, R. A., Hirsch, V. M., Purcell, R. H., and Johnson, P. R., 1989, Nucleotide sequence analysis of feline immunodeficiency virus: Genome organization and relationship to other lentiviruses, *Proc. Natl. Acad. Sci. USA* **86**:8088.

Ou, C.-Y., Ciesielski, C., Myers, G., Bandea, C., Luo, C.-C., Korber, B., Mullins, J., Schochetman, G., Berkelman, R., Economou, N., Witte, J., Furman, L., Curran, J., Jaffe, H., *et al.*, 1992, Molecular epidemiology of HIV transmission in a dental practice, *Science* **256**:1165.

Overbaugh, J., Rudensey, L. M., Papenhausen, M. D., Benveniste, R. E., and Morton, W. R., 1991, Variation in simian immunodeficiency virus *env* is confined to V1 and V4 during progression to simian AIDS, *J. Virol.* **65**:7025.

Pathak, V., and Temin, H., 1990a, Broad spectrum of *in vivo* forward mutations, hypermutations, and mutational hotspots in a retroviral shuttle vector after a single replication cycle: Substitutions, frameshifts, and hypermutations, *Proc. Natl. Acad. Sci. USA* **87**:6019.

Pathak, V., and Temin, H., 1990b, Broad spectrum of *in vivo* forward mutations, hypermutations, and mutational hotspots in a retroviral shuttle vector after a single replication cycle: Deletions and deletions with insertions, *Proc. Natl. Acad. Sci. USA* **87**:6024.

Payne, S. L., Salinovich, O., Nauman, S. M., Issel, C. J., and Montelaro, R. C., 1987, Course and extent of variation of equine infectious anemia virus during parallel persistent infections, *J. Virol.* **61**:1266.

Payne, S. L., Ball, J. M., Issel, C. J., and Montelaro, R. C., 1988, Envelope gene variation in equine infectious anemia virus: Implications for vaccine development, in: *Vaccines 88* (H. Ginsberg *et al.*, eds.), pp. 297–302, Cold Spring Harbor Laboratory, Cold Spring Harbor, New York.

Peeters, M., Fransen, K., Delaporte, E., Van den Haesevelde, M., Gersy-Damet, G. M., Kestens, L., van der Froen, G., and Piot, P., 1992, Isolation and characterization of a new chimpanzee lentivirus (simian immunodeficiency virus isolate cpz-ant) from a wild-captured chimpanzee, *AIDS* **6**:447.

Perrino, F. W., Preston, B. D., Sandell, L. L., and Loeb, L. A., 1989, Extension of mismatched 3' termini of DNA is a major determinant of the infidelity of human immunodeficiency virus type 1 reverse transcriptase, *Proc. Natl. Acad. Sci. USA* **86**:8343.

Pert, C. B., Hill, J. M., Ruff, M. R., Berman, R. M., Robey, W. G., Arthur, L. O., Ruscetti, F. W., and Farrar, W. L., 1986, Octapeptides deduced from the neuropeptide receptor-like pattern of antigen T4 in brain potently inhibit human immunodeficiency virus receptor binding and T-cell infectivity, *Proc. Natl. Acad. Sci. USA* **83**:9254.

Preston, B. D., Poiesz, B. J., and Loeb, L. A., 1988, Fidelity of HIV-1 reverse transcriptase, *Science* **242**:1168.

Querat, G., Audoly, G., Sonigo, P., and Vigne, R., 1990, Nucleotide sequence analysis of SA-OMVV, a visna-related ovine lentivirus: Phylogenetic history of lentiviruses, *Virology* **175**:434.

Rando, R. F., Srinivasan, A., Feingold, J., Gonczol, E., and Plotkin, S., 1990, Characterization of multiple molecular interactions between human cytomegalovirus (HCMV) and human immunodeficiency virus type 1 (HIV-1), *Virology* **176**:87.

Rayfield, M., De Cock, K., Heyward, W., Goldstein, L., Krebs, J., Kwok, S., Lee, S., McCormick, J., Moreau, J. M., Odehouri, K., Schochetman, G., Sninsky, J., and Ou, C.-Y., 1988, Mixed human immunodeficiency virus (HIV) infection in an individual: Demonstration of both HIV type 1 and type 2 proviral sequences by using polymerase chain reaction, *J. Infect. Dis.* **158**:1170.

Reddy, E. P., Sandberg-Wollheim, M., Mettus, R. V., Ray, P. E., DeFreitas, E., and Koprowski, H., 1989, Amplification and molecular cloning of HTLV-I sequences from DNA of multiple sclerosis patients, *Science* **243**:529.

Reiher III, W. E., Blalock, J. E., and Brunck, T. K., 1986, Sequence homology between acquired immunodeficiency virus envelope protein and interleukin 2, *Proc. Natl. Acad. Sci. USA* **83**:9188.

Ricchetti, M., and Buc, H., 1990, Reverse transcriptases and genomic variability: The accuracy of DNA replication is enzyme specific and sequence dependent, *EMBO J.* **9**:1583.

Rimsky, L., Hauber, J., Dukovich, M., Malim, M. H., Langlois, A., Cullen, B. R., and Greene, W. C., 1988, Functional replacement of the HIV-1 *rev* protein by the HTLV-1 *rex* protein, *Nature* **335**:738.

Roberts, J. D., Bebenek, K., and Kunkel, T. A., 1988, The accuracy of reverse transcriptase from HIV-1, *Science* **242**:1171.

Roberts, J. D., Preston, B. D., Johnston, L. A., Soni, A., Loeb, L. A., and Kunkel, T. A., 1989, Fidelity of two retroviral reverse transcriptases during DNA-dependent DNA synthesis *in vitro, Mol. Cell. Biol.* **9**:469.

Saag, M. S., Hahn, B. H., Gibbons, J., Li, Y., Parks, E. S., Parks, W. P., and Shaw, G. M., 1988, Extensive variation of human immunodeficiency virus type-1 *in vivo, Nature* **334**:440.

Salfeld, J., Gottlinger, Sia, R., Park, R., Sodroski, J. G., and Haseltine, W. A., 1990, A tripartite HIV-1 *tat–env–rev* fusion protein, *EMBO J.* **9**:965.

Saltarelli, M., Querat, G., Konings, D. A. M., Vigne, R., and Clements, J. E., 1990, Nucleotide sequence and transcriptional analysis of molecular clones of CAEV which generate infectious virus, *Virology* **179**:347.

Sarin, P. S., Sun, D. K., Thornton, A. H., Naylor, P. H., and Goldstein, A. L., 1986, Neutralization of HTLV-III/LAV replication by antiserum to thymosin alpha-1, *Science* **232**:1135.

Schwartz, S., Felber, B. K., Benko, D. M., Fenyo, E., and Pavlakis, G. N., 1990a, Cloning and functional analysis of multiply spliced mRNA species of human immunodeficiency virus type 1, *J. Virol.* **64**:2519.

Schwartz, S., Felber, B. K., Fenyo, E., and Pavlakis, G. N., 1990b, Env and Vpu proteins of human immunodeficiency virus type 1 are produced from multiple bicistronic mRNAs, *J. Virol.* **64**:5448.

Seiki, M., Hattori, S., Hirayama, Y., and Yoshida, M., 1983, Human adult T-cell leukemia virus: Complete nucleotide sequence of the provirus genome integrated in leukemia cell DNA, *Proc. Natl. Acad. Sci. USA* **80**:3618.

Selik, R. M., Haverkos, H. W., and Curran, J. W., 1984, Acquired immune deficiency syndrome (AIDS) trends in the United States, 1978–1982, *Am. J. Med.* **76**:493.

Sharp, P. M., 1986, What can AIDS virus codon usage tell us?, *Nature* **324**:114.

Sharp, P. M., and Li, W.-H., 1988, Understanding the origins of AIDS viruses, *Nature* **336**:315.

Shpaer, E. G., and Mullins, J. I., 1990, Selection against CpG dinucleotides in lentiviral genes: A possible role of methylation in regulation of viral expression, *Nucleic Acids Res.* **18**:5793.

Sidow, A., and Wilson, A. C., 1990, Compositional statistics: An improvement of evolutionary parsimony and its application to deep branches in the tree of life, *J. Mol. Evol.* **31**:51.

Siekevitz, M., Josephs, S. F., Dukovich, M., Peffer, N., Wong-Staal, F., and Greene, W. C., 1987, Activation of the HIV-1 LTR by T cell mitogens and the transactivator protein of HTLV-I, *Science* **238**:1575.

Simmonds, P., Balfe, P., Ludlam, C. A., Bishop, J. O., and Leigh-Brown, A. J., 1990a,

Analysis of sequence diversity in hypervariable regions of the external glycoprotein of human immunodeficiency virus type 1, *J. Virol.* **64**:5840.

Simmonds, P., Balfe, P., Peutherer, J. F., Ludlam, C. A., Bishop, J. O., and Leigh-Brown, A. J., 1990b, Human immunodeficiency virus-infected individuals contain provirus in small numbers of peripheral mononuclear cells and at low copy numbers, *J. Virol.* **64**:864.

Smith, M. R., and Greene, W. C., 1989, The same 50-kDa cellular protein binds to the negative regulatory elements of the interleukin 2 receptor a-chain gene and the human immunodeficiency virus type 1 long terminal repeat, *Proc. Natl. Acad. Sci. USA* **86**:8526.

Smith, R. F., and Smith, T. F., 1990, Automatic generation of primary sequence patterns from sets of related sequences, *Proc. Natl. Acad. Sci. USA* **87**:118.

Smith, T. F., Marcus, M., and Myers, G., 1988a, Phylogenetic analysis of HIV-1 and HIV-2, in: *Vaccines 88* (H. Ginsberg *et al.*, eds.), pp. 317–321, Cold Spring Harbor Laboratory, Cold Spring Harbor, New York.

Smith, T. F., Srinivasan, A., Schochetman, G., Marcus, M., and Myers, G., 1988b, The phylogenetic history of immunodeficiency viruses, *Nature* **333**:573.

Sodroski, J., Patarca, R., Rosen, C., Wong-Staal, F., and Haseltine, W., 1985, Location of the trans-activating region on the genome of human T-cell lymphotropic virus type III, *Science* **229**:74.

Solomin, L., Felber, B. K., and Pavlakis, G. N., 1990, Different sites of interaction for *rev*, *tev*, and *rex* proteins within the *rev* responsive element of human immunodeficiency virus type 1, *J. Virol.* **64**:6010.

Sonigo, P., Alizon, M., Staskus, K., Klatzmann, D., Cole, S., Danos, O., Retzel, E., Tiollais, P., Haase, A., and Wain-Hobson, S., 1985, Nucleotide sequence of the visna lentivirus: Relationship to the AIDS virus, *Cell* **42**:369.

Spire, B., Sire, J., Zachar, V., Rey, F., Barre-Sinoussi, F., Galibert, F., Hampe, A., and Chermann, J.-C., 1989, Nucleotide sequence of HIV1-NDK: A highly cytopathic strain of the human immunodeficiency virus, *Gene* **81**:275.

Srinivasan, A., York, D., Rangapanhan, P., Ferguson, R., Butler, D., Jr., Feorino, P., Kalyanaraman, V., Jaffe, H., Curran, J., and Anand, R., 1987, Transfusion-associated AIDS donor–recipient human immunodeficiency virus exhibits genetic heterogeneity, *Blood* **69**:1766.

Srinivasan, A., York, D., Butler, D., Jannoun-Nasr, R., Getchell, J., McCormick, J., Ou, C.-Y., Myers, G., Smith, T., Chen, E., Flaggs, G., Berman, P., Schochetman, G., and Kalyanaraman, S., 1989, Molecular characterization of HIV-1 isolated from a serum collected in 1976: Nucleotide sequence comparison to recent isolates and generation of hybrid HIV, *AIDS Res. Hum. Retroviruses* **5**:121.

Steinhauer, D. A., and Holland, J. J., 1987, Rapid evolution of RNA viruses, *Annu. Rev. Microbiol.* **41**:409.

Stephens, C. E., Taylor, P. E., Zang, E. A., Morrison, J. M., Harley, E. J., de Cordoba, S. R., Bacino, C., Ting, R. C. Y., Bodner, A. J., Sarngadharan, M. G., Gallo, R. C., and Rubenstein, P., 1986, Human T-cell lymphotropic virus type III infection in a cohort of homosexual men in New York City, *J. Am. Med. Assoc.* **255**:2167.

Stephens, R. M., Derse, D., and Rice, N. R., 1990, Cloning and characterization of cDNAs encoding equine infectious anemia virus *tat* and putative *rev* proteins, *J. Virol.* **64**:3716.

Sved, J., and Bird, A., 1990, The expected equilibrium of the CpG dinucleotide in vertebrate genomes under a mutation model, *Proc. Natl. Acad. Sci. USA* **87**:4692.

Tadmori, W., Mondal, D., Tadmori, I., and Prakash, O., 1991, Transactivation of human immunodeficiency virus type 1 long terminal repeats by cell surface tumor necrosis factor alpha, *J. Virol.* **65**:6425.

Takahashi, H., Merli, S., Putney, S. D., Houghten, R., Moss, B., Germain, R. N., and Berzofsky, J. A., 1989, A single amino acid interchange yields reciprocal CTL specificities for HIV-1 gp160, *Science* **246**:118.

Takeuchi, Y., Nagumo, T., and Hoshino, H., 1988, Low fidelity of cell-free DNA synthesis by reverse transcriptase of human immunodeficiency virus, *J. Virol.* **62**:3900.

Talbott, R. L., Sparger, E. E., Lovelace, K. M., Fitch, W. M., Pedersen, N. C., Luciw, P. A., and Elder, J. H., 1989, Nucleotide sequence and genomic organization of feline immunodeficiency virus, *Proc. Natl. Acad. Sci. USA* **86**:5743.

Temin, H. M., 1985, Reverse transcription in the eukaryotic genome: Retroviruses, pararetroviruses, retrotransposons, and retrotranscripts, *Mol. Biol. Evol.* **2**:455.

Temin, H. M., 1988a, Evolution of cancer genes as a mutation-driven process, *Cancer Res.* **48**:1697.

Temin, H. M., 1988b, Evolution of retroviruses and other retrotranscripts, in: *Human Retroviruses, Cancer, and AIDS: Approaches to Prevention and Therapy*, UCLA Symposia on Molecular and Cellular Biology (D. Bolognesi, ed.), pp. 1–28, Alan R. Liss, New York.

Temin, H. M., 1989, Is HIV unique or merely different?, *J. Acquired Immune Defic. Syndr.* **2**:1.

Temin, H. M., and Engels, W., 1984, Movable genetic elements and evolution, in: *Evolutionary Theory: Paths into the Future* (J. W. Pollard, ed.), pp. 173–201, Wiley, New York.

Tersmette, M., Gruters, R. A., de Wolf, F., de Goede, R. E., Lange, J. M., Schellekens, P. T., Goudsmit, J., Huisman, H. G., and Miedema, F., 1989, Evidence for a role of virulent human immunodeficiency virus (HIV) variants in the pathogenesis of acquired immunodeficiency syndrome: Studies on sequential HIV isolates, *J. Virol.* **63**:2118.

Thayer, R. M., Power, M. D., Bryant, M. L., Gardner, M. B., Barr, P. J., and Luciw, P. A., 1987, Sequence relationships of type D retroviruses which cause simian acquired immunodeficiency syndrome, *Virology* **157**:317.

Tiley, L. S., Brown, P. H., Le, S.-Y., Maizel, J. V., Clements, J. E., and Cullen, B. R., 1990, Visna virus encodes a post-transcriptional regulator of viral structural gene expression, *Proc. Natl. Acad. Sci. USA* **87**:7497.

Toyoshima, H., Itoh, M., Inoue, J.-I., Seiki, M., Takaku, F., and Yoshida, M., 1990, Secondary structure of the human T-cell leukemia virus type 1 *rex*-responsive element is essential for *rex* regulation of RNA processing and transport of unspliced RNA, *J. Virol.* **64**:2825.

Tristem, M., Marshall, C., Karpas, A., Petrik, J., and Hill, F., 1990, Origin of *vpx* in lentiviruses, *Nature* **347**:341.

Tsujimoto, H., Hasegawa, A., Maki, N., Fukusawa, M., Miura, T., Speidel, S., Cooper, R. W., Moriyama, E. N., Gojobori, T., and Hayami, M., 1989, Sequence of a novel simian immunodeficiency virus from a wild-caught African mandrill, *Nature* **341**:539.

Varmus, H., 1988, Regulation of HIV and HTLV gene expression, *Genes Dev.* **2**:1055.

Vartanian, J.-P., Meyerhans, A., Asjo, B., and Wain-Hobson, S., 1991, Selection, recombination and G→A hypermutation of HIV-1 genomes, *J. Virol.* **65**:1779.

Vega, M. A., Guigo, R., and Smith, T. F., 1990, Autoimmune response in AIDS, *Nature* **345**:26.

Viglianti, G. A., Sharma, P. L., and Mullins, J. I., 1990, Simian immunodeficiency virus displays complex patterns of RNA splicing, *J. Virol.* **64**:4207.

Wain-Hobson, S., 1992, HIV-1 quasispecies *in vivo* and *ex vivo*, in: *Current Topics in Microbiology and Immunology* Vol. 176 (J. J. Holland, ed.) pp. 181–193, Springer-Verlag, Berlin.

Watanabe, T., Seiki, M., Tsujimoto, H., Miyoshi, I., Hayami, M., and Yoshida, M., 1985,

Sequence homology of the simian retrovirus genome with human T-cell leukemia virus type 1, *Virology* **144**:59.

Watanabe, T., Seiki, M., Hirayama, Y., and Yoshida, M., 1986, Human T-cell leukemia virus type I is a member of the African subtype of simian viruses (STLV), *Virology* **148**:385.

Weiss, R., Teich, N., Varmus, H., and Coffin, J., 1985, *RNA Tumor Viruses: Molecular Biology of Tumor Viruses*, 2nd ed., Cold Spring Harbor Laboratory, Cold Spring Harbor, New York.

Werner, T., Ferroni, S., Saermark, T., Brack-Werner, R., Banati, R. B., Mager, R., Steinaa, L., Kreutzberg, G. W., and Erfle, V., 1991, HIV-1 *nef* protein exhibits structural and functional similarity to scorpion peptides interacting with K+ channels, *AIDS* **5**:1301.

Westervelt, P., Gendelman, H. E., and Ratner, L., 1991, Identification of a determinant within the HIV-1 surface envelope glycoprotein critical for productive infection of cultured primary monocytes, *Proc. Natl. Acad. Sci. USA* **88**:3097.

Wolfs, T. F. W., De Jong, J.-J., van den Berg, H., Tijnagel, J. M. G. H., Krone, W. J. A., and Goudsmit, J., 1990, Evolution of sequences encoding the principal neutralization epitope of human immunodeficiency virus 1 is host dependent, rapid, and continuous, *Proc. Natl. Acad. Sci. USA* **87**:9938.

Wolinsky, S. M., Wike, C. M., Korber, B., Hutto, C., Parks, W., Rosenblum, L. A., Kunstman, K. J., Furtado, M. R., and Munoz, J., 1992, Selective transmission of human immunodeficiency virus type 1 variants from mothers to infants, *Science* **225**:1134.

Wong-Staal, F., 1990, Human immunodeficiency viruses and their replication, in: *Virology*, 2nd ed. (B. N. Fields and D. M. Knipe, eds.), pp. 1529–1543, Raven Press, New York.

Yamada, M., Zurbriggen, A., Oldstone, M. B. A., and Fujinami, R. S., 1991, Common immunologic determinant between human immunodeficiency virus type 1 gp-41 and astrocytes, *J. Virol.* **65**:1370.

Yokoyama, S., Chung, L., and Gojobori, T., 1988, Molecular evolution of the human immunodeficiency and related viruses, *Mol. Biol. Evol.* **5**:237.

Zapp, M., and Green, M., 1989, Sequence-specific RNA binding by the HIV-1 *rev* protein, *Nature* **342**:714.

CHAPTER 4

Retroelements in Microorganisms

David J. Garfinkel

I. INTRODUCTION

Retroelements are a diverse, widely distributed group of genetic elements that can replicate through an RNA intermediate. Genes for reverse transcriptase, the key enzyme in this process, have been found associated with retroelements in plants, animals, eukaryotic protists, and, most recently, in prokaryotes. Much progress has been made in understanding the life cycles of retroelements, as is illustrated by several recent reviews on this topic (Boeke, 1988, 1989; Boeke and Garfinkel, 1988; Kingsman and Kingsman, 1988; Temin, 1989; Varmus, 1989; Boeke and Corces, 1989; Sandmeyer *et al.*, 1990; Boeke and Sandmeyer, 1992; Brosius, 1991; Boeke and Chapman, 1991). In addition, detailed comparisons of retroelement coding sequences have yielded insights into their evolution (Johnson *et al.*, 1986; Doolittle *et al.*, 1989; Xiong and Eickbush, 1990).

The aim of this chapter is to describe the biology of retroelements found in microorganisms and their relationships to elements found in multicellular organisms. Specifically, this chapter will attempt to summarize findings on the mechanisms of retrotransposition and replication, structures of the elements, functions of encoded gene products,

DAVID J. GARFINKEL ● National Cancer Institute, Frederick Cancer Research and Development Center, ABL-Basic Research Program, Frederick, Maryland 21701-1201.

The Retroviridae, Volume 1, edited by Jay A. Levy. Plenum Press, New York, 1992.

and interactions with "host" genes that are important for retroelement transposition.

II. DEFINITION OF TERMS

I will follow the conventions for retroelement classification and nomenclature proposed at a recent meeting (Hull and Will, 1989) and presented in Table I. The retroelements presently characterized in eukaryotic microorganisms are retrotransposons and retroposons (Table II). The key feature distinguishing retrotransposons and retroposons from retroviruses is that the first are predominantly noninfectious. The lack of an element-encoded *env* gene in retrotransposons and retroposons is consistent with an intracellular replication cycle. However, retrotransposons and retroposons can be transmitted horizontally as a result of cell fusion or rare uptake of virus-like particles (VLPs). These are also the routes of transmission used by killer virus of *Saccharomyces cerevisiae* (El-Sherbeini and Bostian, 1987).

Similarities between retroviruses, retrotransposons, and retroposons include a genome-length transcript, characteristic target-site duplications, internal and external sequences (e.g., tRNAs) that are required for priming reverse transcription, and transcription signals that are often developmentally or environmentally regulated (Table I). Retrotransposons contain *gag* and *pol* genes that are surrounded by long terminal repeats (LTRs). Retroposons sometimes contain a full complement of *gag* and *pol*, but do not carry LTRs. Instead, these elements usually contain a repeated segment at the beginning of the element, which is required for transcription and possibly for replication/integration, and a

TABLE I. Viral and Nonviral Retroelements[a]

Type	Characteristics[b]				
	Infectivity	LTR	poly d(AT)	RT	IN
I. Viral retroelements					
A. Retrovirus	+	+	−	+	+
B. Pararetrovirus	+	−	−	+	−
II. Nonviral retroelements					
A. Retrotransposon	−	+	−	+	+
B. Retroposon	−	−	+	+	+
C. Retron	?	?	?	+	?
D. Retrosequence (processed pseudogenes)	−	−	+	−	−

[a] Proposed at the meeting on Molecular Biology of Retroid Viruses and Elements, Flumersberg, Switzerland, 3–7 April 1989.
[b] LTR, long terminal repeat; RT, reverse transcriptase; IN, integrase.

TABLE II. Retrotransposons and Retroposons in Eukaryotic Microorganisms

Element name	Organism or genus	Reference
	Retrotransposon	
Ty1	*Saccharomyces cerevisiae*	Clare and Farabaugh (1985)
Ty2	*Saccharomyces cerevisiae*	Warmington *et al.* (1985)
Ty3	*Saccharomyces cerevisiae*	Hansen *et al.* (1988)
Ty4	*Saccharomyces cerevisiae*	Stucka *et al.* (1989)
Tf1	*Schizosaccharomyces pombe*	Levin *et al.* (1990)
Tf2	*Schizosaccharomyces pombe*	Levin *et al.* (1990)
DIRS1	*Dictyostelium*	Capello *et al.* (1985)
DRE	*Dictyostelium*	Marschalek *et al.* (1989)
*Hpa*II	*Physarum*	Pearston *et al.* (1985)
	Retroposon	
Tad	*Neurospora*	Kinsey and Helber (1989)
Tdd-2	*Dictyostelium*	Poole and Firtel (1984)
Tdd-3	*Dictyostelium*	Poole and Firtel (1984)
SLACS/MAE	*Trypanosoma*	Carrington *et al.* (1987), Aksoy *et al.* (1990)
CRE	*Crithidia*	Gabriel *et al.* (1990)
RIME/INGI/TRIS	*Trypanosoma*	Hasan *et al.* (1984)

poly(dA)- or poly(dA-T)-rich region at the 3′ end. In some cases, these elements resemble retrosequences (also called processed pseudogenes) more than they resemble retroviruses [reviewed by Temin (1985) and Weiner *et al.* (1986)]. (See Chapter 1.)

Two additional types of elements that qualify as retroelements have been assigned to the retroposon class (Table I). The first are plasmids found in *Neurospora crassa* mitochondria, and the other are class I introns found in fungal mitochondria. The recently discovered prokaryotic retrons have been categorized separately because of their unique structure and unknown transposition mechanism. Furthermore, telomere synthesis involves a specialized reverse transcriptase (Shippen-Lentz and Blackburn, 1990). This is the first example of an essential cellular process that relies on reverse transcription (Lundblad and Szostak, 1989; Boeke, 1990; Lundblad and Blackburn, 1990). Given the rapid increase in the discovery of new reverse transcription systems, this classification scheme will certainly be expanded in the future.

III. TY ELEMENTS

Ty elements have become a paradigm for studying retrotransposition because they are found in *S. cerevisiae*, an organism with powerful classical and molecular genetic systems. I will focus on the process of Ty retrotransposition and its regulation. Properties of Ty elements as genetic tools and insertional mutagens and their involvement in genome rearrangements will not be extensively covered here. Treatment of these aspects of Ty biology have recently been presented elsewhere (Garfinkel *et al.*, 1988a; Boeke, 1989; Boeke and Sandmeyer, 1991).

A. Ty Element Diversity

The four classes of Ty elements can be placed into two related groups (Fig. 1, Table III). Ty1 and Ty2 elements are closely related and belong to a larger phylogenetic group that includes copia from *Drosophila*, Ta1 from *Nicotiana*, and Tnt1 from *Arabidopsis* (Xiong and Eickbush, 1990). Ty3 elements are more closely related to the *Drosophila* gypsy element superfamily than to Ty1 or Ty2 (Hansen *et al.*, 1988; Xiong and Eickbush, 1990). These observations suggest that Ty3 elements may have entered the yeast genome independently of Ty1 or Ty2.

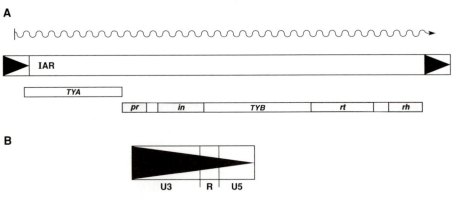

FIGURE 1. Structure of Ty elements. (A) On the top is the genomic Ty transcript. The boxed black triangles represent Ty long terminal repeats (LTRs). Ty1- and Ty2-element LTRs are also called delta (δ) sequences; Ty3-element LTRs are called sigma (σ); Ty4-element LTRs are called tau (τ). The segment between the LTRs contains most of the Ty coding sequence and has been called epsilon (ε). The "enhancer" region of promoter is designated IAR (internal activating region) for Ty1 and Ty2 elements; the promoter regions of Ty3 and Ty4 elements have not been defined. The *TYA* and *TYB* open reading frames are indicated below. The domains of amino acid sequence similarity to retroviral *pol* genes are indicated by the shaded regions and labeled as follows: pr, protease; in, integrase; rt, reverse transcriptase; rh, ribonuclease H. The order shown is for Ty1 and Ty2 elements; the order in Ty3 elements and retroviruses is pr–rt–rh–in; the analysis for Ty4 elements is incomplete. (B) The U_3 (unique 3' region), R (repeat region) and U5 (unique 5' region) of a Ty LTR is presented. The boxed black triangle indicates the direction of transcription. The transcription initiation site defines the U3–R boundary.

TABLE III. Sequence Features of Ty Elements[a]

Ty prototype	Length ε (kb)	LTR (nt)	Terminal inverted repeat	TYA	TYB	Reference
Ty1-912	5.2	334	TG/CA	294–1616 (440 codons)	1576–5562 (1328 codons)	Clare and Farabaugh (1985)
Ty2-117	5.2	332	TG/CA	292–1608 (438 codons)	1562–5608 (1349 codons)	Warmington et al. (1985)
Ty3-1	4.7	340	TGTTGTAT/ ATACAACA	416–1288 (290 codons)	1248–5060 (1270 codons)	Hansen et al. (1988)
Ty4	5.6	371	TGTTG/ CAACA	ND	ND	Genbauffe et al. (1984) Stucka et al. (1989)

[a] ε, central region of Ty that is bracketed by LTRs (refer to Fig. 1); LTR, long terminal repeat; kb, kilobase; nt, nucleotide; ND, not determined

The recently discovered Ty4 elements share properties with both Ty1/ Ty2 and Ty3 element classes (Stucka et al., 1989).

Ty1 (Cameron et al., 1979; Clare and Farabaugh, 1985) and Ty2 (Kingsman et al., 1981; Warmington et al., 1985) elements consist of a 5.3-kilobase (kb) fragment (originally denoted as epsilon, ε) that is bracketed by 350-nucleotide (nt) LTRs (originally known as delta elements, δ) (Fig. 1, Table III). Ty3 elements are slightly smaller (5.4 kb), and their LTRs were initially called sigma (σ) elements (Delrey et al., 1982; Clark et al., 1988; Hansen et al., 1988). Ty4 elements are 6.3 kb long and are bracketed by tau (τ) LTRs (Stucka et al., 1989). Solo LTRs from each type of Ty element can also be found dispersed in the genome (Delrey et al., 1982; Chisholm et al., 1984; Genbauffe et al., 1984).

Ty elements contain two overlapping genes: TYA, which corresponds to gag and specifies a protein that has homology to DNA-binding proteins (Clare and Farabaugh, 1985; Warmington et al., 1985; Hansen et al., 1988), and TYB, which is equivalent to pol and specifies a protein with limited homology to retroviral protease (specified by the domain pr), integrase (in), reverse transcriptase (rt), and ribonuclease H (rh) (Clare and Farabaugh, 1985; Toh et al., 1985; Warmington et al., 1985; Johnson et al., 1986; Stucka et al., 1986; Hansen et al., 1988) (Fig. 1). The order of the pol homology domains within TYB is the same in Ty1 and Ty2 elements, but differs in Ty3: the order in Ty1 and Ty2 elements is pr–in–rt–rh, while the order in Ty3 elements and retroviruses is pr–rt–rh–in.

Ty1 was the first transposable element isolated from yeast (Cameron et al., 1979). It was found as a middle-repeat sequence, one copy of which was associated with a tRNA^{Tyr} gene in some strains. Ty1 is the most abundant class of elements, with 20–25 copies found dispersed in the genome of most haploid *S. cerevisiae* strains (Cameron *et al.*, 1979; Curcio *et al.*, 1990). There are about ten copies of Ty2 (Kingsman *et al.*, 1981; Curcio *et al.*, 1990) and fewer than four copies of Ty3 (Clark *et al.*, 1988; Curcio *et al.*, 1990) and Ty4 (Stucka *et al.*, 1989) in most strains. Strains have been isolated that lack complete Ty1, Ty3, and Ty4 elements, suggesting that these Ty elements are not essential for cell viability (Clark *et al.*, 1988; Curcio *et al.*, 1990; P. Phillipsen, personal communication). Direct evidence for Ty1 and Ty2 transposition was initially provided by the isolation of two Ty-induced mutations in the *HIS4* promoter region (Roeder *et al.*, 1980; Roeder and Fink, 1980). Subsequently, Ty1- and Ty2-induced mutations have been found in many other genes. Ty3 elements have never been isolated as insertional mutagens, perhaps because they have a more restricted target site specificity (Chalker and Sandmeyer, 1990).

Complete Ty1, Ty2, and Ty3 elements have been sequenced (Clare and Farabaugh, 1985; Hauber *et al.*, 1985; Warmington *et al.*, 1985; Boeke *et al.*, 1988a; Hansen *et al.*, 1988; Hansen and Sandmeyer, 1990; P. Farabaugh, personal communication) and many others have been characterized by restriction endonuclease mapping (Boeke *et al.*, 1985, 1986; Natsoulis *et al.*, 1989; Wilke *et al.*, 1989; Curcio *et al.*, 1990). There are numerous single-nucleotide changes among Ty elements, but overall the elements within a group are relatively homogeneous. In contrast to retrotransposons of multicellular organisms, most Ty elements are not grossly rearranged and do not usually contain chain-terminating mutations. Endogenous transposition-defective Ty1 and Ty3 elements that contain mutations in *TYB* have been isolated (Boeke *et al.*, 1988a; Hansen *et al.*, 1988; Hansen and Sandmeyer, 1990). The maintenance of open reading frames and a high level of protein homology suggest that Ty elements may be under positive selection, but the nature of this is unknown (Fulton *et al.*, 1985).

B. Ty Element Transposition

Ty elements can replicate passively along with chromosomal DNA or they can replicate through an RNA intermediate during the transposition process (Fig. 2). Retrotransposition results in insertions of cDNA copies of the element and involves several steps: transcription, translational frameshifting, VLP maturation, reverse transcription, and integration. Ty elements have been studied using two basic approaches. First, Ty elements under the control of their own promoters and integrated at specific genes or chromosomal sites have been analyzed. This

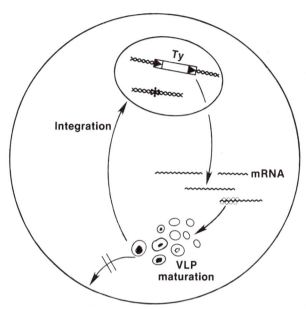

FIGURE 2. Ty element life cycle. Ty elements reside in the nuclear genome where they are transcribed. Ty RNA is terminally redundant because directly repeated LTR (long terminal repeat) sequences present at the ends of the element are transcribed (boxed arrows point in the direction of Ty transcription). Ty RNA directs the synthesis of proteins that are essential for transposition. These proteins assemble around the RNA to form a cytoplasmic Ty-VLP (virus-like particle). Reverse transcription of Ty RNA takes place within the particle via a Ty-encoded particle associated reverse transcriptase. In contrast to infectious retroviruses, Ty-VLPs do not leave the cell. It is likely that an integration intermediate made up of a subparticle protein/nucleic acid complex journeys back to the nucleus where integration takes place.

approach has worked well for studying Ty transcription, but has not yielded much information concerning the rest of the transposition process because Ty transposition is a rare event. The second approach makes use of a set of expression plasmids that contain Ty elements fused to a strong yeast promoter (Boeke *et al.*, 1985; Mellor *et al.*, 1985b; Muller *et al.*, 1987; Hansen *et al.*, 1988). A tremendous induction in the rate of Ty transposition has been achieved by expressing Ty elements from the inducible *GAL1* promoter carried on high-copy plasmids (pGTy plasmids) (Boeke *et al.*, 1985; Hansen *et al.*, 1988). The term "transposition induction" describes this process (Garfinkel *et al.*, 1985). In addition, various marker genes have been introduced into Ty1, Ty2, and Ty3 elements, allowing transposition events to be detected phenotypically (Boeke *et al.*, 1985, Boeke 1988b; Curcio *et al.*, 1988; Eichinger and Boeke, 1988; Garfinkel *et al.*, 1988b; Chalker and Sandmeyer, 1990; Curcio and Garfinkel, 1991). Much of our present knowledge concerning the mechanism of Ty transposition has come from studying transposition-induced cells, because these cells provide the only source of large amounts of Ty proteins.

1. Ty Element RNA

The major Ty1, Ty2, and Ty3 transcripts fulfill the requirements of a retrotransposon replication template (Figs. 1 and 2; Table III). The 5.7-kb Ty1 and Ty2 and 5.4-kb Ty3 transcripts cover almost the complete genome of the element and are bracketed by R–U5 and U3–R motifs required for reverse transcription (Elder *et al.*, 1980; Clark *et al.*, 1988; Hansen *et al.*, 1988). The transcription initiation site defines the beginning of the R (the directly repeated terminus of the RNA). This occurs within the consensus sequence TPyGA for Ty1 and Ty2 elements (Elder *et al.*, 1983) and is preceded by an AT-rich sequence that resembles TATA boxes of many other yeast genes. Transcriptional termination defines the end of the R region. In Ty1 elements, the major 3' end of the RNA creates an R region of about 45 nt. The sequences implicated in 3'-end formation are also found at the 5' end of the transcript, yet are apparently not utilized for termination (Yu and Elder, 1989a). The U_5 (unique 5') region of Ty1 RNA is about 60 nt, and U_3 (unique 3') is about 240 nt. Ty2 RNA is probably similar to Ty1 RNA in structure, because Ty1 and Ty2 LTRs are highly homologous. Ty3 elements have an LTR U_3–R–U_5 structure of 222, 18, and 99 nt, respectively (V. Bilanchone and S. Sandmeyer, personal communication).

Minor Ty1 transcripts of 2.2 and 5 kb have also been observed in wild-type cells (Elder *et al.*, 1980, 1983; Fulton *et al.*, 1988), but their role in transposition is unknown. The 2.2-kb RNA initiates within the upstream LTR and terminates just after entering *TYB* (Fulton *et al.*, 1988). There are conflicting data concerning the initiation site of the 5-kb transcript. R-loop heteroduplex analysis suggested that the 5-kb transcript initiates in the 5' LTR and terminates in ϵ (Elder *et al.*, 1980). Fulton *et al.*, (1988) used nuclease protection to place the initiation site of the 5-kb transcript within ϵ. Winston *et al.*, (1984b) mapped the start of a 5-kb Ty transcript that predominates in *spt3* mutants to a position 800 nt into ϵ.

Ty1 and Ty2 RNAs are among the most abundant RNA polymerase II transcripts in the cell (Elder *et al.*, 1980; Curcio *et al.*, 1990). Ty1 RNA is about tenfold more abundant than Ty2 RNA, even after correction for differences in the element's copy number (Curcio *et al.*, 1990). Recent estimates indicate that Ty1 plus Ty2 RNA makes up about 0.8% of total yeast RNA. Interestingly, M. Nonet and R. Young (personal communication) have shown that Ty RNA is exceptionally stable. When a temperature-sensitive RNA polymerase II mutant defective in initiation is shifted to a nonpermissive temperature, Ty1 mRNA has a half-life of at least 3 hr, which is comparable to that of rRNA. These results further explain how such high levels of Ty1 RNA accumulate in yeast cells. It is tempting to imagine that Ty RNA is protected from degradation by compartmentalization within an immature VLP. However, Ty1 RNA cofractionates with transcripts of several other genes, suggesting that Ty RNA is not in a physically distinct cellular compartment (M. J. Curcio and D. J. Garfinkel, unpublished results).

2. Regulation of Ty Transcription

Basal level transcription of Ty1, Ty2, and Ty3 elements is highest in haploid cells (Elder *et al.*, 1980; Taguchi *et al.*, 1984; S. Sandmeyer, personal communication) and is stimulated by environmental modulators (Rolfe *et al.*, 1986; Van Arsdell *et al.*, 1987; Bradshaw and McEntee, 1989). Unlike retroviruses and most yeast genes, Ty1 and Ty2 elements contain internal activating regions (IARs) downstream of the initiation site that stimulate transcription of the element (Liao *et al.*, 1987; Fulton *et al.*, 1988; Farabaugh *et al.*, 1989; Yu and Elder, 1989b) (Fig. 1). Ty1 LTRs do not contain classical upstream activating sequences (UAS) (Fulton *et al.*, 1988; Yu and Elder, 1989b). Ty2-917 apparently requires an additional UAS upstream of the TATA region for efficient transcription (Liao *et al.*, 1987). Ty2-917 also contains a downstream region, approximately 750 nt into the element, that decreases RNA accumulation (Farabaugh *et al.*, 1989). Since this sequence lowers expression when placed upstream of a heterologous gene, it may function as a "silencer" or "negative enhancer" of gene expression (Brand *et al.*, 1985; Laimins *et al.*, 1986). The Ty2 silencer may also account for the lower overall level of Ty2 RNA (Curcio *et al.*, 1990).

Conflicting results concerning the definition of an IAR have been reported. Some of these problems undoubtedly result from using different Ty elements for transcriptional studies. It is also likely that adjacent cellular sequences affect Ty transcription (Silverman and Fink, 1984). Finally, it is possible that the transcription assays used by different laboratories influence the results. IAR sequences have been defined in the same general region for several different Ty1 and Ty2 elements. In Ty1-D15, an IAR has been identified that can activate Ty transcription when assayed within a Ty1-D15/*his3* fusion element (Yu and Elder, 1989b). The Ty1-D15 IAR is present within a 400-nt *Pvu*II–*Hpa*I restriction fragment (nt 475 and 815, respectively) that is located about 140 nt downstream of the transcription start site (Fig. 1). Immediately upstream of the *Pvu*II site in Ty1-CYC7H2 is a binding site for *STE12*, a factor required for Ty1 transcription (Company *et al.*, 1988; Errede and Ammerer, 1989). Two regions of homology to the SV40 core enhancer have also been reported for Ty1-CYC7H2 (Errede *et al.*, 1985, 1987). Both homology blocks are immediately downstream of the Ty1-D15 IAR (Yu and Elder, 1989b). An IAR called TAS2 has been defined in a comparable segment of another Ty1 element (Fulton *et al.*, 1988). A distinct IAR is present in the same region of a Ty2 element (Liao *et al.*, 1987; Farabaugh *et al.*, 1989). Transcriptional activation by the Ty1-D15 IAR is independent of orientation and the presence of the normal initiation site. Conversely, Ty1-15 (Fulton *et al.*, 1988) and Ty2-917 (Liao *et al.*, 1987) IARs require specific LTR sequences that probably contain a TATA element for efficient transcription.

Definition of Ty promoter regulatory regions and transcription factors has been a very active area of research because of the properties of Ty-induced promoter mutations. These insertion mutations result in either activation and concomitant deregulation (Lemoine *et al.*, 1978; Errede *et al.*, 1980a,b; Ciriacy and Williamson, 1981; Jauniaux *et al.*, 1981, 1982; Williamson *et al.*, 1981, 1983; Dubois *et al.*, 1982; Roeder and Fink, 1982; Scherer *et al.*, 1982; Cooper and Chisholm, 1984; Tschumper and Carbon, 1986; Iida, 1988; Weinstock *et al.*, 1990) or inactivation (Chaleff and Fink, 1980; Roeder and Fink, 1980) of an adjacent chromosomal gene. Ty-induced promoter mutations are generally unstable, allowing the selection of intragenic or extragenic suppressors. Intragenic suppressors have helped to define transcriptional control sequences within the Ty element (Roeder and Fink, 1982; Silverman and Fink, 1984; Roeder *et al.*, 1985b; Coney and Roeder, 1988; Hirschman *et al.*, 1988), while extragenic suppressors have identified important transcription factors (Table IV). Ty insertions within a gene's coding sequence also occur (Eibel and Philippsen, 1984; Rose and Winston, 1984; Simchen *et al.*, 1984; Garfinkel *et al.*, 1988b; Natsoulis *et al.*, 1989; Wilke *et al.*, 1989). In contrast to promoter insertions, these Ty-induced mutations almost never revert.

Ty elements require several types of genes for transcription in haploid cells, many of which were isolated as extragenic suppressors of Ty-induced mutations (Table IV): *STE* (Errede *et al.*, 1980a,b; Taguchi *et al.*, 1984; Company *et al.*, 1988; Errede and Ammerer, 1989), required for haploid gene expression, *SPT* (Winston *et al.*, 1984a, 1987; Clark-Adams *et al.*, 1988; Fassler and Winston, 1988; Eisenmann *et al.*, 1989; G. Natsoulis and J. Boeke, personal communication), *ROC* or *TEC* (Dubois *et al.*, 1982; Laloux *et al.*, 1990), *TYE* (Ciriacy and Williamson, 1981), and *SNF* (Abrams *et al.*, 1986; Neigeborn *et al.*, 1987). Several of these genes are involved in other regulatory networks or have global roles in yeast transcription. *SPT3* is required for normal Ty transcription but does not affect the length or abundance of pGTy transcripts (Winston *et al.*, 1984b; Boeke *et al.*, 1986). *spt3* mutants produce a shorter, less abundant Ty1 transcript that is incapable of acting as a transposition intermediate. *SPT3* also plays a role in mating and sporulation (Winston *et al.*, 1984a; Hirschman and Winston, 1988). Recent genetic analyses suggest that *SPT3* and *SPT15* interact (F. Winston, personal communication). *SPT6* is allelic to *CRE2* and *SSN20*, a regulator of sucrose metabolism (Winston *et al.*, 1984a; Clark-Adams and Winston, 1987; Neigeborn *et al.*, 1987; Denis and Malvar, 1990). In addition, *snf2* (*tye3*), *snf5* (*tye4*), and *snf6* mutants have lower levels of Ty1 transcripts (Neigeborn and Carlson, 1984; Abrams *et al.*, 1986; Happel *et al.*, 1991; M. Ciriacy, personal communication). *SPT13* is allelic to *GAL11*, a gene involved in galactose utilization (Fassler and Winston, 1988). *SPT11* and *SPT12* are allelic to histones H2A (*HTA1*) and H2B (*HTB1*) (Clark-Adams *et al.*, 1988), and *SPT15* is the yeast TATA-binding factor TFIID (Eisenmann *et al.*, 1989). *TEC1* appears to be required specifically for Ty1 element transcription (Laloux *et al.*, 1990). Unidentified DNA-binding activities have also been reported that are specific for Ty2 sequences involved in adjacent gene activation (Goel and Pearlman, 1988).

Ty elements are haploid-specific genes and, therefore, are controlled by the mating-type locus (*MAT*). Ty RNA levels are reduced by about 20-fold in MATa/α diploids (Elder *et al.*, 1980; S. Sandmeyer, personal communication), although this is somewhat dependent on the carbon source (Taguchi *et al.*, 1984). *MAT* regulation probably affects Ty transcription indirectly and directly. Ty elements require the haploid-specific transcription factor *STE12* for maximum transcription levels (Company *et al.*, 1988; Errede and Ammerer, 1989). Since *STE12* is repressed in a/α diploid cells [reviewed by Herskowitz, (1989)], *MAT* control of Ty transcription via *STE12* is indirect. Ty may also be under direct control of the a/α repressor complex. Sequences with partial homology to the a/α consensus binding site have been identified in the Ty1 LTR and in ϵ (Errede *et al.*, 1985; Russell *et al.*, 1986; Errede *et al.*, 1987). A 112-nt segment from Ty1-CYC7H2 containing a potential a/α binding site near the SV40 core enhancer confers *MAT*-regulated expression when placed upstream of *CYC7* (Errede *et al.*, 1987; Company *et al.*, 1988). When placed upstream of a reporter gene, this IAR is orientation- and *STE12*-independent, and can also bring the *STE12*-responsive activator under mating-type control (Company and Errede, 1987). However, activation of a promoterless *PGK1* gene by an SV40 core enhancer-like sequence from Ty1-15 (TAS2) was not regulated by *MAT* (Fulton *et al.*, 1988). *MAT* regulation of Ty1-15 (Rathjen *et al.*, 1987; Fulton *et al.*, 1988) is conferred by a region (TAS1) that best corresponds to the *STE12*-dependent activator of Ty1-CYC7H2 (Company *et al.*, 1988; Errede and Ammerer, 1989).

MAT regulation of Ty expression also influences the phenotype of an interesting class of Ty1- and Ty2-induced promoter-up mutations called *ROAM* mutations (Errede *et al.*, 1980a,b). Several general proper-

TABLE IV. Extragenic Suppressors of Ty-Induced Mutations

Gene	Comments	Reference
SPT1	suppresses HIS4-912 (Ty1)	Winston et al. (1984a)
SPT2	some alleles dominant; suppresses mutations induced by solo-LTR and complete Ty2 elements	Winston et al. (1984a,b), Simchen et al. (1984), Roeder et al. (1985a)
SPT3	Weak sterile; required for Ty1, Ty2, MFa1, MFa2, MFα1 expression	Winston et al. (1984a,b), Simchen et al. (1984), Winston and Minehart (1986), Hirschman and Winston (1988)
SPT4	Some alleles methylmethane sulfonate-sensitive; synthetic lethal with spt5, spt6	Winston et al. (1984a)
SPT5	Synthetic lethal with spt4, spt6; acidic nuclear protein	Winston et al. (1984a), Swanson et al. (1991)
SPT6	Allelic to CRE2 and SSN20	Winston et al. (1984a), Clark-Adams and Winston (1987), Neigeborn et al. (1987), Denis and Malvar (1990)
SPT7	Required for Ty transcription and sporulation	Winston et al. (1984a, 1987)
SPT8	Required for Ty transcription and sporulation	Winston et al. (1984a, 1987)
SPT9	One allele is temperature-sensitive; essential gene	Fassler and Winston (1988)
SPT10	Activates transcription from 3' LTR; synthetic lethal with spt11, spt12; allelic to CRE1	Fassler and Winston, (1988), G. Natsoulis and J. Boeke (personal communication), Denis and Malvar (1990)
SPT11	Synthetic lethal with spt10; allelic to HTA1 (histone H2A)	Fassler and Winston (1988), Clark-Adams et al. (1988)

Gene	Description	Reference
SPT12	Synthetic lethal with spt10; allelic to HTB1 (histone H2B)	Clark-Adams et al. (1988), Fassler and Winston (1988)
SPT13	Required for sporulation; allelic to GAL11	Fassler and Winston (1988)
SPT14	Essential gene	Fassler and Winston (1988)
SPT15	Essential gene; encodes TFIID	Winston et al. (1987), Eisenmann et al. (1989)
SPT16	Essential gene; allelic with CDC68	Malone et al. (1991), Rowley et al. (1991)
SPT21	Activates transcription from 3' LTR; synthetic lethal with spt11, spt12	G. Natsoulis and J. Boeke (personal communication)
SPT23	Gene-dosage suppressor of Ty-induced mutations and ROAM repression	T. Burkett and D. J. Garfinkel (unpublished results)
TEC1 (ROC1)	Required for expression of Ty1-induced mutations at CAR1 and DUR2, 1; required for Ty1 transcription	Dubois et al. (1982), Laloux et al. (1990)
TEC2 (ROC2)	Required for expression of Ty-induced mutations at CAR2	Dubois et al. (1982)
TYE1, 2	Required for expression of Ty-induced mutations at ADH2	Ciriacy and Williamson (1981)
TYE3	Required for expression of Ty-induced mutations at ADH2 and Ty transcription; allelic to SNF2	A. Happel (personal communication), Ciriacy and Williamson, (1981), M. Ciriacy, (personal communication), Neigeborn and Carlson (1984), Abrams et al. (1986)
TYE4	Required for expression of Ty-induced mutations at ADH2 and Ty transcription; allelic to SNF5	A. Happel (personal communication), Ciriacy and Williamson, (1981), M. Ciriacy (personal communication), Neigeborn and Carlson (1984), Abrams et al. (1986)

ties of *ROAM* mutations are as follows: (1) the transcriptional orientation of the Ty is usually opposite to that of the target gene; (2) activation of the target gene is significantly reduced in diploid cells and is dependent on the haploid-specific *STE7*, *STE11*, and *STE12* genes, as well as the *MAT* locus; and (3) the strength of adjacent gene activation depends on the Ty element's transcriptional activity and distance from the target gene (Boeke *et al.*, 1986; Coney and Roeder, 1988; Hirschman *et al.*, 1988). Therefore, it is likely that regulation of adjacent gene and Ty element transcription are controlled by similar Ty IARs.

The strength of the *ROAM* phenotype may involve competition between transcription factors that bind to the promoters of Ty and adjacent genes. If the Ty promoter is preferentially utilized, adjacent gene expression is lowered. If the adjacent gene promoter is used more efficiently, the *ROAM* overexpression phenotype is stronger. Results from several different approaches support this model. Ty insertion sites in the 5' region of the promotorless *his3Δ4* gene are much more dispersed if His$^+$ revertants are obtained in an *spt3* mutant background (Boeke *et al.*, 1986). Since *spt3* mutants produce very little Ty transcript, the *his3Δ4* promoter may now compete more effectively for transcription factors. This would allow the Ty activator to work over a greater distance. More direct evidence for promoter competition is provided by studying different Ty2 elements inserted at the same site in the *HIS4* promoter (Roeder and Fink, 1982; Roeder *et al.*, 1985b; Coney and Roeder, 1988). These Ty insertions confer His$^-$, weak His$^+$, and strong His$^+$ phenotypes that accurately reflect *HIS4* expression (Coney and Roeder, 1988). Maximum *HIS4* expression is correlated with lower levels of expression of the adjacent Ty2 element. This can be brought about by mutations in the TATA regions of Ty or adjacent genes, or in the Ty IAR. Mutations in the TATA region of the Ty1 (Hirschman *et al.*, 1988) or Ty2 (Coney and Roeder, 1988) LTR that lower Ty expression increase the level of *HIS4* expression. Mutations in the *HIS4* TATA region that increase *HIS4* expression also decrease transcription from the LTR promoter (Hirschman *et al.*, 1988). Sequences within the Ty2 IAR are required for weak *HIS4* expression but are not essential for Ty2 transcription (Coney and Roeder, 1988). However, reduction of Ty transcription is not always sufficient for activation of adjacent gene expression. *snf2*, *snf5*, and *snf6* mutants are defective for Ty and solo LTR transcription, but are not suppressors of Ty-induced mutations (Happel *et al.*, 1991).

Finally, Eisenmann *et al.*, (1989) have recently shown that mutant forms of yeast TFIID (Table IV; *SPT15*), a TATA-binding protein required for initiation of *polII* transcription, suppress two different solo-LTR-induced mutations at *HIS4*. These insertion mutations have the capacity to produce two different transcripts. One transcript initiates within the LTR and is nonfunctional (Chaleff and Fink, 1980; Silverman and Fink, 1984), and the other transcript initiates at the normal position in *HIS4*. In Spt15$^+$ cells, transcription initiates within the LTR. In Spt15$^-$ cells, transcription preferentially starts at the normal *HIS4* initiation site and not within the LTR. The simplest explanation for this shift in transcription initiation is that wild-type TFIID preferentially recognizes the LTR TATA, whereas the mutant TFIID recognizes the normal *HIS4* TATA.

An additional form of transcriptional regulation is used by Ty3 elements and their solo-LTR (σ) derivatives. Ty3 and σ transcription is increased dramatically in response to mating pheromones (Van Arsdell *et al.*, 1987; S. Sandmeyer, personal communication). *STE2* (α-pheromone receptor) or *STE5* (*a*-pheromone receptor) is required for pheromone induction. There are several repeats of a sequence shared by several pheromone-inducible genes that are present in σ. It is suspected that pheromone treatment will also stimulate Ty3 transposition. Ty3 element and adjacent tRNA transcription also appear to be interrelated, even though different RNA polymerases apparently act on these promoters (P. Kinsey and S. Sandmeyer, personal communication).

Like other transposable elements and certain cellular genes [reviewed by McClintock (1984)], Ty element transcript levels increase in response to agents that damage the genome (McClanahan and McEntee, 1984; Ruby and Szostak, 1985; Bradshaw and McEntee, 1989). Ty element RNA levels increase tenfold after a 1-hr treatment with UV light (Rolfe *et al.*, 1986; Bradshaw and McEntee, 1989), or a 4- to 6-hr treatment with the mutagen 4-nitroquinoline-1-oxide (4NQO) (McClanahan and McEntee, 1984; Bradshaw and McEntee, 1989). Ty1 and Ty2 transposition into specific target genes (Morawetz, 1987; Bradshaw and McEntee, 1989) or spontaneous transposition into the genome (D. J. Garfinkel, unpublished results) occurs after UV or γ irradiation or methylmethane sulfonate or 4NQO treatment. Since Ty elements belong to a family of yeast genes that increase their expression in response to DNA damage, they may prove to be a useful reporter gene to study this process.

3. Translational Frameshifting

All Ty elements contain a +1 reading frame discontinuity that prevents normal translation of *TYB* (Clare and Farabaugh, 1985; Mellor *et al.*, 1985a; Wilson *et al.*, 1986; Clare *et al.*, 1988; Hansen *et al.*, 1988; Belcourt and Farabaugh, 1990). The mechanism of frameshifting in Ty1 elements has been intensively studied (Belcourt and Farabaugh, 1990). The efficiency of frameshifting for Ty1 is about 20% (Clare *et al.*, 1988; Belcourt and Farabaugh, 1990). Alignment of *TYA* and *TYB* genes from Ty1 and Ty2 elements shows a conserved 14-nt sequence that is sufficient to frameshift a +1 mutation placed in a heterologous coding sequence (Wilson et al., 1986; Clare et al., 1988; Belcourt and Farabaugh, 1990). Recently, Belcourt and Farabaugh (1990) have shown that a 7-nt sequence (CTTAGGC) contains the minimum frameshift region. Furthermore, they demonstrated that frameshifting occurs within two overlapping Leu codons (CUU in the TYA frame and UUA in the TYB frame), tRNA$^{Leu-UAG}$ may be the tRNA able to recognize both Leu codons, and a translational pause at a rare Arg-AGG codon is required for optimum frameshifting. Xu and Boeke (1990b) also have shown that increasing the level of the rarely used tRNA$^{Arg-CCU}$ partially inhibits Ty transposition by altering the stoichiometry of TYA to TYB proteins. Ty3-1 and Ty3-2 contain a putative +1 frameshift, but do not contain the CTTAGGC frameshifting sequence (Hansen et al., 1988; Hansen and Sandmeyer, 1990). It will be interesting to determine if Ty3 elements use a different frameshift mechanism to translate TYB.

Some retroviruses contain a −1 frameshift between gag and pol, which is suppressed in order to translate pol (Jacks and Varmus, 1985; Jacks et al., 1987, 1988a,b; Moore et al., 1987; Wilson et al., 1988). The present results suggest that different mechanisms are used to effect the −1 retroviral and +1 Ty1 element frameshifting event. The minimum frameshift region for Ty1 does not contain homopolymeric runs or stem-loop structures that have been implicated in various retroviral frameshift events. In addition, rabbit reticulocyte lysates do not translate Ty1 RNA containing the +1 frameshift, but do translate Ty1 RNA if the frameshift is corrected (J. Boeke and D. J. Garfinkel, unpublished results).

4. VLP Maturation

Ty-VLP maturation reflects the initial stages of retroviral core assembly (Garfinkel et al., 1985; Mellor et al., 1985c; Adams et al., 1987; L. Hansen and S. Sandmeyer, personal communication). This is a complex process involving processing of Ty precursor proteins by Ty-pr and encapsidation of Ty RNA and the tRNA^{Met} primer required for reverse transcription. Ty1-VLPs are spherical to ovoid in shape and are 60 nm in diameter (Fig. 3). Upon sucrose gradient sedimentation, the particles have an apparent sedimentation value of 175S (Mellor et al., 1985c) and a buoyant density of 1.2 g/ml (S. Youngren and D. J. Garfinkel, unpublished results). Isolated Ty1-VLPs appear to be permeable to macromolecules (Garfinkel et al., 1985; Mellor et al., 1985c; Eichinger and Boeke, 1988, 1990). Exogenous primer/template complexes, integration substrates, pancreatic ribonuclease, and micrococcal nuclease can act on Ty-VLP nucleic acid or protein components without prior solubiliza-

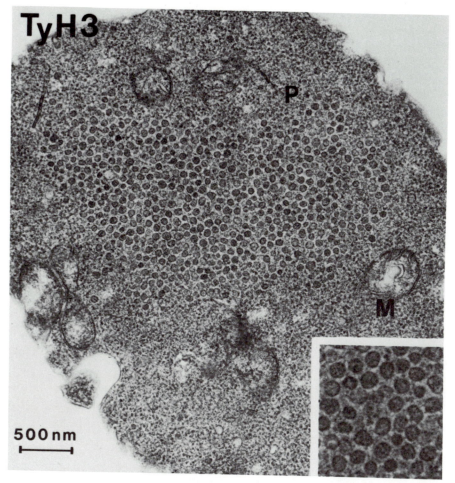

FIGURE 3. Transposition-induced cells contain Ty-VLPs. Strain DG692, which harbors the plasmid pGTy1-H3*NEO*, was transposition-induced and prepared for electron micros-copy. The mitochondria (M) and Ty-VLPs (P) are labeled in the 33,480× magnification. The inset is a 83,700× magnification of the particles. The bar represents 500 nm. K. Nagashima and M. Gonda (PRI, NCI-FCRDC, Frederick, Maryland) helped with the elec-tron microscopy.

tion. It remains to be determined whether this is an intrinsic structural property of the VLPs or an artifact of the isolation procedure.

There is a wealth of indirect evidence suggesting that Ty-VLP for-mation is an essential step in transposition. High levels of transposition are always correlated with the appearance of VLPs (Garfinkel *et al.*, 1985; Mellor *et al.*, 1985c; L. Hansen and S. Sandmeyer, personal communication; A.-M. Hedge and D. J. Garfinkel, unpublished results). Most of the Ty reverse transcriptase activity, *TYA* and *TYB* proteins representing all Ty homology domains, Ty RNA, tRNA^{Met}, and Ty DNA cosediment with the VLPs (Garfinkel *et al.*, 1985; J. Boeke, personal communication). Ty1-*pr* mutations completely inhibit transposition of marked elements (Youngren *et al.*, 1988) and cause morphologically altered Ty-VLPs to form (Adams *et al.*, 1987; Muller *et al.*, 1987; Youngren *et al.*, 1988). Finally, purified Ty1-VLPs catalyze Ty element integration into a known target substrate *in vitro* (Eichinger and Boeke, 1988).

Particle formation could be important for at least two reasons. First, it may protect the cell against potentially deleterious levels of free reverse transcriptase. In yeast, cDNA released from the VLP would be highly recombinogenic and might result in unwanted genome rearrangements (Gargouri *et al.*, 1983; Fink *et al.*, 1986; Fink, 1987). Second, it is likely that Ty-VLPs sequester the necessary components for reverse transcription and integration at the proper concentrations. In this view, the VLP would play a role in regulating the use of RNA for reverse transcription or translation (Fuetterer and Hohn, 1987).

For Ty1 elements, *TYA1* and *TYB1* precursor proteins are proteolytically processed by Ty-*pr* during VLP maturation (Adams *et al.*, 1987; Garfinkel *et al.*, 1991; Muller *et al.*, 1987; Youngren *et al.*, 1988). The major structural core proteins of Ty-VLPs are proteolytically cleaved from the full-length *TYA1* precursor. *TYA1* protein is functionally similar to retroviral *gag* proteins. Unprocessed *TYA1* protein can also form morphologically altered VLPs, suggesting that *TYA1* contains all the information in its primary sequence to form immature VLPs. Covalent protein modifications may also be required for assembly of Ty-VLPs. *TYA1* protein is phosphorylated *in vivo* and binds nucleic acid nonspecifically *in vitro* (Mellor *et al.*, 1985b). *TYA1* and *TYA2* genes have homology to the consensus sequence for prokaryotic DNA-binding proteins (Clare and Farabaugh, 1985; Hauber *et al.*, 1985; Warmington *et al.*, 1985), while *TYA3* has one copy of the Cys–His motif ($CX_2CX_4HX_4C$) that is found in the retroviral *gag* nucleocapsid protein (Hansen *et al.*, 1988).

Some progress has been made in understanding Ty1-VLP maturation by analyzing Ty1-*pr* mutants and by defining the precursor–product relationship of Ty proteins (Adams *et al.*, 1987; Garfinkel *et al.*, 1991; Muller *et al.*, 1987; Youngren *et al.*, 1988). Ty1-*pr* is required for all cleavage events of *TYA* and *TYA–TYB* precursors. Ty-*pr* mutations abolish transposition of genetically marked elements and form immature Ty-VLPs. The mutant particles display reverse transcriptase activity, but do not synthesize Ty DNA *in vitro*. This is because mutant VLPs do not contain normal levels of Ty RNA. It is not known if *pr*-mutant

VLPs fail to encapsidate Ty RNA or if the VLP-associated Ty RNA is more susceptible to nuclease degradation.

To determine the processing pathway of Ty1 proteins, antibodies raised against specific regions of *TYA1* and *TYB1* have been used in immunoblotting and immunoprecipitation experiments (Adams *et al.*, 1987; Garfinkel *et al.*, 1991). *TYA1* encodes a protein of about 50 kilodaltons (kDa) (Mellor *et al.*, 1985b; Adams *et al.*, 1987). Perhaps because of its high proline content (the N-terminal region of *TYA1* is 20% proline), the primary translation product of the gene migrates anomalously on gels as about a 58-kDa polypeptide, p58-*TYA1* (also called PRO-TYA or p1), that is cleaved to form p54-*TYA1* (also called TYA or p2) (Mellor *et al.*, 1985b; Adams *et al.*, 1987; Muller *et al.*, 1987; Boeke and Garfinkel, 1988; Youngren *et al.*, 1988). *TYB1* protein is included in the maturing VLP as part of a 190-kDa *TYA1–TYB1* polyprotein (Clare and Farabaugh, 1985; Mellor *et al.*, 1985a; Adams *et al.*, 1987; Muller *et al.*, 1987; Youngren *et al.*, 1988), which is then processed by Ty-*pr* to form mature *pr*, *in*, and *rt/rh* products (Garfinkel *et al.*, 1991). Two *TYB1*-processing intermediates can be detected from Ty1-H3. Ty1-*pr* cleaves p190-*TYA1–TYB1* to form a p160-*TYB1*-processing intermediate. The p160-*TYB* intermediate is cleaved to form a 23-kDa protease and a precursor of about 140 kDa. The 90-kDa Ty1-*in* and 60-kDa Ty1-*rt/rh* are derived from p140-*TYB1*.

Encapsidation of Ty RNA has been studied using helper-dependent mini-Ty1 elements (Xu and Boeke, 1990a). The smallest mini-Ty1 element capable of transposition contains the 5' R–U5 region, 285 nt of adjacent *TYA1* ϵ sequence, 23 nt of terminal ϵ sequence, and the 3' LTR. The RNA from this mini-Ty1 is packaged into VLPs with wild-type efficiency. Since the mini-Ty1 assay measures all of the steps in transposition required in *cis*, it is likely that the packaging site is located within the remaining Ty1 sequences. An interesting observation made by Xu and Boeke (1990a) is that certain genomic transcripts cofractionate with Ty1-VLPs and may be encapsidated. This may provide a means to form processed pseudogenes in yeast (Fink, 1987; Derr *et al.*, 1991). Most if not all retroviruses also contain two RNA molecules per particle [reviewed by Coffin (1984)]. Genetic evidence suggests that the same is true for Ty elements (Boeke *et al.*, 1986).

5. Reverse Transcription

Ty reverse transcription probably occurs by a retrovirus-like mechanism, since many of the characteristics of this process are similar. Initial experiments on the mechanism of Ty transposition indicated that a complete 5' LTR is regenerated during the process, and sequence polymorphisms are transferred from the 5' end of the transcript to the newly synthesized 3' LTR and from the 3' end of the transcript to the 5' LTR (Boeke *et al.*, 1985; Muller *et al.*, 1991). These patterns of inheritance are exactly what one would predict based on the accepted model for retroviral reverse transcription (Gilboa *et al.*, 1979). Ty elements contain sequences homologous with the 3' end of initiator tRNA^{Met} (Simsek and RajBhandary, 1972; Eibel *et al.*, 1981; Cigan and Donahue, 1986) that are located just inside the 5' LTR and a putative polypurine tract inside the 3' LTR. A minus-strand strong-stop DNA containing an alkali-labile component (presumably tRNA^{Met}) has been identified for Ty1 (A. Bystrom and G. R. Fink, personal communication). tRNA^{Met} is also specifically enriched in Ty-VLP preparations (J. Boeke, personal communication). Plus-strand strong-stop DNA has not been detected (Muller *et al.*, 1991).

Recombination events frequently occur during retroviral replication [reviewed by Linial and Blair (1984) and Hu and Temin (1990)]. This is a consequence of the nature of retroviral reverse transcription and the fact that most retroviruses contain a diploid genome consisting of two RNA molecules. Ty elements also undergo high levels of "transpositional recombination." Genetically marked Ty1 transposition insertions often contain restriction site polymorphisms that are different from those found in the starting element (Boeke *et al.*, 1985). The restriction site changes are also commonly found in genomic Ty elements, suggesting that they are picked up by the marked element during the transposition process (Baltimore, 1985). Genetic evidence supporting transpositional recombination has been obtained by comparing marked transposition events obtained in *SPT3* and *spt3* mutant strains (Boeke *et al.*, 1986). In contrast to the situation in *SPT3* cells, marked Ty1 insertions obtained in *spt3* mutants do not contain restriction site polymorphisms. Since *SPT3* is required for chromosomal Ty transcription but not pGTy transcription, the genomic Ty RNA cannot recombine with *GAL1*-promoted Ty RNA during reverse transcription. These results also support the idea that Ty1-VLPs contain more than one copy of Ty RNA.

Ty1 elements show genetic instabilities that have been attributed to genetic markers placed in the element (Xu and Boeke, 1987) or to mistakes in reverse transcription (Errede *et al.*, 1986; Wilke *et al.*, 1989; Weinstock *et al.*, 1990). In particular, markers that are present as direct repeats often resolve to one copy during the transposition process (Xu and Boeke, 1987). Examination of marked Ty RNA and DNA species in the VLP suggests that deletion events occur during reverse transcription. A process similar to this may also be responsible for creating Ty element insertions with internal deletions (Curcio *et al.*, 1990). Aberrant reverse transcription has been implicated in formation of a 5' LTR with an inverted duplication (U5'–R'–R–U5) (Errede *et al.*, 1986) and in multimeric Ty insertions (Weinstock *et al.*, 1990).

The biochemistry of Ty reverse transcriptase/RNase H has not been intensively studied. Ty1 and Ty2 elements encode a highly related 60-kDa protein that probably contains reverse transcriptase, DNA-dependent DNA polymerase, and RNase H activities (Garfinkel *et al.*, 1991). The Ty1 enzyme may be active when it is part of p190-*TYA1-TYB1*, as well as in its mature form (Youngren *et al.*, 1988). Mutations in the Ty1 *rt* and *rh* homology domains are transposition-defective and result in low levels of reverse transcriptase activity (D. J. Garfinkel, unpublished results). VLP-associated Ty1 reverse transcriptase activity is heat-sensitive (Garfinkel *et al.*, 1985) and is not activated by detergent treatment of the VLPs (J. Boeke and D. J. Garfinkel, unpublished results). A thermolabile enzyme would also explain the low levels of Ty transposition observed at temperatures above 30°C (Paquin and Williamson, 1984, 1986).

6. Integration

Considerable progress has been made in understanding the integration mechanism of Ty1 elements because of the development of a cell-free integration assay (Eichinger and Boeke, 1988, 1990). Similar assays have also been developed to study retrovirus integration (Brown *et al.*, 1987; Fujiwara and Mizuuchi, 1988). The requirements for *in vitro* integration of genetically marked Ty1-H3 elements are simple. Ty1-H3 VLPs purified by sucrose gradient sedimentation and concentrated by ultracentrifugation are incubated with purified bacteriophage λgtWES DNA, then the phage are packaged into λ particles *in vitro*. The phage contain amber mutations in several essential genes that are suppressed by a *supF* gene carried by Ty1-H3. Putative Ty1*supF* transpositions into λ are detected by their ability to form plaques on a *sup*⁰ host. The integration reaction requires a divalent cation and is quite rapid if endogenous marked Ty elements are the integration substrate. Insertions are found in different regions of λ that are nonessential for lytic growth. Transposition events contain a 5-nt target-site duplication characteristic of Ty element insertions *in vivo* (Farabaugh and Fink, 1980; Gafner and Philippsen, 1980). Therefore, the cell-free system faithfully reproduces the first cutting and joining reactions of normal Ty integration.

Recently, Eichinger and Boeke (1990) have shown that exogenously supplied integration substrates can be utilized by Ty1-VLPs in the cell-free integration reaction. These experiments reveal several interesting features of Ty integration. First, integration requires linear DNA containing as little as 12 nt. of Ty sequence from the ends of the LTR. Circular

DNA containing an appropriate circle junction sequence was not an effective substrate in the reaction. Second, a terminal 3'-OH is needed for transposition, but a 5'-PO$_4$ is not. Third, unlike retroviruses, Ty1 elements do not lose a dinucleotide from their ends during integration. Finally, addition of exogenous substrates greatly stimulates the integration activity of the VLPs. This result suggests that formation of integration-competent Ty cDNA is a rate-limiting step in transposition.

How does Ty element cDNA return to the nucleus and cross the nuclear membrane from the cytoplasm? One possibility is that a Ty-VLP returns to the nucleus. But it is unclear how a 60-nm VLP transverses the nuclear membrane. Since the yeast nuclear membrane does not break down during mitosis, it may present a permanent barrier to a VLP. A more likely possibility is that the VLP releases an *in*-Ty cDNA complex as the penultimate step in the transposition pathway. A prediction of this model is that Ty-*in* may be the only protein required for integration. Indeed, *in* proteins from avian leukosis virus (Katz *et al.*, 1990) and Moloney murine leukemia virus (Craigie *et al.*, 1990) have recently been shown to be sufficient to catalyze retroviral integration *in vitro*.

All Ty element classes demonstrate target-site preference. In the case of Ty1 and Ty3 elements, it appears that integration specificity and transcriptional activity of the target site are related. Ty1 element insertion sites have been studied by isolating the sequences that flank chromosomal Ty elements and by studying insertions into specific target genes. *GAL1*-promoted transposition of marked Ty1 and Ty3 elements has greatly facilitated the analysis of Ty insertion sites (Natsoulis *et al.*, 1989; S. Sandmeyer, personal communication). Several features of endogenous Ty insertion sites have become evident from studies over the past several years. First, Ty1 and Ty2 insertions tend to be associated with AT-rich regions (Oyen and Gabrielsen, 1983). Second, clustered insertion events suggest that Ty elements may prefer to integrate near or into other Ty elements (Warmington *et al.*, 1986; Warmington *et al.*, 1987). This feature is also prevalent in retroelements found in *Dictyostelium* (Marschalek *et al.*, 1989, 1990). Third, all Ty elements are frequently found near tRNA genes (Kingsman *et al.*, 1981; Delrey *et al.*, 1982; Eigel and Feldmann, 1982; Sandmeyer and Olson, 1982; Brodeur *et al.*, 1983; Gafner *et al.*, 1983; Stucka *et al.*, 1987; Clark *et al.*, 1988; Hauber *et al.*, 1988; Chalker and Sandmeyer, 1990).

Target-site selection has also been studied by exploiting Ty elements as insertional mutagens. A number of spontaneous and pGTy-induced transposition events have been characterized using a plasmid-based *HIS3* gene (*his3Δ4*), which lacks a promoter (Boeke *et al.*, 1985; Boeke *et al.*, 1986), and several chromosomal genes: *ADH2* (Williamson *et al.*, 1983), *LYS2* (Eibel and Philippsen, 1984; Simchen *et al.*, 1984; Natsoulis *et al.*, 1989), *URA3* (Natsoulis *et al.*, 1989), *CAN1* (Wilke *et* al., 1989), and *SUP4-0*, which is an ochre- or amber-suppressing tRNA^{Tyr} gene (Giroux *et al.*, 1988). Since these Ty insertions are obtained by genetic selections, target-site specificity may have been influenced by the nature of the selection and context of the target gene. Even with these considerations, the location of Ty1 and Ty2 elements in these genes has shown that there are strongly preferred integration regions.

Ty-induced mutants at *ADH2* and *his3Δ4* possess the *ROAM* phenotype and result from insertions in the promoter region. Several *ADH2* insertions occurred near the TATA box, while none was found near UAS elements (Williamson *et al.*, 1983). The *his3Δ4* promoter deletion contains bacteriophage λ sequences in place of the normal *HIS3* promoter region (Scherer *et al.*, 1982). Most Ty insertions occur within 125 nt of the normal *HIS3* transcription start site (Boeke *et al.*, 1985, 1986). Insertions also occur further upstream, but these only confer a His$^+$ phenotype in an *spt3* mutant background (Boeke *et al.*, 1986). There is a pronounced hot spot about 50 nt upstream of the *his3Δ4* transcription start in *SPT3* cells, but this site is not extensively utilized in *spt3* mutants. Presumably, the two patterns of insertion sites are dependent on the His$^+$ selection and not solely on insertion-site specificity.

Selections at *CAN1*, *LYS2*, *URA3*, and *SUP4-0* were for loss of function and therefore presented a much larger target site. Ty insertions have been found near the 5' end of *CAN1*, *LYS2*, and *URA3*, and hot spots are also evident in all three genes. However, no striking target-site consensus sequence emerged from these studies. A loose consensus of (A/C/G)-N-(A/T)-N-(T/G/C) was deduced from the collection of insertions at *URA3* (Natsoulis *et al.*, 1989). Insertion-site preference at *CAN1* appears to depend on the strain background, and possibly the *RAD6* gene (Wilke *et al.*, 1989). These results suggest that Ty recognizes preferred targets by a primary sequence-independent mechanism, which might involve chromatin structure. Evidence for preferred integration has also been obtained from insertions that inactivate *SUP4-0* (Giroux *et al.*, 1988). Of 12 Ty-induced mutants, 10 insertions occurred at the same position in the middle of the gene.

Ty3 elements have a markedly restricted target-site preference that is somehow related to genes transcribed by polymerase III. Both endogenous insertions and *de novo* transposition events are always found near tRNA genes (Brodeur *et al.*, 1983; Chalker and Sandmeyer, 1990). Furthermore, *GAL1*-promoted Ty3 element expression followed by selections for mutations at the *URA3* locus do not result in any Ty3-induced mutations (Chalker and Sandmeyer, 1990). *GAL1*-promoted Ty3 transposition insertions are found within 20 nt upstream of an unoccupied tRNA gene, which is consistent with the location of endogenous Ty3 element and σ insertions. Insertions into tRNA genes already occupied by a σ element are also not preferred targets for *de novo* Ty3 transpositions. Recently, D. L. Chalker and S. B. Sandmeyer (personal communication) developed a plasmid-targeting assay to detect Ty3 transpositions obtained after pGTy3 induction. In this assay, additional polymerase III genes (5S and U6) are used as targets, and target-site preference is stronger if the target gene is transcriptionally active. Since the tRNA transcription start site and the staggered cut initiating Ty3 insertion are also close together (P. Kinsey and S. Sandmeyer, personal communication), it will be interesting to determine what role polymerase III initiation factors play in this process.

C. Regulation of Ty Transposition

An unusual feature that is common to the biology of many retrotransposons but not understood is the low rate of transposition despite high levels of mRNA accumulation (Lueders et al., 1977; Finnegan et al., 1978; Elder et al., 1980; Curcio et al., 1990). Ty RNA is extremely abundant, accounting for an estimated 1% of total RNA (Curcio et al., 1990) and at least 5–10% of polyadenylated RNA in a haploid cell (Elder et al., 1980). In addition, doubling the Ty1 element copy number does not destabilize the genome (Boeke et al., 1991). What makes these observations paradoxical is that increasing the total level of Ty RNA severalfold by expressing a Ty element from the heterologous GAL1 promoter results in a dramatic increase in the frequency of transposition (Boeke et al., 1985).

Controlling the level of transposition is necessary for genetic stability of the yeast cell. As described above (Section III.B.2), regulation of transposition can occur at the transcriptional level. Other types of regulation have been observed or proposed for Ty elements, although the relative importance of these different types of regulation is unknown (Fulton et al., 1985; Fink et al., 1986; Boeke et al., 1988a; Curcio et al., 1988). One hypothesis is that both competent and defective Ty elements are normal residents of the yeast genome, but that at least some of the defective elements are heavily transcribed, whereas the competent elements are poorly transcribed or repressed. A related model is that regulation of Ty transposition occurs by the accumulation of defective copies. In this model, transpositionally nonfunctional elements tend to accumulate in the cell, while functional elements tend to be lost. A strong prediction of this model is that most, if not all, endogenous Ty elements should be inactive in the pGTy system. A third general hypothesis is that there is posttranscriptional regulation of the transposition process. Points at which such regulation could occur include the mode of expression of the important TYB protein, posttranslational modification of

Ty-encoded proteins, which has been demonstrated for a *TYA* protein (Mellor *et al.*, 1985b), and proteolytic processing of *TYA* and *TYA–TYB* primary translation products (Adams *et al.*, 1987; Muller *et al.*, 1987; Youngren *et al.*, 1988), as well as general translational controls.

1. Working Model

Of the three general hypotheses stated previously, it does not appear that regulation of transposition occurs solely by transcriptional repression of competent elements or by accumulation of defective copies (Curcio *et al.*, 1988; Garfinkel *et al.*, 1988a). Curcio *et al.*, (1988) observed that the endogenous elements Ty1-588 and Ty2-117 transpose at levels comparable to Ty1-H3. Therefore, complete active elements exist in the yeast genome. It also appears that Ty1-588 is transcribed at a relatively high level in normal cells (Garfinkel *et al.*, 1988a). This result suggests that a component of transposition regulation occurs posttranscriptionally.

The following working model attempts to explain these and other observations (M. J. Curcio and D. J. Garfinkel, unpublished results). The yeast genome contains both functional and nonfunctional Ty elements, many of which are actively transcribed (Boeke *et al.*, 1988a; Curcio *et al.*, 1988; Garfinkel *et al.*, 1988a). The levels of Ty proteins are low because of a transpositional inhibitor acting at the level of protein synthesis and/or maturation. The inhibitor may be a cellular gene or a chromosomal Ty product. The effects of the inhibitor are overcome by *GAL1*-promoted Ty expression, allowing Ty proteins to accumulate in the cell and catalyze transposition. Predictions of this model can be tested by identifying rate-limiting steps in transposition, and by trying to isolate mutations in the hypothesized "transpositional inhibitor."

2. Distribution of Functional and Nonfunctional Ty Elements in the Genome

Since the yeast genome contains both transpositionally competent and transcriptionally active chromosomal Ty elements, as well as mutant Ty elements, it is possible that *trans*-dominant Ty mutations play a role in regulating transposition. However, the relative numbers of mutant versus competent elements within a given strain are unknown. To survey a large number of different genomic Ty elements for transposition competence, the efficient homologous recombination system of yeast has been used to create recombinant elements from a known pGTy plasmid (M. J. Curcio and D. J. Garfinkel, unpublished results). The recombinant elements contain sequences from a variety of different chromosomal Ty elements that are joined to segments of the transposition-competent plasmid-borne element pGTy1-H3*HIS3* or pGTy2-917*HIS3*. Over 70% of the Ty1 recombinants are transposition-competent when tested, suggesting that most of the Ty elements in the genome are not mutant. It will be interesting to determine if any of the defective Ty elements contain dominant mutations.

3. Chromosomal Genes That Regulate Ty Transposition

In addition to the known regulators of Ty transcription (Section III.B.2) and frameshifting (Section III.B.3), additional cellular genes have been identified that affect Ty transposition. It has been reported that the rate of Ty transposition into the *CAN1* or *URA3* loci increases 100-fold in certain *rad6* mutant backgrounds (Picologlou *et al.*, 1990). Interestingly, the increased level of Ty transposition is not associated with an increase in the level of Ty RNA. These results suggest that *RAD6* may act posttranscriptionally to inhibit Ty transposition. *RAD6* is required for a variety of cellular functions involving DNA repair, induced mutagenesis, and sporulation [reviewed by Haynes and Kunz (1981)]. Jentsch *et al.* (1987) showed that *RAD6* is a ubiquitin-conjugating enzyme and that two of its substrates *in vitro* are histone H2A and H2B. Perhaps *RAD6* restricts Ty insertion specificity by altering the chromatin structure of a target gene. The *RAD52* gene, which is involved in DNA repair and recombination, has also been implicated in regulating Ty transposition. Kunz *et al.* (1989) have reported that although several classes of mutations inactivating *SUP4-0* occur at similar frequencies in wild-type and *rad52* mutants, no Ty-induced mutations have been recovered in *rad52* cells. In wild-type cells, Ty insertional inactivation causes about 10% of the *sup4-0* mutants (Giroux *et al.*, 1988; Kunz *et al.*, 1989). In contrast, Ty-induced mutations occur at *his3Δ4* in a *rad52* mutant background (Scherer *et al.*, 1982; G. Fink, personal communication). These results suggest that a variety of genes may influence Ty transposition, but that target site differences or other mutagenic events complicate further interpretation.

Chapman and Boeke (1991) have recently isolated an unusual gene, *DBR1*, that is required for Ty1 transposition. *GAL1*-promoted Ty1 transposition events drop about tenfold in a *dbr1-1* background. Surprisingly, *DBR1* is also required for intron turnover. It has been found to encode the debranching enzyme required to hydrolyze the 2'–5' phosphodiester linkage at the branch point of excised intron lariats. However, it is not known what role *DBR1* plays in Ty transposition.

4. A Novel Ty Marker Gene for Studying Regulation of Transposition

To study regulation of Ty transposition in greater detail, the *HIS3* marker gene has been modified so that marked transpositions of endogenous Ty elements can be detected (Curcio and Garfinkel, 1991). A unique signal that distinguishes retrotransposition from other DNA-mediated events is the ability to lose an intron during the transposition of a marked element (Boeke *et al.*, 1985; Heidmann *et al.*, 1988). An artificial intron (AI), which is easily manipulated and is completely removed from a gene via splicing (Yoshimatsu and Nagawa, 1989), has been used to conditionally mutate the *HIS3* gene of pGTy1-H3*HIS3*. The 104-nt intron was placed in the *HIS3* coding sequence in the antisense orientation so that it cannot be spliced from the *HIS3* transcript and creates a *his3* null allele. The *HIS3*AI gene was placed within an appropriate pGTy1-H3 plasmid (Garfinkel *et al.*, 1988b), such that *HIS3*AI and Ty are transcribed in opposite directions (denoted by the letter m). This places *AI* in the correct orientation for splicing within the Ty1-H3m*HIS3*AI genome-length transcript. Reverse transcription and transposition of the precisely spliced transcript recreates a functional *HIS3* gene. Therefore, selection for histidine prototrophs identifies cells that contain a transposed copy of Tym*HIS3*AI.

Two features of the system are unique. First, transposed elements can be detected in the presence of the original m*HIS3*AI marker gene, which greatly simplifies and expands the Ty transposition assay (Boeke *et al.*, 1988b; Garfinkel *et al.*, 1988a). Second, since splicing of the AI is not required for transposition, *de novo* unspliced Tym*HIS3*AI insertions can be readily isolated for further analysis. These transposed Tym*HIS3*AI elements are now under control of their own promoter and are inserted in various "unselected" positions in the genome. The genomic Tym*HIS3*AI insertions give rise to spontaneous histidine prototrophs that invariably result from a retrotransposition event with the intron precisely removed. This feature is being exploited to isolate hypertransposition mutants discussed above and to identify new environmental modulators and potential therapeutic agents that affect retrotransposition (M. J. Curcio and D. J. Garfinkel, unpublished data).

The AI transposition assay can be used on any other suspected retrotransposon. Analogous retrotransposon vectors can be built containing an AI within a gene that has a selectable phenotype. These vectors are introduced into cells (or possibly whole organisms) that contain expressed copies of the retrotransposon and, therefore, the potential for *de novo* transposition events. Note that the major requirement for creating a marked element is that all *cis*-acting sequences required for transposition are retained; it is not necessary to construct a helper-independent marked element. Revertants caused by splicing of the intron from the marker gene will be strong candidates for carrying new marked transpositions.

5. Rate of Ty Transposition

Ty transposition into specific target genes occurs at a very low rate, estimated at 10^{-7} to 10^{-8} transpositions per generation (Scherer et al., 1982; Paquin and Williamson, 1984; Boeke et al., 1986; Giroux et al., 1988). Recently, Boeke (1989) used the rate of Ty transposition into his3Δ4 to estimate the rate of random Ty insertions into the genome. After taking several factors into account, a transposition rate in the range of 10^{-3} to 10^{-5} transpositions per generation was obtained. The mHIS3AI marker gene provides another way to estimate the transposition of individual Ty1 elements (Curcio and Garfinkel, 1991). This estimate of individual Ty element transposition is more direct and does not require making the untestable assumption that a particular target sequence is representative of the entire genome. Transposition rates obtained from nine independent TymHIS3AI insertions in four strains are between 3×10^{-7} and 1×10^{-5} transpositions per generation. Three strains containing different TymHIS3AI insertions have comparable rates of transposition per marked Ty element, but the fourth strain tested has a significantly lower rate of His⁺ reversion. A likely explanation for this low rate of transposition is the comparatively low level of marked TymHIS3AI-234 transcript present in this strain.

D. Ty Elements as Retroviruses and Selfish DNA

It is clear that Ty elements and retrotransposons in general are closely related to retroviruses (Tables V and VI). Therefore, they must have a common origin. Two ideas have been put forth to explain how retrotransposons arose. Temin (1974, 1980) proposed that a cellular reverse transcriptase gene was initially incorporated into a retroelement such as Ty. These elements became infectious retroviruses by acquiring gag and envelope genes. It is possible that incorporation of genes for VLP formation (TYA/gag) was also required to prevent reverse transcription of cellular RNAs (see Section III.B.4). The counterargument is that retroviruses are generally parasitic and can become established as endogenous proviruses (Weinberg, 1980). Consequently, it is easy to imagine a Ty element being a degenerate retrovirus that has become fixed in the S. cerevisiae genome. The narrow distribution of Ty elements within yeast and the possibility of interspecies transfer suggest that Ty elements can be transmitted horizontally. But this view does not explain their ulti-

TABLE V. Similarities Between Ty Elements and Retroviral Proviruses

DNA structure
    LTR (long terminal repeat) sequences, direct orientation, similar size
    Small duplication of target DNA
    TG. . .CA terminal dinucleotides
    Homology to tRNA just inside 5′ LTR
    Polypurine stretch just inside 3′ LTR
    Enhancer sequence present
    Integration occurs through a linear DNA intermediate

RNA structure
    Abundant end-to-end transcript
    RNA is terminally repetitious

Open reading frames
    Overlapping reading frames (*TYB* made as readthrough from *TYA*)
    Predicted nucleic acid-binding site in *TYA*
    Homology to retroviral reverse transcriptase/integrase/protease in *TYB*

Virus-like particles
    Contain reverse transcriptase activity
    Contain element-specific RNA and primer
    Contain element-specific proteins
    Mediate integration reaction *in vitro*

mate origin. However, the recent discovery of a cellular reverse transcriptase greatly strengthens the idea that retroelements originally evolved from cellular genes (Shippen-Lentz and Blackburn, 1990).

Do Ty elements perform any useful function for yeast, or are they simply efficient parasites or selfish DNA (Dawkins, 1976; Doolittle and Sapienza, 1980; Orgel and Crick, 1980)? Selfish DNA has two basic properties: (1) the DNA tends to form additional copies of itself and spreads these within the genome. (2) Selfish DNA does not influence

TABLE VI. Differences Between Ty Elements and Retroviruses

Retrovirus	Ty Element
Loss of terminal dinucleotides upon integration	No loss of terminal dinucleotides
Enhancer within LTR (long terminal repeat)	Enhancer outside LTR
Envelope gene required for infectivity	No envelope gene present
Infectious	Noninfectious
Translation of *pol* occurs by nonsense suppression of a −1 frameshift	Translation of *TYB* occurs by a +1 frameshift

phenotype. Ty elements meet the first requirement, but may not fulfill the second. The uncontrolled spread of Ty elements through the genome is limited by mechanisms that regulate transposition and by the deleterious effects of Ty-induced mutations and DNA rearrangements. Ty elements persist in the genome because the negative selection they impose is not strong enough to offset a low basal level of transposition. Furthermore, Ty elements can be lost from the genome by LTR–LTR recombination (Chaleff and Fink, 1980; Roeder and Fink, 1980). This creates a way of eliminating detrimental Ty element insertions. The remaining solo LTRs may provide a substrate for the evolution of new, unique DNA sequences (Fink et al., 1986).

It is notable that Ty-induced mutations far outnumber all other mutagenic events in certain genetic selections. A graphic example of this occurs at ADH4, a cryptic alcohol dehydrogenase gene that can be activated either by Ty insertion (Paquin and Williamson, 1986; Williamson and Paquin, 1987) or chromosomal amplification (Walton et al., 1986). These results indicate that Ty elements are capable of activating genes that are not required for growth under normal laboratory conditions. Therefore, Ty element-mediated insertion mutations or genome rearrangements may increase genetic diversity by their ability to activate cryptic genes (Goebl and Petes, 1986). This idea is consistent with the general concept that all transposable elements function to increase genetic diversity and the potential for rapid genetic change (McClintock, 1984).

IV. RETROELEMENTS IN OTHER EUKARYOTIC MICROBES

A. Other Fungi

Retroelements have been characterized from S. cerevisiae and Saccharomyces kluyveri and quite recently from Schizosaccharomyces pombe (Levin et al., 1990). Saccharomyces species have been surveyed for Ty elements that are related to Ty1 elements (K. Weinstock and J. Strathern, personal communication). Surprisingly, S. kluyveri is the only additional Saccharomyces species that contains Ty1-like elements. However, S. kluyveri Ty elements are clearly diverged from Ty1. This narrow distribution suggests that Ty1 elements may have been acquired recently.

Transposable elements were searched for unsuccessfully in common laboratory strains of S. pombe, using genetic selections that resulted in Ty-induced mutations in S. cerevisiae (Levin et al., 1990). Analysis of moderately repetitive sequences resulted in the isolation of two related retrotransposons, Tf1 and Tf2, which are present in certain *S. pombe* strains but absent in others. Tf elements are transcriptionally active, and are probably transposition-competent (Levin *et al.*, 1990; H. Levin and J. Boeke, personal communication). Tf elements contain one open reading frame, even though there appears to be distinct *gag* and *pol* protein domains. It will be interesting to determine how Tf elements regulate the ratio of *gag* to *pol* proteins in the absence of two separate reading frames. A similar gene organization is found in the *Drosophila* element copia. These elements regulate synthesis of *gag* and *pol* proteins by splicing the genome-length RNA to form a separate *gag* transcript (K. Miller *et al.*, 1989; Brierley and Flavell, 1990; Yoshioka *et al.*, 1990).

Tad elements are the first retroelements to be described in *N. crassa* (Kinsey and Helber, 1989). Tad elements were initially isolated as insertional mutagens in the *am* (glutamate dehydrogenase) locus. Initial genetic selections using several laboratory strains did not yield any transposon-induced mutations (Kinsey and Helber, 1989), a situation similar to that observed with *S. pombe*. To increase the chances of finding a transposable element, a variety of strains were crossed with a standard laboratory stock and the *am* selection was repeated. Tad was isolated in a strain derived from a natural isolate collected in Adiopodoume, Ivory Coast. Indeed, Tad elements have a very narrow distribution, existing only in the strain of origin among 336 isolates tested (Kinsey, 1989).

Two lines of evidence suggest that Tad is a retroposon (Kinsey and Helber, 1989; Kinsey, 1990; J. Kinsey, personal communication). First, preliminary characterization of Tad elements indicates that they lack LTRs and generate rather long and variable target-site duplications (14 and 17 nt). Tad elements also contain an open reading frame with homology to reverse transcriptases. These features are characteristic of retroposons found in a variety of organisms (Xiong and Eickbush, 1990). Second, Tad has been shown to transpose through a cytoplasmic intermediate, which is a fundamental characteristic of a retroelement (Kinsey, 1990). If forced heterokaryons are made between strains that do or do not carry Tad elements, the unoccupied nuclei rapidly gain Tad transposition events. Recently acquired Tad elements can be serially transferred to other naive nuclei, and the total number of elements per nucleus appears to increase during this process. Most importantly, cytoplasmic transfer of the Tad transposition intermediate takes place in the absence of nuclear fusion.

Recently, a solo-LTR-like element, repa, was isolated near the *S* incompatibility locus of the ascomycete *Podospora anserina* (Deleu *et al.*, 1990). Repa elements are strikingly similar in structure and genome location to solo LTRs found in *S. cerevisiae*.

B. Slime Molds

Dictyostelium and *Physarum* apparently contain retroelements, but their exact mode of transposition has not been determined. Several retroelements have been isolated from *Dictyostelium discoidium.* The DIRS1 (also called Tdd-1) element is probably a retroelement, but it has a very unusual structure (Cappello *et al.*, 1985). DIRS1 elements contain inverted instead of directly repeated LTRs, an internal region complementary to the termini of the inverted LTRs, a 3′ AT-rich region, and no duplication of target-site sequences. DIRS1 has a reverse transcriptase open reading frame that spans about 200 residues. Recent alignments place DIRS1 reverse transcriptase within the *Drosophila* gypsy family and most closely related to the Pao elements from *Bombyx* (Xiong and Eickbush, 1990). A detailed model for DIRS1 transposition based entirely on its structure has been proposed (Cappello *et al.*, 1985). The genomic organization of DIRS1 elements suggests that they undergo self-integration events (Cappello *et al.*, 1984; Firtel, 1989). Tdd-2 and Tdd-3 elements do not contain classical LTRs, insertions are flanked by AT-rich sequences, and they tend to integrate into other Tdd elements (Poole and Firtel, 1984). Recently, the *Dictyostelium* DRE elements have been found to be retrotransposons (Marschalek *et al.*, 1989). They are bracketed by LTRs and contain two open reading frames with homology to *gag–pol.* Interestingly, Tdd-3 and DRE elements are closely associated with tRNA genes (Marschalek *et al.*, 1989; Marschalek *et al.*, 1990), much like yeast Ty3 elements and their solo-LTR derivatives.

The *Hpa*II repeat is a major reiterated sequence, comprising up to 20% of the *Physarum* genome (Pearston *et al.*, 1985). *Hpa*II-repeat elements contain LTRs and conserved terminal nucleotides with copia and Ty. There is no evidence for duplication of target-site sequences. These elements have a propensity for integrating into other *Hpa*II copies, allowing the creation of very complex clusters 20–50 kb long.

C. Protozoa

Studies of the Trypanosomatidae have recently uncovered several related retroposons (the generic term retrotransposon-like element, RTnL, is also used to describe these elements). The first group, which is comprised of SLACS (also called MAE) and CRE1 elements, preferentially insert into miniexon gene clusters (Carrington *et al.*, 1987; Aksoy *et al.*, 1990; Gabriel *et al.*, 1990). SLACS and CRE1 elements were first isolated from *T. brucei* subspecies and the insect parasite *Crithidia fasculata*, respectively. They are found within the miniexon coding sequence between nt 11 and 12, and are in the same transcriptional orientation as the miniexon. CRE1 elements apparently undergo high levels of *de novo* transposition, suggesting that these elements could be used to develop a transposition system in *Crithidia*. SLACS and CRE1 elements fit the retroposon paradigm almost perfectly. These elements generate 49- and 29-bp target-site duplications and lack LTRs. They contain variable numbers of a repeated motif at the 5' end, *gag*- and *pol*-like open reading frames, and a long terminal poly(dA) stretch. Their reverse transcriptases are also closely related (Xiong and Eickbush, 1990). Recently, Gabriel and Boeke (1991) demonstrated that CRE1 encodes an enzymatically active reverse transcriptase. Furthermore, CRE1 shows homology to the *in* domain of *pol* (Gabriel *et al.*, 1990).

The second group of retroposons are associated with an element called ribosomal mobile element (RIME), which was found inserted into an rRNA gene of one *T. brucei* stock (Hasan *et al.*, 1984). The initial RIME isolate was internally repetitive, existing as a nearly perfect dimer of a 512-nt sequence. The junction between two monomers contains poly(dA) followed by 6 nt of DNA. The outside ends of the dimer contain a 7-nt direct repeat, which may be a target-site duplication. Two highly related inserts bracketed by RIME monomers have been characterized (Kimmel *et al.*, 1987; Murphy *et al.*, 1987). These composite elements are called INGI and TRS elements. The inserted segments encode reverse transcriptase and presumably other functions required for transposition. The composite elements do not have clearly defined target-site duplications. INGI and TRS insertions are dispersed in the *T. brucei* genome. It will be interesting to determine how the TRS/INGI elements became associated with RIME and how RIME transposition differs from TRS/INGI transposition. A closely related RTnL insertion has recently been recovered in the tubulin gene cluster (Affolter *et al.*, 1989).

V. RETROELEMENTS IN FUNGAL MITOCHONDRIA

Fungal mitochondria have been a rich source of unusual retroelements whose properties, along with those displayed by prokaryotic retrons, may reflect primitive forms of reverse transcription. The best characterized of these elements are mitochondrial group I and II introns of *S. cerevisiae* and plasmids of *Neurospora*.

A. Infectious Introns

The proposal that nuclear introns are mobile (or infectious) came from comparative studies of intron placement and diversity within similar genes from different organisms (Sharp, 1985) [reviewed by Belfort (1989) and Rogers (1989)] and from characterizing a variety of processed pseudogenes [reviewed by Rogers (1985), Vanin (1985), and Weiner *et al.* (1986)] or retrosequences (Table I). Group I and group II introns found in fungal mitochondria have provided an excellent experimental system to study intron insertion and loss. The characteristics of these introns have been recently reviewed (Cech, 1985, 1988; Cech and Bass, 1986; Dujon, 1989; Lambowitz, 1989; Perlman and Butow, 1989). Briefly, group I introns carry a series of short sequence elements that form part of a conserved secondary structure required for splicing. Group II introns have a unique secondary structure and splice by a reaction pathway that more closely resembles nuclear RNA splicing. Although their reaction mechanisms differ, both group I and group II introns are capable of self-splicing *in vitro*. However, genetic analyses of mitochondrial introns indicate that splicing of group I and group II introns requires specific *trans*-acting proteins *in vivo*.

Studies of the group I ω intron found in the large rRNA gene of yeast mitochondria have provided compelling evidence for DNA-mediated intron insertion events (Dujon, 1989; Perlman and Butow, 1989). This intron encodes a site-specific endonuclease that cleaves ω^- intronless DNA, thus promoting the transfer or "homing" of the ω^+ copy by a gene conversion event. This is analogous to the double-strand-break repair mechanism used in yeast mating-type switching [reviewed by Klar (1989)].

Several different RNA-mediated pathways leading to intron insertion or loss have been proposed (Gargouri *et al.*, 1983; Cech, 1985; Sharp, 1985; Dujon, 1989). Introns could potentially insert into other RNA molecules by reversal of the transesterification reaction used for splicing or by proteins that usually promote splicing. Recently, intron insertion was achieved in *trans* using a self-splicing *Tetrahymena* group I intron and β-globin transcript (Woodson and Cech, 1989), or a group II intron bI1 from the *S. cerevisiae* COB gene (Morl and Schmelzer, 1990). Hybrid RNA species could integrate directly into DNA during replication (presumably as part of a primer) or after reverse transcription. Intron insertion could also occur by reverse transcription of an unspliced transcript followed by either illegitimate recombination or an intron-homing mechanism typified by the ω intron. Intron loss could simply involve integration of a cDNA derived from a spliced transcript. This form of intron loss occurs in retrosequences (Rogers, 1985; Vanin, 1985; Weiner *et al.*, 1986; Derr *et al.*, 1991), retroviruses (Shimotohno and Temin, 1982; Sorge and Hughes, 1982; Heidmann *et al.*, 1988), and Ty elements (Boeke *et al.*, 1985; Curcio and Garfinkel, 1991).

Results from several areas permit such fanciful models to be entertained. First, the proteins encoded by group II introns contain characteristic blocks of amino acids that are homologous with retroviral and retrotransposon reverse transcriptases (Michel and Lang, 1985; Xiong and Eickbush, 1990). Interestingly, this includes two highly related introns in the yeast mitochondrial cytochrome oxidase I gene, aI1 and aI2. Intron aI1 has been shown genetically to be a maturase involved in splicing itself and other introns (Bonitz et al., 1980; Levra-Juillet et al., 1989). Second, respiratory-competent revertants obtained from partially defective splicing mutants sometimes lose the relevant intron precisely (Gargouri et al., 1983; Hill et al., 1985; Perea and Jacq, 1985; Seraphin et al., 1988; Levra-Juillet et al., 1989). Other introns in the vicinity, whose function is not considered part of the selection, are also lost. Moreover, the maturase encoded by aI1 and/or aI2 is required for this deletion process. These results suggest that precise intron loss in mitochondria occurs via an RNA intermediate (Gargouri et al., 1983) and that an intron-encoded reverse transcriptase is responsible for cDNA synthesis (Levra-Juillet et al., 1989). A similar phenomenon has also been observed in a *Podospora anserina* gene associated with fungal senescence (Osiewacz and Esser, 1984).

B. Mitochondrial Plasmids

The final argument for involvement of reverse transcription in intron movement comes from mitochondrial plasmids (designated mt-plasmids) found in *Neurospora* strains Mauriceville and Varkud (Nargang et al., 1984; Michel and Lang, 1985; Kuiper and Lambowitz, 1988; Lambowitz, 1989). These plasmids encode a protein that is homologous with reverse transcriptases from retrons and group II introns (Xiong and Eickbush, 1990). A major transcript covers the entire genome, which is expected if the plasmid replicates through an RNA intermediate (Nargang et al., 1984; Akins et al., 1988). In addition, there is a tRNA-like structure at the 3' end of the genomic transcript that may act as a primer for minus-strand DNA synthesis. Similar tRNA-like motifs have also been observed at the ends of plant RNA viruses (W. A. Miller et al., 1986; Dreher and Hall, 1988). The 3' tRNA structures may have been important for replicating RNA molecules early in evolution (Weiner and Maizels, 1987). Furthermore, mt-plasmids also contain sequence motifs characteristic of group I introns (Nargang et al., 1984; Burke et al., 1987; Kuiper and Lambowitz, 1988). These results suggest that mt-plasmids may be an evolutionary "missing link" between catalytically active introns and more advanced retroelements.

Two other results suggest that the mt-plasmids are retroelements. Strains containing mt-plasmids usually show no phenotypic changes. However, senescent mutants have been isolated that result from insertion of the mt-plasmid into the mitochondrial genome (Akins *et al.*, 1986). When the termini of three of these insertions were analyzed, one junction was always at the 5' end of the mt-plasmid genomic transcript. Another series of growth-impaired mutants have been isolated that greatly increase the copy number of the mt-plasmid (Akins *et al.*, 1986, 1989). Several of these mutants now contain a mitochondrial tRNA sequence at or very close to the 5' end of the genomic transcript. These mt-plasmid variants probably arose by alterations in the normal reverse transcription-dependent replication pathway.

Biochemical analysis of the plasmid-encoded reverse transcriptase directly implicates this protein in mt-plasmid replication (Kuiper and Lambowitz, 1988; Kuiper *et al.*, 1990). Reverse transcriptase activity copurifies with mt-plasmid nucleic acid and is highly specific for its endogenous template. The reverse transcriptase synthesizes full-length minus-strand DNAs that begin at the 3' end of the transcript. Antibodies directed against the reverse transcriptase domain cross-react with an 81-kDa protein that has reverse transcriptase activity *in situ*.

VI. BACTERIAL RETRONS

Retrons are retroelements that use a novel reverse transcription mechanism to amplify their copy number (Inouye *et al.*, 1989; Lampson *et al.*, 1989a,b; Lim and Maas, 1989) [reviewed by Temin (1989) and Varmus (1989)] (Fig. 4). They were initially discovered as unusual multicopy, single-stranded DNA (msDNA) satellites in the Gram-negative fruiting bacteria *Myxococcus xanthus* or *Stigmatella aurantiaca* (Yee *et al.*, 1984; Furuichi *et al.*, 1987a,b). *M. xanthus* msDNA exists at a level of over 500 copies per genome without conferring any noticeable phenotype (Dhundale *et al.*, 1988a). Experiments with *M. xanthus* strain DZF1 indicate that distantly related retrons can be found at more than one chromosomal locus (Inouye *et al.*, 1990). Retrons have also been detected in a laboratory *Escherichia coli* strain (Lim and Maas, 1989) and in 5–10% of clinical isolates (Lampson *et al.*, 1989b; Sun *et al.*, 1989).

All retrons have a similar episomal structure, although they show extensive sequence diversity (Dhundale *et al.*, 1988b; Sun *et al.*, 1989; Inouye *et al.*, 1990) (Fig. 4). Retrons usually contain a single-stranded DNA (msDNA) that is covalently attached to an RNA (msdRNA) by a 2'–5' phosphodiester bond (Furuichi *et al.*, 1987a,b). The DNA is joined at its 5' end to an rG residue at nt 19 or 20 of the RNA. The msDNA and msdRNA molecules contain stem-loop structures and base-pair for several nucleotides at their 3' ends.

Characterization of several chromosomal retron loci reveals similar genome organizations (Dhundale *et al.*, 1987; Inouye *et al.*, 1989; Lamp-

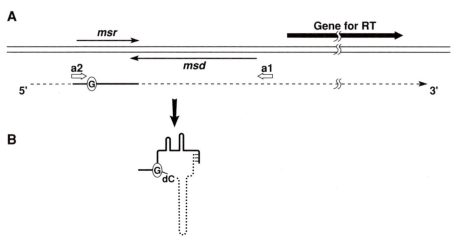

FIGURE 4. (A) Organization of retron chromosomal DNA. The retron chromosomal locus is shown by a double line with the positions of msdRNA (*msr*), msDNA (*msd*) and reverse transcriptase (RT) coding sequences designated with solid arrows. The inverted repeat sequences a1 and a2 are represented by open arrows, the primary retron transcript by the dashed line, msdRNA by the heavy line, and the branch point G is circled. (B) Organization of multicopy single-stranded DNA. DNA is represented by the dotted line and msdRNA by the solid line. [This figure was adapted from Temin (1989).]

son *et al.*, 1989b; Lim and Maas, 1989) (Fig. 4). Each locus contains sequences homologous to msDNA (*msd*), msdRNA (*msr*), and an *rt* open reading frame. These sequences are present in a genome-length transcript that initiates upstream of *msr*. The *msd* and *msr* sequences are on opposite strands of the DNA and their 3′ ends overlap for 6–11 nt (Dhundale *et al.*, 1987). The *rt* gene is located downstream of *msd* and *msr* (Inouye *et al.*, 1989; Lampson *et al.*, 1989b; Lim and Maas, 1989). The N-terminal one-third of the reverse transcriptase amino acid sequence is not homologous with known proteins and also is quite variable between retron isolates (Inouye *et al.*, 1989; Xiong and Eickbush, 1990). The rest of the protein can be aligned with known reverse transcriptases and is more highly conserved among retron isolates. Recently, a reverse transcriptase–msDNA complex has been purified from *E. coli* (Lampson *et al.*, 1990). The predominant protein species copurifying with the complex is 65 kDa, the predicted size for retron *rt*. The complex can synthesize cDNA using an exogenously provided substrates and primers. Genetic studies have shown that *msd*, *msr*, and *rt* are required for msDNA synthesis in *E. coli* (Lampson *et al.*, 1989b; Lim and Maas, 1989).

Initial analyses of the chromosomal retron locus in *E. coli* did not reveal any obvious sequence motifs that might suggest how the transposition/integration occurs. Further definition of the chromosomal locus of retron Ec67 has dramatically changed previously held views on how retrons "transpose" (Hsu *et al.*, 1990). A possible 26-nt target-site duplication has been found for retron Ec67. Surprisingly, 34 kb of novel sequence lies between the presumed target-site duplications. Recently, Inouye *et al.* have shown that a related retron called Ec73 is part of a 12.7-kb P4-like bacteriophage that can be mobilized by a P2 helper phage (1991). Given these results, retron Ec67 could be part of another integrated phage. Retronphage ϕR73 is transmissible and integrates into the *selC* tRNA gene to form stable lysogens. Thus, the target-site duplication is probably produced when ϕR73 integrates into the genome by site-specific homologous recombination. Newly integrated retronphage produce msDNA-Ec73. Although these results show that retrons can be transmitted intercellularly, it is not clear how the retron entered the phage genome, or what its role is in the phage life cycle. Further insights into retron biology will certainly be forthcoming in the future.

A model has been proposed for msDNA synthesis that is based on intermediate structures isolated from cells and the initial characterization of the genomic locus (Dhundale *et al.*, 1987; Lampson *et al.*, 1989a; Lim and Maas, 1989) (Fig. 4). A precursor transcript containing *msr*, *msd*, and *rt* initiates upstream of *msr* and apparently is required for msDNA synthesis. Short inverted repeats bracketing *msr* and *msd* (a1 and a2) may allow the primary transcript to fold into a structure that places the rG in the correct position for priming cDNA synthesis (Dhundale *et al.*, 1987; Lampson *et al.*, 1989a). The priming reaction occurs after the 2′–5′ linkage is made to the first deoxyribonucleotide, but it is not known how this reaction is catalyzed or if the retron reverse transcriptase can prime synthesis from a 2′-OH. There are probably two types of RNA-processing events that take place during or after elongation. The first type trims nucleotides from the ends of the precursor RNA, while the second is an RNase H activity that degrades RNA from the DNA/RNA intermediate.

VII. CONCLUDING REMARKS

Retroelements present in microorganisms are being studied at two levels. First, new and unusual elements continue to be found in a variety of eukaryotes and prokaryotes. A central concern in these studies has been the choice of strains used to search for new elements. Many laboratory stocks may not contain retroelements, because they represent a small part of the genetic diversity of a given species. If retroelements can be transmitted horizontally, recent or selective retroelement acquisition may result in a very limited distribution. Three general approaches used for isolating retroelements are illustrated in this chapter. (1) Genome diversity can be increased by sampling natural populations. This approach has been exploited to isolate Tad elements from *Neurospora*. (2) Elements can be identified from middle-repeat sequences of the genome. (3) Elements can also be isolated because they insert near other genes of interest.

The second area of study concerns the mechanism of retroelement movement. Elements found in microorganisms such as *S. cerevisiae*, *S. pombe*, *N. crassa*, and *E. coli* offer obvious advantages for understanding the detailed mechanism of retroelement replication. Their small size, rapid generation time, and facile genetic systems allow questions to be addressed that are much more difficult to approach in other organisms. Foremost among these are interactions between retroelement and host genes that modulate the transposition process and genome evolution. Answers to these questions may help develop effective therapies for pathogenic retroelements such as human immunodeficiency virus.

ACKNOWLEDGMENTS. I thank T. Burkett, M. J. Curcio, and G. Sharon for critical reading of this manuscript, and the researchers that provided unpublished data. Research was sponsored by the National Cancer Institute, DHHS, under contract No. N01-CO-74101 with ABL. The contents of this publication do not necessarily reflect the views or policies of the Department of Health and Human Services, nor does mention of trade names, commercial products, or organizations imply endorsement by the U. S. Government.

By acceptance of this article, the publisher or recipient acknowledges the right of the U. S. Government and its agents and contractors to retain a nonexclusive, royalty-free license in and to any copyright covering the article.

VIII. REFERENCES

Abrams, E., Neigeborn, L., and Carlson, M., 1986, Molecular analysis of *SNF2* and *SNF5*, genes required for expression of glucose-repressible genes in *Saccharomyces cerevisiae*, *Mol. Cell. Biol.* **6**:3643.

Adams, S. E., Mellor, J., Gull, K., Sim, R. B., Tuite, M. F., Kingsman, S. M., and Kingsman, A. J., 1987, The functions and relationships of Ty-VLP proteins in yeast reflect those of mammalian retroviral proteins, *Cell* **49**:111.

Affolter, M., Rindisbacher, L., and Braun, R., 1989, The tubulin gene cluster of *Trypanosoma brucei* starts with an intact β-gene and ends with a truncated β-gene interrupted by a retrotransposon-like sequence, *Gene* **80**:177.

Akins, R. A., Kelley, R. L., and Lambowitz, A. M., 1986, Mitochondrial plasmids of *Neurospora*: Integration into mitochondrial DNA and evidence for reverse transcription in mitochondria, *Cell* **47**:505.

Akins, R. A., Grant, D. M., Stohl, L. L., Bottorff, D. A., Nargang, F. E., and Lambowitz, A. M., 1988, Nucleotide sequence of the Varkud mitochondrial plasmid of *Neurospora* and synthesis of a hybrid transcript with a 5' leader derived from mitochondrial RNA, *J. Mol. Biol.* **204**:1.

Akins, R. A., Kelley, R. L., and Lambowitz, A. M., 1989, Characterization of mutant mitochondrial plasmids of *Neurospora* spp that have incorporated tRNAs by reverse transcription, *Mol. Cell. Biol.* **9**:678.

Aksoy, S., Williams, S., Chang, S., and Richards, F. F., 1990, SLACS retrotransposon from *Trypanosoma brucei gambiense* is similar to mammalian LINEs, *Nucleic Acids Res.* **18**:785.

Baltimore, D., 1985, Retroviruses and retrotransposons: The role of reverse transcription in shaping the eukaryotic genome, *Cell* **40**:481.

Belcourt, M. F., and Farabaugh, P. J., 1990, Ribosomal frameshifting in the yeast retrotransposon Ty: tRNAs induce slippage on a 7 nucleotide minimal site, *Cell* **62**:339.

Belfort, M., 1989, Bacteriophage introns: Parasites within parasites?, *Trends Genet.* **5**:209.

Boeke, J. D., 1988, Retrotransposons, in: *RNA Genetics, Retroviruses, Viroids, and RNA Recombination* (E. Domingo, J. Holland, and P. Ahlquist, eds.), pp. 59–103, CRC Press, Boca Raton, Florida.

Boeke, J. D., 1989, Transposable elements in *Saccharomyces cerevisiae*, in: *Mobile DNA* (D. E. Berg and M. M. Howe, eds.), pp. 335–374, American Society for Microbiology, Washington, D. C.

Boeke, J. D., 1990, Reverse transcriptase, the end of the chromosome, and the end of life, *Cell* **61**:193.

Boeke, J. D., and Chapman, K. B., 1991, Retrotransposition mechanisms, *Curr. Opin. Cell Biol.* **3**:502.

Boeke, J. D., and Corces, V. G., 1989, Transcription and reverse transcription of retrotransposons, *Annu. Rev. Microbiol.* **43**:403.

Boeke, J. D., and Garfinkel, D. J., 1988, Yeast Ty elements as retroviruses, in: *Viruses of Fungi and Simple Eukaryotes* (Y. Koltin and M. J. Leibowitz, eds.), pp. 15–39, Marcel Dekker, New York.

Boeke, J. D., and Sandmeyer, S. B., 1992, Yeast transposable elements, in: The *Molecular and Cellular Biology of the Yeast Saccharomyces: Genome Dynamics, Protein Synthesis, and Energetics* (J. R. Broach, J. Pringle, and E. Jones, eds.), pp. 193–261, Cold Spring Harbor Laboratory, Cold Spring Harbor, New York.

Boeke, J. D., Garfinkel, D. J., Styles, C. A., and Fink, G. R., 1985, Ty elements transpose through an RNA intermediate, *Cell* **40**:491.

Boeke, J. D., Styles, C. A., and Fink, G. R., 1986, *Saccharomyces cerevisiae SPT3* gene is required for transposition and transpositional recombination of chromosomal Ty elements, *Mol. Cell. Biol.* **6**:3575.

Boeke, J. D., Eichinger, D., Castrillon, D., and Fink, G. R., 1988a, The *Saccharomyces cerevisiae* genome contains functional and nonfunctional copies of transposon Ty1, *Mol. Cell. Biol.* **8**:1431.

Boeke, J. D., Xu, H., and Fink, G. R., 1988b, A general method for the chromosomal amplification of genes in yeast, *Science* **239**:280.

Boeke, J. D., Eichinger, D., and Natsoulis, G., 1991, Doubling the Ty1 element copy number in *Saccharomyces cerevisiae*: Host genome stability and phenotypic effects, *Genetics* **129**:1043.

Bonitz, S., Coruzzi, G., Thalenfeld, B., Tzagoloff, A., and Macino, G., 1980, Assembly of the mitochondrial membrane system. Structure and nucleotide sequence of the gene coding for subunit 1 of yeast cytochrome oxidase, *J. Biol. Chem.* **255**:11927.

Bradshaw, V. A., and McEntee, K., 1989, DNA damage activates transcription and transposition of yeast Ty retrotransposons, *Mol. Gen. Genet.* **218**:465.

Brand, A. H., Breeden, L., Abraham, J., Sternglanz, R., and Nasmyth, K., 1985, Characterization of a "silencer" in yeast: A DNA sequence with properties opposite to those of a transcriptional enhancer, *Cell* **41**:41.

Brierley, C., and Flavell, A. J., 1990, The retrotransposon *copia* controls the relative levels of its gene products post-transcriptionally by differential expression from its two major mRNAs, *Nucleic Acids Res.* **18**:2947.

Brodeur, G. M., Sandmeyer, S. B., and Olson, M. V., 1983, Consistent association between sigma elements and tRNA genes in yeast, *Proc. Natl. Acad. Sci. USA* **80**:3292.

Brosius, J., 1991, Retroposons—Seeds of evolution, *Science* **251**:753.

Brown, P. O., Bowerman, B., Varmus, H. E., and Bishop, J. M., 1987, Correct integration of retroviral DNA *in vitro*, *Cell* **49**:347.

Burke, J. M., Belfort, M., Cech, T. R., Davies, R. W., Schweyen, R. J., Shub, D. A., Szostak, J. W., and Tabak, H. F., 1987, Structural conventions for group I introns, *Nucleic Acids Res.* **15**:7217.

Cameron, J. R., Loh, E. Y., and Davis, R. W., 1979, Evidence for transposition of dispersed repetitive DNA families in yeast, *Cell* **16**:739.

Cappello, J., Cohen, S. M., and Lodish, H. F., 1984, *Dictyostelium* transposable element DIRS-1 preferentially inserts into DIRS-1 sequences, *Mol. Cell. Biol.* **4**:2207.

Cappello, J., Handelsman, K., and Lodish, H. F., 1985, Sequence of *Dictyostelium* DIRS-1: An apparent retrotransposon with inverted terminal repeats and an internal circle junction sequence, *Cell* **43**:105.

Carrington, M., Roditi, I., and Williams, R. O., 1987, The structure and transcription of an element interspersed between tandem arrays of mini-exon donor RNA genes in *Trypanosoma brucei*, *Nucleic Acids Res.* **15**:10179.

Cech, T. R., 1985, Self-splicing RNA: Implications for evolution, *Int. Rev. Cytol.* **93**:3.

Cech, T. R., 1988, Conserved sequence and structures of group I introns: Building an active site for RNA catalysis—A review, *Gene* **73**:259.

Cech, T. R., and Bass, B. L., 1986, Biological catalysis by RNA, *Annu. Rev. Biochem.* **55**:599.

Chaleff, D. T., and Fink, G. R., 1980, Genetic events associated with an insertion mutation in yeast, *Cell* **21**:227.

Chalker, D. L., and Sandmeyer, S. B., 1990, Transfer RNA genes are genomic targets for *de novo* transposition of the yeast retrotransposon Ty3, *Genetics* **126**:837.

Chapman, K. B., and Boeke, J. D., 1991, Isolation and characterization of the gene encoding yeast debranching enzyme, *Cell* **65**:483.

Chisholm, G. E., Genbauffe, F. S., and Cooper, T. G., 1984, Tau, a repeated DNA sequence in yeast, *Proc. Natl. Acad. Sci. USA* **81**:2965.

Cigan, M., and Donahue, T. F., 1986, The methionine initiator tRNA genes of yeast, *Gene* **41**:343.

Ciriacy, M., and Williamson, V. W., 1981, Analysis of mutations affecting Ty-mediated gene expression in *Saccharomyces cerevisiae*, *Mol. Gen. Genet.* **182**:159.

Clare, J., and Farabaugh, P. J., 1985, Nucleotide sequence of a yeast Ty element: Evidence for an unusual mechanism of gene expression, *Proc. Natl. Acad. Sci. USA* **82**:2828.

Clare, J. J., Belcourt, M., and Farabaugh, P. J., 1988, Efficient translational frameshifting occurs within a conserved sequence of the overlap between the two genes of a yeast Ty1 transposon, *Proc. Natl. Acad. Sci. USA* **85**:6816.

Clark, D. J., Bilanchone, V. W., Haywood, L. J., Dildine, S. L., and Sandmeyer, S. B., 1988, A yeast composite element, Ty3, has properties of a retrotransposon, *J. Biol. Chem.* **263**:1413.

Clark-Adams, C. D., and Winston, F., 1987, The *SPT6* gene is essential for growth and is required for delta-mediated transcription in *Saccharomyces cerevisiae*, *Mol. Cell. Biol.* **7**:679.

Clark-Adams, C. D., Norris, D., Osley, M. A., Fassler, J. S., and Winston, F., 1988, Changes in histone gene dosage alter transcription in yeast, *Genes Dev.* **2**:150.

Coffin, J., 1984, Structure of the retroviral genome, in: *RNA Tumor Viruses,* 2nd ed. (R. Weiss, N. Feich, H. Varmus, and J. Coffin, eds.), pp. 261–369, Cold Spring Harbor Laboratory, Cold Spring Harbor, New York.

Company, M., and Errede, B., 1987, Cell-type-dependent gene activation by yeast transposon Ty1 involves multiple regulatory determinants, *Mol. Cell. Biol.* **7**:3205.

Company, M., Adler, C., and Errede, B., 1988, Identification of a Ty1 regulatory sequence responsive to *STE7* and *STE12, Mol. Cell. Biol.* **8**:2545.

Coney, L. R., and Roeder, G. S., 1988, Control of yeast gene expression by transposable elements: Maximum expression requires a functional Ty activator sequence and a defective Ty promoter, *Mol. Cell. Biol.* **8**:4009.

Cooper, T. G., and Chisholm, G., 1984, Position-dependent, Ty-mediated enhancement of *DUR1,2* gene expression, in: *Genome Rearrangement* (M. Simon and I. Herskowitz, eds.), pp. 289–303, Liss, New York.

Craigie, R., Fujiwara, T., and Bushman, F., 1990, The *IN* protein of Moloney murine leukemia virus processes the viral DNA ends and accomplishes their integration *in vitro, Cell* **62**:829.

Curcio, M. J., and Garfinkel, D. J., 1991, Single-step selection for Ty1 element retrotransposition, *Proc. Natl. Acad. Sci. USA* **88**:936.

Curcio, M. J., Sanders, N. J., and Garfinkel, D. J., 1988, Transpositional competence and transcription of endogenous Ty elements in *Saccharomyces cerevisiae:* Implications for regulation of transposition, *Mol. Cell. Biol.* **8**:3571.

Curcio, M. J., Hedge, A.-M., Boeke, J. D., and Garfinkel, D. J., 1990, Ty RNA levels determine the spectrum of retrotransposition events that activate gene expression in *Saccharomyces cerevisiae, Mol. Gen. Genet.* **220**:213.

Dawkins, R., 1976, *The Selfish Gene,* Oxford University Press, New York.

Deleu, C., Turcq, B., and Begueret, J., 1990, *repa,* a repetitive and dispersed DNA sequence of the filamentous fungus *Podospora anserina, Nucleic Acids Res.* **18**:4901.

Delrey, F. J., Donahue, T. F., and Fink, G. R., 1982, Sigma, a repetitive element found adjacent to tRNA genes of yeast, *Proc. Natl. Acad. Sci. USA* **79**:4138.

Denis, C. L., and Malvar, T., 1990, The *CCR4* gene from *Saccharomyces cerevisiae* is required for both nonfermentative and *spt*-mediated gene expression, *Genetics* **124**:283.

Derr, L. K., Strathern, J. N., and Garfinkel, D. J., 1991, RNA-mediated recombination in *S. cerevisiae, Cell* **67**:355.

Dhundale, A., Lampson, B., Furuichi, T., Inouye, M., and Inouye, S., 1987, Structure of msDNA from *Myxococcus xanthus:* Evidence for a long, self-annealing RNA precursor for the covalently linked, branched RNA, *Cell* **51**:1105.

Dhundale, A., Furuichi, T., Inouye, M., and Inouye, S., 1988a, Mutations that affect production of branched RNA-linked msDNA in *Myxococcus xanthus, J. Bacteriol.* **170**:5620.

Dhundale, A., Inouye, M., and Inouye, S., 1988b, A new species of multicopy single-stranded DNA from *Myxococcus xanthus* with conserved structural features, *J. Biol. Chem.* **263**:9055.

Doolittle, W. F., and Sapienza, C., 1980, Selfish genes, the phenotype paradigm of genome evolution, *Nature* **284**:601.

Doolittle, R. F., Feng, D. F., Johnson, M. S., and McClure, M. A., 1989, Origins and evolutionary relationships of retroviruses, *Q. Rev. Biol.* **64**:1.

Dreher, T. W., and Hall, T. C., 1988, Mutational analysis of the sequence and structural requirements in brome mosaic virus RNA for minus strand promoter activity, *J. Mol. Biol.* **201**:31.

Dubois, E., Jacobs, E., and Jauniaux, J. C., 1982, Expression of the *ROAM* mutations in *Saccharomyces cerevisiae:* Involvement of *trans*-acting regulatory elements and relation with the Ty1 transcription, *EMBO J.* 1:1133.

Dujon, B., 1989, Group I introns as mobile genetic elements: Facts and mechanistic speculations—A review, *Gene* 82:91.

Eibel, H., and Philippsen, P., 1984, Preferential integration of yeast transposable element Ty into a promoter region, *Nature* 307:386.

Eibel, H., Gafner, J., Stotz, A., and Philippsen, P., 1981, Characterization of the yeast mobile genetic element Ty1, *Cold Spring Harbor Symp. Quant. Biol.* 45:609.

Eichinger, D. J., and Boeke, J. D., 1988, The DNA intermediate in yeast Ty1 element transposition copurifies with virus-like particles: Cell-free Ty1 transposition, *Cell* 54:955.

Eichinger, D. J., and Boeke, J. D., 1990, A specific terminal structure is required for Ty1 transposition, *Genes Dev.* 4:324.

Eigel, A., and Feldmann, H., 1982, Ty1 and delta elements occur adjacent to several tRNA genes in yeast, *EMBO J.* 1:1245.

Eisenmann, D. M., Dollard, C., and Winston, F., 1989, *SPT15*, the gene encoding the yeast TATA binding factor TFIID, is required for normal transcription initiation *in vivo*, *Cell* 58:1183.

Elder, R. T., St. John, T. P., Stinchcomb, D. T., and Davis, R. W., 1980, Studies on the transposable element Ty1 of yeast. I. RNA homologous to Ty1, *Cold Spring Harbor Symp. Quant. Biol.* 45:581.

Elder, R. T., Loh, E. Y., and Davis, R. W., 1983, RNA from the yeast transposable element Ty1 has both ends in the direct repeats, a structure similar to retrovirus RNA, *Proc. Natl. Acad. Sci. USA* 80:2432.

El-Sherbeini, M., and Bostian, K., 1987, Viruses in fungi: Infection of yeast with the K1 and K2 killer viruses, *Proc. Natl. Acad. Sci. USA* 84:4293.

Errede, B., and Ammerer, G., 1989, *STE12*, a protein involved in cell-type-specific transcription and signal transduction in yeast, is part of protein–DNA complexes, *Genes Dev.* 3:1349.

Errede, B., Cardillo, T. S., Sherman, F., Dubois, E., Deschamps, J., and Wiame, J. M., 1980a, Mating signals control expression of mutations resulting from insertion of a transposable repetitive element adjacent to diverse yeast genes, *Cell* 22:427.

Errede, B., Cardillo, T. S., Wever, G., and Sherman, F., 1980b, Studies on transposable elements in yeast. I. *ROAM* mutations causing increased expression of yeast genes: Their activation by signals directed toward conjugation functions and their formation by insertion of Ty1 repetitive elements, *Cold Spring Harbor Symp. Quant. Biol.* 45:593.

Errede, B., Company, M., Ferchak, J. D., Hutchison, C. A., and Yarnell, W. S., 1985, Activation regions in a yeast transposon have homology to mating type control sequences and to mammalian enhancers, *Proc. Natl. Acad. Sci. USA* 82:5423.

Errede, B., Company, M., and Swanstrom, R., 1986, An anomalous Ty1 structure attributed to an error in reverse transcription, *Mol. Cell. Biol.* 6:1334.

Errede, B., Company, M., and Hutchison, C. A., 1987, Ty1 sequence with enhancer and mating-type-dependent regulatory activities, *Mol. Cell. Biol.* 7:258.

Farabaugh, P. J., and Fink, G. R., 1980, Insertion of the eukaryotic transposable element Ty creates a 5-base pair duplication, *Nature* 286:352.

Farabaugh, P., Liao, X.-B., Belcourt, M., Zhao, H., Kapakos, J., and Clare, J., 1989, Enhancer and silencerlike sites within the transcribed portion of a Ty2 transposable element of *Saccharomyces cerevisiae*, *Mol. Cell. Biol.* 9:4824.

Fassler, J. S., and Winston, F., 1988, Isolation and analysis of a novel class of suppressor of Ty insertion mutations in *Saccharomyces cerevisiae*, *Genetics* 118:203.

Fink, G. R., 1987, Pseudogenes in yeast?, *Cell* **49**:5.

Fink, G. R., Boeke, J. D., and Garfinkel, D. J., 1986, The mechanism and consequences of retrotransposition, *Trends Genet.* **5**:118.

Finnegan, D. J., Rubin, G. S., Young, M. W., and Hogness, D. S., 1978, Repeated gene families in *Drosophila melanogaster, Cold Spring Harbor Symp. Quant. Biol.* **42**:1053.

Firtel, R. A., 1989, Mobile genetic elements in the cellular slime mold *Dictyostelium discoidium*, in: *Mobile DNA* (D. E. Berg and M. M. Howe, eds.), pp. 557–567, American Society for Microbiology, Washington, D. C.

Fuetterer, J., and Hohn, T., 1987, Involvement of nucleocapsids in reverse transcription: A general phenomenon?, *Trends Biochem. Sci.* **12**:92.

Fujiwara, T., and Mizuuchi, K., 1988, Retroviral DNA integration: Structure of an integration intermediate, *Cell* **54**:497.

Fulton, A. M., Mellor, J., Dobson, M. J., Chester, J., Warmington, J. R., Indge, K. J., Oliver, S. G., de la Paz, P., Wilson, W., Kingsman, A. J., and Kingsman, S. M., 1985, Variants within the yeast Ty sequence family encode a class of structurally conserved proteins, *Nucleic Acids Res.* **13**:4097.

Fulton, A. M., Rathjen, P. D., Kingsman, S. M., and Kingsman, A. J., 1988, Upstream and downstream transcriptional control signals in the yeast retrotransposon, Ty, *Nucleic Acids Res.* **16**:5439.

Furuichi, T., Dhundale, A., Inouye, M., and Inouye, S., 1987a, Branched RNA covalently linked to the 5′ end of a single-stranded DNA in *Stigmatella aurantiaca:* Structure of ms DNA, *Cell* **48**:47.

Furuichi, T., Inouye, S., and Inouye, M., 1987b, Biosynthesis and structure of stable branched RNA covalently linked to the 5′ end of multicopy single-stranded DNA of *Stigmatella aurantiaca*, *Cell* **48**:55.

Gabriel, A., and Boeke, J. D., 1991, Reverse transcriptase encoded by a retrotransposon from the trypanosomatid *Crithidia fasciculata, Proc. Natl. Acad. Sci. USA* **88**:9794.

Gabriel, A., Yen, T. J., Schwartz, D. C., Smith, C. L., Boeke, J. D., Sollner-Webb, B., and Cleveland, D. W., 1990, A rapidly rearranging retrotransposon within the miniexon gene locus of *Crithidia fasciculata, Mol. Cell. Biol.* **10**:615.

Gafner, J., and Philippsen, P., 1980, The yeast transposon Ty1 generates duplications of target DNA on insertion, *Nature* **286**:414.

Gafner, J., DeRobertis, E. M., and Philippsen, P., 1983, Delta sequences in the 5′ noncoding region of yeast tRNA genes, *EMBO J.* **2**:583.

Garfinkel, D. J., Boeke, J. D., and Fink, G. R., 1985, Ty element transposition: Reverse transcriptase and virus-like particles, *Cell* **42**:507.

Garfinkel, D. J., Curcio, M. J., Youngren, S. D., and Sanders, N. J., 1988a, The biology and exploitation of the retrotransposon Ty in *Saccharomyces cerevisiae, Genome* **31**:909.

Garfinkel, D. J., Mastrangelo, M. F., Sanders, N. J., Shafer, B. K., and Strathern, J. N., 1988b, Transposon tagging using Ty elements in yeast, *Genetics* **120**:95.

Garfinkel, D. J., Hedge, A.-M., Youngren, S. D., and Copeland, T. D., 1991, Proteolytic processing of *pol-TYB* proteins from the yeast retrotransposon Ty1, *J. Virol.* **65**:4573.

Gargouri, A., Laxowska, J., and Slonimski, P., 1983, DNA-splicing of introns in the gene: A general way of reverting intron mutations, in: *Mitochondria 1983. Nucleo-Mitochondrial Interactions* (R. Schweyen, K. Wolf, and F. Kaudewitz, eds.), pp. 259–268, de Gruyter, Berlin.

Genbauffe, F. S., Chisholm, G. E., and Cooper, T. G., 1984, Tau, sigma, and delta, *J. Biol. Chem.* **259**:10518.

Gilboa, E., Mitra, S. W., Goff, S., and Baltimore, D., 1979, A detailed model of reverse transcription and a test of crucial aspects, *Cell* **18**:93.

Giroux, C. N., Mis, J. R. A., Pierce, M. K., Kohalmi, S. E., and Kunz, B. A., 1988, DNA sequence analysis of spontaneous mutations in the *SUP4-0* gene of *Saccharomyces cerevisiae*, *Mol. Cell. Biol.* **8:**978.

Goebl, M. G., and Petes, T. D., 1986, Most of the yeast genomic sequences are not essential for cell growth and division, *Cell* **46:**983.

Goel, A., and Pearlman, R. E., 1988, Transposable element-mediated enhancement of gene expression in *Saccharomyces cerevisiae* involves sequence-specific binding of a *trans*-acting factor, *Mol. Cell. Biol.* **8:**2572.

Hansen, L. J., and Sandmeyer, S. B., 1990, Characterization of a transpositionally active Ty3 element and identification of the Ty3 integrase protein, *J. Virol.* **64:**2599.

Hansen, L. J., Chalker, D. L., and Sandmeyer, S. B., 1988, Ty3, a yeast retrotransposon associated with tRNA genes, has homology to animal retroviruses, *Mol. Cell. Biol.* **8:**5245.

Happel, A. M., Swanson, M. S., and Winston, F., 1991, The *SNF2, SNF5* and *SNF6* genes are required for Ty transcription in *Saccharomyces cerevisiae*, *Genetics* **128:**69.

Hasan, G., Turner, M. J., and Cordingley, J. S., 1984, Complete nucleotide sequence of an unusual mobile element from *Trypanosoma brucei*, *Cell* **37:**333.

Hauber, J., Nelbock-Hochstetter, P., and Feldman, H., 1985, Nucleotide sequence and characteristics of a Ty element from yeast, *Nucleic Acids Res.* **13:**2745.

Hauber, J., Stucka, R., Kreig, R., and Feldman, H., 1988, Analysis of yeast chromosomal regions carrying members of the glutamate tRNA gene family: Various transposable elements are associated with them, *Nucleic Acids Res.* **16:**10623.

Haynes, R. H., and Kunz, B. A., 1981, DNA repair and mutagenesis in yeast, in: *The Molecular Biology of the Yeast Saccharomyces. Life Cycle and Inheritance* (J. N. Strathern, E. W. Jones, and J. R. Broach, eds.), pp. 371–414, Cold Spring Harbor Laboratory, Cold Spring Harbor, New York.

Heidmann, T., Heidmann, O., and Nicolas, J.-F., 1988, An indicator gene to demonstrate intracellular transposition of defective retroviruses, *Proc. Natl. Acad. Sci. USA* **85:**2219.

Herskowitz, I., 1989, A regulatory hierarchy for cell specialization in yeast, *Nature* **342:**749.

Hill, J., McGraw, P., and Tzagoloff, A., 1985, A mutation in yeast mitochondrial DNA results in a precise excision of the terminal intron of the cytochrome b gene, *J. Biol. Chem.* **260:**3235.

Hirschman, J. E., and Winston, F., 1988, *SPT3* is required for normal levels of a-factor and α-factor expression in *Saccharomyces cerevisiae*, *Mol. Cell. Biol.* **8:**822.

Hirschman, J. E., Durbin, K. J., and Winston, F., 1988, Genetic evidence for promoter competition in *Saccharomyces cerevisiae*, *Mol. Cell. Biol.* **8:**4608.

Hsu, M.-Y., Inouye, M., and Inouye, S., 1990, Retron for the 67-base multicopy single-stranded DNA from *Escherichia coli*: A potential transposable element encoding both reverse transcriptase and Dam methylase functions, *Proc. Natl. Acad. Sci. USA* **87:**9454.

Hu, W.-S., and Temin, H. M., 1990, Retroviral recombination and reverse transcription, *Science*, **250:**1227.

Hull, R., and Will, H., 1989, Molecular biology of viral and nonviral retroelements, *Trends Genet.* **5:**357.

Iida, H., 1988, Multistress resistance of *Saccharomyces cerevisiae* is generated by insertion of retrotransposon Ty into the 5' coding region of the adenylate cyclase gene, *Mol. Cell. Biol.* **8:**5555.

Inouye, S., Hsu, M.-Y., Eagle, S., and Inouye, M., 1989, Reverse transcriptase associated with the biosynthesis of the branched RNA-linked msDNA in *Myxococcus xanthus*, *Cell* **56:**709.

Inouye, S., Herzer, P. J., and Inouye, M., 1990, Two independent retrons with highly diverse reverse transcriptases in *Myxococcus xanthus, Proc. Natl. Acad. Sci. USA* **87**:942.

Inouye, S., Sunshine, M. G., Six, E. W., and Inouye, M., 1991, Retronphage φR73: An *E. coli* phage that contains a retroelement and integrates into a tRNA gene, *Science* **252**:969.

Jacks, T., and Varmus, H. E., 1985, Expression of the Rous sarcoma virus *pol* gene by ribosomal frameshifting, *Science* **230**:1237.

Jacks, T., Townsley, K., Varmus, H. E., and Majors, J., 1987, Two efficient ribosomal frameshifting events are required for synthesis of mouse mammary tumor virus *gag*-related polyproteins, *Proc. Natl. Acad. Sci. USA* **84**:4298.

Jacks, T., Madhani, H. D., Masiarz, F. R., and Varmus, H. E., 1988a, Signals for ribosomal frameshifting in the Rous sarcoma virus *gag–pol* region, *Cell* **55**:447.

Jacks, T., Power, M. D., Masiarz, F. R., Luciw, P. A., Barr, P. J., and Varmus, H. E., 1988b, Characterization of ribosomal frameshifting in HIV-1 *gag–pol* expression, *Nature* **331**:280.

Jauniaux, J. C., Dubois, E., Crabeel, M., and Wiame, J. M., 1981, DNA and RNA analysis of arginase regulatory mutants in *Saccharomyces cerevisiae, Arch. Int. Physiol. Biochim.* **89**:B111.

Jauniaux, J. C., Dubois, E., Vissers, S., Crabeel, M., and Wiame, J. M., 1982, Molecular cloning, DNA structure, and RNA analysis of the arginase gene in *Saccharomyces cerevisiae*. A study of *cis*-dominant regulatory mutants, *EMBO J.* **1**:1125.

Jentsch, S., McGrath, J. P., and Varshavsky, A., 1987, The yeast DNA repair gene *RAD6* encodes a ubiquitin-conjugating enzyme, *Nature* **329**:131.

Johnson, M. S., McClure, M. A., Feng, D.-F., Gray, J., and Doolittle, R. F., 1986, Computer analysis of retroviral *pol* genes: Assignment of enzymatic functions to specific sequences and homologies with nonviral enzymes, *Proc. Natl. Acad. Sci. USA* **83**:7648.

Katz, R. A., Merkel, G., Kulkosky, J., Leis, J., and Skalka, A. M., 1990, The avian retroviral *IN* protein is both necessary and sufficient for integrative recombination *in vitro, Cell* **63**:87.

Kimmel, B. E., Ole-Moiyoi, O. K., and Young, J. R., 1987, Ingi, a 5.2-kb dispersed sequence element from *Trypanosoma brucei* that carries half of a smaller mobile element at either end and has homology with mammalian LINEs, *Mol. Cell. Biol.* **7**:1465.

Kingsman, A. J., and Kingsman, S. M., 1988, Ty: A retroelement moving forward, *Cell* **53**:333.

Kingsman, A. J., Gimlich, R. L., Clark, L., Chinault, A. C., and Carbon, J. A., 1981, Sequence variation in dispersed repetitive sequences in *Saccharomyces cerevisiae, J. Mol. Biol.* **145**:619.

Kinsey, J. A., 1989, Restricted distribution of the Tad transposon in strains of *Neurospora, Curr. Genet.* **15**:271.

Kinsey, J. A., 1990, *Tad*, a LINE-like transposable element of *Neurospora*, can transpose between nuclei in heterokaryons, *Genetics* **126**:317.

Kinsey, J. A., and Helber, J., 1989, Isolation of a transposable element from *Neurospora crassa, Proc. Natl. Acad. Sci. USA* **86**:1929.

Klar, A. J. S., 1989, The interconversion of yeast mating type: *Saccharomyces cerevisiae* and *Schizosaccharomyces pombe*, in: *Mobile DNA* (D. E. Berg and M. M. Howe, eds.), pp. 671–693, American Society for Microbiology, Washington, D. C.

Kuiper, M. T. R., and Lambowitz, A. M., 1988, A novel reverse transcriptase activity associated with mitochondrial plasmids of *Neurospora, Cell* **55**:693.

Kuiper, M. T., Sabourin, J. R., and Lambowitz, A. M., 1990, Identification of the reverse transcriptase encoded by the Mauriceville and Varkud mitochondrial plasmids of *Neurospora, J. Biol. Chem.* **265**:6936.

Kunz, B. A., Peters, M. G., Kohalmi, S. E., Armstrong, J. D., Glattke, M., and Badiani, K., 1989, Disruption of the *RAD52* gene alters the spectrum of spontaneous *SUP4-0* mutations in *Saccharomyces cerevisiae, Genetics* **122**:535.

Laimins, L., Holmgren-Konig, M., and Khoury, G., 1986, Transcriptional *silencer* element in rat repetitive sequences associated with the rat insulin 1 gene locus, *Proc. Natl. Acad. Sci. USA* **83**:3151.

Laloux, I., Dubois, E., Dewerchin, M., and Jacobs, E., 1990, *TEC1*, a gene involved in the activation of Ty1 and Ty1-mediated gene expression in *Saccharomyces cerevisiae:* Cloning and molecular analysis, *Mol. Cell. Biol.* **10**:3541.

Lambowitz, A. M., 1989, Infectious introns, *Cell* **56**:323.

Lampson, B. C., Inouye, M., and Inouye, S., 1989a, Reverse transcriptase with concomitant ribonuclease H activity in the cell-free synthesis of branched RNA-linked msDNA of *Myxococcus xanthus, Cell* **56**:701.

Lampson, B. C., Sun, J., Hsu, M.-Y., Vallejo-Ramierz, J., Inouye, S., and Inouye, M., 1989b, Reverse transcriptase in a clinical strain of *Escherichia coli:* Production of branched RNA-linked msDNA, *Science* **243**:1033.

Lampson, B. C., Viswanathan, M., Inouye, M., and Inouye, S., 1990, Reverse transcriptase from *Escherichia coli* exists as a complex with msDNA and is able to synthesize double-stranded DNA, *J. Biol. Chem.* **265**:8490.

Lemoine, Y., Dubois, E., and Wiame, J.-M., 1978, The regulation of urea amidolyase of *Saccharomyces cerevisiae*. Mating type influence on a constitutivity mutation acting in *cis, Mol. Gen. Genet.* **166**:251.

Levin, H. L., Weaver, D. C., and Boeke, J. D., 1990, Two related families of retrotransposons from *Schizosaccharomyces pombe, Mol. Cell. Biol.* **10**:6791.

Levra-Juillet, E., Boulet, A., Seraphin, B., Simin, M., and Faye, G., 1989, Mitochondrial introns aI1 and/or aI2 are needed for the *in vivo* deletion of intervening sequences, *Mol. Gen. Genet.* **217**:168.

Liao, X.-B., Clare, J. J., and Farabaugh, P. J., 1987, The UAS site of a Ty2 element of yeast is necessary but not sufficient to promote maximal transcription of the element, *Proc. Natl. Acad. Sci. USA* **84**:8520.

Lim, D., and Maas, W. K., 1989, Reverse transcriptase-dependent synthesis of a covalently linked, branched DNA–RNA compound in *E. coli* B, *Cell* **56**:891.

Linial, M., and Blair, D., 1984, Genetics of retroviruses, in: *RNA Tumor Viruses*, 2nd ed. (R. Weiss, N. Feich, H. Varmus, and J. Coffin, eds.), pp. 147–187, Cold Spring Harbor Laboratory, Cold Spring Harbor, New York.

Lueders, K. K., Segal, S., and Kuff, E. L., 1977, RNA sequences specifically associated with mouse intracisternal P particles, *Cell* **11**:83.

Lundblad, V., and Blackburn, E. H., 1990, RNA-dependent polymerase motifs in *EST1:* Tentative identification of a protein component of an essential yeast telomerase, *Cell* **60**:529.

Lundblad, V., and Szostak, J. W., 1989, A mutant with a defect in telomere elongation leads to senescence in yeast, *Cell* **57**:633.

Malone, E. A., Clark, C. D., and Winston, F., 1991, Mutations in *SPT16/CDC68* suppress *cis-* and *trans*-acting mutations that affect promoter function in *Saccharomyces cerevisiae, Mol. Cell. Biol.* **11**:5710.

Marschalek, R., Brechner, T., Amon-Bohm, E., and Dingermann, T., 1989, Transfer RNA genes: Landmarks for integration of mobile genetic elements in *Dictyostelium discoideum, Science* **244**:1493.

Marschalek, R., Borschet, G., and Dingermann, T., 1990, Genomic organization of the transposable element Tdd-3 from *Dictyostelium discoideum, Nucleic Acids Res.* **18**:5751.

McClanahan, T., and McEntee, K., 1984, Specific transcripts are elevated in *Saccharomyces cerevisiae* in response to DNA damage, *Mol. Cell. Biol.* **4**:2356.

McClintock, B., 1984, The significance of responses of the genome to challenge, *Science* **226**:792.

Mellor, J., Fulton, A. M., Dobson, M. J., Roberts, N. A., Wilson, W., Kingsman, A. J., and Kingsman, S. M., 1985a, The Ty transposon of *Saccharomyces cerevisiae* determines the synthesis of at least three proteins, *Nucleic Acids Res.* **13**:6249.

Mellor, J., Fulton, S. M., Dobson, M. J., Wilson, W., Kingsman, S. M., and Kingsman, A. J., 1985b, A retrovirus-like strategy for expression of a fusion protein encoded by yeast transposon Ty1, *Nature* **313**:243.

Mellor, J., Malin, M. H., Gull, K., Tuite, M. F., McCready, S. M., Finnsysesn, T., Kingsman, S. M., and Kingsman, A. J., 1985c, Reverse transcriptase activity and Ty RNA are associated with virus-like particles in yeast, *Nature* **318**:583.

Michel, F., and Lang, B. F., 1985, Mitochondrial class II introns encode proteins related to the reverse transcriptases of retroviruses, *Nature* **316**:641.

Miller, K., Rosenbaum, J., Zorzezna, V., and Pogo, A. O., 1989, The nucleotide sequence of *Drosophila melanogaster copia*-specific 2.1-kb mRNA, *Nucleic Acids Res.* **17**:2134.

Miller, W. A., Bujarski, J. J., Dreher, T. W., and Hall, T. C., 1986, Minus-strand initiation by brome mosiac virus replicase within the 3′ tRNA-like structure of native and modified RNA templates, *J. Mol. Biol.* **187**:537.

Moore, R., Dixon, M., Smith, R., Peters, G., and Dickson, C., 1987, Complete nucleotide sequence of a milk-transmitted mouse mammary tumor virus: Two frameshift suppression events are required for translation of *gag* and *pol*, *J. Virol.* **61**:480.

Morawetz, C., 1987, Effect of irradiation and mutagenic chemicals on the generation of *ADH2*-constitutive mutants in yeast. Significance for the inducibility of Ty transposition, *Mutat. Res.* **177**:53.

Morl, M., and Schmelzer, C., 1990, Integration of group II intron bI1 into a foreign RNA by reversal of the self-splicing reaction *in vitro*, *Cell* **60**:629.

Muller, F., Bruhl, K.-H., Freidel, K., Kowallik, K. V., and Ciriacy, M., 1987, Processing of Ty1 proteins and formation of Ty1 virus-like particles in *Saccharomyces cerevisiae*, *Mol. Gen. Genet.* **207**:421.

Muller, F., Laufer, W., Pott, U., and Ciriacy, M., 1991, Characterization of TY1-mediated reverse transcription in *Saccharomyces cerevisiae*, *Mol. Gen. Genet.* **226**:145.

Murphy, N. B., Pays, A., Tebabi, P., Coquelet, H., Guyaux, M., Steinert, M., and Pays, E., 1987, *Trypanosoma brucei* repeated element with unusual structural and transcriptional properties, *J. Mol. Biol.* **195**:855.

Nargang, F. E., Bell, J. B., Stohl, L. L., and Lambowitz, A. M., 1984, The DNA sequence and genetic organization of a *Neurospora* mitochondrial plasmid suggests a relationship to introns and mobile elements, *Cell* **38**:441.

Natsoulis, G., Thomas, W., Roghmann, M.-C., Winston, F., and Boeke, J. D., 1989, Ty1 transposition in *Saccharomyces cerevisiae* is nonrandom, *Genetics* **123**:269.

Neigeborn, L., and Carlson, M., 1984, Genes affecting the regulation of *SUC2* gene expression by glucose repression in *Saccharomyces cerevisiae*, *Genetics* **108**:845.

Neigeborn, L., Celenza, J. L., and Carlson, M., 1987, *SSN20* is an essential gene with mutant alleles that suppress defects in *SUC2* transcription in *Saccharomyces cerevisiae*, *Mol. Cell. Biol.* **7**:672.

Orgel, L. E., and Crick, F. H. C., 1980, Selfish DNA: The ultimate parasite, *Nature* **284**:604.

Osiewacz, H. D., and Esser, K., 1984, The mitochondrial plasmid of *Podospora anserina*: A mobile intron of a mitochondrial gene, *Curr. Genet.* **8**:299.

Oyen, T. B., and Gabrielsen, O. S., 1983, Non-random distribution of the Ty1 elements within nuclear DNA of *Saccharomyces cerevisiae*, *FEBS Lett.* **161**:201.

Paquin, C. E., and Williamson, V. M., 1984, Temperature effects of the rate of Ty transposition, *Science* **226**:53.

Paquin, C. E., and Williamson, V. M., 1986, Ty insertions at two loci account for most of the spontaneous antimycin A resistance mutations during growth at 15°C of *Saccharomyces cerevisiae* strains lacking *ADH1*, *Mol. Cell. Biol.* **6**:70.

Pearston, D. H., Gordon, M., and Hardman, N., 1985, Transposon-like properties of the major, long repetitive sequence family in the genome of *Physarum polycephalum*, *EMBO J.* **4**:3557.

Perea, J., and Jacq, C., 1985, Role of the 5′ hairpin structure in the splicing accuracy of the fourth intron of the yeast *cob–box* gene, *EMBO J.* **4**:3281.

Perlman, P. S., and Butow, R. A., 1989, Mobile introns and intron-encoded proteins, *Science*, **246**:1106.

Picologlou, S., Brown, N., and Liebman, S. W., 1990, Mutations in *RAD6*, a yeast gene encoding a ubiquitin-conjugating enzyme, stimulate retrotransposition, *Mol. Cell. Biol.* **10**:1017.

Poole, S. J., and Firtel, R. A., 1984, Genomic instability and mobile genetic elements in regions surrounding two discoidin I genes of *Dictyostelium discoideum*, *Mol. Cell. Biol.* **4**:671.

Rathjen, P. D., Kingsman, A. J., and Kingsman, S. M., 1987, The yeast *ROAM* mutation— Identification of the sequences mediating host gene activation and cell-type control in the yeast retrotransposon, Ty, *Nucleic Acids Res.* **15**:7309.

Roeder, G. S., and Fink, G. R., 1980, DNA rearrangements associated with a transposable element in yeast, *Cell* **21**:239.

Roeder, G. S., and Fink, G. R., 1982, Movement of yeast transposable elements by gene conversion, *Proc. Natl. Acad. Sci. USA* **79**:5621.

Roeder, G. S., Farabaugh, P. J., Chaleff, D. T., and Fink, G. R., 1980, The origin of gene instability in yeast, *Science* **209**:1375.

Roeder, G. S., Beard, C., Smith, M., and Keranen, S., 1985a, Isolation and characterization of the *SPT2* gene, a negative regulator of Ty-controlled yeast gene expression, *Mol. Cell. Biol.* **5**:1543.

Roeder, G. S., Rose, A. B., and Perlman, R. E., 1985b, Transposable element sequences involved in the enhancement of yeast gene expression, *Proc. Natl. Acad. Sci. USA* **82**:5428.

Rogers, J. H., 1985, The origin and evolution of retroposons, *Int. Rev. Cytol.* **93**:187.

Rogers, J. H., 1989, How were introns inserted into nuclear genes?, *Trends Genet.* **5**:213.

Rolfe, M., Spanos, A., and Banks, G., 1986, Induction of yeast Ty element transcription by ultraviolet light, *Nature* **319**:339.

Rose, M., and Winston, F., 1984, Identification of a Ty insertion within the coding sequence of the *Saccharomyces cerevisiae URA3* gene, *Mol. Gen. Genet.* **193**:557.

Rowley, A., Singer, R. A., and Johnston, G. C., 1991, *CDC68*, a yeast gene that affects regulation of cell proliferation and transcription, encodes a protein with a highly acidic carboxyl terminus, *Mol. Cell. Biol.* **11**:5718.

Ruby, S. W., and Szostak, J. W., 1985, Specific *Saccharomyces cerevisiae* genes are expressed in response to DNA-damaging agents, *Mol. Cell. Biol.* **6**:4281.

Russell, D. W., Jensen, R., Zoller, M. J., Burke, J., Errede, B., Smith, M., and Herskowitz, I., 1986, Structure of the *Saccharomyces cerevisiae HO* gene and analysis of its upstream regulatory region, *Mol. Cell. Biol.* **6**:4281.

Sandmeyer, S. B., and Olson, M. V., 1982, Insertion of a repetitive element at the same position in the 5'-flanking regions of two dissimilar yeast tRNA genes, *Proc. Natl. Acad. Sci. USA* **79**:7674.

Sandmeyer, S. B., Hansen, L. J., and Chalker, D. L., 1990, Integration specificity of retrotransposons and retroviruses, *Annu. Rev. Genet.* **24**:491.

Scherer, S., Mann, C., and Davis, R. W., 1982, Reversion of a promoter deletion in yeast, *Nature* **298**:815.

Seraphin, B., Simon, M., and Faye, G., 1988, *MSS18*, a yeast nuclear gene involved in the splicing of intron aI5β of the mitochondrial *cox1* transcript, *EMBO J.* **7**:1455.

Sharp, P. A., 1985, On the origin of RNA splicing and introns, *Cell* **42**:397.

Shimotohno, K., and Temin, H. M., 1982, Loss of intervening sequences in genomic mouse alpha-globin DNA inserted in an infectious retrovirus vector, *Nature* **229**:265.

Shippen-Lentz, D., and Blackburn, E. H., 1990, Functional evidence for an RNA template in telomerase, *Science* **247**:546.

Silverman, S. J., and Fink, G. R., 1984, Effects of Ty insertions on *HIS4* transcription in *Saccharomyces cerevisiae*, *Mol. Cell. Biol.* **4**:1246.

Simchen, G., Winston, F., Styles, C. A., and Fink, G. R., 1984, Ty-mediated expression of the *LYS2* and *HIS4* genes of *Saccharomyces cerevisiae* is controlled by the same *SPT* genes, *Proc. Natl. Acad. Sci. USA* **81**:2431.

Simsek, M., and RajBhandary, U. L., 1972, The primary structure of yeast initiator methionine tRNA, *Biochem. Biophys. Res. Commun.* **49**:508.

Sorge, J., and Hughes, S. H., 1982, Splicing of intervening sequences introduced into an infectious retroviral vector, *J. Mol. Appl. Genet.* **1**:547.

Stucka, R., Hauber, J., and Feldmann, H., 1986, Conserved and non-conserved features among the yeast Ty elements, *Curr. Genet.* **11**:193.

Stucka, R., Hauber, J., and Feldmann, H., 1987, One member of the tRNA (Glu) gene family in yeast codes for a minor GAG-tRNA(Glu) species and is associated with several short transposable elements, *Curr. Genet.* **12**:323.

Stucka, R., Lochmuller, H., and Feldmann, H., 1989, Ty4, a novel low-copy number element in *Saccharomyces cerevisiae*: One copy is located in a cluster of Ty elements and tRNA genes, *Nucleic Acids Res.* **17**:4993.

Sun, J., Herzer, P. J., Weinstein, M. P., Lampson, B. C., Inouye, M., and Inouye, S., 1989, Extensive diversity of branched-RNA-linked multicopy single-stranded DNAs in clinical strains of *Escherichia coli*, *Proc. Natl. Acad. Sci. USA* **86**:7208.

Swanson, M. S., Malone, E. A., and Winston, F., 1991, *SPT5*, an essential gene important for normal transcription in *Saccharomyces cerevisiae*, encodes an acidic nuclear protein with a carboxy-terminal repeat, *Mol. Cell. Biol.* **11**:3009.

Taguchi, A. K. W., Ciriacy, M., and Young, E. T., 1984, Carbon source dependence of transposable element-associated gene activation in *Saccharomyces cerevisiae*, *Mol. Cell. Biol.* **4**:61.

Temin, H. M., 1974, On the origin of RNA tumor viruses, *Annu. Rev. Genet.* **8**:155.

Temin, H. M., 1980, Origin of retroviruses from cellular moveable genetic elements, *Cell* **21**:599.

Temin, H. M., 1985, Reverse transcription in the eukaryotic genome: Retroviruses, pararetroviruses, retrotransposons, and retrotranscripts, *Mol. Biol. Evol.* **2**:455.

Temin, H. M., 1989, Retrons in bacteria, *Nature* **339**:254.

Toh, H., Ono, M., Saigo, K., and Miyata, T., 1985, Retroviral protease-like sequence in the yeast transposon Ty1, *Nature* **315**:691.

Tschumper, G., and Carbon, J., 1986, High frequency excision of Ty elements during transformation of yeast, *Nucleic Acids Res.* **14**:2989.

Van Arsdell, S. W., Stetler, G. L., and Thorner, J., 1987, The yeast repeated element sigma contains a hormone-inducible promoter, *Mol. Cell. Biol.* **7**:749.

Vanin, E. F., 1985, Processed pseudogenes: Characteristics and evolution, *Annu. Rev. Genet.* **19**:253.

Varmus, H. E., 1989, Reverse transcription in bacteria, *Cell* **56**:721.

Walton, J., Paquin, C., Kaneko, K., and Williamson, V., 1986, Resistance to antimycin A in yeast by amplification of "ADH4" on a linear, 42 kb palindromic plasmid, *Cell* **46**:857.

Warmington, J. R., Waring, R. B., Newlon, C. S., Indge, K. J., and Oliver, S. G., 1985, Nucleotide sequence characterization of Ty 1-17, a class II transposon from yeast, *Nucleic Acids Res.* **13**:6679.

Warmington, J. R., Anwar, R., Newlon, C. S., Waring, R. B., Davies, R. W., Indge, K. J., and Oliver, S. G., 1986, A 'hot-spot' for Ty transposition on the left arm of yeast chromosome III, *Nucleic Acids Res.* **14**:3475.

Warmington, J. R., Green, R. P., Newlon, C. S., and Oliver, S. G., 1987, Polymorphisms on the right arm of yeast chromosome III associated with Ty transposition and recombination events, *Nucleic Acids Res.* **15**:8963.

Weinberg, R., 1980, Origins and roles of endogenous retroviruses, *Cell* **22**:643.

Weiner, A. M., and Maizels, N., 1987, tRNA-like structures tag the 3' ends of genomic RNA molecules for replication: Implications for the origin of protein synthesis, *Proc. Natl. Acad. Sci. USA* **84**:7383.

Weiner, A. M., Deininger, P. L., and Efstratiadis, A., 1986, Nonviral retroposons: Genes, pseudogenes, and transposable elements generated by the reverse flow of genetic information, *Annu. Rev. Biochem.* **55**:631.

Weinstock, K. G., Mastrangelo, M. F., Burkett, T. J., Garfinkel, D. J., and Strathern, J. N., 1990, Multimeric arrays of the yeast retrotransposon Ty, *Mol. Cell. Biol.* **10**:2882.

Wilke, C. M., Heidler, S. H., Brown, N., and Liebman, S. W., 1989, Analysis of yeast retrotransposon Ty insertions at the *CAN1* locus, *Genetics* **123**:655.

Williamson, V. M., and Paquin, C. E., 1987, Homology of *Saccharomyces cerevisiae* ADH4 to an iron-activated alcohol dehydrogenase from *Zymomonas mobilis*, *Mol. Gen. Genet.* **209**:374.

Williamson, V. M., Young, E. T., and Ciriacy, M., 1981, Transposable elements associated with constitutive expression of yeast alcohol dehydrogenase II, *Cell* **23**:605.

Williamson, V. M., Cox, D., Young, E. T., Russell, D. W., and Smith, M., 1983, Characterization of transposable element-associated mutations that alter yeast alcohol dehydrogenase II expression, *Mol. Cell. Biol.* **3**:20.

Wilson, W., Malim, M. H., Mellor, J., Kingsman, A. J., and Kingsman, S. M., 1986, Expression strategies of the yeast retrotransposon Ty: A short sequence directs ribosomal frameshifting, *Nucleic Acids Res.* **14**:7001.

Wilson, W., Braddock, M., Adams, S. E., Rathjen, P. D., Kingsman, S. M., and Kingsman, A. J., 1988, HIV expression strategies: Ribosomal frameshifting is directed by a short sequence in both mammalian and yeast systems, *Cell* **55**:1159.

Winston, F., and Minehart, P. L., 1986, Analysis of the yeast *SPT3* gene and identification of its product, a positive regulator of Ty transcription, *Nucleic Acids Res.* **14**:6885.

Winston, F., Chaleff, D. T., Valent, B., and Fink, G. R., 1984a, Mutations affecting Ty-mediated expression of the *HIS4* gene of *Saccharomyces cerevisiae*, *Genetics* **107**:179.

Winston, F., Durbin, K. J., and Fink, G. R., 1984b, The *SPT3* gene is required for normal transcription of Ty elements in *S. cerevisiae*, *Cell* **39**:675.

Winston, F., Dollard, C., Malone, E. A., Clare, J., Kapakos, J. G., Farabaugh, P., and Mineheart, P. L., 1987, Three genes required for *trans*-activation of Ty element transcription in yeast, *Genetics* **115**:649.

Woodson, S. A., and Cech, T. R., 1989, Reverse self-splicing of the *Tetrahymena* group I intron: Implication for the directionality of splicing and for intron transposition, *Cell* **57**:335.

Xiong, Y., and Eickbush, T. H., 1990, Origin and evolution of retroelements based upon their reverse transcriptase sequences, *EMBO J.* **9**:3353.

Xu, H., and Boeke, J. D., 1987, High frequency deletion between homologous sequences during retrotransposition of Ty elements in *Saccharomyces cerevisiae*, *Proc. Natl. Acad. Sci. USA* **84**:8553.

Xu, H., and Boeke, J. D., 1990a, Localization of sequences required in 'cis' for yeast Ty1 element transposition near the long terminal repeats: Analysis of mini-Ty1 elements, *Mol. Cell. Biol.* **10**:2695.

Xu, H., and Boeke, J. D., 1990b, Host genes that influence transposition in yeast: The abundance of a rare tRNA regulates Ty1 transposition frequency, *Proc. Natl. Acad. Sci. USA* **87**:8360.

Yee, T., Furuichi, T., Inouye, S., and Inouye, M., 1984, Multicopy single-stranded DNA isolated from a Gram-negative bacterium, *Myxococcus xanthus*, *Cell* **38**:203.

Yoshimatsu, T., and Nagawa, F., 1989, Control of gene expression by artificial introns in *Saccharomyces cerevisiae*, *Science* **244**:1346.

Yoshioka, K., Honma, H., Zushi, M., Knodo, S., Togashi, S., Miyaek, T., and Shiba, T., 1990, Virus-like particle formation of *Drosophila copia* through autocatalytic processing, *EMBO J.* **9**:535.

Youngren, S. D., Boeke, J. D., Sanders, N. J., and Garfinkel, D. J., 1988, Functional organization of the retrotransposon Ty from *Saccharomyces cerevisiae*: Ty protease is required for transposition, *Mol. Cell. Biol.* **8**:1421.

Yu, K., and Elder, R. T., 1989a, Some of the signals for 3'-end formation in transcription of the *Saccharomyces cerevisiae* Ty-D15 element are immediately downstream of the initiation site, *Mol. Cell. Biol.* **9**:2431.

Yu, K., and Elder, R. T., 1989b, A region internal to the coding sequences is essential for transcription of the yeast Ty-D15 element, *Mol. Cell. Biol.* **9**:3667.

CHAPTER 5

Mechanisms of Retrovirus Replication

PAUL A. LUCIW AND NANCY J. LEUNG

I. INTRODUCTION

A. Scope

The retrovirus family encompasses a diverse group of metazoan viruses that have a replication step whereby DNA is synthesized from virion RNA in a process designated reverse transcription (Temin and Baltimore, 1972) (Fig. 1; Table I) (see Chapter 1). Molecular mechanisms in the virus life cycle are reviewed in this chapter, and the focus is on retroviruses containing genes for virion proteins but lacking genes that regulate viral expression. Retroviruses with simple genomes express the polyproteins (i.e., precursor polypeptides) encoded by the following genes: *gag* for group-specific antigen in the virion core, *pol* for RNA-dependent DNA polymerase, and *env* for the viral envelope glycoprotein (Fig. 2). This genome organization is a feature of three genera in the retrovirus family, and both horizontally transmitted exogenous viruses and vertically transmitted endogenous viruses are included (Table II) (see Chapters 1 and 2) (Coffin, 1982b; Coffin and Stoye, 1985). Retroviruses with complex genomes (i.e., lentiviruses, spumaviruses, and certain oncoviruses) encode regulatory genes as well as virion proteins;

PAUL A. LUCIW AND NANCY J. LEUNG • Department of Pathology, School of Medicine, University of California, Davis, California 95616.

The Retroviridae, Volume 1, edited by Jay A. Levy. Plenum Press, New York, 1992.

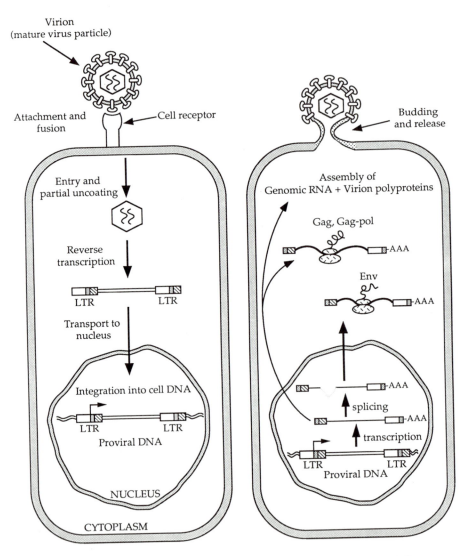

EARLY PHASE   LATE PHASE

FIGURE 1. Major features of the replication cycle of a prototype retrovirus. The early phase of replication involves attachment of a virion to a receptor on the cell surface, entry and uncoating, viral DNA synthesis by reverse transcription in the cytoplasm, transport of unintegrated viral DNA (complexed with viral integrase and *gag* proteins) to the nucleus, and integration into host-cell DNA to produce a provirus. Viral DNA is detected in the cytoplasm at about 2 hr after infection, full-length linear viral DNA molecules start to appear in the nucleus about 4–6 hr later, and proviral DNA is first detected about 12 hr after infection. The late phase starts with synthesis of viral transcripts from the provirus in the nucleus at about 15 hr after infection and continues through to release of progeny virions (about 24 hr after infection). Full-length viral transcripts serve as genomic RNA and as mRNA for *gag*, *prt*, and *pol* polyproteins; *env* gene mRNA is generated from the

replication of these viruses is discussed in other chapters and volumes in this series (see also Green and Chen, 1990; Mergia and Luciw, 1991; Cullen, 1991; Haseltine, 1991). Nonetheless, many aspects of the life cycle of retroviruses with simple genomes are shared by all retroviruses. Investigations of replication mechanisms have depended largely on analysis in both infected cell cultures and in *in vitro* systems based on cell-free extracts and/or defined components (Varmus and Brown, 1989; Coffin, 1990a). Other chapters in this volume describe retroviral infections of humans and other animals (see also Teich *et al.*, 1982; Gardner and Luciw, 1989; Gallo and Wong-Staal, 1990).

Much of the information on the replication of retroviruses with simple genomes has been obtained from extensive studies on the avian sarcoma-leukemia virus (ASLV) as well as the murine leukemia virus (MuLV or MLV) groups see Chapters 6 and 7. Additional retroviruses with simple genomes are reticuloendotheliosis virus (REV) and spleen necrosis virus (SNV). These two highly related avian retroviruses (henceforth designated REV) are not related to any members of the ASLV group but are distantly related to the MuLV group. Primate retroviruses with simple genomes include the gibbon-ape leukemia virus (GALV) and Mason–Pfizer monkey virus (MPMV); the latter agent is in the simian type D retrovirus (SRV) system. Studies of the mouse mammary tumor virus (MMTV) have also contributed to an understanding of basic retroviral replication mechanisms. Members of several retrovirus groups serve as helpers for replication-defective viruses; for example, ASLV and MuLV provide replication functions for avian erythroblastosis virus (AEV) and murine sarcoma virus (MSV), respectively (Coffin, 1982a). Generally, defective retroviruses possess an oncogene(s) (e.g., *erb-A* and *erb-B* in AEV, *ras* in MSV) and are defective in sequences encoding a virion protein(s). Descriptions of the biological properties of all these retroviruses as well as discussions of helper-dependent viruses and endogenous viruses are found in other chapters in this volume (see Chapters 6 and 7).

This review will focus largely on observations in the prototypic

---

full-length viral transcript by splicing. Free cytoplasmic polysomes are the site of translation of mRNA for viral polyproteins derived from the *gag, prt,* and *pol* genes; these polyproteins subsequently associate with genomic viral RNA to form an intracellular core structure. The mRNA for *env* polyprotein is translated on membrane-bound polysomes; *env* polyprotein is modified by glycosylation and proteolytic cleavages mediated by host-cell enzymes and then inserted into the cell plasma membrane. Through the budding process, immature cores are engulfed by the cell plasma membrane and thus acquire a lipid bilayer membrane which contains *env* glycoproteins. Proteolytic processing of *gag, prt,* and *pol* polyproteins takes place during the budding step and continues in newly released particles (immature virions) to produce mature infectious virions. The time course indicated above for major events in a single replication cycle is based on observations in actively growing tissue culture cells that have not been synchronized with respect to the cell cycle.

TABLE I. Characteristics of Retroviruses[a]

Spherical enveloped virion (100–130 nm diameter) containing a nucleoprotein core
  (or capsid)
Dimeric RNA genome (single-stranded, positive polarity) packaged into virions
Three genes for virion polyproteins (*gag, pol,* and *env*)[b]
Replication via reverse transcription and integration into host-cell genome
  (virion-associated reverse transcriptase and integrase encoded by *pol* gene)
Long terminal repeats (LTRs) at each end of integrated viral DNA

---

[a] A replication-competent retrovirus has all the functions for completing the replication cycle and
  yields infectious progeny virions. Replication-deficient retroviruses have one or more mutations in
  essential viral functions and depend on "helper" virus to provide the missing function(s).
[b] Some retroviruses encode additional genes that may be dispensable (e.g., *src* of RSV and *orf* of
  MMTV) or essential (e.g., *tat* of HIV) for viral replication.

ASLV and MuLV systems to establish a conceptual framework. Observations on retroviruses with complex genomes (as well as on retrotransposons) have provided a basis for both modifying previous models and posing new hypotheses (see Chapters 1 and 4). Retroviruses demonstrate diversity with respect to several aspects of replication as well as heterogeneity of biological properties. Nonetheless, an emphasis on common features of diverse retroviruses is essential to maintain both clarity and cohesiveness within this chapter.

This review of retroviral replication emphasizes knowledge acquired within the last 2–3 years, to reduce overlap with previous writings, including the excellent reviews by Varmus and Brown (1989) and Coffin (1990a). In addition, the following articles collectively offer comprehensive coverage of retroviruses: Stephenson (1980), Weiss *et al.* (1982, 1985), Hanafusa *et al.* (1989), Chen (1990), and Swanstrom and Vogt (1990). Thus, citations are largely limited to seminal findings and recent publications; to conserve space, specialized reviews are frequently cited to provide additional information and references.

B. Overview of Retroviral Replication

Retroviruses have a dimeric RNA genome that is reverse-transcribed by the viral-coded RNA-dependent DNA polymerase or reverse transcriptase (RT) (Temin and Mizutani, 1970; Baltimore, 1970) (see Chapter 1). Several features, in addition to sequences encoding RT, unify and distinguish retroviruses from other viruses (Table I) (Varmus and Brown, 1989; Coffin, 1990a). Virions (i.e., virus particles) are spherical, 80–130 nm in diameter, and consist of a nucleoprotein core surrounded by a lipid bilayer membrane. The core contains a dimer of two identical molecules of single-stranded RNA associated with several viral proteins. All retroviruses encode *gag, pol,* and *env* genes, and virion enzymes derived from *gag–pol* polyproteins include protease (PR), RT, RNase-H, and integrase (IN) [see Leis *et al.* (1988) for the nomenclature of viral proteins]. The retroviral RNA genome is reverse-transcribed into double-stranded linear DNA which contains long terminal repeats (LTRs) derived from unique sequences at each end of genomic RNA. Viral DNA is integrated into host-cell DNA, and the integrant (i.e., provirus) begins and ends with short inverted repeats that terminate in the dinucleotides 5' TG . . . CA 3'. Direct repeats of 4–6 base pairs (bp) in host-cell DNA flank the provirus, and the size of the repeat depends on the virus. Although all retroviruses share these features, diversity is noted with respect to sequence, cell tropism, species range, and pathogenic potential (Teich, 1982; Temin, 1989; Katz and Skalka, 1990) (see Chapters 2 and 3).

The retroviral replication cycle is divided into early and late phases, each consisting of several sequential steps (Varmus and Swanstrom, 1982) (Fig. 1). The early phase starts with the binding of a virus particle (i.e., virion) to a cell receptor and continues to the establishment of an integrated provirus. Attachment to a receptor and entry are events largely mediated by the viral *env* glycoprotein. Viral and cellular membranes fuse, and the virion core (i.e., capsid) is released into the cell cytoplasm. Viral enzymes RT and IN remain associated with genomic viral RNA in the form of a nucleoprotein complex. Reverse transcription of genomic viral RNA into double-stranded linear DNA takes place in the nucleoprotein complex in the cell cytoplasm. LTRs flanking the viral genes are produced by strand switches in the reverse transcription process. Linear viral DNA molecules remain in a nucleoprotein complex which is transported to the nucleus, where termini of viral DNA are covalently linked to host-cell DNA in a reaction mediated by IN.

The late phase includes events from the transcription of the provirus to the release of mature virions that are competent to reinitiate the replication cycle. LTRs in the provirus provide signals for cellular factors that control synthesis and processing of viral RNA. All viral transcripts have a 5' methylated cap structure and a 3' poly-A tail and thus resemble eukaryotic mRNA. Spliced as well as full-length viral RNA molecules are transported to the cytoplasm. Both species of viral RNA are translated on host-cell polysomes into virion polyproteins (i.e., precursor polypeptides); however, full-length transcripts also interact with virion polyproteins and are assembled into immature virus particles. By budding through the cell plasma membrane, these particles acquire a lipid bilayer membrane that contains *env* glycoprotein. Final maturation steps in newly released extracellular particles involve processing of assembled polyproteins by the viral PR to yield fully infectious virions (Varmus and Brown, 1989; Coffin, 1990a).

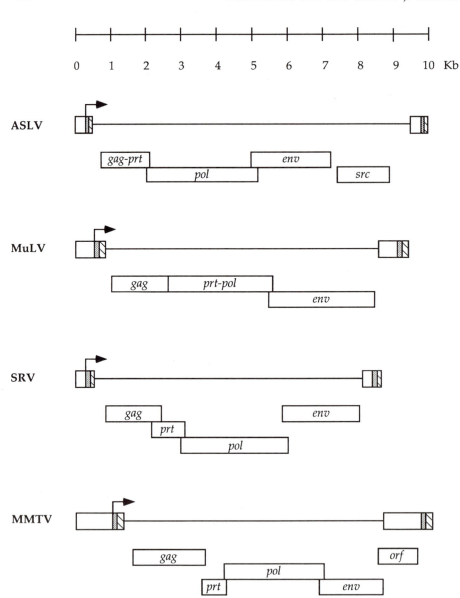

FIGURE 2. Genomes of replication-competent retroviruses. The organization of retroviral DNA genomes is shown for avian sarcoma-leukosis virus complex (ASLV), murine leukemia virus (MuLV), simian type D retroviruses (SRV), and mouse mammary tumor virus (MMTV). Open boxes below each genome indicate the locations of open translation frames for viral genes. Virion proteins are specified by the following genes: *gag* (group-specific antigen), *prt* (protease), *pol* (RNA-dependent DNA polymerase), and *env* (envelope glycoprotein). The *prt* gene of SRV, but not other retroviruses with simple genomes, encodes a domain with homology to deoxyuridine triphosphatase (dUTPase) (Power *et al.*, 1986; Elder *et al.*, 1992). Nonvirion proteins are also encoded by some retroviruses;

C. Significance of Retroviral Replication

Although this review elaborates on molecular mechanisms that characterize events in the retroviral replication cycle, many of these findings have contributed valuable insights into several areas of biology (Varmus, 1988). For example, paradigms based on retroviral genomic organization and reverse transcription were critically important for the analysis of retrotransposons and other transposable elements in diverse organisms as well as certain DNA viruses which replicate through an RNA intermediate (e.g., hepadnaviruses and caulimoviruses) (Garfinkel *et al.*, 1985; Temin, 1985, 1988; Boeke and Corces, 1989; Sandemeyer *et al.*, 1990; Mason *et al.*, 1987) (see Chapter 4). Investigations into mechanisms regulating replication are also essential for understanding virus–host interactions, including patterns of pathogenesis in infected animals and humans (Gardner and Luciw, 1989; Fan, 1990). In addition, a comprehensive knowledge of the retroviral life cycle is required for the development and utilization of these viruses as vectors to transduce heterologous genes; retroviral vector systems will be valuable for pursuing many fundamental issues in both basic biology and medicine (Friedmann, 1989; Soriano *et al.*, 1989; Verma, 1990; Anderson, 1992).

II. VIRION STRUCTURE AND RNA GENOME

A. General Features of Virions

The electron microscope has been a major tool for examining the morphology of both retrovirus particles (i.e., virions) and assembly intermediates in infected cells (Teich, 1982) (see Chapter 2). In addition, information on the properties of specific components of virions has been obtained from extensive biochemical studies on purified virus particles as well as on viral proteins and morphogenesis intermediates in infected cells (Dickson *et al.*, 1982) (see Section VIII on virion assembly). Virions are spherical particles with a diameter ranging from 80 to 130 nm; Fig. 3 is a schematic representation of a prototype virion in the retrovirus family (Bolognesi *et al.*, 1978). Each virus particle contains two identical molecules of the single-stranded RNA genome complexed with viral-coded proteins derived from the *gag* and *pol* genes; this nu-

---

Rous sarcoma virus (RSV) in the ASLV group has the *src* oncogene and MMTV has a large open reading frame designated *orf*. Boxes at the ends of each genome represent the long terminal repeats (LTR); domains in the LTR from left to right are U3 (open box), R (shaded box), and U5 (cross-hatched box) (see Figs. 6 and 11). An arrow above the 5′ LTR marks the start site for viral RNA transcripts. The scale is in kilobases (kb).

TABLE II. Retroviruses with Simple Genomes[a]

Group	Examples	Features, comments
Type B virus	Mouse mammary tumor virus (MMTV)	Exogenous and endogenous; causes mostly carcinoma and some T-lymphomas
Avian type C viruses	Rous sarcoma virus (RSV)[b]	Exogenous; contains *src* oncogene
	Rous-associated virus 1, 2, etc. (RAV-1,2, etc.)[b]	Exogenous; induce neoplastic and nonneoplastic diseases
	Avian erythroblastosis virus (AEV)[b]	Exogenous; requires helper virus (e.g., RAV-1); contains *erb-A* and *erb-B* oncogenes
	RAV-0[b,c]	Endogenous; no disease
	Reticuloendotheliosis virus (REV)[d]	Exogenous; induces nonneoplastic diseases
Mammalian type C viruses	Moloney murine leukemia virus (MoMuLV)	Exogenous; causes T-lymphomas
	Harvey murine sarcoma virus (HaMSV)	Exogenous; requires helper virus; contains oncogene H-*ras*
	AKR-MuLV[c]	Endogenous; no disease
	Human endogenous retrovirus (HERV-K)[c]	Endogenous; not infectious; no disease
	Gibbon-ape leukemia virus (GALV)	Exogenous; leukemia
	Feline leukemia virus (FeLV)	Exogenous; induces neoplastic and nonneoplastic diseases
Type D viruses	Simian type D retrovirus 1 to 5 (SRV-1 to -5)	Exogenous; induces immunodeficiency; prototype: Mason–Pfizer monkey virus (MPMV)–SRV-3
	Squirrel monkey virus (SMRV)[c]	Endogenous; no disease

[a] Information for this table was obtained from Weiss *et al.* (1982) and Chapter 2.
[b] RSV, RAV-1, 2, etc. (formerly identified as avian leukosis viruses, ALV), AEV (and other helper-dependent avian retroviruses containing oncogenes), and RAV-0 are members of the avian sarcoma-leukosis virus (ASLV) complex (see Chapter 2).
[c] Endogenous viruses are discussed in Chapter 2 and other chapters in this series (Coffin, 1982b; Coffin and Stoye, 1985; Krieg *et al.*, 1992).
[d] REV and spleen necrosis virus (SNV) show about 90% DNA sequence homology to each other; these viruses are unrelated to members of the ASLV group. The classification scheme in Chapter 2 positions REV in the subgenera of mammalian type c oncoviruses.

cleoprotein core is also referred to as the capsid. Virion enzymes derived from *gag*-encoded polyproteins are PR, RT, and IN (functions of these enzymes are discussed in several sections below). SRV particles, but not those of other retroviruses with simple genomes, also contain deoxyuridine triphosphatase (dUTPase), and the C-terminal domain of the *prt* gene of SRV is presumed to encode this activity (Elder *et al.*, 1992).

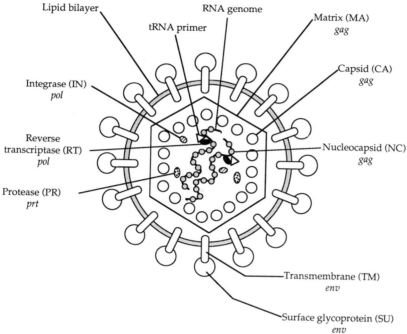

FIGURE 3. Structure of retroviral particles (i.e., virions). The schematic representation of a mature (prototype) virion shows the arrangement of proteins derived from the *gag, pol,* and *env* genes. The capsid or core consists of dimers of genomic RNA, tRNA primers, *gag* cleavage products (MA, CA, and NC), as well as viral enzymes derived from the *prt* and *pol* genes (PR, RT, and IN). Not shown is the virion-associated dUTPase that is derived from the *prt* gene of some retroviruses (e.g., SRV) but not others (e.g., ASLV, MuLV) (Elder *et al.,* 1992). The lipid bilayer surrounding the core contains the *env* glycoprotein, which is a heterodimeric complex of the TM and SU domains. Processing patterns of ASLV and MuLV polyproteins as well as molecular weights for mature virion proteins are shown in Fig. 10.

Viral-coded enzymes are probably part of the core, although the precise stoichiometries and topologies are not known (Figs. 3 and 4). Features of the viral RNA genome are discussed in detail below (Fig. 6). Small amounts of cellular RNA and protein species as well as some cellular DNA are also packaged into virions. The nucleoprotein complex is surrounded by a lipid bilayer envelope, which contains the glycoprotein encoded by the viral *env* gene. This envelope is acquired from host cells during the budding process that releases virus particles. Functions of specific virion proteins derived from the *gag, pol,* and *env* polyproteins are described in detail below and in the sections on viral protein synthesis and assembly. Analysis of the chemical composition of retroviruses shows that virions are (by weight) 60–70% protein, 30–40% lipid, 2–4% protein-associated carbohydrate, and about 1% RNA (Teich, 1982).

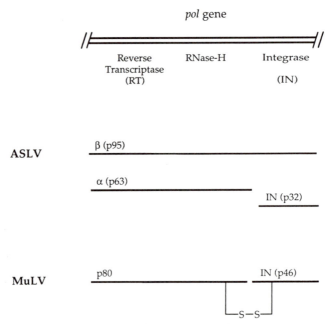

FIGURE 4. Products of the *pol* genes of avian sarcoma–leukosis viruses (ASLV) and murine leukemia viruses (MuLV). The *pol* polyprotein is proteolytically processed within assembled virus particles to produce proteins with enzymatic activities required for viral DNA synthesis and integration. Shown are locations of enzymes encoded by *pol:* reverse transcriptase (RT), RNase H, and integrase (IN). In mature virions, the RT of ASLV is a heterodimeric complex that contains a 63-kDa (α) and a 95-kDa (β) subunit; noncovalent interactions hold the two subunits together. The MuLV RT is isolated as an 80-kDa dimer from virions. IN proteins of ASLV and MuLV are 32-kDa and 46-kDa polypeptides, respectively. The MuLV IN appears to be linked by disulfide bonds to RT in virions. [Redrawn from Kulkosky and Skalka (1990).]

B. Virion Morphology and Intracellular Forms

Unique features of virion ultrastructure revealed by electron microscopy forms the basis for the first classification scheme of retroviruses (Fig. 5) (Teich, 1982, 1985) (see Chapter 2). Ultrastructural observations of intracellular viral forms provide additional criteria for subdividing retroviruses. This classification scheme has limited merit, since genetic and biological properties of retroviruses do not correlate in a simple fashion with morphology (Table II); accordingly, other criteria have recently been established (see Chapter 2). Nonetheless, electron microscopy is important for studying virion assembly, determining the identity of retroviruses, and examining tissues and fluids from infected hosts. Described below are salient features of type A, B, C, and D viral particles from the previously designated oncovirus subfamily (Fig. 5;

Table II); detailed characterizations of these various morphological forms together with electron micrographs have been presented by Teich (1982) (see Chapter 2). A discussion of the morphologies of lentiviruses (e.g., HIV) and complex oncoviruses (e.g., HTLV-I) is beyond the scope of this review (see Gelderblom, 1991, and references therein).

Type A particles, also designated intracellular A-type particles, are strictly intracellular and are not infectious (Fig. 5) [reviewed by Kuff and Lueders (1988); Keshet *et al.* (1991); Dorner *et al.* (1991); Reuss and Schaller (1991)]. These particles are 60–90 nm in diameter and have a relatively clear (i.e., electron-lucent) center surrounded by a bilayer membrane. Type A particles may be located within cisternae or in the cytoplasm; presumably, a topogenic signal on a protein in the type A particle controls intracellular localization. Intracisternal type A particles are sometimes noted in cells infected with type C or type D retroviruses, whereas intracytoplasmic type A particles are observed in cells infected with type B retrovirus. Thus, type A forms appear to represent intermediates in virion assembly (see Section VIII on virion assembly).

The mature form of the type B particle is 125–130 nm in diameter and has an electron-dense nucleoid located eccentrically within the enveloped virion (Fig. 5). During assembly, a toroidal core about 75 nm in diameter is attached to the plasma membrane. In the budding process, these cores acquire a lipid envelope and long viral glycoprotein spikes (Kramarsky *et al.*, 1971; Teich, 1982).

Mature type C particles (80–110 nm) have an electron-dense core concentrically positioned within the virion; immature extracellular forms have an electron-lucent core (Fig. 5). In infected cells, an electron-dense crescent-shaped core is observed when budding has started at an area of the plasma membrane that contains virion envelope spikes; no other intracytoplasmic precursor forms are noted. A topogenic signal in the *gag* polyprotein appears to direct the precursor polypeptides to sites of assembly at the inner surface of the cell plasma membrane (see Section VIII on virion assembly). During budding, the plasma membrane surrounds the electron-lucent viral core to produce newly released immature virions. In contrast to type B virions, the *env* glycoprotein forms relatively short surface spikes on extracellular type C virions (Fig. 5) (Kramarsky *et al.*, 1971; Bolognesi *et al.*, 1978; Teich, 1982).

Extracellular forms of type D retroviruses (100–120 nm in diameter) contain a bar-shaped, electron-dense nucleoid which is often eccentrically positioned within the lipid bilayer envelope (Fig. 5). Type D particles have short glycoprotein surface spikes. Intracellular forms are ring-shaped with a diameter of 60–95 nm; generally, these immature forms are positioned near the plasma membrane (Kramarsky *et al.*, 1971). The *gag* polyprotein contains topogenic signals that initiate particle assembly in the cytoplasm and then direct the immature core to the

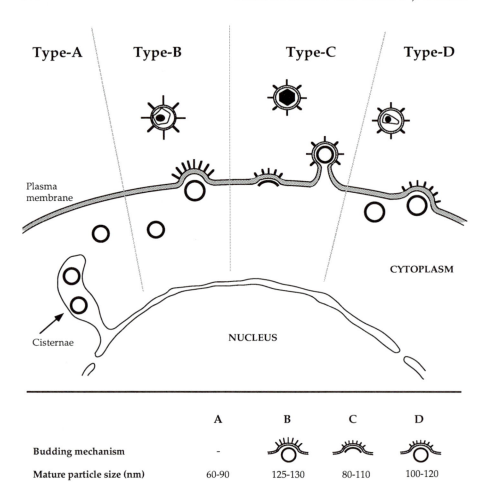

FIGURE 5. Virion morphologies. Schematic representations of retroviral morphologies are based on electron microscopy of mature virions and assembly intermediates in infected cells. Type A forms, depicted as rings, are exclusively intracellular particles observed either within cytoplasmic cisternae or free in the cytoplasm. Intracellular assembly intermediates for type B and type D retroviruses are morphologically similar to type A forms. For type C retroviruses, the major intracellular form is an electron-dense crescent juxtaposed next to the cell plasma membrane. Intracellular forms of retroviral particles are nucleoprotein complexes composed of genomic viral RNA and *gag* polyproteins; these nucleoprotein complexes are designated cores or capsids. Viral-coded *env* glycoproteins are inserted into the cell membrane and appear as knobs on the external surfaces of cells and virions. Immature viral particles are released from cells by the budding process during which viral cores acquire a membrane that contains *env* glycoprotein molecules. Morpho-

plasma membrane (see Section VIII on virion assembly) (Rhee and Hunter, 1990a, 1991).

C. Virion Proteins

This section describes salient features of proteins in mature retroviral particles. Products of the *gag*, *prt* (protease), *pol*, and *env* genes found in virions of several retroviruses are shown in Figs. 3 and 4 (also see Fig. 14 below). The term polyprotein is used henceforth to specify the precursor polypeptide that is the primary translation product of each of these viral genes (Dickson *et al.*, 1982). Patterns of translation and processing of viral polyproteins are described in detail in the sections below on viral protein synthesis and assembly (see also Chapter 2). The nomenclature proposed by Leis *et al.* (1988) has been adopted, and roles of virion proteins in intracellular viral replication events are more extensively discussed in other sections of this review (see also Dickson *et al.* 1982, 1985).

Retroviral *gag* genes encode polyproteins that are cleaved into at least three proteins, and these are present in stoichiometric amounts of about 2000 molecules per virion (Dickson *et al.* 1982; Coffin, 1990a). The three mature *gag* gene products common to all retroviruses are designated matrix (MA), capsid (CA), and nucleocapsid (NC) (Fig. 3). MA protein, ca. 15–20 kilodaltons (kDa), is modified at the N-terminus by acylation in ASLV or myristylation in mammalian retroviruses, and MA is located in the matrix between the capsid and the viral membrane envelope (Schultz *et al.* 1988; Wills and Craven, 1991). CA protein, ca. 24–30 kDa, is the major structural component of the capsid. NC protein, ca. 10–15 kDa, contains a cysteine–histidine motif which resembles metal-binding domains (i.e., zinc fingers) of proteins that interact with nucleic acids, and NC has an affinity for the viral RNA genome (Katz and Jentoft 1989). In many retroviruses, additional proteins are derived from the *gag* polyprotein; however, the functions of these proteins remain to be elucidated (Dickson *et al.*, 1985).

The retroviral protease (PR), ca. 10–15 kDa, is an enzyme distantly related to cellular aspartyl proteases and mediates cleavage of *gag* and *pol* polyproteins during virion assembly (Figs. 3 and 4; see also Fig. 14) (Oroszlan and Luftig, 1990). The location of sequences encoding PR varies depending on the retrovirus: (i) at the end of the *gag* gene in ASLV, (ii) after the stop codon in *gag* but in the same translation frame as the *pol* gene of MuLV, or (iii) in a unique translation frame between the *gag*

---

genetic processes involving cleavage of *gag* polyproteins take place during budding and in newly released particles to produce mature infectious virions. [Redrawn from Gelderblom (1991).]

and *pol* genes in SRV and MMTV (Fig. 2; see also Fig. 14). The amount of PR in virions is governed by the level of polyproteins encoded by either *gag–prt* (for ASLV), *gag–prt–pol* (for MuLV), or both *gag–prt* and *gag–prt–pol* (for SRV and MMTV) (Jacks, 1990).

Reverse transcriptase (RT) is an RNA-dependent DNA polymerase encoded by the *pol* genes of all replication-competent retroviruses (Figs. 3 and 4) (Goff, 1990). Each virion contains about 10–20 molecules of RT which are loosely associated with the nucleoprotein core. The RT of ASLV is a heterodimer that contains a 63-kDa subunit and a 95-kDa subunit; the smaller subunit is derived from the amino terminus of the larger protein (Fig. 4). MuLV has a monomeric RT, ca. 80 kDa (Fig. 4), and the RT of MMTV is also monomeric, with a size of ca. 85–100 kDa. RT also binds the tRNA primer, and a separate domain on this protein functions as a ribonuclease specific for RNA–DNA hybrids (RNase H). Integrase (IN) is a separate protein derived from the carboxy terminus of the *pol* gene (Fig. 4). The functions of RT, RNase H, and IN in viral DNA synthesis and integration are covered below (also see Varmus and Brown, 1989; Coffin, 1990a).

The *env* gene encodes a polyprotein that is posttranslationally modified in the endoplasmic reticulum by cleavages and glycosylation events to yield a surface (SU) domain and a transmembrane (TM) domain (Fig. 3; see also Figs. 16 and 17); these processing events are mediated by host-cell enzymes (Hunter and Swanstrom, 1990; Daar and Ho, 1990). SU glycoprotein is located on the external surface of the viral membrane, where it functions to bind virions to receptors on cells (Figs. 1 and 3). Accordingly, SU plays a critical role in viral host range and patterns of pathogenesis. TM protein is embedded into the lipid bilayer envelope and anchors the SU domain to the membrane (Figs. 1 and 3). TM mediates the fusion of viral and cell membranes during entry (Figs. 1). In virions, SU and TM form a heterodimer that involves either disulfide bonding (i.e., ASLV) or noncovalent interactions (i.e., MuLV, SRV), and two to four heterodimers of SU and TM associate into oligomeric complexes in infected cell plasma membranes and in the virion envelope. Retroviral *env* glycoproteins mediate binding of virions to cell receptors, are targets for antiviral immune responses in infected hosts, and control some cytopathic effects (e.g., syncytium formation) in cell culture. The greatest degree of viral strain variation is observed in the *env* genes of both closely related retroviruses and independent isolates of the same virus (Hunter and Swanstrom, 1990; Daar and Ho, 1990).

D. Virion RNA and *cis*-Acting Elements

Each retroviral particle contains two identical molecules of single-stranded genomic RNA that have positive polarity with respect to translation, and the size of the genome for replication-competent retroviruses ranges from about 8 to 10 kilobases (kb) (Varmus and Swanstrom, 1982). Two copies of genomic RNA are required in each virus particle to accommodate the replication mechanism, although genetic evidence suggests that only one provirus is produced per infectious virus particle (Hu and Temin, 1990a). A pseudodiploid arrangement of the genome provides a means to form heterozygotes and generate recombinant viruses (Weiss *et al.*, 1973; Hu and Temin, 1990a, b; Katz and Skalka, 1990, Coffin, 1990b) (see Chapter 1). Genomic RNA can be extracted from mature virions in a dimeric complex; thus, proteins are not required to hold the two viral RNA molecules together. Both RNA molecules appear to interact near the 5′ ends either through hydrogen bonds formed between molecules aligned in the same polarity or through short antiparallel alignments (Coffin, 1982a; Tounekti *et al.*, 1992). Viral RNA is synthesized by the host-cell transcription apparatus from integrated viral DNA in the cell nucleus. Figure 6 is a schematic representa-

VIRION RNA GENOME

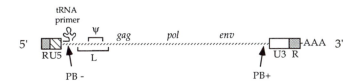

VIRAL DNA GENOME (PROVIRUS)

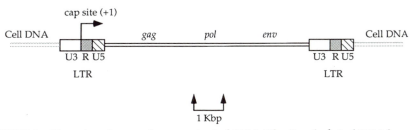

FIGURE 6. *Cis*-acting elements in genomic viral RNA. The 5′ end of viral RNA has a cap structure (7 mG) and the 3′ end has a poly-A tail. R is the short repeat at each end of the genome, U5 is a unique sequence element immediately after the 5′ R sequence, PB− is the primer site for minus-strand DNA synthesis (tRNA binding site), L is the leader region before the start of *gag*, Ψ is the element required for assembly of viral RNA into virions, PB+ is the primer site for plus-strand DNA synthesis, and U3 is a unique sequence at the 3′ end of the genome. Replication-competent retroviruses have genomes about 8 kb in length. This figure also reveals the relationship of the U3, R, and U5 elements in viral RNA with respect to the LTRs in linear viral DNA. Table III lists the lengths of these various *cis*-acting sequence elements for several viruses. Genes encoding virion polyproteins *gag, pol,* and *env* are also shown; the RNA and DNA genomes are co-linear.

tion of a prototype viral RNA genome and shows several *cis*-acting elements that function in reverse transcription, viral gene expression, and virion assembly (Coffin, 1982a, 1985).

The 5' end of virion RNA as well as viral mRNA has a cap structure which is 7-methylguanosine in a 5'–5' linkage via a triphosphate to a second 2'-O-methylated nucleotide. This structure is also designated $m^7G^{5'}ppp^{5'}NmpMp$ (N and M are the first and second nucleotides encoded by the virus). A small number of internal methylations in the form of N^6-methyladenosine (m^6A) residues have been reported for the ASLV genome. The 3' end of all retroviral transcripts has a poly-A tail from about 100 to 200 bases in length (Fig. 6). Cap structures, internal methylations, and poly-A tails are characteristic of cellular mRNA, and all three of these posttranscriptional modifications of viral RNA are carried out by host-cell enzymes (Stoltzfus, 1988).

A short repeated sequence, designated R, is located at each end of the genome, and R varies from 15 to 80 bases, depending on the retrovirus (Haseltine *et al.*, 1977) (Fig. 6; Table III). During reverse transcription, this repeat provides a means to transfer newly initiated DNA strands from the 5' end to the 3' end of viral RNA (see Fig. 8) (Varmus and Brown, 1989). Although R is present in two copies in each genomic viral RNA molecule, only the 5' R element is capped and only the 3' R element has poly-A tails.

A unique sequence, designated U5 (ranging from 80 to 100 bases), is located between R and the site for attachment of a host-cell tRNA which functions as a primer for viral DNA synthesis (Fig. 6; Table III). U5 sequences form stem-loop arrangements with sequences in the leader, with the inverted repeat sequence proximal to the tRNA primer-binding site, and with the TΨC loop in the tRNA primer (see Fig. 9 below) (Cobrinik *et al.*, 1988, 1991; Aiyar *et al.*, 1992). These structural features involving U5 sequences are required for efficient initiation of reverse

TABLE III. *Cis*-Acting Sequence Elements in Retroviral Genomes[a]

Retrovirus group	Virus strain	U3	R	U5	L (SD)	tRNA primer
Type B	MMTV	1194	13	119		Lys
Avian type C	RSV	233	21	102	380 (398)	Trp
	RAV-0	177	21	80		Trp
	REV	396	52			—
Mammalian type C	MoMuLV	449	68	76	621 (206)	Pro
	FeLV					—
Type D	SRV-1	235	13	98	136	Lys

[a] Abbreviations are as follows: U3 is the unique 3' sequence, R is the short repeat at each end of the viral RNA genome, U5 is the unique 5' sequence, L is the leader region (from the cap site +1 to the *gag* initiation codon), and SD designates the position of the splice donor site where known.

transcription (Aiyer *et al.*, 1992). In the mechanism of reverse transcription, R and U5 become part of the flanking LTRs in linear viral DNA (see Fig. 8, below) (Varmus and Brown, 1989).

A specific tRNA molecule is bound near the 5' end of the viral genome and initiates DNA synthesis (Varmus and Swanstrom, 1982). A sequence of 16–19 bases at the 3' end of the tRNA molecule is hydrogen-bonded to the complementary sequence, designated the minus (or negative)-strand primer-binding site (PBS or PB−), in the viral genomic RNA (Fig. 6; also see Fig. 9). Each group of related retroviruses contains a unique tRNA primer (e.g., Trp-tRNA for ASLV, Glu- or Pro-tRNA for members of the MuLV, Lys-tRNA for SRV) (Table III). A genetically engineered MuLV genome with a deletion mutation that removes the Glu-tRNA primer-binding site is defective for replication; however, passage of this clone through mammalian cells yields infectious virus which acquires a PB− specific for Phe-tRNA (Colicelli and Goff, 1986).

Following the PB− is an untranslated leader (L) sequence (ranging from 150 to 200 bases) that precedes the initiation codon for *gag* (Fig. 6; Table III). A sequence element within L, designated the "encapsidation (EN)" or "packaging (Ψ)" sequence, has been shown to play a role in assembling genomic RNA into virions (also see Fig. 9) (Linial and Miller, 1990). Other important *cis*-acting sequences in genomic viral RNA include splice donors and acceptors; these are discussed in more detail in Section VI on the synthesis and processing of viral RNA (Stoltzfus, 1988).

Several *cis*-acting signals that play roles in reverse transcription, integration, and synthesis of viral RNA are located in the 3' portion of the genome. A short sequence about 15 bases long and rich in purines is located downstream from *env*; this element serves to initiate synthesis of plus-strand viral DNA and is designated the plus (or positive)-strand primer (PB+) (Fig. 6, also see Fig. 9). U3 is a unique sequence element extending from PB+ to R at the 3' end of the viral genome (Fig. 6). U3 ranges in length from about 200 bases to 1 kb and contains promoter elements that control transcriptional initiation of the integrated provirus by cellular RNA polymerase II (Table III) (Majors, 1990).

Figure 6 also shows the relationship of genomic viral RNA with proviral DNA; the LTR is produced during viral DNA synthesis and contains the U3 sequence juxtaposed to R–U5. Members within a closely related group of retroviruses show relatively high sequence variation in U3, and examples include endogenous and exogenous viruses in the ASLV and MuLV groups (Coffin, 1990a). Thus, differences in transcriptional control elements in U3 influence not only viral replication rate and cell tropism in culture, but also distribution of virus and patterns of pathogenesis in infected animals (Stoltzfus, 1988; Varmus and Brown, 1989; Coffin, 1990a; Fan, 1990) (see Chapters 6 and 7).

III. VIRAL ATTACHMENT AND ENTRY

A. Overview

The retroviral replication cycle is initiated by a specific interaction of the virion *env* glycoprotein with a cell surface molecule (i.e., receptor) (Fig. 1). Studies in most retrovirus systems have focused largely on biological properties of receptors (Weiss, 1982; Weiss, 1991). Biochemical characterizations have been limited by difficulties in isolating and characterizing receptors which are cell membrane proteins that are hydrophobic and present in low amounts (DeLarco and Todaro, 1976). The first retroviral receptor was identified in the HIV system by a combination of immunological and biochemical methods (Dalgleish *et al.*, 1984; Klatzmann *et al.*, 1984). HIV binds to the CD4 protein, a member of the immunoglobulin supergene family that is found on the surface of the T-helper/inducer subset of T-lymphoid cells (Sattentau and Weiss, 1988; McDougal *et al.*, 1986). Recently, molecular cloning approaches have shown that the cell receptors for ecotropic MuLV and GALV are membrane transport proteins (i.e., permeases) for amino acids and phosphate, respectively (Kim *et al.*, 1991; H. Wang *et al.*, 1991a; Johan *et al.*, 1992).

Entry, or penetration, follows attachment of virions to the cell surface receptor. Although cell surface proteins that bind retroviral *env* glycoproteins have been identified for a few viruses, accessory cellular components appear to be required for efficient entry (Maddon *et al.*, 1986; H. Wang *et al.*, 1991b). Members within a virus family are taken into the cell by either the endocytic (pH-dependent) pathway or by direct fusion of viral and cell membranes at neutral pH [reviewed by White (1990)]. However, retroviruses appear to be an exception to this generalization in that certain members utilize one or the other pathway. The TM domain of *env* plays a major role in the mechanism of viral–cell membrane fusion that is central to entry. In addition, fusion of cell membranes in infected cultures (i.e., formation of syncytia) is also mediated by the retroviral *env* glycoprotein (Pinter *et al.*, 1986; Lifson *et al.*, 1986). The culmination of the entry process is the release of the virion core into the cell cytosol, where additional uncoating events occur and reverse transcription is initiated (Varmus and Swanstrom, 1982).

B. Significance of Viral Attachment and Entry

Viral host range is largely controlled by the recognition process between the SU domain on the virion envelope and the cell surface receptor, although intracellular factors may also influence viral replication

(Hunter and Swanstrom, 1990; Daar and Ho, 1990). The term cell tropism is often used synonymously with host range. Retroviruses demonstrate a high degree of host-range specificity; accordingly, classification schemes have been based on the recognition properties of the *env* glycoprotein (Teich, 1982; Weiss, 1982) (see also Chapters 2, 3, and 7). The ability of a retrovirus to enter specific cell types in an animal host is an important determinant of pathogenesis (Teich *et al.*, 1982, 1985; Fan, 1990). In addition, utilization of retroviral vectors for gene transfer requires an understanding of host-range properties involving cell surface molecules and the entry process.

C. Biological Characterization of Cell Receptors for Retroviruses

Biological aspects of retroviral receptor specificity have been analyzed in cell culture systems that measure resistance as well as susceptibility to infection (Weiss, 1982, 1993). Either of two mechanisms may account for cell surface resistance to viral replication. First, cells may lack the gene(s) for a functional receptor; thus, these cells are genetically resistant to viral entry. In some instances, treatment of cells with a glycosylation inhibitor converts refractory cells to a permissive state (Wilson and Eiden, 1991; Miller and Miller, 1992). Presumably, either glycosylation of the receptor inhibits its function or glycosylation controls the ability of another cell surface protein to block the receptor. Polymorphism of the receptor gene may also account for the capacity of a retrovirus to infect cells of certain species. In the second mechanism, receptors may be present but saturated with viral glycoproteins which are expressed in the cell; thus, virus attachment is blocked (Fig. 7). Either an exogenous or endogenous virus infection of the cell may produce enough *env* glycoprotein to cause receptor interference to superinfection (Rubin, 1960; Weiss, 1982).

Permissivity of a cell for viral replication may be regulated by intracellular mechanisms (e.g., cell specificity of a transcriptional enhancer in the LTR) as well as by attachment and entry (Fan, 1990; Majors, 1990). Therefore, methods have been developed to analyze directly viral host range at the level of these early events [reviewed in Weiss (1982)]. In one procedure, cell fusion assays detect formation of multinucleated syncytia induced by infection of cell cultures with a retrovirus. An alternative method utilizes viral pseudotypes made between the retrovirus under study and a heterologous enveloped virus such as vesicular stomatitis virus (VSV). In coinfected cells, phenotypic mixing of viral glycoproteins yields VSV particles which contain the retroviral *env* glycoprotein as well as the VSV surface glycoprotein; all other components in the pseudotype particle are derived from VSV (Zavada, 1972). The virus

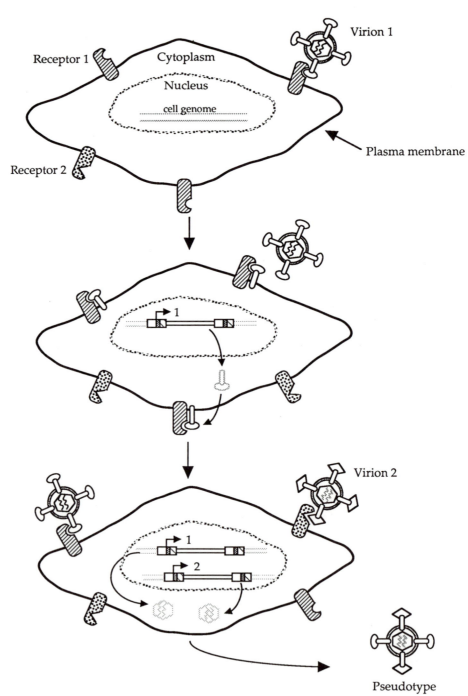

FIGURE 7. Mechanism of retroviral interference at the cell surface. The cell contains surface receptors for retroviruses-1 and -2. After infection with retrovirus-1, the receptors

yield from a coinfected cell culture is treated with an antiserum that neutralizes VSV so that only pseudotype particles remain to establish infection (i.e., plaques). A variation of this method involves pseudotype formation between a test retrovirus glycoprotein and VSV with a conditional mutation in the VSV glycoprotein gene (Boettiger et al., 1975).

Members of the ASLV group isolated from chickens fall into one of five subgroups (A–E) on the basis of host-range properties and interference patterns (Weiss, 1982) (see Chapters 2 and 6). Antibody neutralization properties of ASLV strains also correlate with this classification scheme, and several studies have demonstrated that the env gene plays a major role in cell tropism (Bova et al., 1988; Dorner et al., 1985; Dorner and Coffin, 1986). The ASLVs replicate only in avian cells; the subgroup D virus and the B77 isolate in subgroup C infect mammalian cells, although this infection leads to cell transformation, and progeny virions are not produced. Several strains of REV have been obtained from various avian species; however, sequence comparisons of cloned viral genomes reveal that the REV group has a mammalian origin based on sequence relatedness to mammalian type C retroviruses. Accordingly, the REVs replicate productively in a wide variety of avian cells as well as in certain mammalian cell lines (Koo et al., 1991). ASLV and REV do not cross-interfere and therefore are presumed to utilize different cell receptors. Interference studies indicate that REV and the simian type D retroviruses (SRV) utilize the same receptor (Koo et al., 1992). Indeed, the env genes of REV and SRV show some amino acid sequence similarity, and the positions of cysteine residues in the env genes of both groups of viruses are highly conserved (Thayer et al., 1987).

Viral isolates in the MuLV group are classified into four subgroups on the basis of host range, interference patterns, and antibody neutralization properties (Weiss, 1982; Rein, 1982) (see Chapter 7). Depending on the virus strain, MuLV may replicate only in cells from the host of origin and/or in cells from heterologous species. Designations and host-range properties for the four MuLV subgroups are as follows: ecotropic, murine cells only; xenotropic, nonmurine cells only; amphotropic and dual-tropic, both cell types (Levy, 1978) (Chapter 7). Amphotropic MuLV (e.g., 4070A virus strain) has a very wide species host range, and thus this virus was selected for the development of packaging cell lines for retroviral vectors (Hartley and Rowe, 1976; Rasheed et al., 1976) (reviewed in Miller, 1990a,b). A murine gene encoding the receptor for

---

for this virus are blocked with the env-1 glycoprotein. This cell cannot be superinfected with retrovirus-1 or another retrovirus which utilizes the same receptor. However, a cell preinfected with retrovirus-1 can be superinfected with retrovirus-2, since different receptors are used for attachment. A cell coinfected with retroviruses-1 and -2 produces pseudotyped or mixed progeny virions.

ecotropic MuLV was molecularly cloned and shown to encode a protein related to a cell membrane transport protein for basic amino acids (see below).

The FeLVs are divided into three interference subgroups (A–C) (Weiss, 1982). FeLV-A has the most restricted host range and replicates only in cat cells, whereas FeLV-B and FeLV-C infect cat cells as well as cells from several other mammalian species. Patterns of antibody neutralization also correlate with the subgroup classification of FeLVs (Jarrett et al., 1973).

Host ranges and receptor specificities have been examined for several primate retroviruses (Weiss, 1982). The exogenous simian type D retroviruses naturally infect Asian macaques and are classified into five distinct neutralizing serotypes (SRV-1 to 5) (Marx 1985, cited in Gardner and Luciw, 1989). However, all five viruses are in the same interference group and thus enter cells through the same receptor (Sommerfelt and Weiss, 1990). The SRVs appear to infect only simian and human cells efficiently; both T- and B-lymphoid cells as well as fibroblasts support viral replication (Maul et al., 1988). The endogenous type D retroviruses, squirrel monkey retrovirus (SMRV) and Po-1-Lu from Asian langurs, and the endogenous type C retroviruses from baboons (i.e., BaEV) and domestic cats (i.e., RD114), also show the same receptor specificity as the exogenous SRV strains (Sommerfelt and Weiss, 1990; Kaelbling, et al., 1991). GALV is an exogenous type C retrovirus of primates that infects cells from many mammalian and avian species (Weiss, 1982). Interference studies show that GALV utilizes a different cell receptor than the simian type D retroviruses but that it apparently shares the receptor for FeLV-B (Sommerfelt and Weiss, 1990). A human gene encoding the GALV receptor was molecularly cloned and demonstrated to encode a protein related to a cell membrane transport protein for phosphate (see below).

The finding that several diverse retroviruses utilize the same receptor suggests that a limited set of cell surface proteins function as receptors. Divergent evolution of an ancestral virus may yield distinct viruses that retain common receptor specificity. Alternatively, recombination between viruses may produce envelopes that utilize the same receptor.

Interference patterns of 20 different primate, feline, and murine retroviruses that replicate in human cells were analyzed in one systematic study. These viruses were classified into eight interference groups; thus, human cells express at least eight distinct receptors (Sommerfelt and Weiss, 1990). Each viral subgroup implies the existence of a unique cell receptor; however, this surmise remains to be rigorously established by analyzing receptor specificity at the molecular level.

The genetic basis of receptor specificity has also been studied by classical Mendelian methods and by analysis of viral replication in interspecies somatic cell hybrids (Weiss, 1982). In the avian system, three autosomal loci determine susceptibility to ASLV subgroups A, B, C, and E, and the resistance allele is recessive to the susceptibility allele. In mammalian species, genetic diversity of cell-receptor loci has not been observed; therefore, investigations on mammalian retroviruses have focused on chromosome localization of receptor genes in somatic cell hybrids. Hybrid cell lines are derived from permissive cells of one species and restricted cells of another, each displaying distinguishable karyotypes. Patterns of viral replication are assessed in hybrid cell lines that segregate chromosomes of one or the other parental cell type. Several limitations are inherent in this approach. The receptor must be specific to one cell type of the two parentals used to construct hybrid lines, and the receptor must be constitutively expressed in the hybrid. In addition, receptor specificity must involve a single gene or a closely linked set of genes. With this approach, the receptors for xenotropic and ecotropic MuLV have been assigned to mouse chromosomes 1 and 5, respectively, and the amphotropic MuLV receptor is controlled by a gene encoded in the pericentromeric region of human chromosome 8 (Kozak et al., 1990, and references therein; Garcia et al., 1991). Human chromosome 19 apparently specifies the receptor for type D retroviruses (e.g., SRV), BaEV, and RD114 (Sommerfelt et al., 1990; Sommerfelt and Weiss, 1990; and references therein).

D. Molecular Cloning of Retroviral Receptors

The receptor gene for ecotropic MuLV (ecoR, also designated rec-1) was identified by an elegant molecular cloning procedure that has applicability for the identification of receptors for other retroviruses (Albritton et al., 1989; for review, see Weiss, 1993). Nonpermissive human cells were transfected with DNA from permissive mouse cells. Human cells that acquired and expressed the receptor gene were selected by infection with ecotropic MuLV vectors carrying selectable drug-resistance genes. After selection, mouse DNA was identified in the human cell transfectants by annealing with radioactive probes specific for mouse repeat DNA sequences. Recombinant bacteriophage libraries were then made from these human cell transfectants and screened with probe to mouse repeat DNA. Subsequently, a mouse DNA insert from a recombinant bacteriophage clone was used as a probe to identify a cDNA clone from a recombinant library representing mRNA in permissive mouse cells. Proof that this selection and screening method yields the receptor gene is provided by the observation that nonpermissive human cells transfected with the cDNA clone become susceptible to ecotropic MuLV (Kim et al., 1991; H. Wang et al., 1991a). In addition, the *ecoR* gene product has been shown to mediate attachment of the SU domain (gp70) of the ecotropic MuLV *env* glycoprotein (H. Wang *et al.*, 1991b).

The *ecoR* gene encodes a protein 622 amino acids in length with 14 potential membrane-spanning domains (Kim *et al.*, 1991; H. Wang *et al.*, 1991a). Computer-based sequence comparisons with entries in sequence data banks reveal that the *ecoR* product has structural similarity with an amino acid transport protein (i.e., permease) of yeast. Both *ecoR* and the yeast gene, designated y^+, encode membrane proteins that mediate uptake of cationic amino acids. The protein encoded by *ecoR* has a ubiquitous tissue distribution in the mouse; thus, ecotropic MuLV has the potential to enter a wide variety of cell types in the host. The product of a murine gene designated T-cell early activation gene (*tea*) shows 52% amino acid identity with the *ecoR* product (MacLeod *et al.*, 1990). In addition, a human cDNA clone from a T-cell line encodes a protein with 88% amino acid homology with the *ecoR* product; this gene maps to human chromosome 13 and is expressed in several tissue types (Yoshimoto *et al.*, 1991). Genes related to *ecoR* from both murine and nonmurine species will facilitate the identification of functional domains of the receptor protein.

A receptor gene for GALV (*glvr-1*) was molecularly cloned from permissive human cells through a cloning strategy similar to that described above for *ecoR* (O'Hara *et al.*, 1990). The GALV receptor gene encodes a 679-amino acid protein with several hydrophobic transmembrane domains; this gene is related to a phosphate transport protein in *Neurospora* (Johann *et al.*, 1992). GALV infects several cell types from many species; thus, the functional receptor is widely distributed (see references in Teich 1982). The *glvr-1* gene maps to mouse chromosome 2 and human chromosome 2 (Adamson *et al.*, 1991; Takeuchi *et al.*, 1992).

The significance of membrane transport proteins as receptors for retroviruses remains to be elucidated. The domain on each of these cell receptor proteins that binds the viral *env* glycoprotein is probably separate from the domain that functions to transport metabolites into the cell (Kim *et al.*, 1991; H. Wang *et al.*, 1991a). Cloned cell receptors for these retroviruses are essential for analyzing molecular mechanisms of attachment and early events in viral entry. In addition, studies on the interactions of *env* glycoproteins and cell receptors will provide insight into mechanisms of viral pathogenesis; MuLV *env* is implicated in leukemia, immunosuppression, hemolytic anemia, and neurological disease (Teich *et al.*, 1982, 1985). A role for retroviral *env* gene products in dysfunction of membrane transport proteins is a possible mechanism for cytopathology. Whether other membrane permeases are receptors for additional retroviruses remains to be determined.

E. Virus Entry and Uncoating

In one entry pathway, viral membranes fuse to cellular membranes at low pH conditions within endosomes, and the virion core is released into the cytosol (White, 1990). Evidence for the pH-dependent endocytic pathway has been obtained from infectivity studies in cells treated with lysosomotropic agents (e.g., the weak bases such as ammonium chloride, chloroquine, and amantadine as well as the carboxylic ionophores such as monensin and nigericin). These agents neutralize the acidity of endosomes and thereby inhibit fusion of the viral membrane with the endocytic vesicle membrane. By these criteria, MMTV and the ecotropic strain of MuLV appear to utilize the same endocytic pathway as several other enveloped animal viruses (e.g., orthomyxoviruses) (Portis *et al.*, 1985; McClure *et al.*, 1990, and references therein).

The alternative entry pathway, exemplified by paramyxoviruses, involves direct fusion of the viral membrane with the cell plasma membrane; this process is not affected by a range of pH values (White, 1990). RD114 and SRV utilize the same cell surface receptor and enter cells by this pH-independent mechanism (McClure *et al.*, 1990). Some investigations of HIV-1 show that viral and plasma membranes fuse at the cell surface in a pH-independent fashion similar to the entry process of paramyxoviruses (Stein *et al.*, 1987; McClure *et al.*, 1988); however, other studies of HIV-1 support uptake by receptor-mediated endocytosis (Pauza and Price, 1988). Perhaps some retroviruses (e.g., HIV-1) gain entry via a pH-independent fusion process within endocytic vesicles. Several investigators have shown that the nature and concentration of the cell receptor as well as the cell type may influence the entry mechanism, and these factors may account for the differences reported for HIV entry (McClure *et al.*, 1990; see also Bova-Hill *et al.*, 1991). Under certain circumstances, retroviruses cause plasma membranes of neighboring cells (infected *in vitro*) to fuse and produce multinucleated cells (i.e., syncytia). Formation of syncytia appears to be a reflection of the fusion process by which virions enter cells; accordingly, many retroviruses are titrated by measuring induction of syncytia (Klement *et al.*, 1969; Chatterjee and Hunter, 1980; Weiss, 1982).

The retroviral fusion event is controlled by the hydrophobic domain (i.e., fusion peptide) in the TM domain of the *env* glycoprotein (see below, Fig. 16) [reviewed in White (1990); Hunter and Swanstrom (1990)]. In one model, binding of SU to the receptor causes a conformational change in the *env* glycoprotein so that the fusion peptide inserts itself into the lipid bilayer of the cell membrane. Alternatively, the low pH in an endosome may induce a conformational change in the *env* glycoprotein, exposing the hydrophobic domain of TM, which then mediates fusion of viral and endosomal membranes. Analysis of site-

specific mutations has provided insight into the role of the *env* glycoprotein in fusion and viral entry. Mutations preventing cleavage of SU
and TM abolish infectivity and block formation of syncytia; this cleavage is necessary to produce the hydrophobic end of TM, which merges
into cell membranes during fusion (Granowitz *et al.*, 1991). A small
deletion mutation in the hydrophobic region of TM prevents viral replication at an early step, presumably entry, and placement of charged
amino acids into the hydrophobic peptide of TM blocks both fusion and
infectivity (Hunter and Swanstrom, 1990; Daar and Ho, 1990).

Accessory cellular factors acting in concert with cell surface binding
proteins are hypothesized to facilitate entry and infection. Mouse cells
and some human cells transfected with recombinant human CD4 surface antigen gene cannot be infected with HIV (Maddon *et al.*, 1986;
Ashorn *et al.*, 1990; Chesebro *et al.*, 1990). Similar studies involving
transfection of the *ecoR* gene into certain nonmurine cells also demonstrate that the binding receptor is not sufficient for infection with ecotropic MuLV (H. Wang *et al.*, 1991b). Further studies are required to
identify accessory factors (or second receptors) as well as to elucidate the
precise roles of these factors in early events of viral infection.

The entry process releases the virion core into the cell cytoplasm
(Hunter and Swanstrom, 1990). The extent of uncoating or removal of
virion proteins from the newly entered core is not known. SU, TM, and
presumably MA stay associated with endocytic membranes. Genomic
viral RNA may remain associated with CA, NC, RT, IN, and perhaps PR
proteins in a cytoplasmic nucleoprotein particle. By a mechanism that
remains to be elucidated, the tightly bound NC proteins are removed
from genomic RNA, and subsequently RT is activated to initiate viral
DNA synthesis. Additional studies on early events in infected cells are
required to determine whether viral PR or other factors (e.g., intracellular ionic strength) play a role in freeing viral genomic RNA to serve as a
template for reverse transcription (Roberts and Oroszlan, 1989).

IV. VIRAL DNA SYNTHESIS BY REVERSE TRANSCRIPTION

A. Overview

In 1964, Temin proposed the provirus hypothesis, which held that
RNA tumor viruses (e.g., ASLV) reside in infected cells in the form of
DNA (Temin, 1964, 1976) (see Chapter 1). This hypothesis suggested
the existence of an enzyme that reversed the flow of genetic information
from RNA back to DNA (Crick, 1970). In 1970, Temin, working with
the ASLV system, and Baltimore, working with the MuLV system, inde-

pendently demonstrated that purified preparations of virus particles contained a polymerase that synthesized DNA from viral RNA templates (Baltimore, 1970; Temin and Mizutani, 1970). These findings validated the provirus hypothesis and opened new areas of fundamental research for virology, molecular biology, and medicine (Temin and Baltimore, 1972; Varmus, 1988).

The RNA-dependent DNA polymerase is encoded by the viral *pol* gene and is more commonly designated reverse transcriptase (RT) (Verma, 1977; Varmus and Swanstrom, 1982; Goff, 1990). In addition to a domain for polymerase activity, RT contains a domain that degrades the RNA moiety of RNA–DNA hybrids. This latter activity, designated RNase H, mediates several steps in retroviral replication. Recently, models for RT have been proposed based on results from a combination of structural, enzymological, and genetic studies (Arnold and Arnold, 1991).

Several laboratories have elucidated major steps in reverse transcription, including identification of primers, elucidation of functional domains of RT, and characterization of replicative intermediates (Gilboa *et al.*, 1979a; Varmus and Swanstrom, 1982, 1985; Varmus and Brown, 1989; Coffin, 1990a). Experimental evidence supporting the strand-transfer (or jump) model for reverse transcription has been derived from analysis of early steps in virus-infected tissue culture cells as well as studies utilizing *in vitro* systems with defined components. Investigations in the ASLV and MuLV systems have provided the bulk of experimental evidence for the model of reverse transcription described below (Fig. 8) (Gilboa *et al.*, 1979a; Varmus and Swanstrom, 1982, 1985).

B. Significance of Reverse Transcription

Elucidation of the mechanism of reverse transcription and characterization of the unique enzyme (i.e., RT) mediating this process have been critical not only for understanding the replication and biology of other viruses (e.g., hepadnaviruses and caulimoviruses) and retrotransposons, but also for significant advances in molecular genetics (see Chapter 4) (Temin, 1985; Varmus, 1988; Seeger *et al.*, 1990). Retroviral particles are a source of RT which is essential for synthesizing *in vitro* labeled DNA probes that are used for the detection and identification of nucleic acid sequences in a wide variety of experimental and clinical settings. In a related and critically important development, this enzyme has made it possible to obtain DNA clones representing single-stranded RNA. This latter procedure, designated cDNA cloning, is essential for analysis of splicing patterns of processed RNA transcripts derived from numerous cellular and viral genes. In this fashion, genomes of many RNA viruses

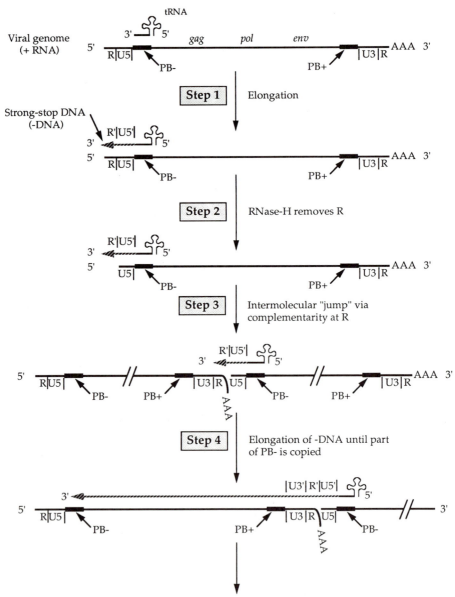

FIGURE 8. Model for retroviral DNA synthesis by reverse transcription. The entire reverse transcription process takes place at the early phase of infection in a nucleoprotein complex in the cell cytoplasm. This schematic shows only the nucleic acid species that participate in the stepwise process. The viral RNA genome has positive-strand polarity and is represented by a heavy line. Each virion contains two molecules of the viral RNA genome and, presumably, the nucleoprotein complex in the cell also has both RNA molecules. *Cis*-acting elements are R and U5 at the 5' end and U3 and R at the 3' end. AAA

have been molecularly cloned in both plasmid and bacteriophage vectors for sequence and genetic analysis.

C. Structure and Function of Reverse Transcriptase (RT)

The *pol* gene specifies a polyprotein which is processed by the viral PR during virion assembly and maturation to yield RT (Fig. 4; also see Fig. 14 below) (Dickson *et al.*, 1982; Goff, 1990). Mechanisms for the synthesis of the *pol* polyprotein for several retroviruses are described in detail below (see Section VII). Several molecules of RT are contained

←

represents the poly-A tail at the 3' end of viral RNA. Step 1: Viral DNA synthesis is initiated by reverse transcriptase (RT) from the host-cell tRNA primer near the 5' end of the RNA genome. A portion of this tRNA is complementary to a sequence in the viral genome that is designated the minus-strand primer-binding site (PB−) (Fig. 9). RT elongates from the 3' end of the tRNA molecule to the 5' end of viral RNA to produce strong-stop DNA that has minus-strand polarity; all minus-strand DNA species are represented by a broken line. Step 2: RNase H (RN) (a functional domain in the RT protein) degrades part or all of R from the 5' end of the viral genome; thus, R' in minus-strand strong-stop DNA is now single-stranded and available for base pairing. Step 3: Although experimental evidence supports both intramolecular as well as intermolecular strand transfer of minus-strand strong-stop DNA, the scheme shown here is provisionally based on an intermolecular event (Panganiban and Fiore, 1988; Hu and Temin, 1990b). The exposed R' hybridizes with the complementary R sequence at the 3' end of the second viral RNA molecule. Step 4: RT elongates from the 3' end of minus-strand strong-stop DNA and copies the viral genome including part of PB−. Step 5: RNase degrades viral RNA in the hybrid at the border with the 3' U3 sequence and generates an oligonucleotide primer designated the plus-strand primer (PB+). PB+ is a stretch of 12–15 purines and has also been designated the polypurine tract (PPT). From PB+, RT copies minus-strand strong-stop DNA through the tRNA sequence complementary to PB−; this species is designated plus-strand strong-stop DNA. All plus-strand DNA species are represented by a shaded line. A methylated adenine in the tRNA molecule marks the termination site for RT in the synthesis of plus-strand strong-stop DNA. The poly-A tail is removed presumably when the 3' R is degraded by RNase H; this may occur shortly after the strand-transfer step. Step 6: The tRNA primer in the hybrid is removed by RNase H and, thus, the PB− sequence at the end of plus-strand strong-stop DNA is exposed. The second strand transfer is an intramolecular (i.e., intrastrand) event in which PB− in plus-strand strong-stop DNA forms a duplex with the complementary sequence at the other end of the minus-strand DNA molecule. Step 7: RT then elongates from PB− in both directions. A strand displacement event separates the duplex between U3, R, U5 and U3', R', U5'. Accordingly, the long terminal repeats (LTRs) are generated as RT continues to elongate through these displaced sequences until the ends of the templates are reached. Synthesis of the plus strand in this model is shown to initiate only at the PB+ site. Internal priming sites have been observed in some instances and, thus, synthesis of plus strands may be discontinuous. The product of reverse transcription is a fully duplex, linear DNA molecule with LTRs and blunt termini. This form of viral DNA is in a cytoplasmic nucleoprotein complex that contains integrase (IN) and other virion proteins. Subsequently, the complex is transported into the nucleus for integration of the linear viral DNA molecule into the host-cell genome to produce the provirus (Fig. 10). [Redrawn from Varmus and Brown (1989).]

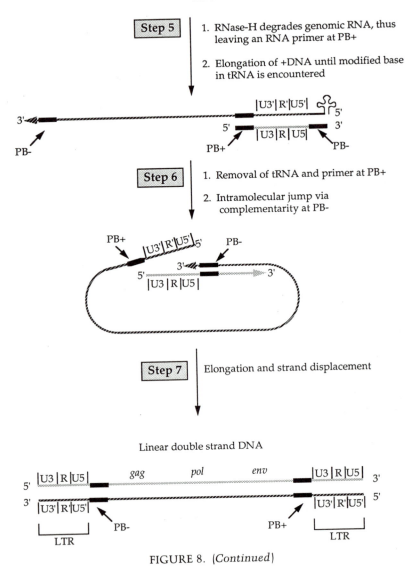

Linear double strand DNA

FIGURE 8. (*Continued*)

within virions, presumably tightly associated with the nucleoprotein core (Fig. 3). The C-terminus of the RT polypeptide also is a separate domain containing RNase H activity which digests the RNA moiety in RNA–DNA hybrids (see below). RT is readily obtained from purified virions for enzymological and structural studies; in addition, expression vectors in genetically engineered bacterial and yeast cells have been used recently to produce large quantities of enzymatically active RT from several retroviruses (Roth *et al.*, 1985; Barr *et al.*, 1987; Framerie *et al.*, 1987). Subunit structures of RT for ASLV (heterodimer) and MuLV (monomer) are shown in Fig. 4.

RT utilizes either single-stranded RNA or single-stranded DNA templates and requires either RNA or DNA primers (Dickson *et al.*, 1982; Goff, 1990). Enzyme activity is demonstrated in preparations of purified virus particles treated with a low concentration of nonionic detergent and incubated with radioactively labeled deoxynucleotide triphosphates (dNTPs) in a solution containing salts and buffer. Different divalent cations are required, depending on the retrovirus, for optimal enzymatic activity (e.g., Mg^{2+} for MMTV, SRV, lentiviruses, and spumaviruses; Mn^{2+} for ASLV and MuLV). Under *in vitro* conditions, virions essentially remain intact, although the detergent permeabilizes the virion membrane, allowing entry of dNTPs. Reaction mixtures are incubated at 37°C, typically for 1–2 h, and activity is determined by measuring the amount of labeled dNTPs incorporated into DNA. In this endogenous reaction, viral genomic RNA is the template, and the primer for the first viral DNA strand (i.e., negative or minus strand) is the host-cell tRNA molecule packaged into the virion and bound to genomic RNA in the previous replication cycle. The primer for the second viral DNA strand (positive or plus strand) is a portion of the RNA genome (i.e., the polypurine track at PB+; see below) and the template is minus-strand viral DNA (Fig. 8). Detergent-activated MuLV particles are capable of producing complete infectious DNA *in vitro* (Rothenberg *et al.*, 1977; Gilboa *et al.*, 1979b).

At higher concentrations of nonionic detergent, virus particles are disrupted, and the solubilized RT will act on templates such as homopolymers of polyadenylate (poly-A). This exogenous reaction requires synthetic primers which may be short oligonucleotides complementary to the templates added to the *in vitro* reaction (e.g., oligo-dT$_{12-18}$ primer and poly-A template). The primer must have a free 3' hydroxyl group to initiate cDNA synthesis in these reactions. RT has the unique ability to utilize the template–primer combination of poly-C and oligo-dG$_{12-18}$, whereas cellular polymerases do not act efficiently on this template–primer combination (Modak and Marcus, 1977; Varmus and Swanstrom, 1982).

In vitro studies with purified enzyme have shown that the DNA synthesis reaction catalyzed by RT follows an ordered mechanism (Jacobo-Molina and Arnold, 1991, and references therein). In the first step, primer attaches to RT and subsequently this complex binds template and substrate. Binding sites on the RT polypeptide have been identified by cross-linking techniques which utilize photoaffinity labels on synthetic templates and primers (Tirumalai and Modak, 1991; see also Majumdar *et al.*, 1989). The divalent cation may be essential for an ionic

interaction that bridges the phosphate groups of the incoming deoxynucleoside triphosphates with the active-site aspartic acid residues in the
highly conserved sequence tyrosine–X–aspartate–aspartate (Kamer and
Argos, 1984; Doolittle *et al.*, 1989). This site also appears to initiate
polymerization on both RNA and DNA templates. RT binds substrate
deoxynucleoside triphosphates at a site near the primer-binding site;
thus, the proximity of these two sites favors phosphodiester bond formation (Nanduri and Modak, 1990; Tirumalai and Modak, 1991).

D. Polymerization Errors

Retroviruses, like several other RNA viruses, acquire mutations at a
high rate (Coffin, 1990b; Katz and Skalka, 1990). RT appears to play an
important role in the generation of diversity in retroviruses largely because this enzyme, like RNA polymerases of other viruses, lacks a 3'-to-
5'-exonuclease proofreading mechanism (e.g., editing). Cellular RNA
polymerase II also does not have a proofreading function; therefore, it is
possible that some mutations may be created in the nucleus during transcription of the integrated viral DNA genome by this cellular enzyme.
Measurements on the fidelity of purified preparations of RT are made
by determining the extent of nucleotide misincorporation in reactions
utilizing defined RNA or DNA templates (Batulla and Loeb, 1974;
Leider *et al.*, 1988). Synthetic homo- and copolymers as well as natural
genomes (e.g., single-stranded bacteriophage ϕX174 DNA) have served
as templates for these assessments. Examples of misincorporation (or
error) rates for RTs are 1 misincorporated nucleotide per 9000–17,000
nucleotides for ASLV and 1 per 30,000 nucleotides for MuLV (Batulla
and Loeb, 1974; Gopinathan *et al.*, 1979; Roberts *et al.*, 1989). In similar
assay systems, HIV-1 RT has a higher misincorporation rate that ranges
from 1 per 1700 to 1 per 4000 nucleotides (Preston *et al.*, 1988; Roberts
et al., 1988). For comparison, the nucleotide error rate for *Escherichia
coli* DNA polymerase, an enzyme that has proofreading capacity, is 1 in
100,000 nucleotides (Battula and Loeb, 1974). Sequence analysis of RT
reaction products made *in vitro* reveals that deletions are also produced
during reverse transcription (Roberts *et al.*, 1989). Accordingly, RT can
extend beyond a mismatch and may be capable of nonprocessive polymerization (i.e., jumping or slippage) under certain circumstances (Bebenek *et al.*, 1989; Yu and Goodman, 1992). Mutations (base-pair substitutions, deletions, and insertions) have been detected during the
course of viral infection in cell culture systems (Dougherty and Temin,
1986; Pathak and Temin, 1990a, b; Pulsinelli and Temin, 1991). Thus,
the analysis of polymerization errors caused by RT *in vitro* is significant

for elucidating mechanisms that generate diversity in retroviruses (Katz and Skalka, 1990; Coffin, 1990b).

E. Inhibitors of Reverse Transcription

Reverse transcription is inhibited by several classes of compounds that function through distinct mechanisms (Mitsuya et al., 1990; Goff, 1990; Tan et al., 1990; Arnold and Arnold, 1991; De Clercq, 1992). Actinomycin-D intercalates between the stacked bases in double-stranded DNA and thus sterically blocks polymerization by RT. This drug inhibits RT reactions utilizing DNA but not those with RNA templates. RT activity is blocked in vitro by substrate analogues such as dideoxynucleotide triphosphates which terminate chain extension. Viral replication is inhibited in tissue culture cells by adding the non-phosphorylated forms of these compounds, including 3'-azido-3'-deoxythymidine (AZT), dideoxycytidine (ddC), and dideoxyinosine (ddI) (Mitsuya et al., 1990). These nucleoside substrate analogues are of particular interest because RT is more sensitive to inhibition than are cellular polymerases. In experimental animal model studies, AZT inhibits infection of mice with MuLV (Ruprecht et al., 1990). Clinical investigations have revealed that AZT reduces virus load in HIV-infected individuals and appears to palliate disease symptoms that progress to AIDS (Mitsuya et al., 1990). Additional substrate analogues which inhibit RT as well as other DNA polymerases are phosphonoformic and phosphonoacetic acids; these agents may inhibit RT because they mimic the phosphate structure of the nucleotide substrate. A wide range of other classes of RT inhibitors have been described; these include polymeric compounds such as dextran sulfates and phosphorothioates, fuchsin, rifabutin, and tetrahydroimidazobenzodiazepinones (TIBO). However, little information is available on the mode of action by which these diverse compounds block the enzyme.

F. RNase H Activity in the Reverse Transcriptase Molecule

RT protein also contains RNase H activity which degrades the RNA moiety in RNA–DNA hybrids in both 5'-to-3' and 3'-to-5' directions (Moelling et al., 1971; Leis et al., 1973; Goff, 1990). Most studies show that the enzyme is largely endonucleolytic (Krug and Berger, 1989); however, some investigators have observed exonuclease activity (Schatz et al., 1990). RNase H plays an essential role in several steps of reverse transcription by removing template RNA during synthesis of viral minus-strand DNA and by creating the ribonucleotide primer for initiation of viral plus-strand DNA. In addition, the 5' capped nucleotide and the 3' poly-A tail are both removed by RNase H from the ends of viral RNA during the process of reverse transcription (see below).

The $\alpha-\beta$ heterodimeric RT complex of ASLV and the monomeric form of MuLV RT both possess RNase H activity (Fig. 4) (Goff, 1990 and references therein). Mutagenesis studies have revealed that the domain responsible for RNase H activity is in the C-terminus of the RT protein and that this activity is required for production of infectious virus (Fig. 4) (Tanese and Goff, 1988; Repaske et al., 1989). These same studies confirmed that the N-terminal portion of the RT protein has the RNA-dependent DNA polymerase activity. A separate primer-binding site has been identified in the RNase H domain (see below). Sequence alignments of the C-termini of many retroviral pol genes reveal that the identity and positions of eight amino acids are conserved in the RNase H domain; six of these amino acids are also conserved with E. coli RNase H (Repaske et al., 1989). Proof that RT and RNase H are separate domains in the RT polypeptide is based on studies in genetically engineered bacteria. Each domain of the MuLV pol gene has been individually expressed in bacteria and each produces the appropriate enzymatic activity (Tanese and Goff, 1988). Thus, RT and RNase H must utilize distinct and nonoverlapping active sites for binding nucleic acids and for catalysis.

G. Structural Models for the Reverse Transcriptase/RNase H Complex

The current understanding of retroviral RT is based largely on biochemical, immunochemical, enzymatic, and genetic analysis (Goff, 1990; Basu et al., 1990; Barber et al., 1990; Jacobo-Molina and Arnold, 1991; Howard et al., 1991). Powerful methods of X-ray crystallography will be required to give a high-resolution structure which accommodates known functions in a three-dimensional context (Jacobo-Molina et al., 1991). Nonetheless, a model of HIV-1 RT in the active reverse transcription complex has recently been proposed. Active HIV-1 RT is a heterodimer consisting of p66 (RT and RNase H domains) and p51 (RT domain) (see references in Barber et al., 1990, and LeGrice et al., 1991). Both polypeptide subunits in the heterodimer may be parallel to each other and the interface of these subunits is hydrophobic and probably forms part of the binding site for the primer–template substrate. An alternative proposal holds that the two subunits in the heterodimer are aligned in opposite orientations. In both models, the axis of symmetry of the heterodimer is oriented in the same direction as the template (Arnold and Arnold, 1991). The RNase H domain is in a position that follows the polymerization site. After reverse transcription, short stretches (7–13 nucleotides) of the template RNA are cleaved by RNase H; the length of these released oligonucleotides is approximately one turn of a double helical hybrid. Additional structural studies are necessary to validate this model as well as to develop a model for retroviruses that have a native RT that is active in the monomeric form (e.g., RT of MuLV). These structural studies will also be relevant for elucidating processing patterns of the *gag–pol* polyprotein during virion morphogenesis. Design and development of antiviral agents will require analysis of structure–activity relationships of these enzymes (i.e., RT and RNase H) unique to retroviruses and other retrotransposons (Barber *et al.*, 1990).

H. Reverse Transcriptase in Other Viruses and Retrotransposons

Reverse transcription plays an essential role in the life cycle of certain DNA viruses and retrotransposons in diverse eukaryotic organisms. Hepadnaviruses (e.g., human hepatitis B virus) and caulimoviruses (e.g., cauliflower mosaic virus) package DNA into virions but replicate via an intracellular RNA intermediate (reviewed by Ganem and Varmus, 1987; Mason *et al.*, 1987; W. S. Robinson, 1990). Accordingly, these viruses encode a *pol* gene that has homology with conserved regions in retroviral RT and RNase H (Toh *et al.*, 1983; Doolittle *et al.*, 1989). Retrotransposons in *Saccharomyces* (e.g., Ty-1), *Nicotiana* (Tnt-1 element), and *Drosophila* (e.g., copia) as well as many other eukaryotic transposable elements also encode *pol* genes that have some sequence homology with the retroviral enzyme (Toh *et al.*, 1985; Mount and Rubin, 1985; Grandbastien *et al.*, 1989; Doolittle *et al.*, 1989; Boeke and Corces, 1989) (see Chapter 4). Eukaryotic cells contain pseudogenes, which are intronless forms of cellular genes that normally contain exons and introns; it is hypothesized that pseudogenes and other retroelements arise from reverse transcription of spliced transcripts of cellular genes (Sharp, 1987; Weiner *et al.*, 1986). Whether a cell enzyme or RT from a retrotransposon mediates the formation of pseudogenes remains to be determined. Long interspersed elements (LINE-1 or L1) in mammalian DNA are repeated sequences with structural similarities to retrotransposons (Hutchison *et al.*, 1989). Recently, a human LINE-1 has been shown to encode a protein with sequence homology to retroviral RT, and this protein exhibits RT activity (Mathias *et al.*, 1991). The sequence conservation noted in all polymerases with RT activity is highest in the N-terminal portion, where the highly conserved motif tyrosine–X–aspartate–aspartate is found (Kamer and Argos, 1984; Yuki *et al.*, 1986; Doolittle *et al.*, 1989). This motif is

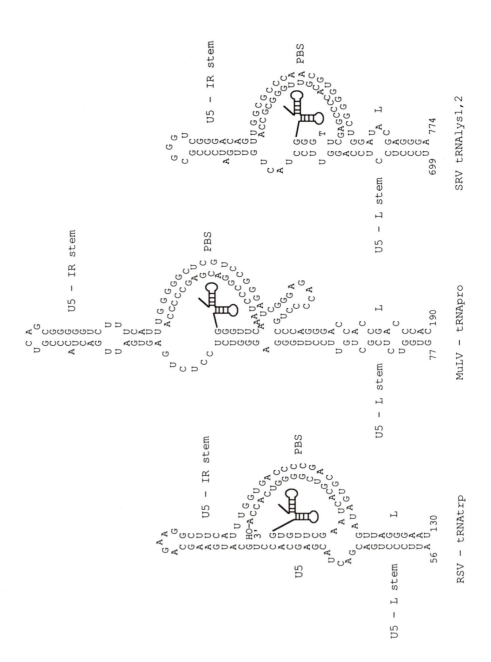

almost certainly part of the enzymatic active site (Goff, 1990; and references therein).

I. Mechanism of Retroviral Reverse Transcription

The model that describes the mechanism of retroviral reverse transcription requires two strand transfers or jumps to produce one molecule of double-stranded linear viral DNA from one or two single-stranded RNA genomes contained within a virion (Gilboa et al., 1979; Varmus and Brown, 1989; Coffin, 1990a). These strand transfers also involve displacement of preexisting sequences. Structural features of the template in virions (i.e., the dimeric complex of genomic viral single-stranded RNA, stem-loops in U5–leader sequences) are described above and illustrated in Figs. 8 and 9. The first DNA strand to be synthesized is complementary to the viral genome and is designated the negative (or minus) viral DNA strand. Subsequently, the positive (or plus) viral DNA strand is synthesized on the negative-strand DNA template.

Shortly after attachment and penetration of a virion into the cell, synthesis of viral DNA by multifunctional RT occurs in the cytoplasm within the uncoated viral core (Fig. 1). The signal that initiates reverse transcription after entry remains to be defined, although secondary structure elements around the PBS may play a role (Cobrinik et al., 1988, 1991). Double-stranded viral DNA is also made in in vitro reactions with purified retroviral particles; thus, the host-cell proteins are not essential for this early stage of viral replication. The initial primer is a specific host tRNA molecule located near the 5' end of the viral RNA genome, at a site designated the minus-strand primer-binding site (PB−) (Fig. 6; Table III) (Sawyer et al., 1974; Taylor and Illmensee, 1975; Taylor, 1977). This tRNA molecule was packaged into the virion during the previous infection cycle. Initiation of reverse transcription requires the interactions of U5 sequences with the leader, the inverted repeat (IR) sequences proximal to the tRNA binding site, and the TΨC loop se-

FIGURE 9. Predicted secondary structures of 5' noncoding regions and potential interactions with tRNA primers for several retroviruses. Numbers are nucleotides from the 5' end of viral RNA. Mutagenesis studies in the ASLV reveal that the interactions between U5 sequences and leader sequences, the inverted repeat proximal to the tRNA primer-binding site, and the TΨC loop in the tRNA primer are required for efficient initiation of reverse transcription (Cobrinik et al., 1988, 1991; Aiyar et al., 1992). ASLV, MuLV, and SRV contain binding sites for tRNA^Trp, tRNA^Pro, and tRNA^Lys, respectively. [Redrawn from Aiyar et al. (1992).]

quence in the tRNA primer (Fig. 9) (Aiyar et al., 1992). NC protein apparently plays a role in reverse transcription, since mutations in the NC domain of gag block viral DNA synthesis (Meric and Goff, 1989). The first viral DNA intermediate observed is a portion of minus-strand DNA that extends from the 3' end of the tRNA primer to the 5' end of the viral genome; this species, designated strong-stop DNA, is the predominant product in in vitro reaction systems (Fig. 8, step 1) (Haseltine et al., 1976; Coffin and Haseltine, 1977).

The first strand transfer requires the short sequence R that is repeated at each end of the retroviral genome, and R has a characteristic length depending on the retrovirus (Table III) (Coffin et al., 1978; Stoll et al., 1977). RNase H removes R from the hybrid structure at the 5' end of viral RNA (Fig. 8, step 2). Thus, the region of strong-stop DNA complementary to R is left unpaired and is available for hybridizing to the R sequence at the 3' end of the viral genome (Fig. 8, step 3). Accordingly, the first strand transfer involves repositioning of newly initiated minus-strand viral DNA from the 5' end to the 3' end of the viral RNA genome. This strand transfer has the potential to involve the end of either the same RNA genome (i.e., intramolecular event) or the second RNA genome copackaged in the virion (i.e., intermolecular event). To distinguish these possibilities, heterozygous virus particles that varied in sequences in U3 and U5 LTR domains were produced by site-directed mutagenesis, and progeny viruses were examined after a single round of infection. Sequence analysis of the marked LTRs revealed that the first strand-transfer step in reverse transcription is an intermolecular event (Panganiban and Fiore, 1988). In a similar study aimed at investigating the mechanism of retroviral recombination, transfer of minus-strand strong-stop DNA was observed to be both intermolecular and intramolecular (Hu and Temin, 1990b). Experimental differences may account for the apparent discrepancy [discussed in Hu and Temin (1990b)]. Although this issue requires resolution, the scheme adopted in this review is provisionally based on an intermolecular transfer event for minus-strand strong-stop DNA (Fig. 8, steps 4 and 5). At this strand-transfer step, the poly-A tail in the RNA genome is removed, presumably as a consequence of RNase H digestion of the 3' R (Fig. 8, steps 4 and 5).

Viruses with mutations in RNase H synthesize strong-stop minus-strand DNA but do not translocate this DNA from the 5' end of the RNA template to the 3' end of a second RNA genome template (Tanese et al., 1991). In vivo studies also show that complete copying of R is not required before this strand transfer occurs; thus, the relative contribution of R from each end of the RNA genome is determined by the length of the R region copied prior to the transfer taking place (Lobel and Goff, 1985). Subsequent to this strand transfer, RT continues to elongate minus-strand viral DNA by copying genomic RNA through the tRNA primer-binding site (i.e., PB−) (Fig. 8, step 4).

Synthesis of viral plus-strand DNA is initiated at a primer site near the 3' end of the viral genome (Fig. 8, step 5). RNase H activity degrades the portion of genomic RNA in RNA–DNA hybrids and generates an oligoribonucleotide primer at the polypurine track (PPT) near the 3' end of the viral genome (Resnick *et al.*, 1984; Smith *et al.*, 1984; Luo *et al.*, 1990). This site is designated the plus-strand primer-binding site (PB+). RT initiates synthesis of plus-strand DNA from the oligoribonucleotide at the PB+ and thus produces the intermediate designated plus-strand strong-stop DNA (Mitra *et al.*, 1979). The enzyme continues to elongate on the minus-strand DNA template and copies the portion of the tRNA molecule that is complementary to the PB– site. A methylated adenine residue in the tRNA molecule blocks RT in the PB– site. Accordingly, this modified nucleotide is the termination point for synthesis of plus-strand DNA. RNase H activity removes the tRNA primer from the 5' end of minus-strand DNA (Omer and Faras, 1982). Elongation of positive-strand DNA then continues to the end of the viral minus-strand template. The second transfer involves base-pairing of the tRNA primer-binding site (PB–) in plus-strand viral DNA to the 3' end of the minus DNA strand (Fig. 8, step 6). Through the use of molecularly cloned heterozygous viruses with distinguishable termini as described above, the second strand transfer was shown to be an intramolecular event (Panganiban and Fiore, 1988; Hu and Temin, 1990b). After the second transfer, RT continues to the ends of the templates and thus generates double-stranded linear viral DNA molecules that have LTRs and blunt termini (Fig. 8, step 7). All steps in reverse transcription take place in a nucleoprotein complex in the cell cytoplasm, and the product viral DNA product in this complex is subsequently transported into the nucleus for integration into the host-cell genome (see Section V below) (Brown, 1990).

J. Alternative Mechanisms for Synthesis of Viral Plus-Strand DNA

Differences among retroviruses are noted with respect to the synthesis of plus-strand DNA. For some retroviruses (e.g., MuLV and MMTV), the plus strand is elongated from one position, the PB+ site (Fig. 8, step 7) (Kung *et al.*, 1981). In contrast, synthesis of ASLV plus strands is discontinuous, and multiple priming sites are utilized to generate overlapping plus-strand DNA molecules (Boone and Skalka, 1981a, b; Kung *et al.*, 1981; Junghans *et al.*, 1982a, b). Breaks in genomic viral RNA may provide internal priming sites for the synthesis of plus strands. Alternatively, RNase H may produce internal oligonucleotide primers at sites that are rich in purines and thus resemble the PB+ site. If the viral plus strand is made in several pieces, then a ligase activity joins these pieces to form fully double-stranded linear viral DNA molecule. The host cell most likely provides the requisite DNA ligase activity.

Certain members of the lentivirus (e.g., visna virus) and spumavirus genera appear to prime efficiently at a sequence approximately in the middle of the genome (i.e., within the *pol* gene) as well as at the PB+ site near the 3' LTR (Mergia and Luciw, 1991, and references therein). The internal priming site for lentiviruses and spumaviruses is rich in purines and is very similar in sequence to the PB+ site of the respective virus. In these examples of discontinuous plus-strand synthesis, DNA ligase activity is required to produce the fully double-stranded linear viral DNA that is the precursor to integrated DNA. The significance of discontinuous synthesis of plus strands remains to be determined. A model for retroviral recombination is based on the notion that plus-strand viral DNA fragments from one viral genome invade a second viral genome that is in the process of reverse transcription (Katz and Skalka, 1990; Hu and Temin, 1990b). The invading DNA may then displace plus strands in the second genome and be assimilated to produce a double-stranded linear viral DNA molecule with a heteroduplex region(s).

V. RETROVIRAL INTEGRATION

A. Overview

The notion that retroviruses integrate into the cell genome is implicit in the provirus hypothesis (Temin, 1976) (see Chapter 1). Direct proof for covalent linkage of retroviral and cell DNA was obtained from molecular cloning and sequencing of viral–cell DNA junctions. In addition, genetic evidence based on analysis of viral mutants demonstrated that integration is required for viral replication. The provirus is co-linear with both unintegrated linear viral DNA and the virion RNA genome, and a short direct repeat of cell DNA is introduced at the site of viral DNA insertion into the cell genome. Many sequences in cell DNA are potential targets for integration, although there may be a preference for some sites (Varmus and Brown, 1989).

Critically important advances on the elucidation of retroviral integration have been made recently in several *in vitro* systems (Brown, 1990). Linear viral DNA (i.e., substrate) is processed by nicking and subsequently joined to target DNA in a reaction system containing purified IN protein; these reactions closely mimic steps in the integration process observed in infected cells. Studies with *in vitro* systems have also ruled out previous notions that circular viral DNA forms are intermediates in the integration pathway (see below). Accordingly, the subsequent elaboration will focus on properties and functions of IN as well as on features of both target DNA and the termini of the integrative precursor (i.e., linear viral DNA). Investigations of bacteriophage mu and other transposable elements have also provided useful paradigms for proposing and testing hypotheses relevant to the mechanism of retroviral integration (Kleckner, 1989, 1990; Pato 1989; Sandemeyer et al., 1990; see Chapter 4).

B. Significance of Retroviral Integration

Retroviruses have evolved a specific integration mechanism because higher eukaryotic cells do not have an efficient mechanism for precisely inserting exogenous DNA into the cell genome (Subramani and Berg, 1983; Miller and Temin, 1986; Bollag et al., 1989; Coffin, 1990c, and references therein). Transfection and microinjection studies with cloned DNA molecules reveal that cells utilize an illegitimate recombination mechanism for integrating foreign DNA, which undergoes random deletions and rearrangements inside the cell (Perucho et al., 1980; Luciw et al., 1983, 1984; Hwang and Gilboa, 1984). Under special circumstances, exogenous DNA is integrated by means of a homologous recombination pathway; however, the frequency of insertion by this means is generally low (Capecchi, 1989; Coffin, 1990c; Gridley, 1991).

Integration has significance for retroviral replication and implications for virus–host interactions. Proviruses are replicated in concert with the host-cell genome; thus, viral genetic information is faithfully passed on to daughter cells during each cell cycle (Temin, 1976). Stable vertical transmission of endogenous retroviruses is a feature of insertion of viral DNA into germline cells (see Chapters 2, 6, and 7). In addition, the integrated form of retroviral DNA is the template for transcription; synthesis of viral RNA transcripts involves the interaction of host-cell RNA polymerase II and other cellular factors with promoter elements in the LTR (Majors, 1990). This is an example of the parasitism of viruses, since retroviruses utilize components of the host-cell transcriptional apparatus for regulating viral gene expression. Integrated retroviruses are able to establish a state of latency and can also produce genetic changes in both infected cells and animals (e.g., activation of proto-oncogenes in somatic cells, mutations in the germ line) (see Chapter 1) (Jaenisch, 1976; Jaenisch et al., 1983; also see Varmus et al., 1981). Transduction of cellular genes (e.g., proto-oncogenes) is an infrequent but important consequence of retroviral integration [reviewed by Kung and Vogt (1991)]. Finally, precise and stable patterns of insertion into host-cell DNA are desirable features of retroviral vectors designed for therapeutic applications (see Section IX).

C. Termini of Linear Viral DNA

Specific sequences in the termini of unintegrated viral DNA play a major role in integration. The outside edges of the LTRs are short inverted repeats that define the retroviral attachment (*att*) site which functions in *cis* in the integration process (Varmus and Brown, 1989). These inverted repeats are generally less than 20 nucleotides in length and are often imperfect (Table IV). The terminal four nucleotides in linear DNA are 5'-AATG. . .CATT-3'; these sequences are highly conserved in the termini of retroviruses and other retrotransposons (Varmus and Brown, 1989; Sandemeyer *et al.*, 1990). Mutagenesis studies in several retroviral systems indicate that these terminal inverted repeats are required for efficient integration and viral replication; mutations in other regions of the LTR do not affect integration (Panganiban and Temin, 1983; Colicelli and Goff, 1988a; Cobrinik *et al.*, 1987; Roth *et al.*, 1989b). Presumably, IN protein forms a multimeric complex (e.g., homodimer or tetramer) and thereby interacts with the *att* sites (i.e., inverted repeats) at both termini of linear viral DNA (Fig. 10).

D. Nucleoprotein Complexes Containing Linear Viral DNA in Infected Cells

Following entry of the virion into the cell, a sequence of reverse transcription events in the cytoplasm produces linear double-stranded viral DNA molecules, and viral proteins remain associated with the newly synthesized DNA (Fig. 8). Biochemical fractionation procedures have been used to isolate this viral nucleoprotein complex from the cytoplasm (Brown *et al.*, 1987; Bowerman *et al.*, 1989; Fujiwara and Mizuuchi, 1988; Brown, 1990, and references therein; Farnet and Haseltine, 1990, 1991). In rate-zonal centrifugation conditions, the complex from MuLV-infected cells sediments at 160S and contains a molecule of linear viral DNA (ca. 20S) that has blunt ends. Viral IN is contained within the nucleoprotein complex, and CA, RT, and NC may also be present (Brown *et al.*, 1989). The nucleoprotein structure is permeable to macromolecules, because viral DNA in purified 160S cytoplasmic particles is sensitive to exogenously added DNase (Bowerman *et al.*, 1989). In cells infected with HIV-1, linear viral DNA is found in a cytoplasmic nucleoprotein complex that contains IN; however, other virion proteins are not detected in this complex (Farnet and Haseltine, 1991). Whether HIV-1 and MuLV are intrinsically different with respect to the composition of the integration complex or whether experimental differences account for the disparity in composition remains to be deter-

TABLE IV. Retroviral *att* Sites at the Termini of Linear Viral DNA[a]

	3' LTR	5' LTR
ASLV	AATGTAGTCTTATGCAATACTCTTG......... TTACATCAGAATACGTTATGAGAAC.........	.........CCTGCATGAAGCAGAAGGCTTCATT .........GGACGTACTTCGTCTTCCGAAGTAA
MuLV	AATGAAAGACCCCACCTGTAGGTTT......... TTACTTTCTGGGGTGGACATCCAAA.........	.........TACCCGTCAGCGGGGTCTTTCATT .........ATGGGCAGTCGCCCCCAGAAAGTAA
REV	AATGTGGGAGGGAGCTCTGGGGGA......... TTACACCCTCCCTCGAGACCCCCCT.........	.........ATCCGTAGTACTTCGGTACAACATT .........TAGGCATCATCAAGCCATGTTGTAA
MMTV	AATGCCGCGCCTGCAGCAGAAATGG......... TTACGGCGCGGAGGTCGTCTTTACC.........	.........CCTCAGGTCAGCCGACTGCGGCATT .........GGAGTCCAGTCGGCTGACGCCGTAA
SRV	AATGTCCGGAGCCCTGCAGCCCGGA......... TTACAGGCCTCGGCACGTCGGGCCT.........	.........TGTTGGTCCCGCGGGACGGGACATT .........ACAACCAGGGCGCGCCCTGCCCTGTAA

[a] Retroviral *att* sites are short inverted (imperfect) repeats at the ends of linear viral DNA. Nucleotides in these repeats are underlined. Integrase recognizes *att* sites in the linear viral DNA molecule contained within the cytoplasmic nucleoprotein complex and subsequently mediates a nicking reaction that removes the terminal TT dinucleotide [Figs. 9 and 10). During integration, the ends of viral DNA 5'-AATG…CATT-3' are converted to 5'-host-TG…CA-host-3'.

mined. The stoichiometry of IN per viral DNA molecule and the topology in the cytoplasmic nucleoprotein complex have not been determined. A specific structural arrangement of IN molecules may be necessary to hold the two viral DNA ends in close proximity to each other and to permit a precise association of the complex with cellular target DNA in the nucleus during integration (see below). The observation that linear retroviral DNA is contained within a nucleoprotein complex is consistent with the emerging view that recombination systems in a variety of organisms involve complex structures formed by DNA–protein and protein–protein interactions (Echols, 1986; Craig, 1988).

The nucleoprotein complex containing linear viral DNA enters the nucleus by a mechanism that remains to be elucidated. Potential nuclear localization signals on viral proteins may participate in active transport of the complex into the nucleus. IN of ASLV has been shown to localize to the nucleus and thus may play a role in the movement of the complex through nuclear pores (Morris-Vasios et al., 1988; Nigg et al., 1991). Other viral proteins (e.g., CA or NC gag proteins) in the complex may mediate this process. Alternatively, the viral nucleoprotein complex may enter the nucleus during mitosis when the nuclear membrane transiently diassembles.

An in vitro system that utilizes these viral nucleoprotein complexes has been devised to investigate the mechanism of retroviral integration (Brown et al., 1987, 1989; Fujiwara and Mizuuchi, 1988; Fujiwara and Craigie, 1989; Brown, 1990, and references therein). The 160S complex, purified from the cytoplasm of cells acutely infected with MuLV, is incubated with naked target DNA (e.g., bacteriophage lambda DNA) and divalent cations. Products of the integration reaction (i.e., recombinant DNA molecules) are scored and selected with sensitive genetic and cloning methods. Integration in this in vitro system is faithful in that proviral structures are obtained. Linear viral DNA loses the terminal dinucleotide from each end, covalent linkage is through the CA 3' dinucleotide in viral DNA termini, and a short duplication is produced in target DNA at the site of insertion (Fig. 10) (Brown et al., 1987, 1989; Fujiwara and Craigie, 1989; see also Hughes et al., 1981; Majors and Varmus, 1981). This in vitro system has been modified to evaluate the effects of topology of the target DNA; depending on experimental conditions, integration into minichromosomes is as efficient as or more efficient than integration into naked DNA (Pryciak et al., 1992) (see below). Supplements of extracts from infected cells to these in vitro systems and an exogenous energy source are not required (Brown, 1990, and references therein). Thus, the 160S complex apparently contains all components necessary for recapitulating the normal pathway of retroviral integration.

E. Genetics, Structure, and Topology of Integrase

IN protein is derived by proteolytic cleavage from the 3' portion of the *gag–pol* polyprotein (Fig. 4) (Leis *et al.*, 1983). Genetic studies with viruses containing site-specific mutations demonstrate the importance of this enzyme for viral replication. Mutations in the IN domain of the *gag–pol* gene do not affect early events up through reverse transcription and entry of the viral nucleoprotein complex into the nucleus (Donehower and Varmus, 1984; Hippenmeyer and Grandgenett, 1984; Schwartzberg *et al.*, 1984a; Panganiban and Temin, 1984b; Quinn and Grandgenett, 1988). Processing of the 3' ends of linear viral DNA in the cytoplasm is blocked, and integration as well as all subsequent steps in the viral life cycle are precluded (Brown *et al.*, 1989; Roth *et al.*, 1989a; see also Hagino-Yamagishi *et al.*, 1987).

Alignment of predicted amino acid sequences of IN proteins from diverse retroviruses reveals high conservation in this portion of the *pol* gene (Doolittle *et al.*, 1989). The N-termini of IN proteins of several diverse retroviruses contain a specific arrangement of histidine and cysteine residues that resembles the metal finger domain of proteins which bind nucleic acids (Johnson *et al.*, 1986; Berg, 1986). ASLV IN proteins with mutations affecting the histidine–cysteine motif at the N-terminus are able to bind DNA *in vitro*, although the cleavage and joining activities are reduced (Khan *et al.*, 1991; Mumm and Grandgenett, 1991). The central region of IN proteins is also highly conserved and shows similarities with sequences deduced for bacterial insertional (IS) elements (Doolittle *et al.*, 1989). Mutations in this central domain of ASLV IN abrogate the nonspecific mode of binding to DNA (Khan *et al.*, 1991).

Secondary structure predictions, which draw on a composite of hydropathy, amphipathicity, and chain flexibility profiles, revealed a highly conserved amphipathic helix in the IN proteins of several diverse retroviruses (Linn and Grandgenett, 1991). The proposed amphipathic helix (coiled coil pattern) may function in cooperative interactions of IN molecules and/or may influence binding to DNA by a mechanism similar to that of several eukaryotic transcriptional factors (Landschulz *et al.*, 1988; O'Shea *et al.*, 1989).

ASLV IN recovered from virions is phosphorylated whereas MuLV IN is not (Tanese *et al.*, 1986; Horton *et al.*, 1991). This modification appears to be mediated by a host cell protein kinase trapped in virions and occurs after particles are released from cells (Eisenman *et al.* 1980). Presumably, phosphorylation directs proteolytic processing (by viral PR) of the *gag–pol* polyprotein to generate mature ASLV IN.

IN may hold the ends of linear viral DNA together and may simultaneously interact with viral and host-cell DNA (Bushman *et al.*, 1990;

Grandgenett and Mumm, 1990) (Fig. 11). Accordingly, models are proposed on the basis of either one or two active sites on the IN molecule. If
the cleavages and strand-transfer reactions involve one molecule of IN,
then each molecule has two active sites. One site recognizes the end of
linear viral DNA and the other interacts with target cell DNA (dimeric
form of IN, Fig. 11). Alternatively, cleavage and strand transfer may
involve independent units whereby one IN molecule cleaves an end of
viral DNA and the other IN molecule acts on target DNA (tetrameric
form of IN, Fig. 11). IN may also interact with CA (i.e., major core
protein from *gag*) and, perhaps, with cellular proteins (Brown, 1990).
Domains of IN that mediate these potential protein–protein interactions remain to be defined. Presumably, different active sites on IN are
involved in nicking of viral DNA termini (in the cytoplasm) and cleavage of target DNA (in the nucleus). In addition, the sequence-dependent
recognition of viral DNA termini and the sequence-independent interactions with target cell DNA are probably mediated by distinct sites on
IN. Analysis of site-specific IN mutations in *in vitro* integration systems together with elucidation of its crystal structure will be essential to
map functional domains and to validate models that describe interactions of IN proteins with substrate and target DNA.

F. Function of Integrase Analyzed *in vitro* with Defined Components

Essential events mediated by IN protein in the retroviral integration
model are shown in Fig. 10 (Varmus and Brown, 1989; Kulkosky and
Skalka, 1990; Grandgenett and Mumm, 1990). In the nucleoprotein integration complex, IN may interact with both (blunt) ends of linear viral
DNA in a sequence-specific fashion by recognizing the short inverted
terminal repeats. A nick is introduced by IN at each terminus of the viral
DNA molecule so that two base pairs (TT) are lost and recessed 3' ends
are generated. After transport of the nucleoprotein complex to the nucleus, IN may mediate sequence-independent interactions between the
complex and target cell DNA. Subsequently, IN makes a staggered cut in
cell DNA and catalyzes single-strand joining of the recessed dinucleotide CA 3' ends of viral DNA to these cell DNA 5' overhangs. Finally,
host-cell enzymes may remove mismatched overhangs (i.e., AA dinucleotide at 5' viral ends) and seal single-stranded regions in target DNA,
thus establishing direct repeats at the site of integration. The size of the
direct repeat is characteristic for the infecting virus and does not depend
on the cell type; therefore, IN controls the size of the staggered cut.

The nicking function of IN has been directly examined in several *in
vitro* systems with purified IN protein and defined DNA substrates

(Katzman *et al.*, 1989; Katz *et al.*, 1990; Vora *et al.*, 1990; Bushman *et al.*, 1990; Sherman and Fyfe, 1990). Both virions and genetically engineered expression systems are the source of IN protein. Substrates that mimic the ends of LTRs are short linear double-stranded oligodeoxynucleotides. IN has endonucleolytic activity and makes nicks two nucleotides from the 3' hydroxyl ends of these synthetic DNA substrates; thus, two nucleotides that extend beyond the highly conserved AC residues are removed (Fig. 10). The complementary strands are not cut and the reaction is specific in that sequences representing termini of a heterologous retrovirus are not nicked at an analogous position. The *in vitro* system has been used to show that nucleotides proximal to the cleavage site are most critical for nicking. In ASLV, the minimum substrate size for efficient cleavage activity corresponds to the 15-bp terminal inverted repeat at the ends of linear viral DNA (i.e., viral *att* site) (Katzman *et al.*, 1989). These findings on substrate requirements for IN activity *in vitro* have been validated by genetic analysis of mutant viral genomes in cells. These genetic studies show that the terminal 10–20 nucleotides in viral DNA are essential for both efficient integration and viral replication in infected cells (Panganiban and Temin, 1983; Colicelli and Goff, 1988a; Cobrinik *et al.*, 1987; Roth *et al.*, 1989b). Thus, *in vitro* studies of IN, together with observations *in vivo*, demonstrate that this protein has appropriate endonucleolytic activity for preparing the integrative precursor. In addition, these *in vitro* findings provide enzymatic evidence that linear viral DNA is the precursor to the integrated provirus.

The joining (i.e., strand-transfer) reaction mediated by IN has also been examined in *in vitro* assay systems (Katz *et al.*, 1990; Craigie *et al.*, 1990; Bushman *et al.*, 1990; Bushman, and Craigie, 1991; Vora *et al.*, 1990). In one assay configuration with purified IN protein, substrate oligodeoxynucleotides with the recessed 3' hydroxyl ends produced by IN (i.e., donor DNA) are covalently linked to various sites in either the plus or minus strands of target oligonucleotides (i.e., acceptor DNA) (Fig. 10). Analysis of the reaction products by DNA sequencing revealed that joining is indeed through the recessed 3' end of the donor molecules (Fig. 10). Thus, this system demonstrates that IN nicks substrate DNA and mediates a strand-transfer reaction between substrate and target oligodeoxynucleotides (Katz *et al.*, 1990; Craigie *et al.*, 1990; Bushman *et al.*, 1990; Bushman and Craigie, 1991).

In another *in vitro* assay configuration, the fidelity of the strand-transfer reaction was evaluated by examining sequences in target DNA (Katz *et al.*, 1990; Bushman *et al.*, 1990; Craigie *et al.*, 1990). Linear plasmid molecules that have recessed 3' ends (i.e., donor DNA) similar to those produced by IN are incubated with purified IN protein and bacteriophage lambda target DNA (i.e., acceptor DNA). Integration products are recombinant phage containing plasmid inserts. Sequence

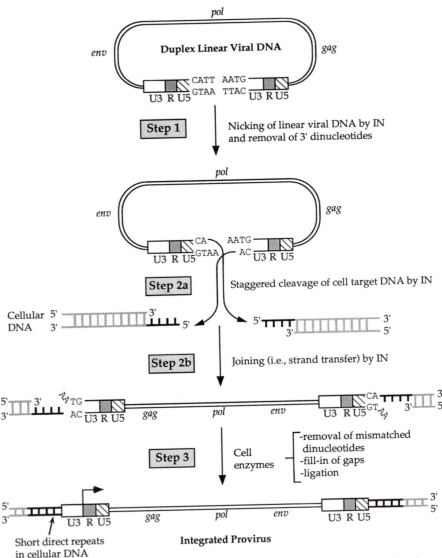

FIGURE 10. Model for integration of viral DNA into the host-cell genome. This figure emphasizes the interactions of the ends of linear viral DNA molecules with host-cell DNA during the integration process. Step 1: Double-stranded linear viral DNA is made by reverse transcription in a cytoplasmic nucleoprotein complex which also contains viral CA, RT, and IN and perhaps other virion core proteins. Either a dimeric or tetrameric complex of IN recognizes and binds to the short inverted repeats at the ends of linear viral DNA (viral *att* sites) (Table IV); consequently, both ends of the linear viral DNA molecule may be held in close proximity within the nucleoprotein complex (Fig. 11). IN nicks the blunt ends of linear viral DNA to remove a dinucleotide from each 3' end and thus generates recessed termini with free 3' hydroxyl groups. This nicking may occur in the cytoplasm

analysis of junctions between plasmid and phage DNA revealed a short direct repeat flanking the termini of the donor plasmid; in addition, joining was through the AC dinucleotide in the donor DNA. Integrated proviruses in cells infected with MuLV and ASLV are flanked by 4- and 6-bp direct repeats in cell DNA, respectively (Varmus and Brown, 1989, and references therein). Accordingly, *in vitro* reactions with IN protein from these two viruses faithfully recapitulated the predicted staggered cut and yielded the appropriate size for the direct repeat in target DNA (Katz *et al.*, 1990; Bushman *et al.*, 1990; Craigie *et al.*, 1990). Thus, reaction systems with defined components reflect major features of retroviral integration observed in infected cells (Fig. 10).

IN-mediated cleavage of target DNA and the joining reaction *in vitro* do not require an exogenous energy source (P. O. Brown, 1990, and references therein). Most viral DNA molecules have a 3' recessed end before entering the nucleus (Fig. 10); therefore, IN-mediated cleavage of the viral termini cannot be coupled to the joining reaction (P. O. Brown *et al.*, 1989; Roth *et al.*, 1989b). The 3' hydroxyl ends of linear viral DNA act as the nucleophile in a transesterification mechanism to attack a phosphodiester bond at the site of insertion into target DNA (Engelman *et al.*, 1991). In the HIV-1 and MuLV systems, the stereochemical course of the cleavage and strand-transfer reactions was determined by incorporating phosphorothioate linkages into the DNA substrate and target DNA. A phosphorothioate linkage has a chiral center; thus, an inversion will be observed in a one-step reaction and no change in configuration will be noted in a two-step mechanism (i.e., via a DNA–IN covalent intermediate). Analysis of integration *in vitro* with IN protein revealed that both the viral DNA cleavage reaction and the strand-transfer reaction involve inversion of the chirality of the participating phosphoro-

←——————————————————————————————————————

and/or during transport of the nucleoprotein complex to the nucleus. Steps 2a and 2b are shown separately for pedagogical reasons; however, the cleavage and joining events probably take place in a concerted reaction. Step 2a: In the nucleus, the nucleoprotein complex attaches to a site in target-cell DNA. IN probably mediates selection of the target site and numerous preferred sites have been identified within the cell genome; however, structural features of chromatin may influence site selection. Cleavage of target cell DNA and joining to termini of linear viral DNA occur in a coordinated reaction. A staggered cut is made by IN in target DNA to generate a short overhang (4–6 bases, depending on the retrovirus) with a phosphorylated 5' end. Step 2b: The 5' phosphorylated end of target DNA is joined by IN to the 3' hydroxyl ends of linear viral DNA. Energy for the joining reaction may come from (i) a transient DNA–IN intermediate that stores energy from the cleaved phosphodiester bond in a transient high-energy DNA–protein bond, or (ii) from a transesterification reaction involving nucleophilic attack of a phosphodiester bond at the site of insertion in target DNA by the 3' hydroxyl end of linear viral DNA. Step 3: The fully integrated provirus is generated after several steps that probably involve host-cell enzymes; these steps include removal of the two mismatched bases at each 5' terminus of viral DNA, completion of the single-stranded gap in target DNA, and ligation of the remaining ends of viral and target DNA. [Redrawn from Varmus and Brown (1989).]

thioate groups; this outcome supports a one-step transesterification mechanism (Engelman *et al.*, 1991). In addition, IN has been shown to mediate exchange of DNA 3' hydroxyl groups *in vitro* with oligomeric substrates that mimic the integration intermediate (Fig. 10, step 2b) (Chow *et al.*, 1992). This latter observation demonstrates that the protein catalyzes a template-guided DNA transesterification reaction. A one-step transesterification mechanism has also been proposed to account for the transposase function of bacteriophage mu (Mizuuchi and Adzuma, 1991).

Additional support for a model in which IN is not a reactant but an enzyme is based on the finding that this protein promotes the integration reaction in reverse via a process designated disintegration (Chow *et al.*, 1991). Oligonucleotides that mimic the integration intermediate (Fig. 10, step 2b) are substrates for monitoring disintegration. IN mediates a DNA splicing reaction analogous to RNA splicing when the viral DNA component of the disintegration substrate is single-stranded; this activity of IN is speculated to be relevant for joining the viral 5' end to target DNA (Fig. 10, step 3) (Chow *et al.*, 1992). Additional analysis of the forward and reverse reactions provides a means to determine physical constraints on viral and target DNA molecules as well as to elucidate the influence of accessory viral and cellular factors on the integration process.

An alternative mechanism that has been suggested to provide energy for retroviral integration involves a step mediated by a DNA–IN covalent intermediate (Katzman *et al.*, 1991). Precedents include site-specific recombinases and topoisomerases which store the energy of a cleaved phosphodiester bond via a transient high-energy DNA–protein phosphodiester bond (e.g., bacteriophage lambda integration) (Wang, 1985; Craig, 1988). Covalent linkage of ASLV IN and target DNA has been detected in *in vitro* reactions; however, this linkage may be adventitious and may not represent an integration intermediate (Katzman *et al.*, 1991).

After the IN-mediated strand-transfer reaction in infected cells, viral and/or cellular enzymes remove the mismatched terminal dinucleotides, fill in the gaps, and ligate the remaining DNA strands (Fig. 10). It is tenable that the gap may be filled in by RT and that IN may cleave the unpaired terminal dinucleotides (at the 5' end of viral DNA) and subsequently join viral and target DNA (Chow *et al.*, 1992). Although IN is sufficient to mediate the cleavage and joining functions for integration, viral and cellular factors may affect the efficiency of these critical steps in retroviral replication. *In vitro* integration systems can be used to investigate the roles of accessory factors in the integration pathway, and these systems offer unique opportunities to test and develop inhibitors of IN function.

G. Cell Target Site for Retroviral Integration (Sequence Specificity and Topology)

Critical issues in retroviral integration are target site specificity and topology. Studies on proviruses in infected cells show that many sites in the cell genome are available for integration and that very little sequence preference is exhibited. To evaluate site specificity, about ten host–viral junctions for several retroviruses were compared; a common target sequence was not disclosed by these comparisons (Dhar et al., 1980; Varmus and Brown, 1989, and references therein). However, results from an extensive analysis of ASLV integrations in avian cells supports the notion of preferred sites or "hot spots" in the cell genome (Shih et al., 1988). A selective cloning procedure was used to obtain a large library of host–viral junctions from a culture of cells infected with ASLV; randomly selected cloned integration regions were used as probes to map the entire recombinant library. Approximately 20% of integrations mapped to 1 of about 1000 target regions in the avian cell genome; strikingly, independent insertions into the same region were identical. Thus, ASLV integrates into a hot spot about 100,000 times more often than expected for a random site (Shih et al., 1988). Whether these cloned target sites from the avian genome will be recognized as preferred sites (in the form of naked DNA substrates) for integration in vitro remains to be determined.

A unique in vivo system for examining target site specificity is based on the observation that retroviruses integrate into the genomes of DNA viruses (i.e., herpes viruses) in coinfected cells. Cultures of avian cells were coinfected with REV and Marek's disease virus (MDV) and, after several passages, MDV minichromosomes were recovered and found to contain integrated REV DNA (Isfort et al., 1992). Sequencing revealed that the REV LTR lost the terminal two nucleotides and was inserted into duplicated (5 bp) MDV target sequences; this is the pattern observed at retrovirus-cell DNA junctions when REV integrates into the cell genome. In similar coinfection experiments, ALV has also been shown to insert into the MDV genome. Examination of natural MDV isolates also demonstrated the presence of REV LTRs that appeared to be correctly integrated (Isfort et al., 1992). The evolutionary and functional significance of retroviral insertion into herpes viruses remains to be determined. Nonetheless, this coinfection system is a means to investigate target site specificity (and topology) in retroviral integration in cells. In addition, nonessential genes of herpes viruses can be mapped by retroviral insertional mutagenesis, a procedure formally analogous to transposon-induced insertional mutagenesis.

Topological features of target cell DNA may influence retroviral integration. The *in vitro* integration systems described above demonstrate that naked DNA is a target for IN-mediated joining reactions (Brown, 1990). This finding reveals that neither supercoiling nor chromatin structure is required for integration; however, these features may affect the efficiency of target site selection. Proviruses are found more frequently in actively transcribed regions and in DNase I hypersensitive sites than in regions of closed chromatin (Robinson and Gagnon, 1986; Vijaya *et al.*, 1986; Rohdewohld *et al.*, 1987; Scherdin *et al.*, 1990; Mooslehner *et al.*, 1990). Actively dividing cells support both synthesis of viral DNA and integration more efficiently than do nondividing cells (Varmus *et al.*, 1977; Humphries *et al.*, 1981; Harel *et al.*, 1981; Miller *et al.*, 1990). Accordingly, cell replication may uncover areas in the target DNA for attachment of the viral nucleoprotein complex. Alternatively, the requirement for cellular DNA replication may mean that cellular functions expressed in the S phases are essential for either entry of the viral nucleoprotein complex into the nucleus or efficient integration (Richter *et al.*, 1984). In the analysis of the integration patterns in ASLV-infected cells described above, the nature of target sites with respect to chromatin topology was not determined (Shih *et al.*, 1988).

To evaluate the role of structural accessibility of target DNA *in vitro*, minichromosomes made from either a yeast plasmid or a mammalian viral genome (e.g., SV40) are used as substrates in integration reactions containing the MuLV 160S nucleoprotein complex from infected cells (see above) (Pryciak *et al.*, 1992). Depending on reaction conditions (e.g., concentration of target, presence of polyamines), minichromosome targets are as efficient as or more efficient than naked DNA targets. Sequencing and cloning revealed that the junctions of retroviral DNA and minichromosome target DNA mimic the authentic integration pattern observed in infected cells (i.e., 2 bp are lost from linear viral DNA and a 6-bp duplication is introduced into the target). A preference was not observed for insertion into exposed regions free of nucleosomes. However, target sites were spaced at about 10 bp. In addition, some sequence selectivity was demonstrated, since many sites contained multiple insertions. Thus, retroviral integration in this *in vitro* system occurs preferentially into the exposed face of the DNA helix in a nucleosome. Perhaps the nucleosomal arrangement presents the major groove of the DNA helix to the integration apparatus, and DNA on the inside face of the nucleosome core may be sterically inaccessible. Alternatively, a bend may be induced in target DNA pointed outward from the nucleosomal core; accordingly, this feature of altered DNA helical structure may influence the retroviral integration process. In either case, rotational orientation of the DNA target appears to control target site specificity (Pryciak *et al.*, 1992; Simpson, 1991). Additional studies are required to identify the factors that govern rotational orientation of

DNA in nucleosomes and to establish whether this topological feature is a determinant of retroviral integration *in vivo* (also see Pryciak and Varmus, 1992)

H. Circular Viral DNA in Infected Cells

Several forms of circular viral DNA forms, in addition to linear viral DNA molecules, are also produced in acutely infected cells (Guntaka *et al.*, 1976; Shoemaker *et al.*, 1980, 1981; see also Lee and Coffin, 1990). Circular forms are found predominantly in the nucleus and are detected at early times after infection. The most abundant circular species is a supercoiled viral DNA molecule with one copy of the LTR; the mechanism that produces this form may be homologous intramolecular recombination between the LTRs of a linear viral DNA molecule. A less abundant form of circular viral DNA has two juxtaposed LTRs (Shank *et al.*, 1978). This form appears to arise from intramolecular ligation of the ends of linear viral DNA. Sequencing has revealed that in a portion of the circular molecules both copies of the LTR are not complete at the point of joining (i.e., circle junction). Extensive studies of circle junctions in unintegrated ASLV DNA recovered from infected cells revealed that multiple genetic determinants (i.e., the *pol* and *gag* genes acting in *cis* and sequences in the LTR acting in *trans*) influence both the proportion of circular molecules with LTR-associated deletions and the relative amounts of circular forms with one or two copies of the LTR (Olsen *et al.*, 1990). Nevertheless, a small proportion of circular forms in cells infected with a variety of retroviruses contain two intact LTRs; thus, the circle junction in this form of viral DNA has two adjacent copies of the viral integration site (i.e., *att*) (Varmus and Brown, 1989).

Previous views of the retroviral integration process held that viral supercoils were the integrative precursor and mechanistic models were proposed on the basis of circular intermediates (Varmus and Swanstrom 1982, 1985, and references therein). In a circular viral DNA molecule with one copy of the LTR, a staggered cut at each end of the LTR was a predicted step in generating a precursor. This cleavage produces an integrative intermediate that can be accommodated by the Shapiro model proposed for intermolecular transposition in bacterial systems (Shapiro, 1979). In an *in vitro* system, the IN protein of ASLV cleaves circular DNA substrates containing two LTRs (i.e., circle junction) at precisely two nucleotides 5′ to the circle junction, thus yielding sites in viral DNA that are joined to cell DNA in the integration pathway (Grandgenett and Vora, 1985; Duyk *et al.*, 1985). This and other observations based on analysis of integration patterns of genetically engineered viral mutants have supported the role of circular viral DNA molecules as integrative precursors (Panganiban and Temin, 1984a). However, recent results obtained largely from extensive studies in several *in vitro* systems clearly favor the linear form of viral DNA as the direct precursor to the integrated provirus (see above) (Brown *et al.*, 1987, 1989; Fujiwara and Mizuuchi, 1988). Accordingly, viral supercoils with one and two copies of the LTR probably do not play an essential role in the viral life cycle.

Additional circular viral DNA forms have been observed in infected cells (Varmus and Brown, 1989). Multimers that consist of head-to-tail monomers may be by-products of the reverse transcription process (Goubin and Hill, 1979; Kung *et al.*, 1980). Some circular molecules appear to be due to autointegration; the viral *att* sites at the end of a linear viral DNA molecule are inserted into an internal target sequence in the same viral DNA molecule. In some instances, these autointegrants show the hallmarks of retroviral integration (i.e., removal of a dinucleotide at the *att* sites and a short duplication in the target sequence) (Shoemaker *et al.*, 1980, 1981). A cell-free system from ASLV-infected cells demonstrates autointegration as well as normal integration into exogenous target DNA (Y. M. H. Lee and Coffin, 1990). The significance (if any) of all these circular DNA forms for retroviral replication is not known.

I. Role of Host Cell in Retroviral Integration

Retroviral DNA synthesis and integration are influenced by events in the host cell cycle and by specific cellular factors. Nondividing cells in the stationary phase of the cell cycle do not support efficient synthesis of unintegrated viral DNA (Varmus *et al.*, 1977; Fritsch and Temin, 1977; Humphries *et al.*, 1981). A stable intermediate(s) may persist in the form of viral RNA or a mixture of partially reverse-transcribed forms, and subsequent stimulation of cell division by a variety of agents (e.g., serum, mitogens) produces physiological changes in cells so that all phases of viral replication can occur (Varmus *et al.*, 1977; Harel *et al.*, 1981; Zack *et al.*, 1990). Genetic studies have also implicated cellular factors in the control of retroviral DNA synthesis and integration. An X-linked mouse gene controlling cell DNA synthesis also affects production of unintegrated linear MuLV DNA. Defective forms of unintegrated linear MuLV DNA molecules accumulate at the nonpermissive temperature in a cell line that contains a temperature-sensitive defect in cell DNA replication; the precise nature of the cellular defect remains to be determined (Richter *et al.*, 1984). Another example of cellular control of retroviral integration involves the mouse Fv-1 restriction gene. A determinant mapping to the *gag* gene of MuLV influences integration in cells of different Fv-1 genotypes (Boone *et al.*, 1983; Ou *et al.*, 1983). In a proposed model, the Fv-1 gene product interacts with the CA protein of an infecting subviral core particle and prevents either integration or entry into the nucleus (Jolicoeur and Rassart, 1980, 1981; Yang et al., 1980). Further studies are required to elucidate the potential role of host-cell factors in maturation, transport, and integration of retroviral DNA.

J. Summary of the Model for Retroviral Integration

A summary of the major events in the retroviral integration pathway is outlined below and schematically presented in Fig. 10. Background information for this model, reviewed above, is drawn from *in vivo* (i.e., cell culture) as well as *in vitro* studies on several different retroviruses (Varmus and Brown, 1989; Kulkosky and Skalka, 1990; Grandgenett and Mumm, 1990; Katz et al., 1990; Engelman et al., 1991; Sandemeyer et al., 1990). Not all aspects of the integration pathway have been rigorously demonstrated for any one retrovirus and many important issues remain to be elucidated. Accordingly, the proposed model may be modified and revised as new findings are made. Nonetheless, this pathway provides an operational framework for additional studies on the components (both viral and cellular) and mechanisms that control retroviral integration.

Reverse transcription yields a double-stranded linear viral DNA molecule which is part of a cytoplasmic nucleoprotein complex that contains and IN protein. The termini of viral DNA are short inverted repeats that define the *cis*-acting *att* site, and these ends are blunt. Presumably, a multimeric complex of IN holds the ends of the linear viral DNA molecule in proximity to each other (Fig. 11). Within the nucleoprotein complex, IN protein nicks a dinucleotide from each 3' end of the viral DNA to generate recessed termini with free 3' hydroxyl groups (Fig. 10, step 1). This processing event may occur in the cytoplasm and/or during migration of the complex into the nucleus. The nucleoprotein complex enters the nucleus and attaches to target cell DNA (Fig. 10, step 2). IN introduces a staggered cut in the target DNA to generate a short overhang (4–6 nucleotides, depending on the retrovirus) with a phosphorylated 5' end (Fig. 10, step 2). These 5' phosphorylated ends are joined by IN to the recessed 3' hydroxyl ends of linear viral DNA (Fig. 10, step 3). Cleavage and joining take place in a coordinated reaction that derives energy from a transesterification step. Cellular and/or viral enzymes may remove the two mismatched nucleotides at each 5' terminus of viral DNA, fill the single-stranded gap, and nick-seal (i.e., ligate) the remaining ends to generate the fully integrated provirus (Fig. 10, step 4).

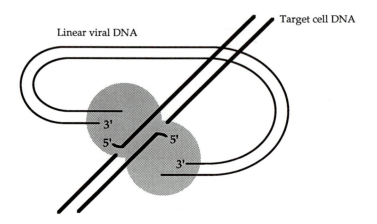

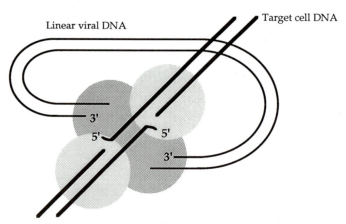

FIGURE 11. Models for the interaction of IN protein with the ends of linear viral DNA and target cell DNA. Each shaded sphere represents a molecule of IN protein. The double lines represent double-stranded viral and cellular DNA. In the cytoplasmic nucleoprotein, either a dimeric or tetrameric complex of IN binds to the short inverted repeats at the termini of linear viral DNA (i.e., viral *att* sites). The complex migrates to the nucleus and attaches to target cell DNA. IN probably mediates sequence-independent binding to target DNA. These models predict at least four functional domains on the IN protein: (i) a site(s) for binding and nicking the termini of linear viral DNA, (ii) a site(s) for IN–IN interactions to generate either the dimeric or tetrameric complex, (iii) a domain for recognizing and nicking target cell DNA, and (iv) a site for joining the 3′ ends of viral DNA to 5′ ends of target DNA.

Nicking and joining steps mediated by retroviral IN proteins are very similar to the transposase functions of bacterial transposons mu and Tn10 (Craigie and Mizuuchi, 1987; Pato, 1989; Kleckner, 1989). Both IN and transposase proteins have site-specific endonucleolytic activities which generate recessed 3' ends in the mobile DNA element, both catalyze a staggered cut in target DNA, and both join the recessed 3' ends of the element to the 5' ends of nicked target DNA. Thus, these common features support the notions that the basic biochemical mechanisms are similar for IN and transposase and that these processes are evolutionarily conserved.

VI. SYNTHESIS OF RETROVIRAL RNA

A. Overview of Viral RNA Synthesis

Synthesis of RNA from viral DNA (i.e., transcription) takes place in the cell nucleus and is mediated by the multisubunit complex of cellular RNA polymerase II (Young, 1991; Sawadogo and Sentenac, 1990). Regulation of viral transcription involves the interplay of specific cellular protein factors and viral sequence elements (Johnson and McKnight, 1989; Mitchell and Tjian, 1989), and the LTR contains cis-acting elements that control initiation and processing (i.e., polyadenylation) of viral transcripts, although internal sequences (between the LTRs) govern splicing (Stoltzfus, 1988; Majors, 1990). Transport of viral RNA to the cytoplasm appears to be closely associated with splicing and/or poly-A addition. Retroviruses with a simple genetic organization (e.g., ASLV, MuLV) depend entirely on host-cell components and cis-acting viral sequences for controlling RNA synthesis (Majors, 1990). In contrast, the complex retroviruses (e.g., HTLV-I, HIV) encode transcriptional and posttranscriptional transactivators which play important control functions in synthesis and processing of viral RNA (Green and Chen, 1990; Haseltine, 1991).

The amount of retroviral-specific RNA in infected cells is small; generally, less than 1% of total RNA is viral (Varmus and Swanstrom, 1982, 1985; Stoltzfus, 1988). Retroviral transcription is closely linked to regulatory events in the cell. Dividing cells support viral gene expression more efficiently than do nondividing cells (Humphries and Temin, 1974; Humphries et al., 1981). In addition, some retroviruses utilize cellular activation mechanisms to modulate levels of viral RNA (Nabel and Baltimore, 1987; Tong-Starksen et al., 1987; Tong-Starksen and Peterlin, 1990). An example is the induction of MMTV transcription by glucocorticoid hormones, which act on regulatory sequences in the viral LTR (Majors, 1990).

B. Significance of Retroviral Transcription Mechanisms

Biological properties such as cell tropism are determined, at least in part, by the levels of cellular factors that interact with viral regulatory sequences in the LTR. These transcriptional regulatory sequences control viral replication rate, organ tropism, and patterns of disease in infected animals (Fan, 1990; Kung and Vogt, 1991). Thus, an analysis of viral transcription is essential not only for elucidating mechanisms regulating viral gene expression, but also for understanding pathogenesis and for designing retroviral vectors for gene transfer.

C. Basal Promoter Elements in the LTR

The retroviral LTR flanks the viral DNA genome and consists of sequences derived from both the 3′ end (i.e., U3) and 5′ end (R–U5) of viral RNA (Fig. 6). This transposition of sequences is a consequence of reverse transcription of viral genomic RNA into DNA (Fig. 8). LTRs vary from about 300 to 1200 bp (Table III). In the integrated provirus, both LTRs are covalently linked to host-cell DNA and the 5′ LTR serves as the promoter to initiate viral RNA synthesis (Figs. 1, 11, and 12). Promoters are DNA sequence elements, located at the 5′ ends of cellular and viral genes, that contain binding sites for various protein factors that initiate and regulate transcription (Breathnach and Chambon, 1981; Mitchell and Tjian, 1989; Lewin, 1990). Figure 12 is a schematic drawing that shows *cis*-acting sequence elements in the LTRs that control viral transcription of ASLV and MuLV (Speck and Baltimore, 1987; Stoltzfus, 1988; Majors, 1990).

Many promoters for viral genes as well as cellular genes contain a TATA box, which is a sequence element recognized by a cellular factor designated TATA-binding protein (TBP) (Greenblatt, 1991, and references therein). This factor is a basic part of the cellular transcriptional apparatus that involves cellular RNA polymerase II, a multisubunit enzyme (Sawadogo and Sentenac, 1990; R. A. Young, 1991). Generally, the TATA box, located 20–30 bp upstream from the initiation site, is required for accurate and efficient initiation (Lewin, 1990; Greenblatt, 1991). All known retroviruses have a TATA box in the U3 domain of the LTR (Fig. 12). In the model for transcription in eukaryotic cells, TBP binds to the TATA box and interacts with specific cellular factors to form a preinitiation complex. Subsequently, components of the multisubunit RNA polymerase II system interact with the preinitiation complex to form a stable initiation complex, which then incorporates ribonucleotide triphosphates into RNA in a template-dependent reaction (Lewin, 1990; Sawadogo and Sentenac, 1990; Greenblatt, 1991). Viral transcripts are modified by cellular capping enzymes which add an inverted guanyl nucleotide to the 5' ends of newly initiated RNA (Stoltzfus, 1988); the retroviral cap site, designated +1, also defines the 5' end of the R region in the LTR (Figs. 6 and 12).

Regulation of transcription initiation also involves a variety of other cellular factors that interact with sequence elements in the U3 domain upstream from the TATA box in the LTR (Majors, 1990, and references therein). One class of cis-acting regulatory sequences, designated upstream elements, are usually positioned within 150 bp of the TATA sequence (Muller et al., 1988; Lewin, 1990). Upstream elements are generally 8–20 bp long and serve as binding sites for specific cellular transcription factors (Mitchell and Tjian, 1989; P. F. Johnson and McKnight, 1989); examples for the ASLV and MuLV systems are shown in Fig. 12 (Majors, 1990). The TATA box together with upstream elements is often referred to as the basal promoter (Lewin 1990). Transcription initiation in the LTR is regulated by precise combinations and arrangements of these binding sites, and changes in upstream elements have significant effects on rates of RNA synthesis (Lewin, 1990; Ptashne and Gann, 1990; Majors, 1990). Analysis of transcriptional regulatory elements in the U3 region of the ASLV LTR have not revealed negative-acting sequences (Luciw et al., 1983; Cullen et al., 1985b); however, the U3 domains of REV, MuLV, and HIV-1 contain negative regulatory elements (Rosen et al., 1985b; Hirano and Wong, 1988; Flanagan et al., 1989, 1992). Although transcription of retroviruses with simple genomes is generally regulated through cis-acting elements upstream from the start site of RNA synthesis (i.e., cap site at +1), in some instances R-region sequences influence viral gene expression in a positive fashion either at initiation or at a posttranscriptional step(s) (Ridgway et al., 1989; Cupelli and Lenz, 1991).

Retroviral promoters play a direct role in oncogenesis mediated by replication-competent retroviruses that do not encode transforming genes (i.e., oncogenes). This observation was first made in chickens developing B-lymphoid tumors after infection with ALV (see Chapter 6). Analysis of proviruses in these tumors revealed that the virus had integrated next to the c-myc proto-oncogene (Hayward et al., 1981; Payne et al., 1981). In normal cells, c-myc is transcribed from its native promoter at low levels; however, ALV tumors contain high levels of c-myc transcripts directed by the proviral LTR inserted upstream from the c-myc promoter. This mechanism of retroviral neoplasia is termed promoter insertion. Studies in several additional avian retroviral systems as well as in MuLV, MMTV, and FeLV tumors have also provided examples of direct LTR control of cellular proto-oncogenes adjacent to integrated proviruses (see Chapters 6 and 7) (Kung, 1991; Nusse, 1991; Neil et al., 1991; Tsichlis and Lazo, 1991).

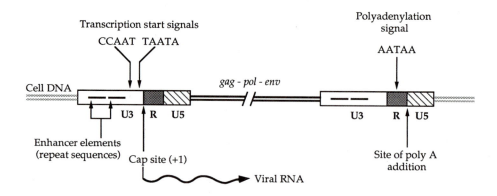

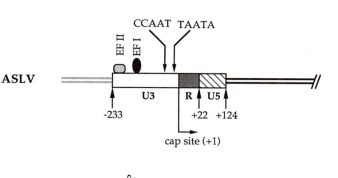

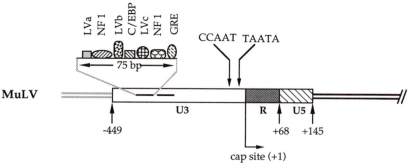

FIGURE 12. Transcriptional control elements in the retroviral LTR. The integrated form of the provirus for a prototype retrovirus is shown together with several *cis*-acting elements in the U3, R, and U5 domains of the LTR that play roles in regulating viral transcription. In addition, binding sites for cellular transcription factors are shown for the 5′ LTRs of ASLV and MuLV. The cap site is the location of the 5′ end of viral transcripts. Cellular transcription factor TBF binds the TATA box and controls the formation of an initiation complex with cellular RNA polymerase II. An upstream element recognized by the CCAAT transcription factor/nuclear factor 1 (CTF/NF-1) is also shown. The U3 domain of the ASLV LTR binds cellular enhancer factors 1 and 2 (EF-I and EF-II). Two 75-bp direct repeats are located in the MuLV LTR. These repeats are complex enhancer elements

D. Enhancer Elements Regulate Viral Transcription

Another important class of *cis*-acting sequences that regulate viral as well as cellular promoters are enhancers (Levinson *et al.*, 1982; Luciw *et al.*, 1983; Gruss and Khoury, 1983; Maniatis *et al.*, 1987). These elements, generally positioned 100–200 bp upstream of the TATA box, bind specific cellular factors that enhance or activate transcription from the basal promoter (Atchison, 1988; Kriegler, 1990). In some cellular genes (e.g., immunoglobulin light- and heavy-chain genes), the enhancer is located in an intron downstream from the start site of transcription (Lenardo *et al.*, 1987). Locations of the ASLV and MuLV enhancers in the U3 domains of the respective LTRs for each virus are shown in Fig. 12. A segment within the ASLV *gag* gene also has enhancer activity (Arrigo and Beemon, 1987; Karnitz *et al.*, 1987; Stoltzfus *et al.*, 1987b). Some enhancer elements appear to function only in specific cell types from certain species, whereas others have a broad range (Kriegler, 1990). Thus, the nature of the enhancer is a fundamental determinant of biological properties of retroviruses, including cell tropism and species range as well as pathogenic potential (see below).

Transient expression assays with plasmids containing cloned transcription units have revealed that enhancers function in an orientation-independent and position-independent fashion. An enhancer usually has a core domain, about 10–15 bases long, that binds a specific cellular factor, and different enhancers do not show extensive sequence homology (Maniatis *et al.*, 1987; Kriegler, 1990). The fully competent enhancer contains a hierarchy of binding domains and thus may have a modular structure that is up to 70 bases long (Herr and Clarke, 1986) (Fig. 12). Promoters may have duplicated copies of the entire enhancer and, in some instances, a functional enhancer is created by multimerization of enhancer core domains. The mechanism(s) by which these transcriptional elements regulate initiation is not fully resolved, although several possibilities have been proposed (Muller *et al.*, 1988; Dynan, 1989). An enhancer-binding factor may change the conformation of chromatin and thereby make proximal as well as distal promoters more accessible to other cellular transcription factors and RNA polymerase II.

ASLV has an enhancer in the U3 domain of the LTR that is active in a variety of cell types (Luciw *et al.*, 1983; Laimins, 1984a; Ju and Cullen, 1985). The enhancer in U3 appears to be composed of three domains,

---

which contain binding sites for several cellular transcription factors: leukemia virus proteins a, b, and c (LV-a, -b, and -c), CTF/NF-1, core enhancer binding protein (C/EBP), and the glucocorticoid response element (GRE). Both CTF/NF-1 and C/EBP play roles in developmental and tissue-specific regulation of MuLV gene expression. [Redrawn from Majors (1990).]

and thus resembles other enhancers in that multiple sequence elements are involved (Laimins *et al.*, 1984a; Stoltzfus, 1988). One domain is similar to an SV40 core enhancer and another has homology to the octamer sequence found in the immunoglobulin gene enhancer (Parslow *et al.*, 1984; Herr and Clarke, 1986).

Several studies on protein factors in avian cells present a complex picture of regulation through enhancer sequences and other *cis*-acting elements in the ASLV LTR. Uninfected avian cells contain proteins (i.e., EF-I and EF-II) which recognize sequences in the LTR (Fig. 12) (Sealy and Chalkley, 1987; Goodwin, 1988). EF-I recognizes a CCAAT motif at −129 to −133 and binds to a second CCAAT motif at −65 to −69 with lower affinity (Ryden and Beemon, 1989; Faber and Sealy, 1990). Detailed biochemical analysis of EF-I protein and characterization of cDNA clones of EF-I from avian cells reveal that this factor is either identical to or highly related to the cellular transcription factors NF-Y and CBF (Ozer *et al.*, 1990; Faber and Sealy, 1990; Boulden and Sealy, 1990). Binding studies and mutational analysis show that additional distinct factors in avian cells (E2BP and E3BP) recognize sequences between −203 to −196 and −169 to −158 (Kenny and Guntaka, 1990). Footprinting experiments indicate that avian erythroid cells contain factors, designated F-III and F-I, which recognize sequences −225 to −214 and −195 to −171, respectively (Goodwin, 1988). In RSV-transformed cells, v-*src* appears to increase the level of the cellular factor(s) that recognizes the CCAAT element, thereby stimulating transcription directed by the viral LTR (Dutta *et al.*, 1990).

A region within the ASLV *gag* gene, between nucleotides 788 and about 1000, functions as a relatively weak enhancer in transient expression assays (Arrigo *et al.*, 1987; Stoltzfus *et al.*, 1987; Carlberg *et al.*, 1988). In infected cells, proviral sequences within *gag* are also nuclease-hypersensitive sites, and thus this region of ASLV may influence chromatin structure on the provirus (Chiswell *et al.*, 1982; Conklin and Groudine, 1986; also see Groudine *et al.*, 1981). Although proteins from hamster cells have been shown to interact with sequences in this part of *gag* in *in vitro* binding assays, the precise relationship between binding and enhancer activity has not been established (Karnitz *et al.*, 1987, 1989). Sequences in *gag* also function to negatively regulate splicing of viral mRNA; however, mutational analysis shows that enhancer function is separable from the precise region that controls the processing of viral transcripts (Carlberg *et al.*, 1988; McNally *et al.*, 1991) (see below). The enhancer in *gag* may play a role in insertional activation of oncogenes and possibly in the process that selects the 5′ LTR for transcription initiation (Stoltzfus, 1988) (see below).

LTRs of endogenous avian proviruses (*ev-1* and *ev-2*) are 10- to 100-fold less active than exogenous virus LTRs in transient expression assays

(Cullen *et al.*, 1983, 1985a; Norton and Coffin, 1987), and additional studies demonstrated that these LTRs lack enhancer activity when tested for activity on promoters from a cellular gene and a heterologous virus in transient expression assays (Cullen, 1985b; Weber and Schaffner, 1985). Proviral forms of *ev-1* and *ev-2* are not expressed *in vivo*; however, these proviruses can be activated by treating cells with nucleotide analogues (Robinson, 1979; Conklin *et al.*, 1982) (see Chapter 6). Also, molecular clones of the endogenous viruses are expressed after transfection into tissue culture cells (Cullen *et al.*, 1983, 1985a). To investigate the role of transcriptional control elements with respect to the regulation of endogenous virus gene expression, chimeric LTRs were constructed by replacing the native U3 domain of the RSV LTR with U3 regions from the endogenous viruses. Although the endogenous virus U3 regions lacked detectable enhancer activity in the natural context, these elements behaved as strong enhancers in the chimeric LTRs containing promoter elements from RSV (Conklin *et al.*, 1991). In addition, enhancer function in the U3 regions of the endogenous viruses was not dependent on orientation in these chimeric LTRs (Conklin *et al.*, 1991). Presumably, both the arrangement of modules that compose an enhancer and the precise sequences proximal to the enhancer are critical for activity.

The MuLV system is an example of a retroviral LTR with a complex enhancer structure (Speck and Baltimore, 1987; Majors, 1990). Two tandem direct repeats (75 bp) that have enhancer function are located about 150 bp upstream from the TATA box in the U3 domain (Levinson *et al.*, 1982; Laimins *et al.*, 1984b) (Fig. 12). Positive selection may account for the presence of duplicated enhancers in MuLV and several other retroviruses; a MuLV derivative with one enhancer replicates less well than wild-type virus and has a longer latency period for leukemogenesis in mice (Li *et al.*, 1987; also see Hanecak *et al.*, 1986). Assays utilizing the gel retardation method have detected binding sites (i.e., core elements) for at least six host-cell factors in the direct repeats (Fig. 12) (Speck and Baltimore, 1987; Speck *et al.*, 1990b; Boral *et al.*, 1989; Thornell *et al.*, 1988a, 1988b, 1991). Several proteins that bind the core site in the MoMuLV enhancer have been purified by core-oligonucleotide affinity chromatography from calf thymus extracts. These proteins, designated core-binding factors (CBFs), recognize enhancers of both viral and cellular genes that are specifically transcribed in lymphocytes (Wang and Speck, 1992). Additional investigations are required to determine whether the multiple polypeptides in affinity-purified preparations of CBFs represent a family of transcription factors, products of differential splicing of a single gene, or proteolytic cleavage of a single CBF. Two additional proteins distinct from CBF that bind to the MoMuLV core enhancer are the CAAT/enhancer-binding protein (C/EBP) and activating protein 3 (AP-3) (P. F. Johnson *et al.*, 1987; Mercurio and Karin, 1989).

The number of binding sites that are occupied within the enhancer at any one time remains to be determined. Possibly, these binding proteins also interact with each other to regulate transcriptional activity of the LTR. One or more of these factors appears to be specific for certain lymphoid cell types, and a glucocorticoid response element within the enhancer confers conditional regulation (see below) (Celander and Haseltine, 1987; Speck and Baltimore, 1987; Speck *et al.*, 1990b). Accordingly, cell-specific factors that bind core elements in the MuLV enhancer may be a critical determinant of tissue specificity and pathogenicity in infected animals (see below).

Retroviral enhancers are positive regulators of viral transcription in infected cells (Maniatis *et al.*, 1987; Atchison, 1988). Accordingly, these enhancers may have the potential to regulate viral expression when integration occurs in areas of the cell genome that are not transcriptionally active (see Section on V integration above). However, several studies show that retroviruses integrate into regions of the cell genome that appear to be transcriptionally active (Vijaya *et al.*, 1986; Rohdewohld *et al.*, 1987; Scherdin *et al.*, 1990; Mooslehner *et al.*, 1990). Proviruses with mutations in enhancers are poorly expressed relative to wild-type proviruses (Majors, 1990). Therefore, integration into (apparent) transcriptionally active regions of host chromatin does not compensate for attenuated transcription in viral genomes with enhancer mutations.

E. Biological Properties Controlled by Retroviral Enhancers

Regulatory effects of the viral transcriptional enhancer underlie a mechanism of neoplasia induced by nonacute retroviruses that do not encode oncogenes. Seminal studies in ALV-induced B-lymphomas revealed that in some tumors, the provirus was located near the c-*myc* proto-oncogene but upstream and in the opposite transcriptional orientation relative to the c-*myc* promoter (Hayward *et al.*, 1981; Payne *et al.*, 1981). In these tumors, levels of c-*myc* RNA from its native promoter are increased (i.e., activated) by the transcriptional function of the enhancer in the adjacent provirus. In addition, a provirus inserted downstream from a proto-oncogene may also cause overexpression of the proto-oncogene promoter. RAV-0 is an endogenous avian retrovirus that contains an LTR lacking enhancer activity; accordingly, absence of an enhancer may account for the nonpathogenic nature of RAV-0 infections (Luciw *et al.*, 1983; Weber and Schaffner, 1985; see also Brown *et al.*, 1987). Several studies on other avian retroviral systems as well as in several mammalian retroviral systems (e.g., MMTV, MuLV, FeLV) have provided additional examples of enhancer activation of proto-oncogenes (see Chapters 6 and 7) (Kung and Vogt, 1991). (In addition to promoter insertion and enhancer activation, proto-oncogenes may be activated by either the leader-insertion or the poly-A-insertion mechanism see (Kung et al., 1991) (Chapter 6).

Transcriptional enhancers of several retroviruses function in specific cell types (LoSardo et al., 1989; Spiro et al., 1988) [reviewed in Majors (1990) and in Tsichlis and Lazo (1991)]. Accordingly, the enhancer in the LTR controls the pattern of tumor induction (Fan, 1990; Coffin, 1990a) and influences the neuropathic potential of MuLV (Des-Groseillers et al., 1985). This conclusion has been validated through direct analysis of LTRs in transient expression assays together with studies of both site-specific viral mutants and recombinant viruses (constructed from parental viruses with distinct phenotypes). The most extensive investigations on the biological role of enhancers have been performed on several distinct MuLV strains, including T-lymphomagenic MoMuLV, erythroleukemogenic Friend-MuLV (F-MuLV), nonleukemogenic Akv, and T-lymphomagenic viruses derived from Akv, SL3-3, Gross passage A, and recombinant mink cell focus-forming viruses (MCFs) (Chatis et al., 1983; Celander and Haseltine, 1984; Des-Groseillers and Jolicoeur, 1984; Lenz et al., 1984; Short et al., 1987; Li et al., 1987; Ishimoto et al., 1987; Holland et al., 1989; Fan, 1990; Majors, 1990) (see Chapter 7).

Two examples from the MuLV system are selected to illustrate the role of the transcriptional enhancer in the LTR with respect to replication in animals and disease specificity. In inoculated mice, MoMuLV causes T-lymphoblastic lymphomas, whereas F-MuLV induces erythroleukemias (Fan, 1990). Tumor induction involves the activation of specific proto-oncogenes by each virus in the target tissue. In tissue culture systems, MoMuLV replicates more efficiently in T-lymphoid cells, whereas erythroid cells preferentially support F-MuLV replication. Results of transient expression assays reveal that the M-MuLV and F-MuLV enhancers function efficiently in T-lymphoid cells and erythroid cells, respectively (Bosze et al., 1986; Cone et al., 1987). To focus on the enhancer in the LTR, recombinant viruses were constructed by exchanging small blocks of sequences (i.e., core elements) within the enhancers of both viruses (Golemis et al., 1989; Fan, 1990). The pattern of tumor induction (i.e., leukemias versus erythroid tumors) mapped to the enhancers, however, disease specificity was not controlled by a single core element. In another study, MoMuLV strains with site-specific mutations in either of two elements within the enhancer (i.e., C/EBP and LVb) induced erythroleukemias in mice (Fig. 12) (Speck et al., 1990a). Thus, these findings demonstrate that both individual core elements as well as interactions among core elements contribute to tumor specificity (Hollon and Yoshimura, 1989).

The precise context of the enhancer in a transcriptional unit may affect function. In the MuLV system, the importance of LTR sequences outside the enhancers is exemplified by studies on a mutant of M-MuLV which lacks a 23-bp GC-rich region flanking the enhancer (Fig. 12) (Hanecak et al., 1991). Mice infected with the deletion mutant virus show a longer latency period for disease and a wider repertoire of hematologic tumor types (e.g., acute myeloid leukemia, erythroleukemia, and B-cell lymphoma as well as lymphoblastic lymphoma typical of wild type M-MuLV). With respect to transcription mechanisms to account for the altered tissue specificity, the deletion may have altered critical spatial arrangements of binding sites for promoter elements and the enhancer region or, alternatively, a novel tissue-specific enhancer may have been generated at the junction of the ends of the deletion. Direct binding studies on the mutant LTR are required to determine altered interactions of known transcriptional factors and to identify putative new recognition sites.

The function of retroviral transcriptional enhancers with respect to development has been studied in murine embryonal carcinoma (EC) and embryonal stem (ES) cells maintained in culture. Undifferentiated EC and ES cells do not support MuLV replication, whereas these cells become permissive for MuLV infection after treatment with an agent that induces differentiation (Gorman et al., 1985; Teich et al., 1977; Speers et al., 1980; also see Taketo et al., 1985). Sequences in the MuLV 5' untranslated region near the tRNA primer-binding site appear to inhibit expression in undifferentiated EC cells; however, the LTR also plays a major role in regulating transcription in this system (Barklis et al., 1986; Loh et al., 1987, 1988; Weiher et al., 1987). The differentiation process is accompanied by changes in levels and/or activities of cellular transcriptional factors (Tsukiyama et al., 1989), and transient expression assays revealed that the enhancers in the LTR are not active in undifferentiated EC cells but acquire activity after differentiation (Linney et al., 1984; Feuer et al., 1989; Loh et al., 1987). MuLV mutants have been selected for expression in EC and ES cells (Franz et al., 1986; Akgun et al., 1991; Grez et al., 1990; Hilberg et al., 1987). A single point mutation in the enhancer activates the LTR in ES cells; this point mutation creates a recognition site for the murine equivalent of the basic cellular transcription factor SP1 (Grez et al., 1991). Additional studies with chimeric MuLV LTR containing heterologous enhancers (viral and cellular) also support the conclusion that the MuLV enhancers, together with the leader, are targets of regulatory factors in EC cells (Feuer et al., 1989; Taketo and Tanaka, 1987).

The role of the retroviral enhancer for controlling tissue-specific expression has also been investigated by replacing the viral enhancer with regulatory elements from a cellular gene. The mouse transthyretin gene contains an upstream-distal enhancer and a promoter-proximal region that control expression of the gene in liver and brain (Costa et al., 1986). Wild-type MuLV (MoMuLV strain) is normally not efficiently expressed in these organs in infected mice. A chimeric MuLV LTR was constructed by replacing the native enhancers with a DNA fragment (approximately 200 bp in length) that contains both of the transthyretin regulatory elements (Feuer and Fan, 1991). A cloned viral genome with chimeric LTRs was infectious in tissue culture cells and in mice. In situ hybridization analysis of mouse organs demonstrated that the chimeric virus was expressed in liver and brain as well as in tissues that support replication of wild-type MoMuLV. These and other experiments with genetically engineered MuLVs (containing chimeric LTRs constructed with regulatory elements from heterologous viruses such as SV40 or polyoma virus) also point to the critical role of enhancer elements in tissue-specific transcription (Hanecak et al., 1988; Fan et al., 1988). Potentially, retroviral vectors can be targeted for expression in specific cell types in animals by substituting the viral enhancers with regulatory components from cellular genes or heterologous viruses (see below).

F. Conditional Regulation by the LTR

LTRs of several MuLV strains (e.g., the T-lymphomagenic SL3-3 virus) are responsive to glucocorticoid hormones. The viral enhancers contain the conserved glucocorticoid-responsive element (GRE); however, a precise role for the glucocorticoid receptor in the induction process has not been established (Fig. 12) (DeFranco and Yamamoto, 1986; Celander and Haseltine, 1987; Celander et al., 1988). A nuclear factor specific to T-cells has been shown to interact with the GREs in the retroviral enhancer, and the predicted amino acid sequence of a molecular clone reveals that this protein is member of the helix–loop–helix transcriptional activator family of mammalian DNA-binding proteins (Corneliussen et al., 1991).

MMTV is a murine retrovirus (unrelated to MuLV) that causes primarily mammary carcinomas, and, infrequently, T-lymphomas and renal tumors are also observed in MMTV-infected mice (Teich et al., 1982; Morris, 1991) (see Chapter 7). Steroid hormones, including glucocorticoids and progesterone, regulate transcription directed by the viral LTR. A hormone-responsive element (HRE) was identified by analysis of LTR deletions and chimeric heterologous promoters in transient expression assays in tissue culture cells (Majors and Varmus, 1983;

Chandler *et al.*, 1983; Hynes *et al.*, 1983; Onts *et al.*, 1985; Majors, 1990, and references therein). These experiments, together with DNase foot-printing analysis, revealed multiple binding sites for the glucocorticoid receptor and the progesterone receptor proteins in the U3 domain of the LTR (Payvar *et al.*, 1983). The core recognition sequence (TGTTCT) for both receptors is present four times (at -175, -119, -98, and -83) upstream from the TATA box, and a site for the cellular transcriptional factor NF-1 (an inverted repeat of two TGGA motifs) is located between -75 and -64 (Miksicek *et al.*, 1987; Buetti and Kuhnel, 1986; Majors, 1990 and references therein).

Even though the sequence TGTTCT is the major recognition motif for both the glucocorticoid and progesterone receptors in the MMTV LTR, detailed binding studies and functional assessments in transient expression assays revealed some differences between these two hormones (Cato *et al.*, 1988; Chalepakis *et al.*, 1988). Precise recognition patterns of the receptors for sequences in the LTR were analyzed by DNase I footprinting and methylation protection methods. Although both receptors interact with the TGTTCT motif, differences in binding patterns were noted within the region (between -130 and -100) that encompasses these elements. LTRs with site-specific mutations in the HREs were constructed and evaluated for responsiveness to each hormone in cells in transient expression assays. Differences were observed with respect to the requirement of individual HREs for induction by either glucocorticoid or progesterone.

The gene for the glucocorticoid receptor from several mammalian species has been molecularly cloned and sequenced (Evans, 1988; Beato, 1989, and references therein). Structure–function studies show that the N-terminal portion is an activation domain, the central portion has a zinc-finger motif (i.e., DNA-binding domain), and the C-terminal portion is the ligand-binding domain. Both the DNA-binding function and the transcriptional regulatory function are regulated directly by hormone. In cells treated with hormone, the surface glucocorticoid receptor binds the hormone and the complex translocates to the nucleus. Subsequently, regions of cellular chromatin near the glucocorticoid response elements (GREs) become DNase I hypersensitive. These exposed regions are more accessible to cellular transcription factors, and thus initiation by RNA polymerase II is activated from promoters containing a GRE(s) (Evans, 1988; Beato, 1989).

This scenario for hormone regulation of the MMTV LTR has been validated by additional investigations in cells and in *in vitro* systems. These studies have also provided insight on the effects of nucleosome structure on modulating retroviral transcription (Grunstein, 1990). Mutational analysis of the LTR reveals that the cellular transcription factor NF-1 (as well as TBP) is required for efficient transcription of the viral promoter after hormone induction in cells (Buetti and Kuhnel, 1986; Miksicek *et al.*, 1987). Ternary complexes between the glucocorticoid receptor and NF-1 on DNA have not been detected in hormone-treated cells (Cordingly *et al.*, 1987). *In vitro*, purified glucocorticoid receptor and NF-1 individually bind DNA containing the respective target sequences; however, cooperative or synergistic binding of these two proteins to their targets is not observed (Pina *et al.*, 1990). Thus, these *in vitro* binding systems with naked DNA do not reproduce the effects of hormone induction in cells.

To investigate the role of chromatin structure on hormone regulation of the MMTV LTR, minichromosomes were made *in vitro* by reconstituting nucleosomes with purified histone octamers and a DNA fragment encompassing the viral promoter (Perlmann and Wrange, 1988; Pine *et al.*, 1990). These minichromosomes have a nucleosome-like structure covering the four GREs and single NF-1 site; thus, the sequence between −202 and −30 in the LTR contains the information for precisely positioning the histone octamer (Pina *et al.*, 1990). This phased nucleosome pattern observed *in vitro* mimics the pattern found *in vivo* by analysis of hypersensitive sites with DNase I (Perlmann and Wrange, 1988; Pina *et al.*, 1990). In the proposed model, glucocorticoid receptors first bind the exposed GREs in LTR sequences in fully packaged chromatin. The regulatory nucleosome is destabilized by this interaction so that the two hidden GREs and the NF-1 site are made accessible. Subsequently, all the appropriate components of the transcriptional apparatus enter and activate initiation of viral RNA synthesis from the promoter. In this model, the histone octamer functions as a repressor to mediate negative control of transcription. Thus, nucleosome positioning is a mechanism of viral gene regulation in addition to mechanisms involving protein–protein interactions between transcriptional factors and components of the multisubunit RNA polymerase II complex (Ptashne and Gann, 1990; Lewin, 1990).

The role of the conditional enhancer in the MMTV LTR has been investigated with respect to tissue-specific expression and oncogenesis in mice (Nusse, 1991; Majors, 1990, and references therein). Virus replicates primarily in mammary and salivary glands of females as well as in accessory genital glands and salivary glands of males. Studies in transgenic mouse systems have validated the notion that the LTR regulates tissue-specific transcription. Analysis of mice with transgenes containing the LTR linked to either c-*myc* or v-Ha-*ras* revealed that the tissue pattern of LTR-directed expression resembled that of viral replication in normal mice (R. A. Stewart *et al.*, 1988; Sinn *et al.*, 1987). MMTV tumor induction involves an enhancer-activation mechanism that, in principle, is similar to the ALV and MuLV tumor systems described above (Nusse *et al.*, 1984; Dickson *et al.*, 1985) (see Chapter 7). Analysis of integrated proviruses in mammary carcinomas reveals that the MMTV generally inserts near the cellular genes *wnt-1* and *wnt-2*, although proviruses are sometimes found at other regions of the cell genome (Nusse, 1991; Morris, 1991). Several patterns of proviral insertion relative to the *wnt-1* locus have been described for different mammary carcinomas. A provirus may be upstream from *wnt-1* and in the opposite transcriptional orientation. In some instances, the provirus is downstream from *wnt-1* and in the same transcriptional orientation. In the rare thymic lymphomas and renal carcinomas associated with MMTV, integrated proviruses have rearranged LTRs; however, the mammary-specific enhancer is intact (Majors, 1990, and references therein). Further studies are required to elucidate the mechanism(s) of oncogenesis in these infrequent MMTV tumors.

G. Capping and Methylation of Retroviral RNA

The 5′ ends of newly synthesized viral transcripts are capped and methylated by host-cell enzymes (Banerjee, 1980; Stoltzfus, 1988). Both full-length and subgenomic retroviral RNAs have a cap structure which is 7-methylguanosine in a 5′–5′ linkage via a triphosphate to a second 2′-O-methylated nucleotide (Furuichi *et al.*, 1975; Keith and Fraenkel-Conrat, 1975). This structure is designated $m^7G^5ppp^5NmpMp$ (N and M are the first and second nucleotides encoded by the virus) or, more simply, 7mG.

Methylation of internal A bases has been reported for the ASLV genome (Furuichi *et al.*, 1975; Beemon and Keith, 1977). Approximately 10–15 N^6-methyladenosine residues (m^6A) are found in the 3′ half of virion RNA and clustering of methylated sites is noted (Kane and Beemon, 1985). These modifications take place in the nucleus and may occur before splicing (Chen-Kiang *et al.*, 1979). Analysis of modification patterns reveals that Pum^6ACU is a consensus sequence, and the preference for Pu is G. Although this site is found many times in the ASLV genome, only a portion of the potential methylation sites are modified (Dimock and Stoltzfus, 1977; Kane and Beemon, 1985). In addition, certain sites are methylated in some viral RNA molecules and not in others. Factors that influence the distribution of m^6A residues and the stoichiometry of modification are not known; however, the secondary structure of the viral transcript may control the accessibility of certain sites to cellular methylases. A low level of m^6A modification has also been detected in cellular messages and in transcripts of other viruses (e.g., SV40, adenovirus) (Kane and Beemon, 1985, and references therein). A role for m^6A residues in splicing has been proposed on the basis of experiments involving methylation inhibitors in ASLV-infected tissue culture cells; however, additional studies are required to substantiate this notion (Stoltzfus, 1988).

H. Processing of Viral RNA by Splicing

The 5' LTR initiates the synthesis of the full-length viral transcript that enters one of three pathways: (i) genomic RNA for packaging into virions, (ii) messenger RNA for translation into *gag*-encoded polyproteins, and (iii) precursor for subgenomic viral transcripts (Fig. 1) (Varmus and Swanstrom, 1982, 1985; Stoltzfus, 1988). These subgenomic messages are made by splicing a leader sequence from the 5' end of viral RNA to an acceptor sequence within the viral genome; thus, all spliced viral messages utilize a common splice donor and share 5' and 3' ends (Fig. 13). The *env* gene of all retroviruses is expressed from a spliced transcript (Fig. 13). For MuLV and MMTV, the splice donor is between the 5' LTR and the start of *gag*; and the splice acceptor is immediately upstream from the initiation codon for the *env* gene. In contrast, in the ASLV system, the spliced leader contains the first six codons of *gag*, which are then attached to the 5' end of the *env* gene; thus, the ASLV *env* gene product is a fusion protein which contains six amino acids encoded by a sequence spliced from the beginning of the *gag* gene (Hunter *et al.*, 1983). Several strategies have been evolved by retroviruses with simple genomes to control the relative levels of unspliced viral transcripts required for virion assembly and subgenomic mRNA required for protein synthesis (e.g., *env* and *src* of ASLV and *orf* of MMTV).

Splice donor and acceptor sites, the branchpoint sequence, and sequences within the flanking exons are *cis*-acting elements which control the selection of splice sites as well as the efficiency of splicing (Stoltzfus, 1988; Kranier and Maniatis, 1988). Splice signals in retroviral RNA generally conform to those described for many eukaryotic genes and genes of eukaryotic DNA viruses which contain introns (Sharp, 1987). Several observations in both the ASLV and MuLV systems support the notion that cellular factors act on the viral precursor transcript in the nucleus to carry out the functions of splicing. For instance, in the ASLV system, the proportion of full-length viral RNA (i.e., precursor transcript and genomic RNA) to *env* mRNA is about 2:1 in productively infected avian cells. However, in nonpermissive mammalian cells harboring ASLV, the predominant viral mRNA is the subgenomic message for *src*; genomic and *env* transcripts are present in very low amounts (Quintrell *et al.*, 1980). In addition, *env* mRNA is spliced aberrantly in mammalian cells (Berberich *et al.*, 1990).

Mutagenesis studies have shown that alterations in the ASLV genome affect the ratio of spliced to unspliced transcripts. A sequence

ASLV

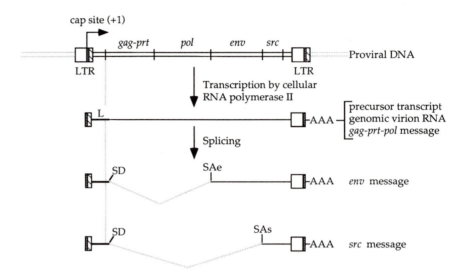

MuLV

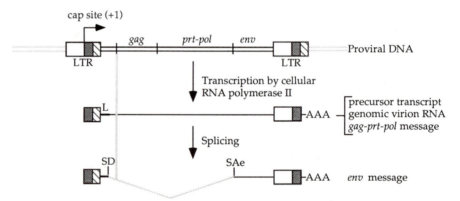

FIGURE 13. Synthesis of full-length and spliced (i.e., subgenomic) retroviral RNA. Integrated proviruses that serve as templates for viral RNA synthesis in the nuclei of infected cells are shown for ASLV and MuLV. The boxes at each end represent LTRs composed of the U3, R, and U5 domains. Transcription mediated by cellular RNA polymerase II initiates in the 5' LTR (designated by an arrow), and a sequence in the 3' LTR is the signal for addition of poly-A tails by a cellular transferase (Fig. 12). The full-length transcript serves as the precursor for subgenomic viral transcripts, genomic virion RNA, and message for *gag–prt–pol*. Leader regions (L) for both viruses are drawn as heavy lines. The splice donor (SD) for subgenomic ASLV messages is within the *gag* gene; consequently, attachment of L to the splice acceptor for *env* (SAe) yields a message in which the translation initiation codon and the first six amino acids of *gag* specify the N-terminus of the *env* polyprotein. The same L is attached to the splice acceptor for *src* (SAs). The *src* gene contains an initiation codon downstream from SAs; consequently, the subgenomic message for *src* is bicistronic. In the MuLV system, the SD in L is upstream from the *gag* initiation codon.

about 300 bases long within the *gag* gene (positions 707–1006) functions as a negative regulator of splicing (NRS) (Arrigo *et al.*, 1987; Stoltzfus and Fogarty, 1989). Accordingly, removal of this NRS from the native setting within the *env* and *src* introns results in an increase of spliced to unspliced RNA. In addition, this element inhibits splicing only in the sense orientation; inhibition is observed when the NRS is inserted into heterologous introns in both avian and mammalian cells (McNally *et al.*, 1991). Highest levels of inhibitory activity are observed when the NRS is positioned in the intron near the 5′ splice site. Intriguingly, the transcriptional enhancer element within the ASLV *gag* gene partly overlaps the NRS, although mutation analysis has demonstrated that the enhancer and NRS are distinct (Arrigo and Beemon, 1988; Karnitz *et al.*, 1987; McNally *et al.*, 1991) (see above). A computer-predicted secondary structure model for the NRS has been proposed; however, further mutational analysis is required to determine whether splicing is inhibited by the intrinsic secondary structure of the NRS or whether sequences within the NRS anneal to and sequester splice site sequences.

Additional investigations on ASLV splicing have focused on the role of sequences near the splice acceptors for *env* and *src*. Insertion of a synthetic oligonucleotide upstream from the *env* 3′ splice site increases levels of spliced *env* transcript due to utilization of a new branchpoint for splicing (Katz *et al.*, 1988; Fu *et al.*, 1991). Pseudorevertant viruses have mutations either in the inserted sequence or in the *env* exon downstream from the *env* splice acceptor site; thus, the wild-type ASLV splice acceptor appears to be suboptimal and sequences that influence splicing are located within *env*. Wild-type ASLV RNA is spliced inefficiently *in vitro*, and mutations in the pseudorevertants have been shown to function directly at the level of splicing (Fu *et al.*, 1991). Mutagenesis studies and analysis of chimeric constructs in transient expression assays revealed that sequences upstream from the *src* 3′ splice site inhibit the level of splicing at this site (Berberich and Stoltzfus, 1991; McNally and Beemon, 1992). This region (−71 to −185 relative to the *src* 3′ splice site) exerts an inhibitory effect only in the sense orientation and in heterologous contexts, although the inhibitory effect is stronger for the *src* acceptor than for heterologous splice acceptors. Secondary structure analysis reveals that the inhibitory segment immediately upstream of *src* is complementary to sequences encompassing the *src* acceptor site; however, the precise role of this predicted secondary structure feature in splicing remains to be defined (McNally and Beemon, 1992). In summary, these investigations demonstrate that different mechanisms control the underutilization of the *env* and *src* acceptor sites in ASLV.

Examples from studies on MuLV also demonstrate that *cis*-acting sequences in the viral genome influence splicing. Analysis of MuLV genomes with deletion mutations revealed that efficient formation of spliced *env* message requires a region located upstream from the 3' splice site and a second region located in the middle of the intron (Hwang *et al.*, 1984; Armentano *et al.*, 1987). In addition, the presence of *gag* sequences in an MuLV vector (designated N2) augmented expression of the spliced message for the cDNA gene for human adenosine deaminase (ADA) (Armentano *et al.*, 1987).

Taken together, these and other findings show that inherent features in retroviral RNA regulate the efficiency of splicing (Stoltzfus, 1988; Coffin, 1990a, and references therein). Retroviruses with simple genomes do not appear to encode proteins which govern ratios of unspliced to spliced transcripts. Splicing controls provided by *cis*-acting sequences are essential to regulate relative levels of genomic viral RNA for encapsidation and subgenomic viral transcripts for protein synthesis. In contrast, retroviruses with complex genomes encode transactivators (e.g., *rex* in HTLV-1 and *rev* in HIV-1) which control the balance of full-length transcripts to subgenomic RNA at the level of splicing and/ or transport (Greene and Cullen, 1990; Green and Chen, 1990).

I. Addition of Poly-A Tails to the 3' End of Viral Transcripts

Poly-A tails from 100 to 200 residues long are added to the 3' end of cellular and viral RNA molecules transcribed by the cellular RNA polymerase II complex (Figs. 6 and 13); this modification involves cellular functions and occurs in the nucleus (Manley, 1988; Stoltzfus, 1988; Proudfoot, 1991). The precise role of poly-A tails in eukaryotic mRNA remains to be defined; potential functions involve transport from the nucleus to the cytoplasm, stability in the cytoplasm, and/or modulation of translation (Jackson and Standart, 1990; Munroe and Jacobson, 1990; Atwater, 1990). The LTR contains the hexanucleotide sequence AAUAAA which signals the addition of poly-A tails to viral transcripts, and a less conserved 3' GU or U-rich element 3' of the poly-A-addition site also plays a role in processing (Figs. 12 and 13) (Proudfoot, 1991). Addition of poly-A generally occurs about 20 bases downstream from the hexanucleotide signal element. Processing by polyadenylation appears to be a prerequisite for the subsequent termination of transcription by RNA polymerase II (Proudfoot, 1989).

In ASLV, 3'-terminal cleavage and addition of poly-A tails in response to the signal in the 3' LTR is not completely efficient, since a small proportion (about 15%) of viral transcripts initiated in the 5' LTR read through the 3' LTR and into flanking host-cell sequences (Herman and Coffin, 1986). Studies in genetically engineered mutations in the ASLV genome have revealed that two distinct ASLV sequences, one in the *gag* gene and another spanning the *env* splice acceptor site, facilitate

3'-end processing of viral transcripts (Miller and Stoltzfus, 1992). The significance of these long hybrid (i.e., viral–host) transcripts is not established; these can be packaged into virions and may have played a role in the capture of oncogenes by retroviruses (Herman and Coffin, 1987; Varmus and Brown, 1989). Discussed below are mechanisms that may account for preferential utilization of the poly-A signal in the 3' LTR rather than in the identical 5' LTR.

J. Selection of an LTR for Initiation or Poly-A Addition

Both LTRs in the provirus are identical; however, during retroviral transcription, the 5' LTR has promoter activity and the addition of poly-A tails occurs at the 3' LTR (Figs. 12 and 13). Thus, initiation of viral RNA synthesis involves activation of the promoter in the 5' LTR and suppression of the (potential) promoter in the 3' LTR. Similarly, processing of 3' ends of viral transcripts by polyadenylation requires suppression of the (potential) poly-A signal in the 5' LTR and activation of this signal in the 3' LTR (Imperiale and DeZazzo, 1991).

Viral sequences downstream from the 5' LTR may play a role in transcription initiation. In ASLV, it is tenable that the enhancer element in the *gag* gene is critical for preferential utilization of the 5' LTR as a promoter. This notion is supported by the observation that the 3' LTR is activated in an integrated ASLV provirus after (natural) deletion of sequences in the 5' portion of the provirus but with the 5' LTR left intact (Goodenow and Hayward, 1987). Consequently, a cellular gene (e.g., proto-oncogene) downstream of the integrated provirus may be activated by transcription initiated in the 3' LTR (Kung *et al.*, 1991).

An alternative explanation for preferential promoter activity of the 5' LTR is based on the interference (or occlusion) of one promoter by another promoter located immediately upstream (Cullen *et al.*, 1984; Imperiale and DeZazzo, 1991). Accordingly, removal of the upstream promoter is predicted to relieve the block to initiation by the second promoter. In the ASLV system, the occlusion model is supported by studies in cells containing proviruses integrated upstream from a proto-oncogene (Payne *et al.*, 1981; Goodenow and Hayward, 1987). Removal of the 5' LTR leads to activation of the promoter in the 3' LTR; consequently, a hybrid message is produced which contains both viral (i.e., R–U5 sequences from the 3' LTR) and cellular (e.g., proto-oncogene) sequences. Additional support for the occlusion model is derived from experiments in which a poly-A processing signal is positioned between two promoters. In cells transfected with a plasmid containing two ASLV LTRs, the upstream LTR has greater activity than the downstream LTR (Cullen *et al.*, 1984). A derivative plasmid was constructed by cloning a

3' processing site between the two LTRs; analysis of transcripts in transfected cells revealed that the inhibition of the downstream promoter (3' LTR) by the upstream promoter (5' LTR) is alleviated.

Specific mechanisms must also be utilized to either suppress the poly-A signal (AAUAAA) in the 5' LTR or activate it in the 3' LTR (Figs. 12 and 13). In the case of ASLV, this regulatory problem has been obviated because the poly-A signal is located in the U3 domain of the LTR, upstream from the initiation site. Thus, the signal AAUAAA appears only once, at the 3' end of ASLV transcripts (Fig. 12). For most other retroviruses, the poly-A signal is downstream from the initiation site for transcription (e.g., cap site); accordingly, viral transcripts contain the signal AAUAAA at each end (Figs. 12 and 13). For these viruses, cis-acting sequences in the U3 portion of the 3' LTR (i.e., upstream from R) may be required for efficient addition of poly-A tails. Several studies primarily in the REV and HIV-1 systems have demonstrated the importance of U3 sequences in processing of 3' ends of viral transcripts (Dougherty and Temin, 1987; Brown et al., 1991; Valsamakis et al., 1991); however, a clear picture has not yet emerged as to whether specific sequences and structural features as well as position in U3 influence the poly-A addition process (Imperiale and DeZazzo, 1991).

Alternatively, the poly-A site may be occluded if positioned near an active promoter (i.e., 5' LTR). Experimental support for this latter model, also designated the promoter proximity model, has been obtained in the REV and HIV-1 systems (Iwasaki and Temin, 1990a, b; Weichs an der Glon, et al., 1991). Polyadenylation was increased if the target viral poly-A site was positioned at increasing distances from the active promoter. On a mechanistic level, cellular factors required for poly-A processing may not be able to compete with the transcription complex if the poly-A site is close to the initiation site. In summary, it is tenable that both upstream sequences (in U3) and promoter proximity may influence the selection and use of the retroviral poly-A site (Imperiale and DeZazzo, 1991, and references therein). The relative contribution of each mechanism may depend on the retrovirus and may vary with cell type and physiological conditions.

VII. TRANSLATION AND PROCESSING OF VIRAL PROTEINS

A. Overview and Significance of Retroviral Protein Synthesis

In a productively infected cell, retroviral protein synthesis accounts for a small proportion (generally less than 1%) of total cell protein synthesis (Dickson et al., 1982). Translation of retroviral proteins takes place on cytoplasmic polysomes and is regulated by the host-cell translational apparatus. Retroviruses do not encode either tRNA molecules or protein factors which modify the host-cell translation system. The full-length viral transcript is translated on free ribosomes into *gag*-containing polyproteins, and a spliced subgenomic mRNA is translated on membrane-bound ribosomes into *env* polyprotein. Although the synthesis of the *gag, prt,* and *pol* polyproteins is initiated at one start codon at the beginning of the *gag* open reading frame, retroviruses have unique translational mechanisms to regulate the levels of each of these polyproteins (Vogt *et al.,* 1975; Eisenman and Vogt, 1978; Dickson *et al.,* 1982). Mechanisms that control retroviral polyprotein synthesis are suppression of a stop codon and frameshifting at the ribosomal level (Jacks, 1990). Translational frameshifting is also a feature of avian coronavirus and several retrotransposons (e.g., Ty elements in *Saccharomyces*) (Brierly *et al.,* 1987, 1989; Clare and Farabaugh, 1985; Clare *et al.,* 1988; Mellor *et al.,* 1985) [reviewed by Atkins *et al.* (1990)]. Posttranslational processing of *gag*-containing polyproteins is mediated by the viral-coded PR, and these cleavages are tightly coordinated with virion assembly (Oroszlan and Luftig, 1990). Extensive biochemical, genetic, and structural studies have been used to establish a structural model for retroviral PR. Accordingly, this viral-specific enzyme is a target for the design and development of antiviral inhibitors. The retroviral *env* polyprotein is processed by specific proteolytic cleavages and modified by glycosylation (Hunter and Swanstrom, 1990). These processing events appear to conform to mechanisms that regulate the production of cell surface glycoproteins as well as the glycoproteins of other enveloped viruses. Insertion of *env* glycoprotein into the cell plasma membrane may lead to important changes in surface recognition properties of the infected cell. Thus, synthesis of viral polyproteins is mediated by host-cell polysomes, although novel translational mechanisms are utilized, and posttranslational processing events involve both viral-coded (i.e., PR) and cell functions (e.g., glycosylation).

B. Multicistronic mRNAs for Viral Proteins

A multicistronic (or polycistronic) mRNA specifies more than one translation product by using independent initiation and termination codons for protein synthesis (Samuel, 1989). In the mRNA that encodes ASLV *gag* (i.e., full-length viral transcript), three short translation frames precede the actual AUG codon that initiates the *gag* polyprotein (AGC<u>AUG</u>G) (Swanstrom *et al.,* 1982; Schwartz *et al.,* 1983). These translation frames start with an AUG codon, and a termination codon is located at the end of each frame. The sequence at the beginning of the

gag gene is the first start codon in the ASLV transcript to lie in a nucleotide context that is preferred for initiation (i.e., gccA/GccAUGG) (Kozak, 1991). Protein is synthesized from one of these short translation
frames; consequently, the message for ASLV *gag* is bicistronic (Hackett
et al., 1986). The scanning hypothesis holds that a ribosomal initiation
complex formed at the 5' end of mRNA must migrate along the non-
translated region to the initiating AUG (Kozak, 1991). However,
ribosomes can reinitiate translation at a second AUG codon after previously initiating, and subsequently terminating, at an upstream site
(Samuel, 1989).

Subgenomic mRNAs for ASLV *env* and *src* are also multicistronic.
The mRNA for ASLV *env* has the 5' leader region containing the three
short translation frames upstream from *gag*; in addition, the spliced
leader on *env* mRNA attaches the first six codons of *gag* in-frame to the
env gene (Hackett *et al.*, 1982; Schwartz *et al.*, 1983). Thus, the ASLV
env gene product is a fusion protein with six amino acids at the N-
terminus encoded by a sequence spliced from the beginning of the *gag*,
gene (Ficht *et al.*, 1984). The mRNA for ASLV *src* has the same 5' leader
as *env* mRNA; thus, the splicing process produces a transcript that contains the three short translation frames upstream from *gag*, and the *gag*
AUG codon is positioned 90 bases upstream from the AUG codon that
initiates translation of *src* protein (ACCAUGG) (Swanstrom *et al.*,
1982; Takeya and Hanafusa, 1983; Schwartz *et al.*, 1983). The first four
AUG codons in *src* mRNA are each followed by an in-frame termination
triplet. Accordingly, initiation at the *src* start codon (fifth AUG in *src*
mRNA) can occur because terminators for the other translation frames
are located upstream (Hughes *et al.*, 1984; Samuel, 1989). ASLV is not a
unique retrovirus with respect to translation mechanisms involving
multicistronic mRNA; HIV also specifies multicistronic transcripts
(Schwartz *et al.*, 1992).

C. Synthesis of *gag* Gene Products

Polyproteins derived from the *gag* gene range in size from ca. 60 to
80 kDa, depending on the retrovirus (Figs. 2 and 14) (Wills and Craven,
1991). The message for *gag* is the full-length viral transcript which is
also the monomeric form of virion RNA. Generally, the first AUG encountered in viral RNA is the *gag* start codon (Kozak, 1991). An exception is ASLV, which has three short translation frames preceding the
initiation codon for *gag* (see above) (Schwartz *et al.*, 1983; Hackett,
1986). In the ASLV system, *gag* and *prt* sequences are in the same translation frame; accordingly, PR is derived from the C-terminus of the

ASLV *gag* polyprotein (Fig. 2). In contrast, the *gag* translation frame of other retroviruses does not include *prt* (Fig. 2).

Synthesis of MuLV *gag* from the full-length viral transcript involves two pathways. In the major pathway, translation of *gag* initiates at the first AUG codon (AAUAUGG) in a favorable translation context (gccA/GccAUGG) to produce a 65-kDa polyprotein (Pr65gag) that is subsequently assembled into virions and proteolytically processed during particle morphogenesis to produce mature core proteins (Shinnick *et al.*, 1981; Kozak, 1991). In the second pathway, a CUG triplet upstream from the *gag* start codon and in-frame with the *gag* translation frame is recognized as an initiation codon for protein synthesis. The CUG codon is in a favorable context for translational initiation (ACCCUGG) and directs the synthesis of a *gag* polyprotein that has an N-terminal hydrophobic leader peptide (Edwards and Fan, 1979; Prats *et al.*, 1989; Mehdi *et al.*, 1990). This form of the MuLV *gag* polyprotein is glycosylated to produce an 85-kDa glycoprotein (gp85gag) that is inserted into the cell plasma membrane (Pillemer *et al.*, 1986). Subsequently, gp85gag is cleaved into two glycoproteins, with molecular weights of 55 and 40 kDa; these are released into the cell culture medium. The glycosylated form of *gag* accounts for about 5–10% of total *gag* protein in an infected cell and is not incorporated into virions (Henderson *et al.*, 1983). Mutants in MuLV that specifically abrogate synthesis of gp85gag are infectious in tissue culture cells but show delayed replication kinetics relative to wild-type virus (Fan *et al.*, 1983; Prats *et al.*, 1989; also see Schwartzberg *et al.*, 1983). A glycosylated form of *gag* polyprotein has been demonstrated on the surface of HIV-1-infected cells; however, the significance of this modification of *gag* is not known (Shang *et al.*, 1991). A role for secreted forms of glycosylated *gag* protein in MuLV (and HIV-1) has not yet been determined.

A common feature of all retroviruses is that the N-terminus of MA, the first domain in the *gag* polyprotein, is modified (Fig. 14). In the ASLV system, the methionine specified by the initiation codon is acylated by a host-cell enzyme (Palmiter *et al.*, 1978). In *gag* polyproteins of mammalian retroviruses, the N-terminal methionine is removed, and myristic acid is added to the glycine residue that immediately follows the methionine (Schultz and Oroszlan, 1983; Rhee and Hunter, 1987; Schultz *et al.*, 1988; Weaver and Panganiban, 1990). This latter modification occurs in a cotranslational step during *gag* polyprotein synthesis (Wilcox *et al.*, 1987). The host-cell enzyme N-myristyl transferase links myristic acid, donated by myristoyl coenzyme-A, to the α-carbon of the N-terminal glycine acceptor (Olson and Spizz, 1986; Towler and Glaser, 1986). These modifications are necessary for targeting subunits (i.e., *gag* polyproteins and immature cores) to the inner surface of the cell plasma

ASLV

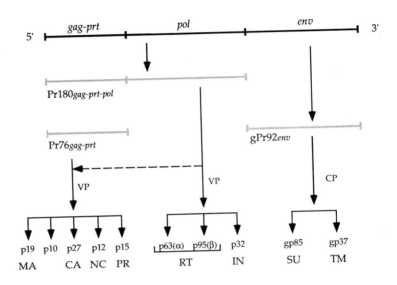

MuLV

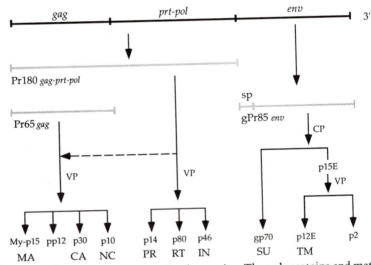

FIGURE 14. Processing patterns of retroviral polyproteins. The polyproteins and mature processed virion proteins derived from the *gag, pol,* and *env* genes of ASLV and MuLV are shown. Proteins are designated according to molecular weights in kilodaltons (kDa). The following designations are used: pr, polyprotein or precursor; p, protein; pp, phosphoprotein; gp, glycoprotein. The nomenclature for virion proteins is based on Leis *et al.* (1988): MA, *gag* matrix; CA, *gag* capsid; NC, *gag* nucleocapsid; PR, protease from *prt*; RT, *pol* reverse transcriptase; IN, *pol* integrase; SU, *env* surface domain; TM, *env*, transmembrane domain. VP and CP denote proteolytic cleavages by viral PR or a cellular protease, respec-

membrane during virion assembly (see Section VIII). Mutations that alter the penultimate glycine residue in the *gag* polyprotein block myristylation and abolish viral infectivity (Rhee *et al.*, 1987; Rein *et al.*, 1986; Gottlinger *et al.*, 1989; Bryant *et al.*, 1989). In type C retroviruses, these mutant *gag* polyproteins are not processed and remain unassembled in the cytoplasm (Wills *et al.*, 1984; Rein *et al.*, 1986; Schultz and Rein, 1989; Wills *et al.*, 1989). In type D retroviruses, mutant *gag* polyproteins that lack the N-terminal myristic acid assemble into immature core particles, but these particles do not associate with the cell plasma membrane to initiate the budding process (Rhee *et al.*, 1987). Mechanisms of transport of *gag* polyproteins to sites of viral morphogenesis in the cell are discussed below in Section VIII on virion assembly.

The *gag* polyproteins are cleaved to mature forms (i.e., MA, CA, and NC proteins) by viral PR during and/or immediately after the budding process (Fig. 14; also see Fig. 18) (Oroszlan and Luftig, 1990; Wills and Craven, 1991). Additional proteins are derived from the *gag* polyprotein of several retroviruses; however, the role of these cleavage products in the viral life cycle is not well defined (Dickson *et al.*, 1985). A detailed discussion of the recognition and enzymatic properties of retroviral PRs is presented below.

D. Translational Frameshifting and Suppression for Synthesis of Retroviral Enzymes PR, RT, and IN

Figure 2 shows the organization of the *gag*, *prt*, and *pol* genes of several retroviruses. These patterns have implications for the synthesis of *prt* (PR) and *pol* (RT and IN) gene products. Retroviral enzymes are derived from polyprotein precursors which are translated from full-length viral transcripts (Dickson *et al.*, 1982). Translation is initiated at the AUG codon for *gag*, and stop codons are located either at the end of *gag* (for MuLV), at the end of *gag–prt* (ASLV), or at the ends of both *gag* and *prt* (MMTV and SRV) (Fig. 2). Previously, RNA splicing was proposed to account for synthesis of the *prt* and *pol* gene products (Dickson *et al.*, 1982). However, it has been established that different translational mechanisms are utilized by each virus to obviate stop codons so that *gag–prt* and *gag–prt–pol* polyproteins are produced (Fig. 15). Results from direct sequencing of viral proteins together with studies in *in vitro* translation systems have shown that stop-codon suppression (for

---

tively. The dashed lines and arrows indicate that parts of certain polyproteins may be cleaved into products represented by other polyproteins (e.g., Pr180 of ASLV is cleaved into gene products derived from *prt* and *gag* as well as *pol*). [Redrawn from Orozslan and Luftig (1990).]

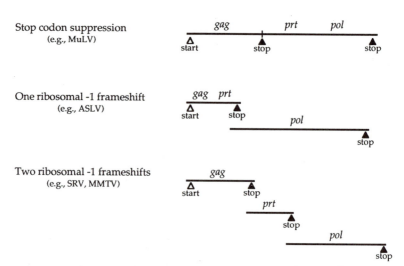

FIGURE 15. Translation mechanisms (i.e., suppression and frameshifting) to generate polyproteins for *gag*, *prt*, and *pol*. Solid lines represent translation reading frames for the *gag*, *prt*, and *pol* genes. Start (initiation) and stop (termination) codons are shown. In all cases, translation initiates at the start codon that specifies the N-terminus of the *gag* polyprotein. [Redrawn from Jacks (1990).]

MuLV) and ribosomal frameshifting (for RSV, MMTV, SRV, and HIV) account for synthesis of these retroviral polyproteins (Fig. 15) (S. Yoshinaka *et al.*, 1985; Jacks and Varmus, 1985; Jacks *et al.*, 1988a; Feng *et al.*, 1989; Atkins *et al.*, 1990; see references in Jacks, 1990, and Hatfield *et al.*, 1992). Posttranslational cleavages of viral polyproteins are mediated by viral PR during virion assembly to produce mature forms of the viral enzymes.

ASLV *gag* and *pol* translation frames overlap 58 bases (Figs. 2 and 14), and the *pol* frame is translated at an efficiency of about 5% in the −1 direction with respect to *gag* (Jacks and Varmus, 1985). For retroviruses that have a *prt* frame which is separate from both *gag* and *pol* (e.g., MMTV and SRV, Figs. 2 and 15), higher efficiencies of −1 frameshifting are required to read through each of two −1 overlaps. In the case of MMTV, the *gag–prt* and the *prt–pol* overlaps have been shown to be translated, respectively, by about 23% and 8% of the ribosomes that initiate at the *gag* AUG (Jacks *et al.*, 1987; Moore *et al.*, 1987). Accordingly, MMTV *pol* polyprotein is synthesized at about 3–5% of the level of the *gag* polyprotein. This ratio of precursor polyproteins conforms to the relative amounts of *gag* gene products and RT measured in mature MMTV particles (Dickson and Atterwill, 1979).

The frameshift event for each retrovirus is at a unique position within the overlapping translation frames involving the *gag*, *prt*, and *pol* genes (Figs. 2 and 15) (Jacks, 1990, and references therein). Studies on several retroviral systems have shown that ribosomal frameshifting requires a seven-base sequence at the frameshift site and a secondary structure element (i.e., stem-loop) immediately downstream of the site. Examples of shifty sequences are A AAU UUA (ASLV), U UUU UUA (HIV-1), and A AAA AAC (MMTV gag–prt). Thus, the general form of the shift site is the sequence X XXY YYZ, in which the triplets are the initial (or "0") translation frame and X may be identical to Y (Chamorro et al., 1992). The base at the 3' end of the frameshift signal is translated once in the 0 frame and once in the −1 frame (Hizi et al., 1987). The stem-loop after the shift site is part of a potential pseudoknot structure that may form when bases in an RNA loop form base pairs with a sequence outside the loop (Schimmel et al., 1989; Ten Dam et al., 1990; Le et al., 1991; Chamorro et al., 1992; also see Wills et al., 1991).

In the simultaneous slippage model for −1 translational frameshifting, a specific sequence (i.e., shift site) allows slippage of tRNA in the ribosomal acceptor site and the stem-loop structure retards the movement of ribosomes so that the probability of frameshifting is increased (Jacks et al., 1988b). Ribosomes shift into the −1 frame when an amino acyl tRNA decodes the YYZ codon and peptidyl tRNA recognizes the XXY triplet in the P site. The anticodons in both tRNA molecules recognize two of three base pairs at the shift site after slipping into the −1 frame. Translation of gag–pol in HIV-1 also requires a −1 frameshift; however, studies with site-specific mutants and chimeric sequences at shift sites have not revealed a clear structural feature for frameshifting in HIV-1 (Madhani et al., 1988). Accordingly, further analysis is essential to elucidate fully the significance of pseudoknots and other structural elements with respect to −1 frameshifting.

Several retroviruses and retrotransposons are predicted to utilize translational frameshifting for protein synthesis. Sequence analysis of lentiviruses (in addition to HIV-1), and the complex oncoviruses (e.g., HTLV), reveals that these retroviruses require a −1 frameshifting mechanism to translate gag-containing polyproteins [reviewed in Jacks (1990)]. Analysis of the gag and pol gene sequences of mammalian intracisternal A particles and Drosophila retrotransposons also implies a −1 frameshift. Additional entities unrelated to retroviruses which utilize a −1 frameshifting event at a pseudoknot are yeast dsRNA and coronaviruses (Brierly et al., 1989, 1991; Dinman, 1991; also see Ten Dam et al., 1990; Le et al., 1991; and Hatfield et al., 1992).

In MuLV, the gag and prt–pol genes are separated by a termination codon (i.e., amber) but remain in the same translation frame (Figs. 2 and 15). Host-cell translational suppression mechanisms utilize a tRNA charged with glutamine to suppress the amber termination codon (i.e., UAG) thus, allowing ribosomes to read through (S. Yoshinaka et al.,

1985; also see Hatfield *et al.*, 1989). Signals for suppression involve sequences near the *gag* stop codon (Panganiban, 1988; Hatfield *et al.*, 1992, and references therein); analysis in cell-free translation systems shows that a pseudoknot located eight nucleotides downstream of the UAG codon in MuLV *gag* enhances readthrough (Wills *et al.*, 1991). Structural features in RNA also appear to play a role in stop-codon suppression in tobacco mosaic virus and mammalian alphaviruses (e.g., Sindbis virus) (Ishikawa *et al.*, 1986; Li and Rice, 1989; Skuzeki *et al.*, 1991); however, sequence comparisons reveal no obvious homology in the region encompassing the suppressible termination codon (Feng *et al.*, 1989; Ten Dam *et al.*, 1990). A precise function for pseudoknot and stem-loop structures in the movement of ribosomes along mRNA remains to be determined. The efficiency of suppression in the MuLV *gag* gene in infected cells as well as in *in vitro* translation systems is about 5%; thus, the *gag–prt–pol* polyprotein is about 20 times less abundant than the *gag* polyprotein (Dickson *et al.*, 1982; Yoshinaka *et al.*, 1985).

E. Structural and Enzymatic Properties of the Retroviral PR

PR cleaves *gag*-containing polyproteins into mature virion proteins during particle morphogenesis (reviewed by Skalka, 1989; Oroszlan and Luftig, 1990; Swanstrom *et al.*, 1990; Fitzgerald and Springer, 1991). This enzyme is required for retroviral replication since noninfectious, immature particles containing uncleaved viral polyproteins are produced if PR is rendered defective by site-specific mutagenesis (Crawford and Goff, 1985; Stewart *et al.*, 1990, and references therein; Bennett *et al.*, 1991). Patterns for the proteolytic processing of virion polyproteins for several retroviruses are shown in Fig. 14. (Proteolytic processing of *env* polyprotein involves a host-cell enzyme localized in the Golgi apparatus; see below.) Mature retroviral PRs range in size from 99 to 126 amino acids, and sequence comparisons reveal that these enzymes are distantly related to cellular aspartyl proteases (Toh *et al.*, 1985). The sequence aspartate–threonine/serine–glycine is conserved in the active sites of both viral and cellular enzymes. Recently, vectors have been used to express large quantities of active PR from several retroviruses in genetically engineered bacteria and yeast. In addition, active HIV-1 PR has been chemically synthesized (Wlodawer *et al.*, 1989).

Similar three-dimensional models for ASLV and HIV-1 PR have been derived from crystallographic analysis together with predictions of structure based on amino acid sequences (Pearl and Taylor, 1987; M. Miller *et al.*, 1989; Navia *et al.*, 1989; H. Weber *et al.*, 1990; Skalka, 1989; Oroszlan and Luftig, 1990; Jaskolski *et al.*, 1990; Arnold and Arnold, 1991). Structures of HIV-1 PR complexed with inhibitory substrate analogues have also been determined (Jaskolski *et al.*, 1991; La-patto *et al.*, 1989; Fitzgerald *et al.*, 1990; Swain *et al.*, 1990; Erickson *et al.*, 1990), and a structure for ASLV PR complexed with an inhibitor has been proposed (Grinde *et al.*, 1992a). A crystal structure has been elucidated for the chemically synthesized 99-residue HIV-1 PR (Wlodawer *et al.*, 1989). Taken together, these structural studies, largely focused on the PR of ASLV and HIV-1, support the notion that the functional form of the retroviral enzyme is a homodimer (Miller *et al.*, 1989; Lapatto *et al.*, 1989; Meek *et al.*, 1989; Navia *et al.*, 1989; Wlodawer *et al.*, 1989; Jaskolski *et al.*, 1990; Weber, 1990).

The PR monomer consists of several β-strands, a long α-helix, and a partial α-helix; this monomer is similar in structure to a single domain of the bilobal cellular aspartyl proteases (Oroszlan and Luftig, 1990; Swanstrom *et al.*, 1990). A single chain of these cellular enzymes is folded into a structure that has an internal pseudodyad of symmetry with both halves of the pseudodimeric active site contributed by different parts of the same polypeptide chain. In the dimeric form, N- and C-termini of both PR monomers are intertwined to form a four-stranded antiparallel β-sheet. The conserved sequence aspartate–threonine/serine–glycine of each retroviral PR monomer is positioned in a loop that forms a part of the catalytic site; the same tripeptide sequence is in a similar loop configuration within the catalytic site of all cellular aspartic proteases (Suguna *et al.*, 1987; Sali *et al.*, 1989). Located above the active site in the retroviral PR dimer is the binding cleft. This cleft is large enough to accommodate a substrate that is seven amino acid residues in length (see below). Two large flaps in the dimer cover the entrance to the binding cleft and appear to move to allow substrate or inhibitor to enter. Topological studies of cellular aspartyl proteases also show that these enzymes have only one flap, which covers the processing complex (Suguna *et al.*, 1987; Sali *et al.*, 1989). Major features of this model for the retroviral PR have been confirmed by mutational analysis (Loeb *et al.*, 1988, 1989; Leis *et al.*, 1989; Bizub *et al.*, 1991). These structural studies provide a basis for analyzing catalytic mechanisms and elucidating the configuration (and processing patterns) of retroviral polyproteins with PR domains. In addition, knowledge of the structure of this viral enzyme can be used to design and evaluate inhibitors that block polyprotein processing and hence viral replication.

Cleavage site specificity of PR has been analyzed (i) by comparing sequences of known recognition sites in retroviral *gag* and *pol* polyproteins (Table V), (ii) in *in vitro* studies with synthetic peptides 7–10 residues long that mimic authentic cleavage sites, and (iii) by constructing and testing site-specific mutations in *gag* and *pol* polyprotein substrates as well as in the catalytic site of PR (Loeb *et al.*, 1989; Pettit *et al.*, 1991; Grinde *et al.*, 1992b). These investigations reveal that PR has specificity

TABLE V. Protease Cleavage Sites in *gag* Polyproteins[a]

Virus	Cleavage junction	P4	P3	P2	P1	—	P1'	P2'	P3'	P4'
ASLV	MA–x	T	S	C	Y	—	H	C	G	T
	x–p10	P	Y	V	G	—	S	G	L	Y
	p10–CA	V	V	A	M	—	P	V	V	I
	CA–x	A	A	A	M	—	S	S	A	I
	x–NC	P	L	I	M	—	A	V	V	N
	NC–PR	P	A	V	S	—	L	A	M	T
	RT–IN	F	Q	A	Y	—	P	L	R	E
MuLV	MA–pp21	S	S	L	Y	—	P	A	L	T
	p12–CA	S	Q	A	F	—	P	L	R	A
	CA–NC	S	K	L	L	—	A	T	V	V
	PR–RT	L	Q	V	L	—	T	L	N	I
	p15E–p2E	V	Q	A	L	—	V	L	T	Q
MMTV	MA–pp21	D	L	V	L	—	L	S	A	E
	pp21–p3	S	K	A	F	—	L	A	T	D
	p3–p8	E	L	I	L	—	P	V	K	R
	p8–n	P	V	G	F	—	A	G	A	M
	n–CA	T	F	T	F	—	P	V	V	F
	CA–NC	G	M	A	Y	—	A	A	A	M
	p30–PR	S	H	V	H	—	W	V	Q	E
SRV	MA–pp21	F	P	V	L	—	L	T	A	Q
	pp21–p12	P	T	V	M	—	A	V	V	N
	p12–CA	K	D	I	F	—	P	V	T	E
	CA–NC	G	L	A	M	—	A	A	A	F

Header spanning P4–P4': Cleavage site sequence

[a] Four amino acids on either side of the PR cleavage sites are shown for *gag* polyproteins of several retroviruses. P1 to P4 and P1' to P4' denote positions of amino acids relative to the cleavage site. PR has specificity for more than one substrate sequence. The amino acid at P1 is always hydrophobic and unbranched at the β-carbon. The majority of cleavage sites fall into two classes, depending on the P1' amino acid. Type I sites have proline at P1' and type II sites have alanine, leucine, or valine at P1'. Adapted from Pettit *et al.*, 1991.

for more than one cleavage site; accordingly, both sequence and structure of the substrate affect enzymatic activity. Specificity is influenced by four amino acids upstream and three amino acids downstream from the scissile bond in the target substrate, and the amino acid immediately upstream from the cleavage site (P1 position) is always hydrophobic and unbranched at the β-carbon (Drake *et al.*, 1988; Kotler *et al.*, 1988, 1989; Pettit *et al.*, 1991). The majority of cleavage site sequences have been grouped into two classes on the basis of the amino acid immediately downstream from the scissile bond (P1' position) (Table V) (Pettit *et al.*, 1991). In type I sites, proline is in the P1' position; CA proteins from *gag* of most retroviruses have a proline residue at the N-terminus (Table V). In type II sites, either alanine, leucine, or valine occupies the P1' position; the cleavage site that generates the C-terminus of CA proteins is a type II site (Table V). The catalytic rate of the HIV-1 PR is about tenfold greater than that of the ASLV enzyme (Kotler et al., 1989), and mutagenesis studies have identified amino acids in the substrate-binding pocket of the ASLV PR that influence the catalytic rate (Grinde et al., 1992a). Rates of PR cleavage of natural substrates may be controlled not only by sequence composition, but also by structural context and surface accessibility of the cleavage site.

F. Processing of Viral Polyproteins by PR

For all retroviruses, PR is translated as part of *gag*-containing polyproteins (Figs. 2 and 14); thus, PR is part of the polyprotein and may be considered a proenzyme or zymogen (Skalka, 1989; Oroszlan and Luftig, 1990; Wills and Craven, 1991). The mechanism of activation of the enzyme appears to be autocatalytic. In the proposed model for processing, the *gag*-containing polyprotein dimerizes and then the first proteolytic cleavage occurs either in *cis* (intramolecular) or in *trans* (intermolecular) to release the mature PR dimer (Skalka, 1989; Kotler et al., 1989). Analysis of the crystal structure of ASLV and HIV-1 PR supports the notion that PR cleaves itself out of the precursor by a *trans* mechanism. The sites at which PR is cleaved from the *gag–pol* precursor polyprotein are on the opposite side of the molecule from the catalytic site; thus, cleavage in *cis* is not likely for steric reasons (Wlodawer et al., 1989; Lapatto et al., 1989; Miller et al., 1989). Formation of the PR dimer under normal conditions is an intermolecular event dependent on the concentration of the precursor.

To test the role of dimerization on processing and assembly, ASLV *gag* polyproteins containing linked PR dimers were constructed and evaluated in transfected cells. These *gag* precursors with dimeric PR were incorporated into core-like particles and proteolytically processed to mature *gag* proteins, whereas *gag* precursors with only one PR domain were not efficiently processed, although core-like particles are produced and released from cells (Burstein et al., 1991). Thus, dimerization is necessary for initiating maturation of *gag*-containing polyproteins. The ASLV PR retains optimal activity in the *gag* polyprotein only if its carboxy terminus is free (Bennett et al., 1991). In addition, the ASLV *gag–pol* polyprotein displays no detectable *cis*- or *trans*-acting PR activity; this processing step is required for releasing active RT. A *gag* polyprotein rescues the *gag–pol* polyprotein, and particles containing active RT are produced (Stewart and Vogt, 1991; also see Craven et al., 1991). Proteolytic processing is also not observed when the MuLV *gag–pol* polyprotein is expressed in the absence of *gag* polyprotein (Felsenstein and Goff, 1991). Studies in the ASLV system have shown that PR can not initiate the processing reaction while it is part of the *gag* precursor (Burstein *et al.*, 1992). Thus, the protease must first be released from its precursor before it can attack other sites in the *gag* and *gag–pol* polyproteins. In the pathway of virion assembly, PR cleavage of *gag*-containing polyproteins is initiated during the budding process and continues in newly released immature particles (see Section VIII below).

Members of the ASLV are distinct from other retroviruses in that the PR domain is at the carboxy terminus of the *gag* polyprotein (Fig. 2). Thus, a 1:1 stoichiometry describes the ratio of ASLV PR to other *gag* proteins. Perhaps this enzyme has a structural role in ASLV assembly; however, virions (noninfectious) are assembled and released from cells harboring viral genomes with deletion mutations in PR (Stewart *et al.*, 1990). One possibility is that PR may be required on the end of every *gag* polyprotein molecule to mediate a *cis* cleavage(s). An alternative explanation is based on the observation that the ASLV PR has a relatively low level of activity as compared to other proteases; accordingly, larger quantities of ASLV may be required for efficient processing (Kotler *et al.*, 1989). All other retroviral proteases as well as cellular aspartyl proteases have the sequence aspartate–threonine–glycine in the active site, whereas the ASLV enzyme substitutes serine for the threonine. Whether this substitution affects the activity of proteases is an issue that can be addressed by site-directed mutagenesis.

A potential role for PR in early stages of the retroviral life cycle is proposed on the basis of experiments in an animal lentivirus system. In purified particles of equine infectious anemia virus (EIAV) incubated *in vitro*, PR cleaves NC protein into smaller polypeptides (Roberts and Oroszlan, 1989; Roberts *et al.*, 1991). Thus, cleavage of NC (mediated by PR) is hypothesized to be a step in the entry and uncoating processes. Removal of NC which is tightly bound to virion RNA may be necessary for reverse transcription to proceed in the cytoplasm to generate double-stranded viral DNA. Presumably, the NC proteins of ASLV and MuLV as well as other retroviruses may play a similar role in early events in replication.

G. Synthesis and Processing of *env* Gene Products

Synthesis and processing of retroviral *env* glycoproteins take place in the cellular secretory pathway which is used for the production of membrane and secreted proteins of both the host cell and numerous enveloped viruses (Figs. 16 and 17) (White, 1990; Hunter and Swanstrom, 1990; Daar and Ho, 1990). The retroviral *env* polyprotein is translated from subgenomic spliced mRNA on membrane-bound polysomes. A signal peptide characterized by a high proportion of hydrophobic amino acid residues is located in the N-terminal leader, which varies

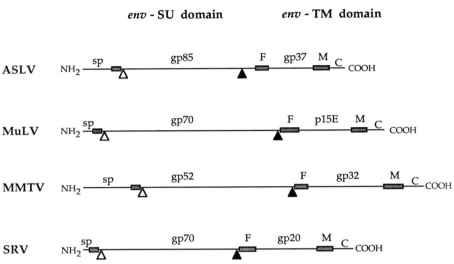

FIGURE 16. Functional domains and proteolytic processing sites in retroviral *env* glycoproteins. Each line represents the retroviral *env* polyprotein; widened areas are hydrophobic regions. Shown are the signal peptide (sp) cleavage site (open triangle) and the cleavage site that separates the surface (SU) and transmembrane (TM) domains (filled triangle). SU and TM domains for each retrovirus are labeled according to the molecular weights of the mature cleavage products. The TM domain contains a hydrophobic fusion peptide (F), a hydrophobic membrane-spanning region (M), and a cytoplasmic anchor (C). [Redrawn from Hunter and Swanstrom (1990).]

from about 30 to 100 amino acids. After translocation of the nascent *env* polyprotein into the lumen of the endoplasmic reticulum, a host endoprotease cleaves the leader peptide; the specificity of this proteolytic cleavage is apparently determined by sequences within the signal peptide. Translocation through the endoplasmic reticulum is halted at a stretch of hydrophobic amino acids (i.e., stop-transfer sequence) located near the C-terminus of the TM domain of *env*; accordingly, the *env* polyprotein remains anchored in the lipid bilayer membranes of the cell. The region of the TM domain extending from the hydrophobic anchor to the natural termination codon is designated the cytoplasmic anchor (Fig. 17). Site-directed mutagenesis has been used to introduce a translational termination codon immediately before the anchor sequence; consequently, the truncated form of the *env* polyprotein lacks an anchor and is secreted into the extracellular medium. These mutagenesis studies revealed that the relatively hydrophilic cytoplasmic anchor portion of TM is dispensable for ASLV replication (Perez *et al.*, 1987). In contrast, similar mutagenesis studies show that the cytoplasmic anchor of HIV is essential for viral infectivity (Kowalski *et al.*, 1987).

During cotranslational transfer of the nascent *env* polyprotein into the endoplasmic reticulum, cellular enzymes add oligosaccharide side

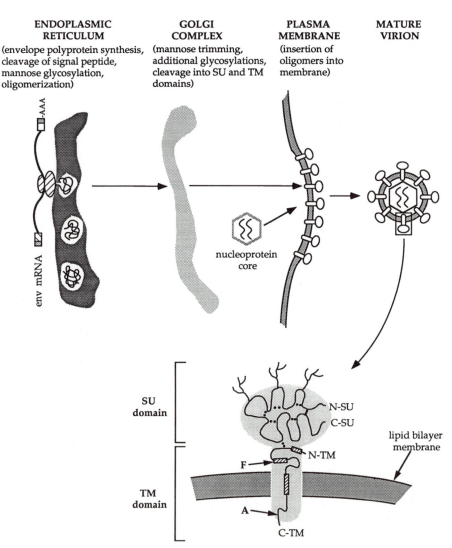

ENDOPLASMIC RETICULUM	GOLGI COMPLEX	PLASMA MEMBRANE	MATURE VIRION
(envelope polyprotein synthesis, cleavage of signal peptide, mannose glycosylation, oligomerization)	(mannose trimming, additional glycosylations, cleavage into SU and TM domains)	(insertion of oligomers into membrane)	

FIGURE 17. Intracellular pathways of *env* glycoprotein synthesis and assembly into membranes. Messenger for *env* is translated on membrane-bound polysomes and the nascent *env* polypeptide is directed into the lumen of the endoplasmic reticulum. Tertiary structure, acquired by oligomerization of *env* polyproteins in the endoplasmic reticulum, is necessary for transport to the Golgi complex where mannose-rich carbohydrate side chains are added. A cellular endoprotease located in the Golgi compartment cleaves the *env* polyprotein into the SU and TM domains. Subsequently, the processed *env* oligomers are inserted into the plasma membrane of the cell. Nucleoprotein complexes composed of genomic viral RNA and *gag* polyproteins bud through areas of the cell plasma membrane that contain a high density of *env* glycoproteins to produce extracellular virions. The inset drawing shows the surface (SU) and transmembrane (TM) domains of the *env* glycoprotein heterodimer. Carbohydrate branches are indicated for both domains. Hydrophobic

chains, composed predominantly of mannose residues esterified into long chains (Fig. 17) (Hunter and Swanstrom, 1990, and references therein). The tripeptide sequences asparagine–X–serine and asparagine–X–threonine are predicted sites of N-linked glycosylations on asparagine. Within the endoplasmic reticulum, *env* polyproteins interact and form oligomers. In the ASLV system, the *env* polyprotein assembles into a trimeric complex (Einfeld and Hunter, 1988). Various reports indicate that the MuLV *env* oligomer contains from three to six subunits, although experimental differences may account for this apparent discrepancy (Pinter, 1989; Kamps *et al.*, 1991; see also Schawaller *et al.*, 1989). Detailed analysis of MuLV revealed that several points of contact between the *env* glycoprotein monomers contribute to stabilization of the oligomer, and a region in TM is critical for maintaining quarternary structure (Tucker *et al.*, 1991). A proposed model for TM shows that a predicted α-helix involves the "leucine zipper"-like motif $LX_6LX_6NX_6LX_6L$, and this structural element may function to maintain protein–protein interactions in the *env* oligomer (Delwart and Mosialos, 1990; Gallaher *et al.*, 1989). In an alternative model, the predicted α-helix is presumed to form a coiled coil that is the basis for subunit interactions in the *env* oligomer (Tucker *et al.*, 1991). Mutagenesis and biochemical studies show that oligomerization is necessary for the transport of *env* from the endoplasmic reticulum into the Golgi compartment (Einfeld and Hunter, 1988; White, 1990, and references therein).

After transport of the *env* precursor into the Golgi complex, many mannose residues are trimmed from the glycosylated side chains; other carbohydrates (i.e., N-acetylglucosamine, galactose, fucose) are added to produce an *env* glycoprotein that contains both complex and hybrid carbohydrate side chains (Fig. 17) (Hunter and Swanstrom, 1990, and references therein). The function(s) of glycosylation is not clear; these oligosaccharide side groups may be necessary for transporting the *env* glycoprotein through various intracellular compartments during synthesis; in addition, the stability of the *env* glycoprotein as well as folding pattern may be influenced by carbohydrate side groups. Studies with site-specific *env* gene mutants and with glycosylation inhibitors reveal that certain carbohydrate modifications are essential for viral infectivity. Carbohydrate side chains also affect the antigenicity of the mature *env* glycoprotein by occluding peptide epitopes. O-linked sugars have also been found in the *env* gene products of several retroviruses; the

---

stretches in the TM domain are depicted as cross-hatched boxes; additional features of TM are the fusion peptide (F) and the cytoplasmic anchor (C). Intramolecular disulfide bonds (shown as filled double dots) hold portions of the *env* polypeptide of SU and TM together to form loops; in addition, the SU and TM domains of some retroviruses interact via disulfide bonds (e.g., ASLV).

significance of O-linked glycosylation remains to be determined (Pinter and Honnen, 1988).

Another important event in *env* biosynthesis in the Golgi compartment is cleavage of the polyprotein into the SU and TM domains by a host-cell endoprotease (Fig. 17) (Hunter and Swanstrom, 1990, and references therein). This cleavage occurs immediately after a stretch of 2–3 basic amino acid residues in the SU domain. Site-specific mutations which alter or delete the amino acids at this cleavage site yield an uncleaved *env* glycoprotein that folds and assembles normally and is inserted into the cell membrane; however, production of infectious virus is blocked (Perez and Hunter, 1987; McCune *et al.*, 1988; Earl *et al.*, 1991). Mammalian subtilisins belonging to the serine protease family cleave after dibasic residues in a variety of cellular and viral proteins (Barr, 1991). Cleavage of the *env* polyprotein in the Golgi compartment may result in a conformational change whereby the hydrophobic fusion peptide in the TM domain is buried within the glycoprotein oligomer. In a new infectious cycle, attachment of virions to a cell receptor via the SU domain is essential for the fusion peptide to insert into the cell plasma membrane (see Section III on virus entry).

The *env* polyprotein, modified by proteolytic processing and glycosylation events and assembled into oligomeric complexes, is transported out of the Golgi compartment and inserted into the cell plasma membrane. Epithelial cells are polar, with apical and basolateral surface membranes; retroviral *env* glycoproteins are transported to the basolateral membrane (Roth *et al.*, 1983; Stephens and Compans, 1986). Retroviral determinants controlling polarized transport were analyzed in epithelial cells infected with vaccinia virus vectors expressing HIV-1 genes. The HIV-1 *env* glycoprotein directed virus maturation and release at the basolateral membrane (Owens *et al.*, 1991). In all cell types (i.e., fibroblasts, lymphoid and epithelial cells), the viral nucleoprotein core buds through an area of the cell plasma membrane that contains *env* oligomers, and thus the core acquires a lipid bilayer membrane (Figs. 5 and 18).

VIII. VIRION ASSEMBLY

A. Overview and Significance of Retroviral Assembly (i.e., Morphogenesis)

The assembly of virions involves noncovalent intermolecular interactions of specific viral components in the cell (i.e., genomic RNA, host-cell tRNA, and *gag, pol,* and *env* gene products) and proteolytic cleavages of viral polyproteins (Figs. 3 and 14). Studies of these complex events in retroviral morphogenesis have involved biochemical and genetic analysis and electron microscopy of assembly intermediates in infected cells (Fig. 5) (Bernhard *et al.*, 1958; Bernhard, 1960; Fine and Schochetman, 1978; Teich, 1982; Gelderblom, 1991, and references therein). Natural viral variants as well as viral mutants constructed by site-specific mutagenesis procedures have been examined (Goff, 1990, and references therein). These studies reveal two pathways of particle morphogenesis; the salient difference is whether the immature core (or capsid) is formed in the cytoplasm or at the plasma membrane (Fig. 5) (Fine and Schochetman, 1978; Rhee and Hunter, 1990a, b). For both pathways, retroviral assembly presumably initiates with the interaction of the NC domain of the *gag* polyprotein with *cis*-acting packaging (i.e., encapsidation) signals in genomic viral RNA (Fig. 6) (Linial and Miller, 1990). In addition, virus particles in both morphogenetic routes acquire a lipid bilayer membrane by budding through the cell plasma membrane (Fig. 18). Although the major features of virion assembly are known, many important details remain to be elucidated.

The retroviral assembly process yields a stable extracellular vehicle that not only protects and transmits viral genetic information, but also participates in early events in an infectious cycle. Reverse transcription takes place in a ribonucleoprotein core complex derived from virions immediately after entry into cells (see above). In addition, the integrative precursor (i.e., linear viral DNA) is contained within a nucleoprotein complex that may consist of the major core protein (i.e., CA) and other virion proteins. Replication intermediates of hepadnaviruses and caulimoviruses are also encapsidated into viral-like particles (Summers and Mason, 1982; Marsh and Guilfoyle, 1987; Mason *et al.*, 1987). Retrotransposons that have only an intracellular life cycle (e.g., Ty-1 and Ty-2 elements in *Saccharomyces*, copia in *Drosophila*, intracisternal type A particles in mammalian cells) also require particle-associated assemblages for transposition (Eichinger and Boeke, 1988; Fuetterer and Hohn, 1987; Boecke and Corces, 1989) (see Chapter 4). Although the packaging of cellular RNA into viral particles is generally an inefficient process, retroviruses may play roles in transposing sequences within the host genome (Baltimore, 1985; Linial, 1987) (see Chapter 1). Studies on assembly are significant for identifying specific mechanisms and viral components that may be targeted for the development of antiviral inhibitors. Also, an understanding of the factors that control virion formation is essential for the design of efficient retroviral vectors for gene transfer (see Section IX).

B. Packaging: *cis*-Acting Viral RNA Sequences

Retroviruses have mechanisms to select genomic viral RNA out of a large cytoplasmic pool of cellular mRNA and subgenomic viral mRNA

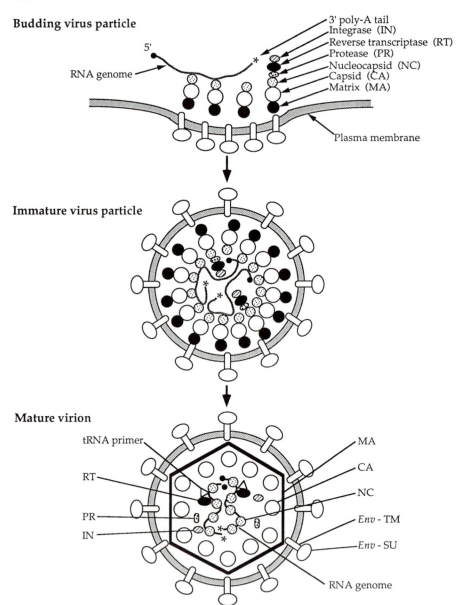

FIGURE 18. Release of type C retroviral particles from cells by budding and subsequent maturation. Myristylation (or acylation) of the N-terminus of the MA domain in the *gag* polyprotein generates a topogenic signal, and the polyprotein is subsequently transported to the inner surface of the cell plasma membrane. Genomic viral RNA molecules interact with the NC domain and perhaps other portions in the *gag* polyprotein molecules located near the plasma membrane. Intracellular portions of the TM domains may also recognize MA, although studies with mutants indicate that TM is not required for release of parti-

for assembly into virions (Linial and Miller, 1990; Aronoff and Linial, 1992, and references therein). Genomic viral RNA as well as cellular mRNA and subgenomic viral mRNA have 5′m7G cap structures and poly-A tails (Figs. 6 and 13) (Coffin, 1985; Stoltzfus, 1988). Therefore, selectivity must involve internal viral sequences which are absent from spliced viral and cellular messages. Cis-acting signals (designated Ψ) in genomic viral RNA as well as trans-acting virion core proteins (i.e., NC) govern the mechanism of encapsidation. Packaging of intact genomic viral RNA is about 1000 times more efficient than incorporation of the same RNA molecules lacking these signals (Adam and Miller, 1988; Dornburg and Temin, 1988; Aronoff and Linial, 1992).

Extensive studies aimed at defining Ψ in genomic viral RNA have been based on the analysis of both naturally occurring viral variants which show defects in assembly and genetically engineered viral mutants constructed by site-specific mutagenesis of cloned proviral DNA (Linial et al., 1978; Watanabe and Temin, 1982; Linial and Miller, 1990, and references therein). Some differences with respect to locations of assembly elements in genomic RNA are observed in comparisons of ASLV with MuLV and REV. In the ASLV system, efficient assembly requires three cis-acting elements: (i) a noncoding segment at least 150 bases long between the PB− and the start codon for gag (i.e., Ψ) (Fig. 6), (ii) sequences within the 5′ end of the gag gene, and (iii) a segment about 100 bases long that forms a direct repeat flanking the src gene (designated DR) (Linial and Miller, 1990, and references therein). The L segment is upstream of the splice donor; thus, subgenomic viral messages encoding env and src contain L as well as the direct repeat element flanking src; however, these messages are not efficiently packaged. Presumably, the absence of an important packaging element within the ASLV gag sequence precludes the assembly of subgenomic (i.e., spliced) viral mRNA.

For both MuLV and REV, the respective Ψ is a sequence about 300 bases long that is located in the L region immediately downstream from

---

cles from cells. Formation of the dimer linkage structure between two genomic viral RNA molecules may occur either at very early stages in the interaction of the gag polyprotein with the genome or during the budding process. The precise point in particle assembly at which the primer tRNA molecule associates with the genome is not known. A high concentration of gag polyprotein–viral RNA complexes in specific sites on the inner face of the plasma membrane produces the electron-dense crescents that are observed in electron micrographs of infected cells; these crescents are the precursors to virion cores. A protrusion in the plasma membrane forms around the gag polyprotein–viral RNA complexes, and the membrane distends until it completely engulfs the core and is released from the cell. Proteolytic processing of gag polyproteins is probably initiated by the budding process and continues in newly released immature particles. Morphological changes accompany these processing events until a mature virion is produced which is fully infectious.

the splice donor in each virus (Fig. 6) (Mann *et al.*, 1983; Mann and Baltimore, 1985; Sorge *et al.*, 1983; Watanabe and Temin, 1979; Embretson and Temin, 1987b; Linial and Miller, 1990). Thus, subgenomic *env* mRNA for these viruses is not assembled into virions, because the *cis*-acting assembly element is in the *env* intron. MuLV Ψ is functional for packaging viral RNA if moved from the normal location to a position downstream of the *env* gene, and this element retains packaging activity only in the sense orientation (Mann and Baltimore, 1985). REV is an avian retrovirus that shows some homology with MuLV *gag* and *pol* genes; transcripts containing MuLV Ψ are efficiently packaged by REV (Embretson and Temin, 1987b; Dornburg and Temin, 1990).

Retroviral Ψ elements required for efficient packaging have been localized by mutational analysis of viral genomes as described above; however, sequences sufficient for packaging have been defined by other approaches. Selectable markers have been cloned into MuLV and REV genomes with deletions that remove the coding genes (i.e., *gag*, *pol*, and *env*) (Fig. 19) (Mann *et al.*, 1983; Cepko *et al.*, 1984; Emerman and Temin, 1984; Kriegler, 1990, and references therein). These constructs yield transcripts that are assembled into particles either in cells infected with replication-competent helper viruses or in genetically engineered cells expressing virion proteins (i.e., packaging cell lines) (Fig. 20) (Miller, 1990a). Constructs containing the LTR and Ψ together with a selectable marker (or other heterologous gene) are designated retroviral vectors (see below) (Coffin, 1985; McLachlin *et al.*, 1990).

An extension of this approach to the construction of retroviral vectors has defined the minimum sequences required for packaging (Adam and Miller, 1988; also see Murphy and Goff, 1989). Segments of MuLV encompassing Ψ were inserted into a nonretroviral transcription unit containing a cytomegalovirus promoter, the neomycin (*neo*) selectable marker, and a poly-A signal from SV40. The MuLV DNA fragment containing sequences from bases 215–1038 (L and part of *gag*) was cloned between the end of the *neo* gene and the SV40 poly-A site in this heterologous transcription unit (Figs. 6 and 20). In transfected packaging cells, transcripts containing the 823-bp MuLV segment were assembled into virions as well as those from the parent virus. In an identical vector, heterologous transcripts containing the region from bases 215–563 (L only) were encapsidated 40-fold less efficiently. Accordingly, the 823-bp region that includes L and the N-terminal sequences for *gag* has been designated the extended packaging signal (Ψ⁺). Other investigators have found that sequences within the MuLV *gag* gene also contribute to the encapsidation efficiency of viral RNA (Armentano *et al.*, 1987; Bender *et al.*, 1987). A similar approach utilizing a nonretroviral vector system has been used to demonstrate that the Ψ⁺ of ASLV is specified by a 683-base segment that includes L and sequences from the N-terminus of

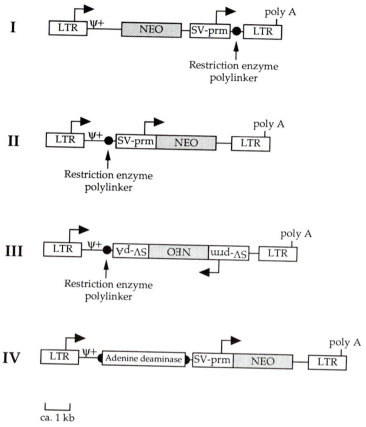

FIGURE 19. Retroviral vectors for transduction of heterologous genes. The proviral configurations shown are based on retroviral vectors containing MuLV LTRs and packaging
signals (Ψ). neo is a gene from the bacterial transposon Tn5 which encodes the selectable
marker neomycin that functions in mammalian cells. ada represents the cDNA copy of the
human adenosine deaminase gene. SV-prm is the promoter from the simian virus 40
(SV40) early gene; this heterologous promoter is located between the retroviral LTRs and
regulates initiation of neo (configurations II, III, and IV). The 5' LTR regulates transcription of full-length transcripts as well as neo (configuration I) and ada (configuration IV).
Arrows indicate transcription start sites and direction of transcription for each vector.
Filled circles are synthetic oligonucleotides containing restriction enzyme sites for inserting heterologous genes (i.e., polylinkers). The 3' LTR provides the signal for addition of
poly-A to full-length transcripts and subgenomic transcripts in configurations I and II.
The SV-prm–neo transcription unit in configuration III is in the antisense orientation
relative to the LTRs; thus, a poly-A site (shown in the figure as SV-pA) is provided by a
sequence element from the 3' end of the SV40 early gene. Thus, in configuration III the
full-length transcript (packaged into virus particles) encodes the LTRs and the neo gene in
the sense orientation. The SV-prm–neo–SV-pA sequence is in the antisense orientation in
the full-length transcript, but the subgenomic transcript will be in the sense orientation
with respect to the neo gene.

Packaging Cell

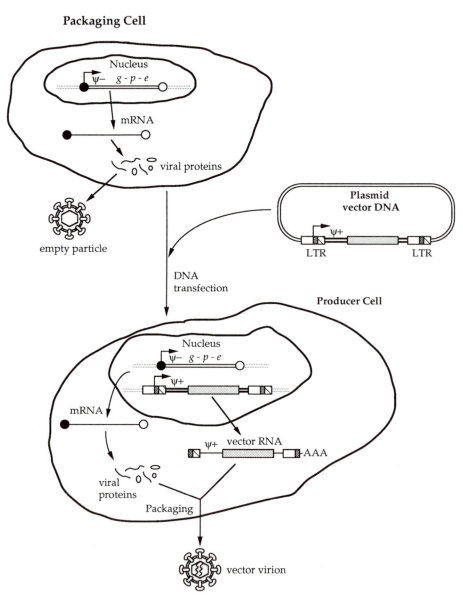

FIGURE 20. Packaging cell lines for retroviral vectors. This packaging cell line contains an integrated transcription unit that expresses MuLV virion polyproteins (*gag, pol,* and *env*) under the control of a promoter (filled circle with arrow) and poly-A site (open circle) from a heterologous virus. In addition, the packaging signal (Ψ) has been deleted from the transcription unit expressing the MuLV polyproteins. A plasmid containing the cloned DNA form of a retroviral vector (Fig. 19) is introduced into the packaging cell line by DNA transfection. Full-length transcripts from the transfected plasmid contain Ψ and are packaged into virions. Accordingly, these virions contain all functions required for one

gag (Fig. 6) (Aronoff and Linial, 1991). The Ψ^+ of ASLV (and MuLV) functions efficiently at the 3′ end of a gene, but only in the sense orientation (Adam and Miller, 1988; Aronoff and Linial, 1992).

Potential RNA secondary structures in MuLV Ψ have been identified by a combination of analysis with chemical probes and computer modeling (Alford *et al.*, 1991; Tounekti *et al.*, 1992). A proposed model involves a predicted pseudoknot together with conventional stem-loop structures. Alignment of packaging regions of REV, MuLV, and VL30 (i.e., endogenous retrovirus-related sequences) revealed only a short stretch (12 bases) of homology; intriguingly, this element is part of a deduced stem-loop structure that is identical for all three viruses (Aronoff and Linial, 1992; see also Tounekti *et al.*, 1992). The Ψ^+ of ASLV also has several predicted stem-loop structures (Aronoff and Linial, 1992). Additional mutational studies in all of these retroviral systems are required to assess the validity of the proposed RNA structures with respect to packaging and/or formation of the dimer linkage structure (Bieth *et al.*, 1990).

Packaging of cellular RNAs into retroviral particles is generally a low-frequency event (Aronoff and Linial, 1992, and references therein). An avian cell line (SE21Q1b) harbors a defective ASLV genome and produces noninfectious particles containing a random sampling of cellular mRNAs (Linial *et al.*, 1978; Gallis *et al.*, 1979). Less than 1% of the RNA in these particles is viral-specific. The defective provirus has a 179-bp deletion extending from the 5′ LTR into L (position 96–274; see Fig. 6) (Anderson *et al.*, 1992). In addition, several differences are noted in sequence comparisons of the NC domain of the defective provirus and wild-type ASLV; however, analysis of site-specific mutations reveals that the changes in the mutant proviral NC protein are not sufficient to account for the altered packaging specificity in the SE21Q1b cell line (Anderson *et al.*, 1992) (see below). Nonetheless, studies with recombinant viral genomes show that the mutation(s) that favors random packaging of cellular RNAs maps to the NC region of *gag*. After superinfection of SE21Q1b cells with wild-type virus, packaging of cellular RNA continues at a relatively high level; thus, the *cis*-acting defect is dominant (Linial *et al.*, 1978). The SE21Q1b cell system was the first example of transduction of randomly packaged cellular RNAs via reverse transcription and integration into the host genome. This process is designated retrofection and the cDNAs representing cellular mRNAs without introns are termed retrogenes (Linial, 1987; Aronoff and Linial, 1991).

round of infection and integration. Virions are harvested from the extracellular medium and used to infect permissive recipient cells which acquire the proviral form of the vector. The host-cell transcription apparatus then mediates expression of the transduced gene(s) in the recipient cell. [Redrawn from J. R. McLachlin *et al.* (1990).]

C. Packaging: *trans*-Acting Viral Proteins

The presence of sites in genomic RNA required for assembly implies that a *trans*-acting factor(s) recognizes these sites (Coffin, 1985). The *pol* and *env* gene products are not required for the formation of virus particles; noninfectious virions containing genomic viral RNA are produced from viral genomes harboring large deletions in either of these genes (Linial and Blair, 1982). However, similar mutagenesis studies show that *gag* is essential for encapsidating viral RNA and particle assembly (Goff, 1990, and references therein; Wills and Craven, 1991, and references therein). Examination of virions revealed that the *gag*-derived NC protein is tightly associated with the genome in the core (Darlix and Spahr, 1982).

Retroviral NC proteins are small, very basic, and hydrophilic (Meric *et al.*, 1984; Karpel *et al.*, 1987). In addition, NC proteins have a cysteine–histidine motif that appears to be similar to the metal finger domains of several proteins which bind nucleic acids (Dickson *et al.*, 1985; Berg 1986). This motif consists of one or two copies of the sequence cysteine–X2–cysteine–X4–histidine–X4–cysteine, and most retroviruses encode an NC protein which contains this motif (Katz and Jentoft, 1989). Also, cauliflower mosaic virus and several retrotransposons (e.g., *Saccharomyces* Ty-1 element) encode proteins with this sequence motif. In contrast, the *gag* genes of several primate spumaviruses do not specify a domain with a cysteine–histidine motif; instead, the NC analogue of this group of retroviruses is presumed to be a *gag* cleavage product that has a high proportion of basic amino acids (Maurer *et al.*, 1988; Mergia and Luciw, 1991; Renne *et al.*, 1992).

The significance of the cysteine–histidine motifs in NC for packaging has been tested by constructing and evaluating viral mutants (Jentoft *et al.*, 1988; Fu *et al.*, 1988; Meric *et al.*, 1988; Gorelic *et al.*, 1988; Meric and Goff, 1989; Dupraz *et al.*, 1990). Molecular clones of ASLV and MuLV genomes with site-specific mutations in NC (i.e., in the cysteine–histidine motifs) have been shown to form particles which do not package viral RNA efficiently. These studies also suggested that the N-terminal cystidine–histidine motif in viruses with two copies is more critical for NC recognition of RNA than the C-terminal motif.

In addition to encapsidation of genomic viral RNA (and initiation of viral DNA synthesis; see Section V), NC may play a role in the formation of the dimer linkage structure and appears to facilitate the annealing of the tRNA primer to the viral genome (Meric and Spahr, 1986; Bieth *et al.*, 1990; Prats *et al.*, 1988). Thus, mutations in NC may have a complex phenotype. Accordingly, the cysteine–histidine motif may be essential for one or more of these functions ascribed to NC (i.e., encapsidation, genome dimerization, and tRNA annealing). Also, this motif may enable NC to form dimers or higher multimers which then bind viral RNA during the assembly process (Aronoff and Linial, 1992).

In vitro, NC proteins of ASLV and MuLV have an affinity for homologous and heterologous RNA (Leis and Jentoft, 1983; Meric et al., 1984; Fu et al., 1985; Karpel et al., 1987). Thus, this interaction is not sequence-specific in these two viral systems; however, recent studies on HIV-1 have established conditions that appear to demonstrate specific recognition of HIV-1 NC with the homologous packaging signal (Luban and Goff, 1991; see below). No evidence of RNA binding via the cysteine–histidine motif has been demonstrated in vitro (Meric et al., 1984; Fu et al., 1985; Karpel et al., 1987). A single molecule of NC covers 4–6 bases and stoichiometric considerations suggest that each molecule of genomic RNA is entirely coated with NC (Karpel et al., 1987). Phosphorylation of NC purified from ASLV particles increases the affinity for RNA in a nonspecific fashion (Fu et al., 1985).

Recent analysis in an in vitro system has revealed that the HIV-1 gag polyprotein specifically recognizes viral sequences near the 5' end of the genome (Luban and Goff, 1991). For this system, gag polyprotein produced in genetically engineered bacteria was fixed to a matrix (e.g., nitrocellulose membranes), and measurements were made on the extent of binding of radioactively labeled HIV-1 RNA. Viral RNA molecules lacking sequences required for efficient packaging (i.e., Ψ) bound less well than wild-type viral RNA, and mutagenesis of the cysteine–histidine motif in the NC domain of the gag polyprotein abrogated binding. Utilization of this blotting procedure may facilitate the in vitro analysis of packaging for other viral systems. Sequences and structural features of NC protein (embedded within the gag polyprotein) as well as viral RNA can be evaluated with respect to binding specificity.

Analysis of viral assembly in infected cells has demonstrated that the packaging process is influenced by other domains in the gag polyprotein in addition to NC. UV cross-linking experiments on freshly isolated ASLV particles revealed that the NC domain embedded within the gag polyprotein interacts directly with genomic viral RNA (Stewart et al., 1990). Studies with site-specific mutations in various regions of the ASLV gag gene show that the NC protein participates in packaging viral RNA only as a part of the full-length gag polyprotein (Oertle and Spahr, 1990). In addition, these same investigations demonstrated that the PR domain influences dimerization of the viral genome. The precise basis for recognition of the viral genome remains to be established; either a structural element in viral RNA is required or another viral gag protein acts in concert with NC to confer specific recognition properties.

D. Pathways for Virion Assembly

In one pathway for retroviral morphogenesis, genomic viral RNA and *gag*-containing polyproteins are assembled into immature cores in the cell cytoplasm; cells infected with the type B and type D retroviruses follow this pattern of morphogenesis (Fig. 5) (Fine and Schochetman, 1978; Rhee and Hunter, 1990a, b; Rhee *et al.*, 1990). These intracellular cores are spherical with a double-walled appearance that is similar to the structure of intracellular type A particles. The MA domain of *gag* has a topogenic signal that targets viral polyproteins to a cytoplasmic compartment where assembly is initiated (Rhee and Hunter, 1990a). A specific point mutation in the MA domain alters the pattern of SRV morphogenesis; the mutant *gag* polyprotein assembles into a core structure at the plasma membrane, much like type C retroviruses (see below), instead of within the cytoplasm. Analysis of additional site-specific mutations reveals that the MA domain of the *gag* polyprotein precursor is also essential for (i) transport of the core from the site of assembly in the cytoplasm to the plasma membrane and (ii) budding through the plasma membrane (Rhee and Hunter, 1991). Myristylation of the amino terminus of the *gag* polyprotein is necessary for the association of the core with the inner surface of the cell membrane (Rhee and Hunter, 1987). The preassembled core protrudes through the plasma membrane, a particle is released from the cell, and proteolytic processing of viral polyproteins continues in newly budded particles. The *env* glycoprotein oligomers are inserted into the plasma membrane of infected cells (Fig. 17). Virions acquire *env* glycoprotein by budding through areas of the plasma membrane that contain the viral glycoprotein. Mutational analysis has revealed that preassembled intracellular cores bud through the plasma membrane in the absence of *env* glycoprotein (Rhee *et al.*, 1990). The core of a mature type D retrovirus particle has a characteristic bar shape (Fig. 5). Electron microscopy shows that type B retroviruses also assemble a core structure within the cell cytoplasm; however, the core in mature type B particles is round and eccentrically positioned within the mature virion (Fig. 5) (Fine and Schochetman, 1978).

In the other morphogenesis pathway, characteristic of type C retroviruses, *gag*-containing polyproteins are individually transported to the cytoplasmic side of the plasma membrane containing *env* glycoprotein (Figs. 5 and 18). The *gag* polyproteins of MuLV and REV must be myristylated to associate with the cell membrane (Schultz and Rein, 1989; Weaver and Panganiban, 1990; Wills and Craven, 1991, and references therein). Studies on the effects of the ionophore monensin (i.e., an inhibitor of vesicular traffic) together with analysis of site-specific mutations in MA of MuLV *gag* revealed that myristylated *gag* polyproteins travel via vesicles to the plasma membrane (Hansen *et al.*, 1990). Additional investigations of *gag–β–gal* fusion proteins also showed that the MA domain is essential for targeting *gag* polyproteins to the plasma membrane (Jones *et al.*, 1990; Hansen *et al.*, 1990). Assembly of immature cores (i.e., complexes of *gag*-containing polyproteins and genomic viral RNA) is concomitant with the budding process, and this core first appears as an electron-dense crescent on the interior surface of the plasma membrane at the site of budding (Bernhard *et al.*, 1958; Bernhard, 1960; Gelderblom, 1991) (Fig. 5). A semicircular protrusion of the cell membrane forms above the immature core. Viral *env* glycoprotein oligomers are inserted into the plasma membrane, and virions acquire membranes with *env* glycoproteins through the budding process (Fig. 17). In the newly budded virus particle, the core appears as an electron-dense outer ring that surrounds an electron-lucent nucleoid (Fig. 5). Subsequently, the immature core condenses into a central, electron-dense structure to yield mature infectious virions (Yoshinaka *et al.*, 1977; Dickson *et al.*, 1982, 1985) (Fig. 5).

At early stages of the budding process, the MA portion of the *gag* polyprotein appears to interact with the cytoplasmic anchor region in the TM domain of *env* (Gebhart *et al.*, 1984; Perez and Hunter, 1987). However, the interaction of TM and MA for virion assembly is not required for budding, since cells harboring proviruses with deletion mutations in the *env* gene release bald particles which lack *env* glycoprotein and are not infectious (Kawai and Hanafusa, 1973; Dickson *et al.*, 1982). In addition, genetically engineered cells transfected with vectors that express only the *gag* polyprotein also release noninfectious particles that resemble retroviral cores (Wills and Craven, 1991, and references therein). Taken together, these observations show that the cytoplasmic anchor region of TM does not influence particle assembly and budding although it is required for efficient incorporation of *env* oligomers into virion membranes (Fig. 18) (Hunter and Swanstrom, 1990, and references therein).

Incorporation of the specific tRNA primer molecule into virions appears to involve an interaction with the RT domain in the *pol* polyprotein, although other studies suggest that NC protein may facilitate the interaction of the tRNA primer with genomic viral RNA (Sawyer and Hanafusa, 1979; Peters and Hu, 1980; Tirumalai and Modak, 1991; Prats *et al.*, 1988). Some host-cell proteins (e.g., protein kinase, ubiquitin) and several species of small cellular RNA molecules are also found in virions; these host components may be trapped in virus particles during the budding process (Eisenman and Vogt, 1978; Putterman *et al.*, 1990).

In both assembly pathways, proteolytic cleavages of viral polyproteins in immature cores are initiated by the viral PR during budding; full maturation of viral cores via these cleavages occurs in newly released virions (Katch *et al.*, 1985; Skalka, 1989; Oroszlan and Luftig, 1990). These specific proteolytic cleavages ensure that the assembly process is unidirectional and that morphogenetic events occur in a stepwise fashion. In addition, viral enzymes (i.e., RT and IN) in the *pol* polyprotein are not active; this proenzyme form prevents inappropriate activity inside the cell during virion morphogenesis.

IX. CONCLUSIONS AND PROSPECTS

Analysis of retroviral replication mechanisms has continued at a vigorous pace for the last two decades. Discovery of RT was a seminal and critical event that provoked intense and profound interest in a group of viruses that was already recognized for fascinating biological properties, namely tumorigenesis and cell transformation. Within the last decade, studies on basic retroviral replication mechanisms have escalated with the discovery that these agents are linked to serious human pathologies (e.g., AIDS). Many of the lessons learned from basic studies on the replication of retroviruses with simple genomes are directly applicable to retroviruses with complex genomes (e.g., HIV, HTLV). In addition, an understanding of these replication mechanisms is relevant not only for providing insight into the molecular biology of eukaryotic cells, but also for establishing the roles of specific viral genes and regulatory elements in virus–host relationships (e.g., latency and persistent infection, mechanisms of pathogenesis).

Bishop, Varmus, and co-workers discovered that retroviruses incorporate cellular proto-oncogenes into their genomes, and this observation substantiated the notion that retroviruses have the capacity to transduce heterologous (i.e., nonretroviral) genes into vertebrate cells (Bishop and Varmus, 1982, 1985) (see Chapter 1). Indeed, retroviruses can be viewed as packages which deliver genes into cells in culture and in animals (Tabin *et al.*, 1982; Joyner *et al.*, 1983; Bandyopadhyay and Temin, 1984; A. D. Miller *et al.*, 1983). Several unique features based on retroviral structure and replication events have been exploited for designing efficient vectors for gene delivery (Coffin, 1985a). These vectors are generally replication-defective and therefore may be limited to only one round of infection (Mann *et al.*, 1983; Cone and Mulligan, 1984; Miller 1990a), although replication-competent vectors have also been developed (Hughes and Kosik, 1984; Petropoulos and Hughes, 1991; Gelinas and Temin, 1986). Host range is controlled by selection of the *env* glycoprotein used for packaging (Fig. 20), and these vectors may accommodate one or more heterologous genes (Fig. 19) (McLachlin *et al.*, 1990). Expression is regulated by cell-specific enhancer elements in the LTR, internal heterologous promoters can be incorporated into the vector to control expression (Fig. 19), and the provirus is integrated into the host-cell genome in a stable fashion in low copy number.

Retroviral vectors have many potential clinical applications (Anderson, 1992; Miller, 1990b), and studies on therapeutic efficacy have been performed in animal model systems (Williams *et al.*, 1986; McLachlin *et al.*, 1990; Kay *et al.*, 1992). Recently, the feasibility of retroviral gene transduction for cancer therapy in humans has been demonstrated (Rosenberg *et al.*, 1990). In addition, a vector system has been constructed to express adenine deaminase (ADA) for treating severe combined immunodeficiency in humans (McLachlin *et al.*, 1990; Anderson, 1992) (Fig. 19). The knowledge of the retroviral replication cycle and virion structure reviewed above provides an essential basis for designing additional vector systems as well as for exploring potential applications and limitations. The converse also applies; hypotheses and issues that deal with mechanisms of retroviral replication and biology can be addressed by constructing and evaluating retroviral vectors (Ellis and Bernstein, 1989; Jones *et al.*, 1990; Pulsinelli and Temin, 1991; Reddy *et al.*, 1991; Hevezi and Goff, 1991; Dillon *et al.*, 1991; Soriano *et al.*, 1991).

On several occasions throughout this chapter, specific gaps in current knowledge on issues dealing with various retrovirus replication steps have been cited. Tools, techniques, and concepts of molecular biology as well as contributions from both immunology and cell biology have been essential for elucidating viral replication mechanisms. New findings in these disciplines will continue to provide a basis for learning more about retroviruses as well as mechanisms which control cells and determine the outcome of infection in animals and humans.

Postscript

The authors leave the reader with these two thoughts:

> *In expanding the field of knowledge we but increase the horizon of ignorance.*
> —Henry Miller

> *Science baits laws with stars to catch telescopes.*
> —e. e. cummings

X. REFERENCES

Adam, M. A., and Miller, A. D., 1988, Identification of a signal in a murine retrovirus that is sufficient for packaging of nonretroviral RNA into virions, *J. Virol.* **62:**3802.

Adamson, M. C., Silver, J., and Kozak, C. A., 1991, The mouse homolog of the gibbon ape leukemia virus receptor: genetic mapping and a possible receptor function in rodents. *Virology* **183**:778.

Aiyar, A., Cobrinik, D., Ge, Z., Kung, H. J., and Leis, J., 1992, Interaction between retroviral U5 viral RNA and the TYC loop of the tRNAtrp primer are required for efficient initiation of reverse transcription, *J. Virol.* **66**:2464.

Akgun, E., Ziegler, M., and Grez, M., 1991, Determinants of retrovirus gene expression in embryonal carcinoma cells, *J. Virol.* **65**:382.

Albritton, L. M., Tseng, L., Scadden, D., and Cunningham, J. M., 1989, A putative murine ecotropic retrovirus receptor gene encodes a multiple membrane-spanning protein and confers susceptibility to virus infection, *Cell* **57**:659.

Alford, R. L., Honda, S., Lawrence, C. B., and Belmont, J. W., 1991, RNA secondary structure analysis of the packaging signal for Moloney murine leukemia virus, *Virology* **183**:611.

Anderson, D. J., Lee, P., Levine, K. L., Sang, J., Shah, S. A., Yang, O. O., Shank, P. R., and Linial, M. L., 1992, Molecular cloning and characterization of the RNA packaging-defective retrovirus SE21Q1b, *J. Virol.* **66**:204.

Anderson, W. F., 1984, Prospects for human gene therapy, *Science* **226**:401.

Anderson, W. F., 1992, Human gene therapy, *Science* **258**:808.

Armentano, D., Yu, S. F., Kantoff, P. W., Von Ruden, T., Anderson, W. F., and Gilboa, E., 1987, Effect of internal viral sequences on the utility of retroviral vectors, *J. Virol.* **61**:1647.

Arnold, E., and Arnold, G. F., 1991, Human immunodeficiency virus structure: Implications for antiviral design, *Adv. Virus Res.* **39**:1.

Aronoff, R., and Linial, M., 1991, Specificity of retroviral RNA packaging, *J. Virol.* **65**:71.

Aronoff, R., and Linial, M. L., 1992, Retroviral RNA encapsidation and retrofection, in: *Techniques and Applications of Genome Research* (K. W. Adolph, ed.), Academic Press, San Diego, California.

Arrigo, S., and Beemon, K., 1988, Regulation of Rous sarcoma virus RNA splicing and stability, *Mol. Cell. Biol.* **18**:4858.

Arrigo, S., Yun, M., and Beemon, K., 1987, *Cis*-acting regulatory elements within the *gag* genes of avian retroviruses, *Mol. Cell. Biol.* **7**:388.

Ashorn, P. A., Berger, E. A., and Moss, B., 1990, Human immunodeficiency virus envelope glycoprotein/CD4-mediated fusion of nonprimate cells with human cells, *J. Virol.* **64**:2149.

Atchison, M. L., 1988, Enhancers: Mechanisms of action and cell specificity, *Annu. Rev. Cell Biol.* **4**:127.

Atkins, J. F., Weiss, R. B., and Gesteland, R. F., 1990, Ribosome gymnastics—Degree of difficulty 9.5, style 10.0, *Cell* **62**:413.

Atwater, J. A., Wisdom, R., and Verma, I. M., 1990, Regulated mRNA stability, *Annu. Rev. Genet.* **24**:519.

Baltimore, D., 1970, RNA-dependent DNA polymerase in virions of RNA tumor viruses, *Nature* **226**:1209.

Baltimore, D., 1985, Retroviruses and retrotransposons: The role of reverse transcription in shaping the eukaryotic genome, *Cell* **40**:481.

Bandyopadhay, P. K., and Temin, H. M., 1984, Expression of a complete chicken thymidine kinase gene inserted in a retrovirus vector, *Mol. Cell. Biol.* **4**:749.

Banerjee, A. K., 1980, 5'-terminal cap structure in eucaryotic messenger ribonucleic acids, *Microbiol. Rev.* **44**:175.

Barber, A. M., Hizi, A., Maizel, J. R., and Hughes, S. H., 1990, HIV-1 reverse transcriptase: Structure predictions for the polymerase domain, *AIDS Res. Hum. Retroviruses* **9**:1061.

Barklis, E., Mulligan, R. C., and Jaenisch, R., 1986, Chromosomal position or virus mutation permits retrovirus expression in embryonal carcinoma cells, *Cell* **47**:391.

Barr, P. J., 1991, Mammalian subtilisins: The long-sought dibasic processing endopro-
teases, *Cell* **66**:1.

Barr, P. J., Power, M. D., Lee-Ng, C. T., Gibson, H., and Luciw, P. A., 1987, Expression of
active human immunodeficiency virus reverse transcriptase in *Saccharomyces cerevisiae*, *Bio/Technology* **5**:486.

Basu, A., Basu, S., and Modak, M. J., 1990, Site-directed mutagenesis of Moloney murine
leukemia virus reverse transcriptase, *J. Biol. Chem.* **265**:17162.

Battula, N., and Loeb, L. A., 1974, The infidelity of avian myeloblastosis virus deoxyribonucleic acid polymerase in polynucleotide replication, *J. Biol. Chem.* **249**:4086.

Beato, M., 1989, Gene regulation by steroid hormones, *Cell* **56**:335.

Bedenek, K., Abbots, J., Roberts, J. D., Wilson, S. H., and Kunkel, T. A., 1989, Specificity
and mechanism of error-prone replication by human immunodeficiency virus-1 reverse transcriptase, *J. Biol. Chem.* **264**:16,948.

Beemon, K., and Keith, J., 1977, Localization of N6-methyladenosine in the Rous sarcoma
virus genome, *J. Mol. Biol.* **113**:165.

Bender, M. A., Palmer, T. D., Gelinas, R. E., and Miller, A. D., 1987, Evidence that the
packaging signal of Moloney murine leukemia virus extends into the *gag* region, *J.
Virol.* **61**:1639.

Bennett, R. P., Rhee, S., Craven, R. C., Hunter, E., and Wills, J. W., 1991, Amino acids
encoded downstream of *gag* are not required by Rous sarcoma virus protease during
gag-mediated assembly, *J. Virol.* **65**:272.

Berberich, S. L., and Stoltzfus, C. M., 1991, Mutations in the regions of the Rous sarcoma
virus 3′ splice sites: Implications for regulation of alternative splicing, *J. Virol.*
65:2640.

Berberich, S. L., Macias, M., Zhang, L., Turek, L. P., and Stoltzfus, C. M., 1990, Comparison of Rous sarcoma virus RNA processing in chicken and mouse fibroblasts: Evidence for double-spliced RNA in nonpermissive mouse cells, *J. Virol.* **64**:4313.

Berg, J. M., 1986, Potential metal-binding domains in nucleic acid binding proteins,
Science **232**:485.

Bernhard, W., 1960, The detection and study of tumor viruses with the electron microscope, *Cancer Res.* **20**:712.

Bernhard, W., Bonar, R. A., Beard, D., and Beard, J. W., 1958, Ultrastructure of viruses of
myeloblastosis and erythroblastosis isolated from plasma of leukemic chickens, *Proc.
Soc. Exp. Biol. Med.* **97**:48.

Bieth, E., Gabus, C., and Darlix, J. E., 1990, A study of the dimer formation of Rous
sarcoma virus RNA and of its effect on viral protein synthesis *in vitro*, *Nucleic Acids
Res.* **18**:119.

Bishop, J. M., and Varmus, H. E., 1982, Functions and origins of retroviral transforming
genes, in: *RNA Tumor Viruses* (R. Weiss, N. Teich, H. Varmus, and J. Coffin, eds.),
pp. 999–1108, Cold Spring Harbor Laboratory, Cold Spring Harbor, New York.

Bishop, J. M., and Varmus, H. E., 1985, Functions and origins of retroviral transforming
genes, in: *RNA Tumor Viruses—Supplements and Appendixes* (R. Weiss, N. Teich,
H. Varmus, and J. Coffin, eds.), pp. 249–256, Cold Spring Harbor Laboratory, Cold
Spring Harbor, New York.

Bizub, D., Weber, I. T., Cameron, C. E., Leis, J. P., and Skalka, A. M., 1991, A range of
catalytic efficiencies with avian retroviral protease subunits genetically linked to
form single polypeptide chains, *J. Biol. Chem.* **266**:4951.

Boeke, J. D., and Corces, V. G., 1989, Transcription and reverse transcription of retrotransposons, *Annu. Rev. Microbiol.* **43**:403.

Boettiger, D., Love, D. N., and Weiss, R. A., 1975, Virus envelope markers in mammalian tropism of avian RNA tumor viruses, *J. Virol.* **15**:108.

Bollag, R. J., Waldman, A. S., and Liskay, R. M., 1989, Homologous recombination in mammalian cells, *Annu. Rev. Genet.* **23**:199.

Bolognesi, D. P., Montelaro, R. C., Frank, H., and Schafer, W., 1978, Assembly of type-C oncornaviruses: A model, *Science* **199**:183.

Boone, L. R., and Skalka, A. M., 1981a, Viral DNA synthesized *in vitro* by avian retrovirus particles permeabilized with mellitin. I. Kinetics of synthesis and size of minus- and plus-strand transcripts, *J. Virol.* **37**:109.

Boone, L. R., and Skalka, A. M., 1981b, Viral DNA synthesized *in vitro* by avian retrovirus particles permeabilized with mellittin. II. Evidence for a strand displacement mechanism in plus-strand synthesis, *J. Virol.* **37**:117.

Boone, L. R., Myer, F. E., Yang, D. M., Ou, C. Y., Koh, C. K., Roberson, L. E., Tennant, R. W., and Yang, W. K., 1983, Reversal of Fv-1 host range by *in vitro* restriction endonuclease fragment exchange between molecular clones of N-tropic and B-tropic murine leukemia virus genomes, *J. Virol.* **48**:110.

Boral, A. L., Okenquist, S. A., and Lenz, J., 1989, Identification of the SL3-3 virus enhancer core as a T lymphoma cell-specific element, *J. Virol.* **63**:76.

Bosze, Z., Theisen, H., and Charnay, P., 1986, A transcriptional enhancer with specificity for erythroid cells is located in the long terminal repeat of the Friend leukemia virus, *EMBO J.* **5**:1615.

Boulden, A., and Sealy, L., 1990, Identification of a third protein factor which binds to the Rous sarcoma virus LTR enhancer: Possible homology with the serum response element, *Virology* **174**:204.

Bova, C. A., Olsen, J. C., and Swanstrom, R., 1988, The avian retrovirus *env* gene family: Molecular analysis of host range and antigenic variants, *J. Virol.* **62**:75.

Bova-Hill, C., Olsen, J. C., and Swanstrom, R., 1991, Genetic analysis of Rous sarcoma virus subgroup D *env* gene: Mammal tropism correlates with temperature sensitivity of gp85, *J. Virol.* **65**:2073.

Bowerman, B., Brown, P. O., Bishop, J. M., and Varmus, H. E., 1989, A nucleoprotein complex mediates the integration of retroviral DNA, *Genes Dev.* **3**:469.

Breathnach, R., and Chambon, P., 1981, Organization and expression of eukaryotic split genes coding for proteins, *Annu. Rev. Biochem.* **50**:349.

Brierly, I., Boursnell, M., Birns, M., Bilmoria, B., Block, V., Brown, T., and Inglis, S., 1987, An efficient ribosomal frameshifting signal in the polymerase-encoding region of the coronavirus IBV, *EMBO J.* **6**:3779.

Brierly, I., Digard, P., and Inglis, S. C., 1989, Characterization of an efficient ribosomal frameshifting signal: Requirement for an RNA psuedoknot, *Cell* **57**:543.

Brierly, I., Rolley, N. J., Jenner, A. J., and Inglis, S. C., 1991, Mutational analysis of the RNA pseudoknot component of a coronavirus ribosomal frameshifting signal, *J. Mol. Biol.* **220**:889.

Brown, D. W., Blais, B. P., and Robinson, H. L., 1988, Long terminal repeat (LTR) sequences, env, and a region near the 5′ LTR influences the pathogenic potential of recombinants between Rous-associated virus types 0 and 1, *J. Virol.* **62**:3431.

Brown, P. H., Tiley, L. S., and Cullen, B. R., 1991, Effect of RNA secondary structure on polyadenylation site selection, *Genes Dev.* **5**:1227.

Brown, P. O., 1990, Integration of retroviral DNA, in: *Retroviruses–Strategies of Replication* (R. Swanstrom and P. K. Vogt, eds.), pp. 19–48, Springer-Verlag, Berlin.

Brown, P. O., Bowerman, B., Varmus, H. E., and Bishop, J. M., 1987, Correct integration of retroviral DNA *in vitro*, *Cell* **49**:349.

Brown, P. O., Bowerman, B., Varmus, H. E., and Bishop, J. M., 1989, Retroviral integration: Structure of the initial covalent product and its precursor, and a role for the IN protein, *Proc. Natl. Acad. Sci. USA* **86**:2525.

Bryant, M. L., Heuckeroth, R. C., Kimata, J. T., Ratner, L., and Gordon, J. I., 1989, Replication of human immunodeficiency virus 1 and Moloney murine leukemia virus is inhibited by different heteroatom-containing analogs of myristic acid, *Proc. Natl. Acad. Sci. USA* **86**:865.

Buetti, E., and Kuhnel, B., 1986, Distinct sequence elements involved in the glucocorticoid regulation of the mouse mammary tumor virus promoter identified by linker scanning mutagenesis, *J. Mol. Biol.* **190**:379.

Burstein, H., Bizub, D., and Skalka, A. M., 1991, Assembly and processing of avian retroviral gag polyproteins containing linked protease dimers, *J. Virol.* **65**:6165.

Burstein, H., Bizub, D., Kotler, M., Schatz, G., Vogt, V. M., and Skalka, A. M., 1992, Processing of avian retroviral gag polyprotein precursors in blocked by a mutation at the NC–PR cleavage site, *J. Virol.* **66**:1781.

Bushman, F. D., and Craigie, R., 1991, Activities of human immunodeficiency virus (HIV) integration protein *in vitro:* Specific cleavage and integration of HIV DNA, *Proc. Natl. Acad. Sci. USA* **88**:1339.

Bushman, F. D., Fujiwara, T., and Craigie, R., 1990, Retroviral integration directed by HIV integration protein *in vitro, Science* **249**:1555.

Capecchi, M. R., 1989, Altering the genome by homologous recombination, *Science* **244**:1288.

Carlberg, K., Ryden, T. A., and Beemon, K., 1988, Localization and foot-printing of an enhancer within the avian sarcoma virus *gag* gene, *J. Virol.* **62**:1617.

Cato, A. C. B., Skroch, P., Weinmann, J., Butkeraitis, P., and Ponta, H., 1988, DNA sequences outside the receptor-binding sites differentially modulate the responsiveness of the mouse mammary tumor virus promoter to various steroid hormones, *EMBO J.* **7**:1403.

Celander, D., and Haseltine, W. A., 1984, Tissue-specific transcription preference as a determinant of cell tropism and leukaemogenic potential of murine retroviruses, *Nature* **312**:159.

Celander, D., and Haseltine, W. A., 1987, Glucocorticoid regulation of murine leukemia virus transcription elements is specified by determinants within the viral enhancer region, *J. Virol.* **61**:269.

Celander, D., Hsu, B. L., and Haseltine, W. A., 1988, Regulatory elements within the murine leukemia virus enhancer regions mediate glucocorticoid responsiveness, *J. Virol.* **62**:1314.

Cepko, C. L., Roberts, B. E., and Mulligan, R. C., 1984, Construction and applications of a highly transmissible murine retroviral shuttle vector, *Cell* **37**:1053.

Chalepakis, G., Arnemann, J., Slater, E., Bruller, H. J., Gross, B., and Beato, M., 1988, Differential gene activation by glucocorticoids and progestins through the hormone regulatory element of mouse mammary tumor virus. *Cell* **53**:371.

Chandler, V. L., Maler, B. A., and Yamamoto, K., 1983, DNA sequences bound specifically by glucocorticoid receptor *in vitro* render a heterologous promoter hormone responsive *in vivo, Cell* **33**:489.

Chaterjee, S., and Hunter, E., 1980, Fusion of normal primate cells: A common biological property of the D-type retroviruses, *Virology* **107**:100.

Chatis, P. A., Holland, C. A., Hartley, J. W., Rowe, W. P., and Hopkins, N., 1983, Role for the 3' end of the genome in determining disease specificity of Friend and Moloney murine leukemia viruses, *Proc. Natl. Acad. Sci. USA* **80**:4408.

Chen, I. S. Y., 1990, *Retrovirus Genome Organization and Gene Expression*, Saunders, Philadelphia, Pennsylvania.

Cheng-Kiang, S., Nevins, J., and Darnell, J. E., 1979, N-6-methyl-adenosine in adenovirus type 2 nuclear RNA is conserved in the formation of messenger RNA, *J. Mol. Biol.* **135**:733.

Chesebro, B., Buller, R., Portis, J., and Wehrly, K., 1990, Failure of human immunodeficiency virus entry and infection in CD4-positive human brain and skin cells, *J. Virol.* **64**:215.

Chiswell, D. J., Gillespie, D. A., and Wyke, J. A., 1982, The changes in proviral chromatin that accompany morphological variation in avian sarcoma virus-infected rat cells, *Nucleic Acids Res.* **10**:3976.

Chamorro, M., Parkin, N., and Varmus, H. E., 1992, An RNA pseudoknot and an optimal heptameric shift site are required for highly efficient ribosomal frameshifting on a retroviral messenger RNA, *Proc. Natl. Acad. Sci. USA* **89**:713.

Chow, S. A., Vincent, K. A., Ellison, V., and Brown, P. O., 1992, Reversal of integration and DNA splicing mediated by integrase of human immunodeficiency virus, *Science* **255**:723.

Clare, J., and Farabaugh, P., 1985, Nucleotide sequence of a yeast Ty element: Evidence for an unusual mechanism of gene expression, *Proc. Natl. Acad. Sci. USA* **82**:2829.

Clare, J. J., Belacort, M., and Farabaugh, P. J., 1988, Efficient translational frameshifting occurs within a conserved sequence of the overlap between the two genes of a yeast Ty1 transposon. *Proc. Natl. Acad. Sci. USA* **85**:6816.

Cobrinik, D., Katz, R., Rerrey, R., Skalka, A., and Leis, J., 1987, Avian sarcoma and leukosis virus pol-endonuclease recognition of the tandem repeat junction: Minimum site required for cleavage is also required for viral growth, *J. Virol.* **61**:1999.

Cobrinik, D., Skoskey, L., and Leis, J., 1988, A retroviral RNA secondary structure required for efficient initiation of reverse transcription, *J. Virol.* **62**:3622.

Cobrinik, D., Aiyar, A., Ge, Z., Katzman, M., Huang, H., and Leis, J., 1991, Overlapping retrovirus U5 sequence elements are required for efficient integration and initiation of reverse transcription, *J. Virol.* **65**:3864.

Coffin, J., 1982a, Structure of the retroviral genome, in: *RNA Tumor Viruses* (R. Weiss, N. Teich, H. Varmus, and J. Coffin, eds.), pp. 261–368, Cold Spring Harbor Laboratory, Cold Spring Harbor, New York.

Coffin, 1982b, Endogenous viruses, in: *RNA Tumor Viruses* (R. Weiss, N. Teich, H. Varmus, and J. Coffin, eds.), pp. 1109–1203, Cold Spring Harbor Laboratory, Cold Spring Harbor, New York.

Coffin, J., 1985, Genome structure, in: *RNA Tumor Viruses—Supplements and Appendixes* (R. Weiss, N. Teich, H. Varmus, and J. Coffin, eds.), pp. 17–73, Cold Spring Harbor Laboratory, Cold Spring Harbor, New York.

Coffin, J. M., 1990a, Retroviridae and their replication, in: *Virology* (B. N. Fields, D. M. Knipe, R. M. Chanock, *et al.*, eds.), pp. 645–708, Raven Press, New York.

Coffin, J. M., 1990b, Genetic variation in retroviruses, in: *Applied Virology Research* (E. Kurstak, R. G. Marusyk, F. A. Murphy, and M. H. V. Van Regenmortel, eds.), pp. 11–34, Plenum Press, New York.

Coffin, J. M., 1990c, Molecular mechanisms of nucleic acid integration, *J. Med. Virol.* **31**:43.

Coffin, J. M., and Haseltine, W. A., 1977, Terminal redundancy and the origin of replication of Rous sarcoma virus RNA, *Proc. Natl. Acad. Sci. USA,* **74**:1908.

Coffin, J., and Stoye, J., 1985, Endogenous viruses, in: *RNA Tumor Viruses* (R. Weiss, N. Teich, H. Varmus, and J. Coffin, eds.), pp. 357–404, Cold Spring Harbor Laboratory, Cold Spring Harbor, New York.

Coffin, J. M., Hageman, T. C., Maxam, A. M., and Haseltine, W. A., 1978, Structure of the genome of Moloney murine leukemia virus: A terminally redundant sequence, *Cell* **13**:761.

Colicelli, J., and Goff, S. P., 1986, Isolation of a recombinant murine leukemia virus utilizing a new primer tRNA, *J. Virol.* **57**:37.

Colicelli, J., and Goff, S. P., 1988a, Sequence and spacing requirements of a retrovirus integration site, *J. Mol. Biol.* **199**:47.

Colicelli, J., and Goff, S. P., 1988b, Isolation of an integrated provirus of Moloney murine leukemia virus with long terminal repeats in inverted orientation: Integration utilizing two U3 sequences, *J. Virol.* **62**:633.

Cone, R., and Mulligan, R., 1984, High efficiency gene transfer into mammalian cells: Generation of helper-free recombinant retrovirus with broad mammalian host range, *Proc. Natl. Acad. Sci. USA* **81**:6349.

Cone, R. D., Weber-Benarous, A., Baorto, D., and Mulligan, R. C., 1987, Regulated expression of a complete human β-globin gene encoded by a transmissible retrovirus vector, *Mol. Cell Biol.* **7**:887.

Conklin, K. F., and Groudin, M., 1986, Varied interactions between proviruses and adjacent host chromatin, *Mol. Cell. Biol.* **6**:3999.

Conklin, K. F., Coffin, J. M., Robinson, H. L., Groudin, M., and Eisenman, R., 1982, Role of methylation in the induced and spontaneous expression of the avian endogenous virus ev-1: DNA structure and gene products, *Mol. Cell. Biol.* **2**:638.

Conklin, K. F., Coffin, J. M., Robinson, H. L., Groudine, M., and Eisenman, R., 1991, Activation of an endogenous retrovirus enhancer by insertion into a heterologous context, *J. Virol.* **65**:2525.

Cordingly, M. G., Riegel, A. T., and Hager, G. L., 1987, Steroid-dependent interaction of transcription factors with the inducible promoter of mouse mammary tumor virus *in vivo*, *Cell* **48**:261.

Corneliussen, B., Thornell, A., Hallberg, B., and Grundstrom, T., 1991, Helix-loop-helix transcriptional activators bind to a sequence in glucocorticoid response elements of retrovirus enhancers, *J. Virol.* **65**:6084.

Costa, R. H., Lai, E., and Darnell, J. E., 1986, Transcriptional control of the mouse prealbumin (transthyretin) gene: Both promoter sequences and a distinct enhancer are cell specific, *Mol. Cell. Biol.* **6**:4697.

Craig, N. L., 1988, The mechanism of conservative site-specific recombination, *Annu. Rev. Genet.* **22**:77.

Craigie, R., and Mizuuchi, K., 1987, Transposition of the Mu DNA: Joining of Mu to target DNA can be uncoupled from cleavage at the ends, *Cell* **51**:493.

Craigie, R., Fujiwara, T., and Bushman, F., 1990, The IN protein of Moloney murine leukemia virus processes the viral DNA ends and accomplishes their integration *in vitro*, *Cell* **62**:829.

Craven, R. C., Bennett, R. P., and Wills, J. W., 1991, Role of the avian retroviral protease in the activation of reverse transcriptase during virion assembly, *J. Virol.* **65**:6205.

Crawford, S., and Goff, S. P., 1985, A deletion mutation in the 5' part of the *pol* gene of Moloney murine leukemia virus blocks proteolytic processing of the gag and pol polyproteins, *J. Virol.* **53**:899.

Crick, F., 1970, Central dogma of molecular biology, *Nature* **227**:561.

Cullen, B. R., 1991, Human immunodeficiency virus as a prototypic complex retrovirus, *J. Virol.* **65**:1053.

Cullen, B. R., Skalka, A. M., and Ju, G., 1983, Endogenous avian retroviruses contain deficient promoter and leader sequences, *Proc. Natl. Acad. Sci. USA* **80**:2946.

Cullen, B. R., Lomedico, P. T., and Ju, G., 1984, Transcriptional interference in avian retroviruses: Implications for the promoter insertion model of leukemogenesis, *Nature* **307**:241.

Cullen, B. R., Raymond, K., and Ju, G., 1985a, Transcriptional activity of avian retroviral long terminal repeats directly correlates with enhancer activity, *J. Virol.* **53**:515.

Cullen, B. R., Raymond, K., and Ju, G., 1985b, Functional analysis of the transcription control region located within the avian retroviral long terminal repeat, Mol. Cell. Biol. 5:438.

Cupelli, L. A., and Lenz, J., 1991, Transcriptional initiation and postinitiation effects of murine leukemia virus long terminal repeat R-region sequences, J. Virol. 65:6961.

Daar, E. S., and Ho, D. D., 1990, The structure and function of retroviral envelope glycoproteins, in: Retrovirus Genome Organization and Gene Expression (I. S. Y. Chen, ed.), pp. 205–214, Saunders, Philadelphia, Pennsylvania.

Dalgleish, A. G., Beverly, P. C. L., Clapham, P. R., Crawford, D. H., Greaves, M. F., and Weiss, R. A., 1984, The CD4 (T4) antigen is an essential component of the receptor for the AIDS retrovirus, Nature 312:763.

Darlix, J. L., and Spahr, P. F., 1982, Binding sites of viral protein p19 onto Rous sarcoma virus RNA and possible controls of viral functions, J. Mol. Biol. 160:147.

De Clerq, E., 1987, Perspectives for the chemotherapy of AIDS, Anticancer Res. 7:1023.

DeFranco, D., and Yamamoto, K., 1986, The two different factors act separately or together to specify functionally distinct activities at a single transcriptional enhancer, Mol. Cell. Biol. 6:993.

DeLarco, J., and Todaro, G. J., 1976, Membrane receptors for murine leukemia viruses: Characterization using the purified viral envelope glycoprotein gp71, Cell 82:365.

Delwart, E. L., and Mosialos, G., 1990, Retroviral envelope glycoproteins contain a "leucine zipper"-like repeat, AIDS Res. Hum. Retroviruses 6:703.

DesGroseillers, L., and Jolicoeur, P., 1984, The tandem direct repeats within the long terminal repeat of murine leukemia viruses are the primary determinant of their leukemogenic potential, J. Virol. 52:945.

DesGroseillers, L., Rassart, E., Robitaille, Y., and Jolicoeur, P., 1985, Retrovirus-induced spongiform encephalopathy: The 3' end long terminal repeat-containing viral sequences influence the incidence of the disease and the specificity of the neurological syndrome, Proc. Natl. Acad. Sci. USA 82:8818.

Dhar, R., McClements, W. L., Enquist, L. W., and Vande Woude, G. F., 1980, Nucleotide sequences of integrated Moloney sarcoma provirus long terminal repeats and their host and viral junctions, Proc. Natl. Acad. Sci. USA 77:3937.

Dickson, C., and Atterwill, M., 1979, Composition, arrangement and cleavage of the mouse mammary tumor virus polyprotein precursor Pr77gag and p110gag, Cell 17:1003.

Dickson, C., Eisenman, R., Fan, H., Hunter, E., and Teich, N., 1982, Protein biosynthesis and assembly, in: RNA Tumor Viruses (R. Weiss, N. Teich, H. Varmus, and J. Coffin, eds.), pp. 513–648, Cold Spring Harbor Laboratory, Cold Spring Harbor, New York.

Dickson, C., Eisenman, R., and Fan, H., 1985, Protein biosynthesis and assembly, in: RNA Tumor Viruses—Supplements and Appendixes (R. Weiss, N. Teich, H. Varmus, and J. Coffin, eds.), pp. 135–185, Cold Spring Harbor Laboratory, Cold Spring Harbor, New York.

Dillon, P. J., Lenz, J., and Rosen, C. A., 1991, Construction of a replication-competent murine retrovirus vector expressing the human immunodeficiency virus type 1 tat transactivator protein, J. Virol. 65:4490.

Dimock, K., and Stoltzfus, C. M., 1977, Sequence specificity of internal methylation in B77 avian sarcoma virus RNA subunits, J. Virol. 5:2298.

Dinman, J. D., Tateo, I., and Wickner, R. B., 1991, A −1 ribosomal frameshift in a double-stranded RNA virus of yeast forms a gag–pol fusion, Proc. Natl. Acad. Sci. USA 88:174.

Donehower, L. A., and Varmus, H. E., 1984, A mutant murine leukemia virus with a single missense codon in pol is defective in a function affecting integration, Proc. Natl. Acad. Sci. USA 81:6461.

Doolittle, R. F., Feng, D., Johnson, M. S., and McClure, M. A., 1989, Origins and evolutionary relationships of retroviruses, Q. Rev. Biol. 64:1.

Dornburg, R., and Temin, H. M., 1988, Retroviral vector system for the study of cDNA gene formation, Mol. Cell. Biol. 8:2328.

Dornburg, R., and Temin, H. M., 1990, Presence of a retroviral encapsidation sequence in nonretroviral RNA increases the efficiency of formation of cDNA genes, J. Virol. 64:886.

Dorner, A. J., and Coffin, J. M., 1986, Determinants for receptor interaction and cell killing on the avian retrovirus glycoprotein gp85, Cell 45:365.

Dorner, A. J., Stoye, J. P., and Coffin, J. M., 1985, Molecular basis of host range variation in avian retroviruses, Mol. Cell. Biol. 6:4387.

Dorner, A. J., Bonneville, F., Kriz, R., Kelleher, K., Bean, K., and Kaufman, R. J., 1991, Molecular cloning and characterization of a complete Chinese hamster provirus related to intracisternal A particle genomes, J. Virol. 65:4713.

Dougherty, J. P., and Temin, H. M., 1986, High mutation rate of a spleen necrosis virus-based vector, Mol. Cell. Biol. 6:4387.

Dougherty, J. P., and Temin, H. M., 1987, A promoterless retroviral vector indicates that there are sequences in U3 required for 3' processing, Proc. Natl. Acad. Sci. USA 84:1197.

Drake, P. L., Nutt, R. F., Brady, S. F., Garsky, V. M., Ciccarone, F. M., Leu, C. T., Lumma, P. K., Freidinger, R. M., Veber, D. F., and Sigal, I. S., 1988, HIV-1 protease specificity of peptide cleavage is sufficient for processing of gag and pol polyproteins, Biochem. Biophys. Res. Commun. 156:297.

Dupraz, P., Oertle, S., Meric, C., Damay, P., and Spahr, P. F., 1990, Point mutations in the proximal Cys–His box of Rous sarcoma virus nucleocapsid protein, J. Virol. 64:4978.

Dutta, A., Stoeckle, M. Y., and Hanafusa, H., 1990, Serum and v-src increase the level of a CCAAT-binding factor required for transcription from a retroviral long terminal repeat, Genes Dev. 4:243.

Duyk, G., Longiaru, M., Cobrinik, D., Kowal, R., deHaseth, P., Skalka, A. M., and Leis, J., 1985, Circles with two tandem long terminal repeats are specifically cleaved by pol gene-associated endonuclease from avian sarcoma and leukosis viruses: Nucleotide sequences required for site-specific cleavage, J. Virol. 56:589.

Dynan, W. S., 1989, Modularity in promoters and enhancers, Cell 58:1.

Earl, P. L., Koenig, S., and Moss, B., 1991, Biological and immunological properties of human immunodeficiency virus type 1 glycoprotein: Analysis of proteins with truncations and deletions expressed by recombinant vaccinia viruses, J. Virol. 65:31.

Echols, H., 1986, Multiple DNA–protein interactions governing high-precision DNA transactions, Science 233:1050.

Edwards, S. A., and Fan, H., 1979, Gag-related polyproteins of Moloney murine leukemia virus: Evidence for independent synthesis of glycosylated and unglycosylated forms, J. Virol. 30:551.

Eichinger, D. J., and Boeke, J. D., 1988, The DNA intermediate in yeast Ty1 element transposition copurifies with virus-like particles: Cell-free Ty1 transposition, Cell 54:955.

Einfeld, D., and Hunter, E., 1988, Oligomeric structure of a prototype retrovirus glycoprotein, Proc. Natl. Acad. Sci. USA 85:8688.

Eisenman, R. N., and Vogt, V. M., 1978, The biosynthesis of oncovirus proteins, Biochim. Biophys. Acta 473:187.

Eisenman, R. N., Mason, W. S., and Linial, M., 1980, Synthesis and processing of polymerase proteins of wild-type and mutant avian retroviruses, J. Virol. 36:62.

Elder, J. H., Lerner, D. L., Hasselkus-Light, C. S., Fontenot, D. J., Hunter, E., Luciw, P. A.,

Montelaro, R. C., and Phillips, T. R., 1992, Distinct subsets of retroviruses encode deoxyuridine triphosphatase, *J. Virol.* **66**:1791.

Ellis, J., and Bernstein, A., 1989, Gene targeting with retroviral vectors: Recombination by gene conversion into regions of nonhomology, *Mol. Cell. Biol.* **9**:1621.

Embretson, J. E., and Temin, H. M., 1987a, Transcription from a spleen necrosis virus 5′ long terminal repeat is suppressed in mouse cells, *J. Virol.* **61**:3454.

Embretson, J. E., and Temin, H. M., 1987b, Lack of competition results in efficient packaging of heterologous murine retroviral RNAs and reticuloendotheliosis virus encapsidation-minus RNAs by the reticuloendotheliosis virus helper cell line, *J. Virol.* **61**:2675.

Emerman, M., and Temin, H. M., 1984, High-frequency deletion in recovered retrovirus vectors containing endogenous DNA with promoters, *J. Virol.* **50**:42.

Engelman, A., Mizuuchi, K., and Craigie, R., 1991, HIV-1 DNA integration: Mechanism of viral DNA cleavage and DNA strand transfer, *Cell* **67**:1211.

Erickson, J., Neidhart, D. J., VanDrie, J., *et al.*, 1990, Design, activity, and 2.8A crystal structure of a c2 symmetric inhibitor complexed to HIV-1 protease, *Science* **249**:527.

Evans, R. M., 1988, The steroid and thyroid hormone receptor superfamily, *Science* **240**:889.

Faber, M., and Sealy, L., 1990, Rous sarcoma virus enhancer factor I is a ubiquitous CCAAT transcription factor highly related to CBF and NF-Y, *J. Biol. Chem.* **265**:22243.

Fan, H., 1990, Influences of the long terminal repeats on retrovirus pathogenicity, in: *Retrovirus Genome Organization and Gene Expression* (I. S. Y. Chen, ed.), pp. 165–174, Saunders, Philadelphia, Pennsylvania.

Fan, H., Chute, H., Chao, E., and Feuerman, M., 1983, Construction and characterization of Moloney murine leukemia virus mutants unable to synthesize glycosylated gag protein, *Proc. Natl. Acad. Sci. USA* **80**:5965.

Fan, H., Chute, H., Chao, E., and Pattengale, P. K., 1988, Leukemogenicity of Moloney murine leukemia virus carrying polyoma enhancer sequences in the long terminal repeat is dependent on the nature of the inserted polyoma sequences, *Virology* **166**:58.

Farmerie, W. G., Loeb, D. D., Casavant, N. C., Hutchison, C. A. I., Edgell, M. H., and Swanstrom, R., 1987, Expression and processing of the AIDS virus reverse transcriptase in *Escherichia coli*, *Science* **236**:305.

Farnet, C. M., and Haseltine, W. A., 1991, Determination of viral proteins present in the human immunodeficiency virus type 1 preintegration complex, *J. Virol.* **65**:1910.

Felsenstein, K. M., and Goff, S. P., 1988, Expression of the gag–pol fusion protein of Moloney murine leukemia virus without gag protein does not include virion formation or proteolytic processing, *J. Virol.* **62**:2179.

Feng, Y. X., Levin, J. G., Hatfield, D. L., Schaefer, T. S., Gorelick, R. J., and Rein, A., 1989, Suppression of UAA and UGA termination codons in mutant murine leukemia viruses, *J. Virol.* **63**:2870.

Feuer, G., and Fan, H., 1990, Substitution of murine transthyretin (prealbumin) regulatory sequences into the Moloney murine leukemia virus long terminal repeat yields infectious virus with altered biological properties, *J. Virol.* **64**:6130.

Feuer, G., Taketo, M., Hanecak, R. C., and Fan, H., 1989, Two blocks in Moloney murine leukemia virus expression in undifferentiated F9 embryonal carcinoma cells as determined by transient expression assays, *J. Virol.* **63**:2317.

Ficht, T. A., Chang, L. J., and Stoltzfus, C. M., 1984, Avian sarcoma virus *gag* and *env* gene structural protein precursors contain common amino-terminal sequence, *Proc. Natl. Acad. Sci. USA* **81**:362.

Fine, D., and Schochetman, G., 1978, Type D primate retroviruses: A review, *Cancer Res.* **38**:3123.

Fitzgerald, P. M. D., and Springer, J. P., 1991, Structure and function of retroviral proteases, *Annu. Rev. Biophys. Biophys. Chem.* **20:**299.

Fitzgerald, P. M. D., McKeever, B. M., VanMiddlesworth, J. F., Springer, J. P., Heimbech, J. C., Leu, C. T., Herber, W. K., Dixon, R. A. F., and Drake, P. L., 1990, Crystallographic analysis of a complex between human immunodeficiency virus type 1 protease and acetyl-pepstatin at 2.0A resolution, *J. Biol. Chem.* **265:**14209.

Flanagan, J. R., Kreig, A. M., Max, E. E., and Khan, A. S., 1989, Negative control region at the 5' end of the murine leukemia virus long terminal repeats, *Mol. Cell. Biol.* **8:**739.

Flanagan, J. R., Becvker, K. G., Ennist, D. L., Gleason, S. L., Driggers, P. H., Levi, B. Z., Appella, E., and Ozato, K., 1992, Cloning of a negative transcription factor that binds to the upstream conserved region of Moloney murine leukemia virus, *Mol. Cell. Biol.* **12:**38.

Franz, T., Holberg, F., Seliger, B., Stocking, S., and Ostertag, W., 1986, Retroviral mutants efficiently expressed in embryonal carcinoma cells, *Proc. Natl. Acad. Sci. USA* **83:**3292.

Friedmann, T., 1989, Progress toward human gene therapy, *Science* **244:**275.

Fritsch, E., and Temin, H. M., 1977, Inhibition of viral DNA synthesis in stationary chicken embryo fibroblasts infected with avian retroviruses, *J. Virol.* **24:**461.

Fu, X. D., Phillips, N., Jentoft, J., Tuazon, P. T., Traugh, J. A., and Leis, J., 1985, Site-specific phosphorylation of avian retrovirus nucleocapsid protein pp12 regulates binding to RNA, *J. Biol. Chem.* **260:**9941.

Fu, X. D., Katz, R. A., and Skalka, A. M., 1988, Site-directed mutagenesis of the avian retrovirus nucleocapsid proteins pp12: Mutation which affects RNA binding *in vitro* blocks viral replication, *J. Biol. Chem.* **263:**2134.

Fu, X. D., Katz, R. A., Skalka, A. M., and Maniatis, T., 1991, The role of branchpoint and 3'-exon sequences in the control of balanced splicing of avian retrovirus RNA, *Genes Dev.* **5:**211.

Fuetterer, J., and Hohn, T., 1987, Involvement of nucleocapsids in reverse transcription: A general phenomenon, *TIBS* **12:**92.

Fujiwara, J., and Craigie, R., 1989, Integration of mini-retroviral DNA: A cell-free reaction for biochemical analysis of retroviral integration, *Proc. Natl. Acad. Sci. USA* **86:**3065.

Fujiwara, J., and Mizuuchi, K., 1988, Retroviral DNA integration: Structure of an integration intermediate, *Cell* **54:**497.

Furuichi, Y., Shatkin, A. J., Stavnezer, E., and Bishop, J. M., 1975, Blocked, methylated 5'-terminal sequence in avian sarcoma virus RNA, *Nature* **257:**618.

Gallaher, W. R., Ball, J. M., Garry, R. F., Griffin, M. C., and Montelaro, R. C., 1989, A general model for the transmembrane proteins of HIV and other retroviruses, *AIDS Res. Hum. Retroviruses* **5:**431.

Gallis, B., Linial, M., and Eisenman, R., 1979, An avian oncovirus mutant deficient in genomic RNA: Characterization of the packaged RNA as cellular messenger RNA, *Virology* **94:**146.

Gallo, R. C., and Wong-Staal, R., 1990, in: *Retrovirus Biology and Human Disease,* Marcel Dekker, New York.

Ganem, D., and Varmus, H. E., 1987, The molecular biology of the hepatitis B viruses, *Annu. Rev. Biochem.* **56:**651.

Garcia, J. V., Jones, C., and Miller, A. D., 1991, Localization of the amphotropic murine leukemia virus receptor gene to the pericentric region of human chromosome 8, *J. Virol.* **65:**6316.

Gardner, M. B., and Luciw, P. A., 1989, Animal models of AIDS, *FASEB J.* **3:**2593.

Garfinkel, D. J., Boeke, J. D., and Fink, G. R., 1985, Ty element transposition: Reverse transcription and virus-like particles, *Cell* **42:**507.

Gebhardt, A., Bosch, J. V., Ziemiecki, A., and Friis, R. R., 1984, Rous sarcoma virus p19 and gp35 can be chemically crosslinked to high molecular weight complexes: An insight into viral association, *J. Mol. Biol.* **174:**297.

Gelderblom, H. R., 1991, Assembly and morphology of HIV: Potential effect of structure on viral function, *AIDS* **5:**617.

Gelinas, C., and Temin, H. M., 1986, Nondefective spleen necrosis virus-derived vectors define the upper size limit for packaging reticuloendotheliosis viruses, *Proc. Natl. Acad. Sci. USA* **83:**9759.

Gilboa, E., Mitra, S. W., Goff, S., and Baltimore, D., 1979a, A detailed model of reverse transcription and tests of curcial aspects, *Cell* **18:**93.

Gilboa, E., Goff, S., Shields, A., Yoshimura, F., Mitra, S., and Baltimore, D., 1979b, *In vitro* synthesis of a 9-kbp terminally redundant DNA carrying the infectivity of Moloney murine leukemia virus, *Cell* **16:**863.

Goff, S. P., 1990, Retroviral reverse transcriptase: synthesis, structure, and function, *J. Acquired Immune Defic. Syndr.* **3:**817.

Golemis, E., Li, Y., Frederickson, T. N., Hartley, J. W., and Hopkins, N., 1989, Distinct segments within the enhancer region collaborate to specify the type of leukemia induced by nondefective Friend and Moloney virus, *J. Virol.* **63:**328.

Golemis, E. A., Speck, N. A., and Hopkins, N., 1990, Alignment of U3 region sequences of mammalian type C viruses: Identification of highly conserved motifs and implications for enhancer design, *J. Virol.* **64:**534.

Goodenow, M. M., and Hayward, W. S., 1987, 5' long terminal repeats of myc-associated proviruses appear structurally intact but are functionally impaired in tumors induced by avian leukosis viruses, *J. Virol.* **61:**2489.

Goodwin, G. H., 1988, Identification of three sequence-specific DNA-binding proteins which interact with the Rous sarcoma virus enhancer and upstream elements, *J. Virol.* **62:**2186.

Gopinathan, K. P., Weymouth, L. A., Kunkel, T. A., and Loeb, L. A., 1979, Mutagenesis *in vitro* by DNA polymerase from an RNA tumor virus, *Nature* **278:**857.

Gorelick, R. J., Henderson, L. E., Hanser, J. P., and Rein, A., 1988, Point mutants of Moloney murine leukemia virus that fail to package viral RNA: Evidence for specific RNA recognition by a "zinc finger-like" protein sequence, *Proc. Natl. Acad. Sci. USA* **85:**8420.

Gorman, C. M., Rigby, P. W. J., and Lane, D. P., 1985, Negative regulation of viral enhancers in undifferentiated embryonic stem cells, *Cell* **42:**519.

Gottlinger, H. G., Sodroski, J. G., and Haseltine, W. A., 1989, Role of capsid precursor processing and myristoylation in morphogenesis and infectivity of human immunodeficiency virus type 1, *Proc. Natl. Acad. Sci. USA* **86:**5781.

Goubin, G., and Hill, M., 1979, Monomer and multimer covalently closed circular forms of Rous sarcoma virus DNA, *J. Virol.* **29:**799.

Grandbastien, M., Spielmann, A., and Caboche, M., 1989, Tnt1, a mobile retroviral-like transposable element of tobacco isolated by plant cell genetics, *Nature* **337:**376.

Grandgenett, D. P., and Mumm, S. R., 1990, Unraveling retrovirus integration, *Cell* **60:**3.

Grandgenett, D. P., and Vora, A. C., 1985, Site-specific nicking at the avian retrovirus LTR circle junction by the viral pp32 endonuclease, *Nucleic Acids Res.* **13:**6205.

Granowitz, C., Colicelli, J., and Goff, S. P., 1991, Analysis of mutations in the envelope gene of Moloney murine leukemia virus: Separation of infectivity from superinfection resistance, *Virology* **183:**545.

Green, P. L., and Chen, I. S. Y., 1990, Regulation of human T cell leukemia virus expression, *FASEB J.* **4:**169.

Greenblatt, J., 1991, Roles of TFIID in transcriptional initiation by RNA polymerase II, *Cell* **66:**1067.

Greene, W. C., and Cullen, B. R., 1990, The rev–rex connection: Convergent strategies for the posttranscriptional regulation of HIV-1 and HTLV-1 gene expression, in: *Retrovirus Genome Organization and Gene Expression* (I. S. Y. Chen, ed.), pp. 195–204, Saunders, Philadelphia, Pennsylvania.

Grez, M., Akgun, E., Hilberg, F., and Ostertag, W., 1990, Embryonic stem cell virus, a recombinant murine retrovirus with expression in embryonic stem cells, *Proc. Natl. Acad. Sci. USA* **87**:9202.

Grez, M., Zornig, M., Nowock, J., and Ziegler, M., 1991, A single point mutation activates the Moloney murine leukemia virus long terminal repeat in embryonal stem cells, *J. Virol.* **65**:4691.

Gridley, T., 1991, Insertional versus targeted mutagenesis in mice, *New Biol.* **3**:1025.

Grinde, B., Cameron, C. E., Leis, J., Weber, I., Wlodawer, A., Burstein, H., Bizub, D., and Skalka, A. M., 1992a, Mutations that alter the activity of the Rous sarcoma virus protease, *J. Biol. Chem.* **267**:9481.

Grinde, B., Cameron, C. E., Leis, J., Weber, I., Wlodawer, A., Burstein, H., and Skalka, A. M., 1992b, Analysis of substrate interactions of the Rous sarcoma virus and human immunodeficiency virus-1 proteases using a set of systematically altered peptide substrates, *J. Biol. Chem.* **267**:9491.

Groudine, M., Eisenman, R., and Weintraub, H., 1981, Chromatin structure of endogenous retroviral genomes and activation by an inhibitor of DNA methylation, *Nature* **292**:311.

Grunstein, M., 1990, Histone function in transcription, *Annu. Rev. Cell Biol.* **6**:643.

Gruss, P., and Khoury, G., 1983, Enhancer elements, *Cell* **33**:313.

Guntaka, R. V., Richards, O. C., Shank, P. R., Kung, H. J., Davidson, N., Fritsch, E., Bishop, J. M., and Varmus, H. E., 1976, Covalently closed circular DNA of avian sarcoma virus: Purification from nuclei of infected quail tumor cells and measurement by electron microscopy, *J. Mol. Biol.* **106**:337.

Hackett, P. B., Swanstrom, R., Varmus, H. E., and Bishop, J. M., 1982, The leader sequence of the subgenomic mRNAs of Rous sarcoma virus is approximately 390 nucleotides, *J. Virol.* **41**:527.

Hackett, P. B., Petersen, R. B., Hensel, C. H., Albericio, F., Gunderson, S. I., Palmenberg, A. C., and Barany, G., 1986, Synthesis *in vitro* of a seven amino acid peptide encoded in the leader RNA of Rous sarcoma virus, *J. Mol. Biol.* **190**:45.

Hagino-Yamagishi, K., Donehower, L. A., and Varmus, H. E., 1987, Retroviral DNA integrated during infection by an integration-deficient mutant of murine leukemia virus is oligomeric, *J. Virol.* **61**:1964.

Hanafusa, H., Pinter, A., and Pullman, M. E., 1989, *Retroviruses and Disease*, Academic Press, San Diego, California.

Hanecak, R., Mittal, S., Davis, B. R., and Fan, H., 1986, Generation of infectious Moloney murine leukemia viruses with deletions in the U3 portion of the long terminal repeat, *Mol. Cell. Biol.* **6**:4634.

Hanecak, R., Pattengale, P. K., and Fan, H., 1988, Addition or substitution of simian virus 40 enhancer sequences into the Moloney murine leukemia virus (M-MuLV) long terminal repeat yields infectious M-MuLV with altered biological properties, *J. Virol.* **62**:2427.

Hanecak, R., Pattengale, P. K., and Fan, H., 1991, Deletion of a GC-rich region flanking the enhancer element within the long terminal repeat sequences alters the disease specificity of Moloney murine leukemia virus, *J. Virol.* **65**:5357.

Hansen, M., Jelinek, L., Whiting, S., and Barklis, E., 1990, Transport and assembly of gag proteins into Moloney murine leukemia virus, *J. Virol.* **64**:5306.

Harel, J., Rassart, E., and Jolicoeur, P., 1981, Cell cycle dependence of synthesis of unintegrated viral DNA in mouse cells newly infected with avian retroviruses, *Virology* **110**:202.

Hartley, J. W., and Rowe, W. P., 1976, Naturally occurring murine leukemia viruses in wild mice: Characterization of a new "amphotropic" class, *J. Virol.* **19**:19.

Haseltine, W. A., 1991, Molecular biology of the human immunodeficiency virus type 1, *FASEB J.* **5**:2349.

Haseltine, W. A., Kleid, D. G., Panet, A., Rothenberg, E., and Baltimore, D., 1976, Ordered transcription of RNA tumor virus genomes, *J. Mol. Biol.* **106**:109.

Haseltine, W. A., Maxam, A. M., and Gilbert, W., 1977, Rous sarcoma virus genome is terminally redundant: The 5' sequence, *Proc. Natl. Acad. Sci. USA* **74**:989.

Hatfield, D., Feng, Y. Y., Lee, B. J., Rein, A., Levin, J. G., and Oroszlan, S., 1989, Chromatographic analysis of the aminoacyl-tRNAs which are required for translation of codons at and around the ribosomal frameshift sites of HIV, HTLV-1, and BLV, *Virology* **173**:736.

Hatfield, D. L., Levin, J. G., Rein, A., and Oroszlan, S., 1992, Translational suppression in retroviral gene expression, *Adv. Virus Res.* **41**:193.

Hayward, W. S., Neel, B. G., and Astrin, S. M., 1981, Activation of a cellular *onc* gene by promoter insertion in ALV-induced lymphoid leukosis, *Nature* **290**:465.

Henderson, L. E., Krutzsch, H. C., and Oroszlan, S., 1983, Myristyl amino-terminal acylation of murine retrovirus proteins: An unusual post-translational protein modification, *Proc. Natl. Acad. Sci. USA* **80**:339.

Herman, S. A., and Coffin, J. M., 1986, Differential transcription from the long terminal repeats of integrated avian leukosis virus DNA, *J. Virol.* **60**:497.

Herman, S. A., and Coffin, J. M., 1987, Efficient packaging of readthrough RNA in ALV: Implications for oncogene transduction, *Science* **236**:845.

Herr, W., and Clarke, J., 1986, The SV40 enhancer is composed of multiple functional elements that can compensate for one another, *Cell* **45**:461.

Hevezi, P., and Goff, S. P., 1991, Generation of recombinant murine retroviral genomes containing the v-*src* oncogene: Isolation of a virus inducing hemangiosarcomas in the brain, *J. Virol.* **65**:5333.

Hilberg, F., Stocking, C., Ostertag, W., and Grez, M., 1987, Functional analysis of a retroviral host-range mutant: Altered long terminal repeat sequences allow expression in embryonal carcinoma cells, *Proc. Natl. Acad. Sci. USA* **84**:5232.

Hippenmeyer, P. J., and Grandgenett, D. P., 1984, Requirement of the avian retrovirus pp32 DNA binding protein domain for replication, *Virology* **137**:358.

Hirano, A., and Wong, T., 1988, Functional interaction between transcriptional control elements in the long terminal repeat of reticuloendotheliosis virus: Cooperative DNA binding of promoter- and enhancer-specific factors, *Mol. Cell. Biol.* **8**:5232.

Hizi, A., Henderson, L. E., Copeland, T. D., Sowder, R. C., Hixson, C. V., and Oroszlan, S., 1987, Characterization of mouse mammary tumor virus gag-pro gene products and the ribosomal frameshift site by protein sequencing, *Proc. Natl. Acad. Sci. USA* **84**:7041.

Holland, C. A., Thomas, C. Y., Chattopadhyay, S. K., Koehne, C., and O'Donnell, P. V., 1989, Influence of enhancer sequences on thymotropism and leukemogenicity of mink cell focus-forming viruses, *J. Virol.* **63**:1284.

Hollon, T., and Yoshimura, F. K., 1989, Mapping of functional regions of murine retrovirus long terminal repeat enhancers: Enhancer domains interact and are not independent in their contributions to enhancer activity, *J. Virol.* **63**:3353.

Horton, R., Mumm, S. R., and Grandgenett, D. P., 1991, Phosphorylation of the avian retrovirus integration protein and proteolytic processing of its carboxy terminus, *J. Virol.* **65**:1141.

Howard, K. J., Frank, K. B., Sim, I. S., and LeGrice, S. F. J., 1991, Reconstitution and properties of homologous and chimeric HIV-1/HIV-2 p66/p51 reverse transcriptase, *J. Biol. Chem.* **266**:23,003.

Hu, W. S., and Temin, H. M., 1990a, Genetic consequences of packaging two RNA genomes in one retroviral particle: Pseudodiploidy and high rate of genetic recombination, *Proc. Natl. Acad. Sci. USA* **87**:1556.

Hu, W. S., and Temin, H. M., 1990b, Retroviral recombination and reverse transcription, *Science* **250**:1227.

Hughes, S. H., and Kosik, E., 1984, Mutagenesis of the region between *env* and *src* of the SR-A strain of Rous sarcoma virus for the purpose of constructing helper-independent vectors, *Virology* **136**:89.

Hughes, S. H., Mutschler, A., Bishop, J. M., and Varmus, H. E., 1981, A Rous sarcoma virus provirus is flanked by short direct repeats of a cellular DNA sequence present in only one copy prior to integration, *Proc. Natl. Acad. Sci. USA* **78**:4299.

Hughes, S. H., Mellstrom, K., Kosik, E., Tamanoi, T., and Brugge, T., 1984, Mutation of a termination codon affects *src* initiation, *Mol. Cell. Biol.* **4**:1738.

Humphries, E. H., and Temin, H. M., 1974, Requirement for cell division for initiation of transcription of Rous sarcoma virus RNA, *J. Virol.* **14**:531.

Humphries, E. H., Glover, C., and Reichmann, M. E., 1981, Rous sarcoma virus infection of synchronized cells establishes provirus integration during S phase DNA synthesis prior to cell division, *Proc. Natl. Acad. Sci. USA* **78**:2601.

Hunter, E., and Swanstrom, R., 1990, Retrovirus envelope glycoproteins, in: *Retroviruses—Strategies of Replication* (R. Swanstrom and P. K. Vogt, eds.), pp. 187–253, Springer-Verlag, Berlin.

Hunter, E., Hill, E., Hardwick, M., Bhown, A., Schwartz, D. E., and Tizard, R., 1983, Complete sequence of the Rous sarcoma virus *env* gene: Identification of structural and functional regions of its product, *J. Virol.* **46**:920.

Hutchison, C. A., Hardies, S. C., Loeb, D. D., Shehee, W. R., and Edgell, M. H., 1989, LINEs and related retroposons: Long interspersed repeated sequences in the eucaryotic genome, in: *Mobile DNA* (D. E. Berg and M. M. Howe, eds.), pp. 593–617, American Society for Microbiology, Washington, D.C.

Hwang, J. V., and Gilboa, E., 1984, Expression of genes introduced into cells by retroviral infection is more efficient than that of genes introduced into cells by DNA transfection, *J. Virol.* **50**:414.

Hwang, L. H. S., Park, J., and Gilboa, E., 1984, Role of intron-contained sequences in formation of Moloney murine leukemia virus *env* mRNA, *J. Virol.* **4**:2289.

Hynes, N., Van Ooyen, A. J. J., Kennedy, N., Herrlich, P., Ponta, H., and Groner, B., 1983, Subfragments of the large terminal repeat cause glucocorticoid-responsive expression of mouse mammary tumor virus and of an adjacent gene, *Proc. Natl. Acad. Sci. USA* **80**:3637.

Imperiale, M. J., and DeZazzo, J. D., 1991, Poly(A) site choice in retroelements: *Deja vu* all over again, *New Biol.* **3**:531.

Isfort, R., Jones, D., Kost, R., Witter, R., and Kung, H. J., 1992, Retrovirus insertion into herpesvirus *in vitro* and *in vivo*, *Proc. Natl. Acad. Sci. USA* **89**:991.

Ishikawa, M., Meshi, T., Motoyoshi, F., Takamatsu, N., and Okada, Y., 1986, *In vitro* mutagenesis of the putative replicase genes of tobacco mosaic virus, *Nucleic Acids Res.* **14**:8291.

Ishimoto, A., Takimoto, M., Adachi, A., Kakuyama, M., Kato, S., Kakimi, K., Fukuoka, K., Ogiu, T., and Matsuyama, M., 1987, Sequences responsible for erythroid and lymphoid leukemia in the long terminal repeats of Friend-mink cell focus-forming and Moloney murine leukemia virus, *J. Virol.* **61**:1861.

Iwasaki, K., and Temin, H. M., 1990a, The U3 region is not necessary for 3′ end formation of spleen necrosis virus RNA, *J. Virol.* **64**:6329.

Iwasaki, K., and Temin, H. M., 1990b, The efficiency of RNA 3′-end formation is determined by the distance between the cap site and the poly(A) site in spleen necrosis virus, *Genes Dev.* **4:**2299.

Jacks, T., 1990, Translational suppression in gene expression in retroviruses and retrotransposons, in: *Retroviruses—Strategies of Replication* (R. Swanstrom and P. K. Vogt, eds.), pp. 93–124, Springer-Verlag, Berlin.

Jacks, R., and Varmus, H. E., 1985, Expression of the Rous sarcoma virus *pol* gene by ribosomal frameshifting, *Science* **230:**1237.

Jacks, T., Townsley, K., Varmus, H. E., and Majors, J., 1987, Two efficient ribosomal frameshift events are required for synthesis of mouse mammary tumor virus gag-related polypeptides, *Proc. Natl. Acad. Sci. USA* **84:**4298.

Jacks, T., Power, M. D., Masiarz, F. R., Luciw, P. A., Barr, P. J., and Varmus, H. E., 1988a, Characterization of ribosomal frameshifting in HIV-1 gag–pol expression, *Nature* **231:**280.

Jacks, T., Madhani, H. D., Masiarz, F. R., and Varmus, H. E., 1988b, Signals for ribosomal frameshifting in the Rouse sarcoma virus *gag–pol* region, *Cell* **55:**447.

Jackson, R. J., and Standart, N., 1990, Do the poly(A) tail and 3′ untranslated region control mRNA translation? *Cell* **62:**15.

Jacobo-Molina, A., and Arnold, E., 1991, HIV reverse transcriptase structure–function relationship, *Biochemistry* **30:**6351.

Jacobo-Molina, A., Clark, A. D., Williams, R. L., Nanni, R. G., Clark, P., Ferris, A. L., Hughes, S. H., and Arnold, E., 1991, Crystals of a ternary complex of human immunodeficiency virus type 1 reverse transcriptase with a monoclonal antibody Fab fragment and double-stranded DNA diffract x-rays to 3.5A resoluction, *Proc. Natl. Acad. Sci. USA* **88:**10895.

Jaenisch, R., 1976, Germ line integration and Mendelian transmission of the exogenous Moloney leukemia virus, *Proc. Natl. Acad. Sci. USA* **73:**1260.

Jaenisch, R., Harbers, K., Schnicke, A., Lohler, J., Chumakov, I., Jahner, D., Grotkopp, D., and Hoffmann, E., 1983, Germ line integration of Moloney murine leukemia virus at the Mov13 locus leads to recessive lethal mutation and early embryonic death, *Cell* **32:**209.

Jarrett, O., Laird, H. M., and Hay, D., 1973, Determinants of the host range of feline leukaemia viruses, *J. Gen. Virol.* **20:**169.

Jaskolski, M., Miller, M., Rao, J. K. M., Leis, J., and Wlodawer, A., 1990, A structure of the aspartic protease from Rous sarcoma virus retrovirus defined at 2-angstrom resolution, *Biochemistry* **29:**5889.

Jaskolski, M., Tomasselli, A. G., Sawyer, T. K., Staples, D. G., Heinrikson, R. L., Schneider, J., Kent, S. B. H., and Wlodawer, A., 1991, Structure at 2.5A resolution of chemically synthesized human immunodeficiency virus type 1 protease complexed with a hydroxyethylene-based inhibitor, *Biochemistry* **30:**1600.

Jentoft, J. E., Smith, L. M., Fu, X., Johnson, M., and Leis, J., 1988, Conserved cysteine and histidine residues of the avian myeloblastosis virus nucleocapsid protein are essential for viral replication but are not "zinc-binding fingers," *Proc. Natl. Acad. Sci. USA* **85:**7094.

Johann, S. V., Gibbons, J. J., and O'Hara, B., 1992, GLVR1, a receptor for gibbon ape leukemia virus, is homologous to a phosphate permease of *Neurospora crassa* and is expressed at high levels in the brain and thymus, *J. Virol.* **66:**1635.

Johnson, M. S., McClure, M. A., Feng, D., Gray, J., and Doolittle, R. F., 1986, Computer analysis of retroviral *pol* genes: Assignment of enzymatic functions to specific sequences and homologies with nonviral enzymes, *Proc. Natl. Acad. Sci. USA* **83:**7648.

Johnson, P. F., and McKnight, S. L., 1989, Eukaryotic transcriptional regulatory proteins, *Annu. Rev. Biochem.* **58:**799.

Johnson, P. F., Landschulz, W. H., Graves, B. J., and McKnight, S. L., 1987, Identification of a rat liver nuclear protein that binds to the enhancer core element of three animal viruses, *Genes Dev.* **1**:133.

Jolicoeur, P., and Rassart, E., 1980, Effect of Fv-1 gene product on synthesis of linear and supercoiled viral DNA in cells infected with murine leukemia virus, *J. Virol.* **33**:183.

Jolicoeur, P., and Rassart, E., 1981, Fate of unintegrated viral DNA in Fv-1 permissive and resistant mouse cells infected with murine leukemia viruses, *J. Virol.* **37**:609.

Jones, T. A., Blaug, G., Hansen, M., and Barklis, E., 1990, Assembly of gag–B-galactosidase proteins into retrovirus particles, *J. Virol.* **64**:2265.

Joyner, A., Keller, G., Phillips, R. A., and Bernstein, A., 1983, Retrovirus mediated transfer of a bacterial gene into mouse haematopoietic progenitor cells, *Nature* **305**:556.

Ju, G., and Cullen, B. R., 1985, The role of avian retroviral LTRs in the regulation of gene expression and viral replication, *Adv. Virus Res.* **30**:179.

Junghans, R. P., Boone, L. R., and Skalka, A. M., 1982a, Products of reverse transcription in avian retrovirus analyzed by electron microscopy, *J. Virol.* **43**:544.

Junghans, R. P., Boone, L. R., and Skalka, A. M., 1982b, Retroviral DNAseH structures: Displacement-assimilation model of recombination, *Cell* **30**:53.

Kaelbling, M., Eddy, R., Shows, T. B., Copeland, N. G., Gilbert, D. J., Jenkins, N. A., Klinger, H. P., and O'Hara, B., 1991, Localization of the human gene allowing infection by gibbon ape leukemia virus to human chromosome region 2q11–q14 and to the homologous region on mouse chromosome 2, *J. Virol.* **65**:1743.

Kamer, G., and Argos, P., 1984, Primary structural comparison of RNA-dependent polymerases from plant, animal, and bacterial viruses, *Nucleic Acids Res.* **12**:7269.

Kamps, C. A., Lin, Y. C., and Wong, P. K. Y., 1991, Oligomerization and transport of the envelope protein of Moloney murine leukemia virus-TB and of ts1, a neurovirulent temperature-sensitive mutant of MoMuLV-TB, *Virology* **184**:687.

Kane, S. E., and Beemon, K., 1985, Precise localization of m6A in Rous sarcoma virus RNA reveals clustering of methylation sites: Implications for RNA processing, *Mol. Cell. Biol.* **5**:2298.

Karnitz, L., Faber, S., and Chalkley, R., 1987, Specific nuclear proteins interact with the Rous sarcoma virus internal enhancer and share a common element with the enhancer located in the long terminal repeat of the virus, *Nucleic Acids Res.* **15**:9841.

Karnitz, L., Poon, D., Weil, A., and Chalkley, R., 1989, Purification and properties of the Rous sarcoma virus internal enhancer binding protein, *Mol. Cell. Biol.* **9**:1929.

Karpel, R. L., Henderson, L. E., and Oroszlan, S., 1987, Interactions of retroviral structural proteins with single-stranded nucleic acids, *J. Biol. Chem.* **262**:4961.

Katoh, I., Yoshinaka, Y., Rein, A., Shibuya, M., Odaka, T., and Oroszlan, S., 1985, Murine leukemia virus maturation: Protease region required for conversion from "immature" to "mature" core form and for virus infectivity, *Virology* **145**:280.

Katz, R. A., and Jentoft, J. E., 1989, What is the role of the Cys–His motif in retroviral nucleocapsid (NC) proteins? *BioEssays* **11**:176.

Katz, R. A., and Skalka, A. M., 1988, A C-terminal domain in the avian sarcoma-leukosis virus *pol* gene product is not essential for viral replication, *J. Virol.* **62**:528.

Katz, R. A., and Skalka, A. M., 1990, Generation of diversity in retroviruses, *Annu. Rev. Genet.* **24**:409.

Katz, R. A., Kotler, M., and Skalka, A. M., 1988, *Cis*-acting intron mutations that affect the efficiency of avian retroviral RNA splicing: Implications for mechanisms of control, *J. Virol.* **62**:2686.

Katz, R. A., Merkel, G., Kulkosky, J., Leis, J., and Skalka, A. M., 1990, The avian retroviral IN protein is both necessary and sufficient for integrative recombination *in vitro*, *Cell* **63**:87.

Katzman, M., Katz, R. A., Skalka, A. M., and Leis, J., 1989, The avian retroviral integration protein cleaves the terminal sequences of linear viral DNA at the *in vivo* sites of integration, *J. Virol.* **63**:5319.

Katzman, M., Mack, J. P. G., Skalka, A. M., and Leis, J., 1991, A covalent complex between retroviral integrase and nicked substrate DNA, *Proc. Natl. Acad. Sci. USA* **88**:4695.

Kawai, S., and Hanafusa, H., 1973, Isolation of defective mutant of avian sarcoma virus, *Proc. Natl. Acad. Sci. USA* **70**:3493.

Kay, M. A., Baley, P., Rothenberg, S., Leland, F., Fleming, L., Ponder, K. P., Liu, T. J., Finegold, M., Darlington, G., Pokorny, W., and Woo, S. L. C., 1992, Expression of human α1-antitrypsin in dogs after autologous transplantation of retroviral transduced hepatocytes, *Proc. Natl. Acad. Sci. USA* **89**:93.

Keith, J., and Fraenkel-Conrat, H., 1975, Identification of the 5′ end of Rous sarcoma virus RNA, *Proc. Natl. Acad. Sci. USA* **72**:3347.

Kenny, S., and Guntaka, R. V., 1990, Localization by mutational analysis of transcription factor binding sequences in the U3 region of Rous sarcoma virus LTR, *Virology* **176**:483.

Keshet, E., Schiff, R., and Itin, A., 1991, Mouse retrotransposons: A cellular reservoir of long terminal repeat (LTR) elements with diverse transcriptional specificities, *Adv. Cancer Res.* **56**:215.

Khan, E., Mack, J. P. G., Katz, R. A., Kulkosky, J., and Skalka, A. M., 1991, Retroviral integrase domains: DNA binding and the recognition of LTR sequences, *Nucleic Acids Res.* **19**:851.

Kim, J. W., Closs, E. I., Albritton, L. M., and Cunningham, J. M., 1991, Transport of cationic amino acids by the mouse ecotropic retrovirus receptor, *Nature* **352**:725.

Klatzmann, D., Champagne, E., Chamaret, S., Gruest, J., Guetard, D., Hercend, T., Gluckman, J. C., and Montagnier, L., 1984, T-lymphocyte T4 molecule behaves as the receptor for human retrovirus LAV, *Nature* **312**:767.

Kleckner, N., 1989, Transposon Tn10, in: *Mobile DNA* (D. E. Berg and M. M. Howe, eds.), pp. 227–268, American Society for Microbiology, Washington, D.C.

Kleckner, N., 1990, Regulation of transposition in bacteria, *Annu. Rev. Genet.* **6**:297.

Klement, V., Rowe, W. P., Hartley, J. W., and Pugh, W. E., 1969, Mixed culture cytopathogenicity: A new test for growth of murine leukemia viruses in tissue culture, *Proc. Natl. Acad. Sci. USA* **63**:753.

Koo, H. M., Brown, A. M. C., Ron, Y., and Dougherty, J. P., 1991, Spleen necrosis virus, an avian retrovirus, can infect primate cells, *J. Virol.* **65**:4769.

Koo, H. M., Gu, J., Varela-Echavarria, A., Ron, Y., and Dougherty, J. P., 1992, Reticuloendotheliosis type C and primate type D oncoretroviruses are members of the same receptor interference group, *J. Virol.* **66**:3448.

Kotler, M., Katz, R. A., Danho, W., Leis, J., and Skalka, A. M., 1988, Synthetic peptides as substrates and inhibitors of a retroviral protease, *Proc. Natl. Acad. Sci. USA* **85**:4185.

Kotler, M., Danho, W., Katz, R. A., Leis, J., and Skalka, A. M., 1989, Avian retroviral protease and cellular aspartic proteases are distinguished by activities on peptide substrates, *J. Biol. Chem.* **264**:3428.

Kowalski, M., Potz, J., Basiripour, L., Dorfman, T., Goh, W. C., Terwilliger, E., Dayton, A., Rosen, C., Haseltine, W., and Sodroski, J., 1987, Functional regions of the envelope glycoprotein of human immunodeficiency virus type 1, *Science* **237**:1351.

Kozak, M., 1991, Structural features in eucaryotic mRNAs that modulate the initiation of translation, *J. Biol. Chem.* **266**:19867.

Kozak, C. A., Albritton, L. M., and Cunningham, J. M., 1990, Genetic mapping of a cloned sequence responsible for susceptibility to ecotopic murine leukemia viruses, *J. Virol.* **64**:3119.

Kramarsky, B., Sarkar, N. H., and Moore, D. H., 1971, Ultrastructural comparison of a virus from a rhesus monkey mammary carcinoma with four oncogenic RNA viruses, *Proc. Natl. Acad. Sci. USA* **68**:1603.

Kranier, A., and Maniatis, T., 1988, RNA splicing, in: *Frontiers in molecular biology,* (B. D. Hames and D. M. Glover), pp. 131–296, IRL Press, Oxford/Washington D.C.

Kriegler, M., 1990, *Gene Transfer and Expression—A Laboratory Manual,* Stockton Press, New York.

Krug, M. S., and Berger, S. L., 1989, Ribonuclease H activities associated with viral reverse transcriptases are endonucleases, *Proc. Natl. Acad. Sci. USA* **86**:3539.

Kuff, E. L., and Lueders, K. K., 1988, The intracisternal A-particle gene family: Structure and functional aspects, *Adv. Cancer Res.* **51**:183.

Kulkosky, J., and Skalka, A. M., 1990, HIV DNA integration: Observations and inferences, *J. Acquired Immune Defic. Syndr.* **3**:839.

Kung, H. J., and Vogt, P. K., 1991, in: *Retroviral Insertion and Oncogene Activation,* (H. J. Kung and P. K. Vogt), Springer-Verlag, Berlin.

Kung, H. J., Shank, P. R., Bishop, J. M., and Varmus, H. E., 1980, Identification and characterization of dimeric and trimeric circular forms of avian sarcoma virus-specific DNA, *Virology* **103**:425.

Kung, H. J., Fung, Y. K., Majors, J. E., Bishop, J. M., and Varmus, H. E., 1981, Synthesis of plus strands of retroviral DNA in cells infected with avian sarcoma virus and mouse mammary tumor virus, *J. Virol.* **37**:127.

Kung, H. J., Boerkoel, C., and Carter, T. H., 1991, Retroviral mutagenesis of cellular oncogenes: A review with insights into the mechanisms of insertional activation, in: *Retroviral Insertion and Oncogene Activation,* (H. J. Kung and P. K. Vogt), pp. 1–25, Springer-Verlag, Berlin.

Laimins, L. A., Tsichlis, P. N., and Khoury, G., 1984a, Multiple enhancer domains in the 3' terminus of the Prague strain of Rous sarcoma virus, *Nucleic Acids Res.* **12**:6427.

Laimins, L. A., Gruss, P., Pozzatti, R., and Khoury, G., 1984b, Characterization of enhancer elements in the long terminal repeat of Moloney murine sarcoma virus, *J. Virol.* **49**:183.

Landschulz, W. H., Johnson, P. F., and McKnight, S. L., 1988, The leucine zipper: A hypothetical structure common to a new class of DNA binding proteins, *Science* **240**:1759.

Lapatto, R., Blundell, T., Hemmings, A., Overington, J., Wilderspin, A., Wood, S., Merson, J. R., Whittle, P. J., Danley, D. E., Geoghegan, K. F., Hawrylik, S. J., Lee, S. J., Scheld, K. G., and Hobart, P. M., 1989, X-ray analysis of HIV-1 proteinase at 2.7A resolution confirms structural homology among retroviral enzymes, *Nature* **342**:299.

Le, S. Y., Shapiro, B. A., Chen, J. H., Nussinov, R., and Maizel, J. V., 1991, RNA pseudoknots downstream of the frameshift sites of retroviruses, *GATA* **8**:191.

Lee, Y. M. H., and Coffin, J. M., 1990, Efficient autointegration of avian retrovirus DNA *in vitro, J. Virol.* **64**:5958.

Lee, F., Hall, C. V., Ringold, G. M., Dobson, D. E., Luh, J., and Jacob, P. E., 1984, Functional analysis of the steroid hormone control region of mouse mammary tumor virus, *Nucleic Acids Res.* **12**:4191.

LeGrice, S. F. J., Naas, T., Wohlgensinger, B., and Schatz, O., 1991, Subunit-selective mutagenesis indicates minimal polymerase activity in heterodimeric-associated p51 HIV-1 reverse transcriptase, *EMBO J.* **10**:3905.

Leider, J. M., Palese, P., and Smith, F., 1988, Determination of the mutation rate of a retrovirus, *J. Virol.* **62**:3084.

Leis, J., and Jentoft, J., 1983, Characteristics and regulation of interaction of avian retrovirus pp12 protein with viral RNA, *J. Virol.* **48**:361.

Leis, J. P., Berkower, I., and Hurwitz, J., 1973, Mechanism of action of ribonuclease H isolated from avian myeloblastosis virus and *Escherichia coli, Proc. Natl. Acad. Sci. USA* **70**:466.

Leis, J., Duyk, G., Johnson, S., Longiaru, M., and Skalka, A., 1983, Mechanism of action of the endonuclease associated with $\alpha\beta$ and $\beta\beta$ forms of avian RNA tumor virus reverse transcriptase, *J. Virol.* **45**:727.

Leis, J., Baltimore, D., Bishop, J. M., Coffin, J., Fleissner, E., Goff, S. P., Oroszlan, S., Robinson, H., Skalka, A. M., Temin, H. M., and Vogt, V., 1988, Standardized and simplified nomenclature for proteins common to all retroviruses, *J. Virol.* **62**:1808.

Leis, J., Bizub, D., Weber, I., Cameron, C., Katz, R., Wlodawer, A., and Skalka, A., 1989, Structure–function analysis of retroviral aspartyl proteases, in: *Current Communications in Molecular Biology: Viral Proteases as Targets for Chemotherapy* (H. Krausslich, S. Oroszlan, and E. Wimmer, eds.), pp. 235–243, Cold Spring Harbor Laboratory, Cold Spring Harbor, New York.

Lenardo, M., Pierce, J. W., and Baltimore, D., 1987, Protein binding sites in Ig gene enhancers determine transcriptional activity and inducibility, *Science* **236**:1573.

Lenz, J. D., Celander, D., Crowther, R. L., Patarca, R., Perkins, D. W., and Haseltine, W. A., 1984, Determination of the leukemogenicity of a murine retrovirus by sequences within the long terminal repeat, *Nature* **308**:467.

Levinson, B., Khoury, G., Vande Woude, G., and Gruss, P., 1982, Activation of SV40 genome by 72-base pair tandem repeats of Moloney sarcoma virus, *Nature* **295**:568.

Levy, J. A., 1978, Xenotropic type C viruses. *Curr. Top. Microbiol. Immunol.* **79**:111.

Lewin, B. L., 1990, Commitment and activation at pol II promoters: A tail of protein–protein interactions, *Cell* **61**:1161.

Li, G., and Rice, C. M., 1989, Mutagenesis of the in-frame opal termination codon preceding nsP4 of Sindbis virus: Studies of translational readthrough and its effect on virus replication, *J. Virol.* **63**:1326.

Li, Y., Colemis, E., Hartley, J. W., and Hopkins, N., 1987, Disease specificity of nondefective Friend and Moloney murine leukemia viruses is controlled by a small number of nucleotides, *J. Virol.* **61**:693.

Lifson, J. D., Feinberg, M. B., Reyes, G. R., Rabin, L., Banapour, B., Chakrabarti, S., Moss, B., Wong-Staal, F., Steimer, K. S., and Engelman, E. G., 1986, Induction of CD4-dependent cell fusion by HTLV-III/LAV envelope glycoproteins, *Nature* **323**:725.

Lin, T. H., and Grandgenett, D. P., 1991, Retrovirus integrase: Identification of a potential leucine zipper motif, *Protein Eng.* **4**:435.

Linial, M., 1987, Creation of a processed pseudogene by retroviral infection, *Cell* **49**:93.

Linial, M., and Blair, D., 1985, Genetics of retroviruses, in: *RNA Tumor Viruses* (R. Weiss, N. Teich, H. Varmus, and J. Coffin, eds.), pp. 650–783, Cold Spring Harbor Laboratory, Cold Spring Harbor, New York.

Linial, M. L., and Miller, A. D., 1990, Packaging: Sequence requirements and implications, in: *Retroviruses—Strategies of Replication*, (R. Swanstrom and P. K. Vogt, eds.), pp. 125–152, Springer-Verlag, Berlin.

Linial, M., Medeiros, E., and Hayward, W., 1978, An avian oncovirus mutant (SE21Q1b) deficient in genomic RNA: Biological and biochemical characterization, *Cell* **15**:1371.

Linney, E., Dvais, B., Overhauser, J., Chao, E., and Fan, H., 1984, Nonfunction of a Moloney murine leukemia virus regulatory sequence in F9 embryonal carcinoma cells, *Nature* **308**:470.

Lobel, L. I., and Goff, S. P., 1985, Reverse transcription of retroviral genomes: Mutations in the terminal repeat sequences, *J. Virol.* **53**:447.

Loeb, D. D., Hutchinson, C. A., Edgell, M. H., Farmerie, W. G., and Swanstrom, R., 1988, Mutational analysis of the HIV-1 protease suggests functional homology with aspartic proteases, *J. Virol.* **63**:111.

Loeb, D. D., Swanstrom, R., Everitt, L., Manchester, M., Stamper, S. E., and Hutchison, C. A., 1989, Complete mutagenesis of the HIV-1 protease, *Nature* **340**:397.

Loh, T. P., Sievert, L. L., and Scott, R. W., 1987, Proviral sequences that restrict retroviral expression in mouse embryonal carcinoma cells, *Mol. Cell. Biol.* **7**:3775.

Loh, T. P., Sievert, L. L., and Scott, R. W., 1988, Negative regulation of retrovirus expression in embryonal carcinoma cells mediated by an intragenic domain, *J. Virol.* **62**:4086.

LoSardo, J. E., Cupelli, L. A., Short, M. K., Berman, J. W., and Lenz, J., 1989, Differences in activities of murine retroviral long terminal repeats in cytotoxic T lymphocytes and T lymphoma cells, *J. Virol.* **63**:1087.

Luban, J., and Goff, S. P., 1991, Binding of human immunodeficiency virus type 1 (HIV-1) RNA to recombinant HIV-1 gag polyprotein, *J. Virol.* **65**:3203.

Luciw, P. A., Bishop, J. M., Varmus, H. E., and Capecchi, M., 1983, Location and function of retroviral and SV40 sequences that enhance biochemical transformation after microinjection of DNA, *Cell* **33**:705.

Luciw, P. A., Oppermann, H., Bishop, J. M., and Varmus, H. E., 1984, Integration and expression of several forms of Rous sarcoma virus DNA used for transfection of mouse cells, *Mol. Cell. Biol.* **4**:1260.

Luo, G., Sharmeen, L., and Taylor, J., 1990, Specificities involved in the initiation of retroviral plus-strand DNA, *J. Virol.* **64**:592.

MacLeod, C. L., Finley, K., Kakuda, D., Kozak, C. A., and Wilkinson, M. F., 1990, Activated T cells express a novel gene on chromosome 8 that is closely related to the murine ecotropic retroviral receptor, *Mol. Cell. Biol.* **10**:3663.

Maddon, P. J., Dalgleish, A. G., McDougal, J. S., Clapham, P. R., Weiss, R. A., and Axel, R., 1986, The T4 gene encodes the AIDS virus receptor and is expressed in the immune system and the brain, *Cell* **47**:333.

Madhani, H. D., Jacks, T., and Varmus, H. E., 1988, Signals for the expression of the HIV pol gene by ribosomal frameshifting, in: *Control of HIV Gene Expression* (R. Franza, B. Cullen, and F. Wong-Staal, eds.), pp. 119–125), Cold Spring Harbor Laboratory, Cold Spring Harbor, New York.

Majors, J., 1990, The structure and function of retroviral long terminal repeats, in: *Retroviruses—Strategies of Replication* (R. Swanstrom and P. K. Vogt, eds.), pp. 49–92, Springer-Verlag, Berlin.

Majors, J. E., and Varmus, H. E., 1981, Nucleotide sequences at host–proviral junctions for mouse mammary tumour virus, *Nature* **289**:253.

Majors, J. E., and Varmus, H. E., 1983, A small region of the mouse mammary tumor virus long terminal repeat confers glucocorticoid hormone regulation on a linked heterologous gene, *Proc. Natl. Acad. Sci. USA* **80**:5866.

Majumdar, C., Stein, C. A., Cohen, J. S., Broder, S., and Wilson, S. H., 1989, Stepwise mechanism of HIV reverse transcriptase: Primer function of phosphorothioate oligodeoxynucleotide, *Biochemistry* **28**:1340.

Maniatis, T., Goodbourn, S., and Fischer, J. A., 1987, Regulation of inducible and tissue-specific gene expression, *Science* **236**:1237.

Manley, J. L., 1988, Polyadenylation of mRNA precursors, *Biochim. Biophys. Acta* **950**:1.

Mann, R., and Baltimore, D., 1985, Varying the position of a retrovirus packaging sequence results in the encapsidation of both unspliced and spliced mRNAs, *J. Virol.* **54**:401.

Mann, R. S., Mulligan, R., and Baltimore, D., 1983, Construction of a retrovirus packaging mutant and its use to produce helper-free selective retrovirus, *Cell* **32**:871.

Marsh, L. E., and Guilfoyle, T. J., 1987, Cauliflower mosaic virus replication intermediates are encapsidated into virion-like particles, *Virology* **161**:129.

Mason, W. S., Taylor, J. M., and Hull, R., 1987, Retroid virus genome replication, *Adv. Virus Res.* **32**:35.

Mathias, S. L., Scott, A. F., Kazazian, H. H., Boeke, J. D., and Gabriel, A., 1991, Reverse transcriptase encoded by a human transposable element, *Science* **254**:1801.

Maul, D. H., Zaiss, C. P., Mackenzie, M. R., Shiigi, S. M., Marx, P. A., and Gardner, M. B., 1988, Simian retrovirus D subgroup 1 has a broad cellular tropism for lymphoid and nonlymphoid cells, *J. Virol.* **62**:1768.

Maurer, B., Bannert, H., Darai, G., and Flugel, R. M., 1988, Analysis of the primary structure of the long terminal repeat and the *gag* and *pol* genes of the human spumaretrovirus, *J. Virol.* **62**:1590.

McClure, M. O., Marsh, M., and Weiss, R. A., 1988, Human immunodeficiency virus infection of CD4-bearing cells occurs by a pH-independent mechanism, *EMBO J.* **7**:513.

McClure, M. O., Sommerfelt, M. A., Marsh, M., and Weiss, R. A., 1990, The pH independence of mammalian retrovirus infection, *J. Gen. Virol.* **71**:767.

McCune, J. M., Rabin, L. B., Feinberg, M. B., Lieberman, M., Kosek, J. C., Reyes, J. R., and Weissman, I. L., 1988, Endoproteolytic cleavage of gp160 is required for activation of human immunodeficiency virus, *Cell* **53**:55.

McDougal, J. S., Kennedy, M. S., Sligh, J. M., Cort, S. P., Mawle, A., and Nicholson, J. K. A., 1986, Binding of HTLV-III/LAV to T4+ cells by a complex of the 110K viral protein and the T4 molecule, *Science* **231**:382.

McLachlin, J. R., Cornetta, K., Eglitis, K., and Anderson, W. F., 1990, Retroviral-mediated gene transfer, *Prog. Nucleic Acid Res.* **38**:91.

McNally, M. T., and Beemon, K., 1992, Intronic sequences and 3' splice sites control Rous sarcoma virus RNA splicing, *J. Virol.* **66**:6.

McNally, M. T., Gontarek, R. R., and Beemon, K., 1991, Characterization of Rous sarcoma virus intronic sequences that negatively regulate splicing, *Virology* **185**:99.

Meek, T. D., Dayton, B. D., Metcalf, B. W., *et al.*, 1989, Human immunodeficiency virus 1 protease expressed in *Escherichia coli* behaves as a dimeric aspartic protease, *Proc. Natl. Acad. Sci. USA* **86**:1841.

Mehdi, H., Ono, E., and Gupta, K. C., 1990, Initiation of translation at CUG, GUG, and ACG codons in mammalian cells, *Gene* **91**:173.

Mellor, J., Fulton, S. M., Dobson, J., Wilson, K., Kingsman, S. M., and Kingsman, A. J., 1985, A retrovirus-like strategy for expression of a fusion protein encoded by the yeast transposon Ty1, *Nature* **313**:243.

Mercurio, F., and Karin, M., 1989, Transcription factors AP-3 and AP-2 interact with the SV 40 enhancer in a mutually exclusive manner, *EMBO J.* **8**:1455.

Mergia, A., and Luciw, P. A., 1991, Replication and regulation of primate foamy viruses, *Virology* **184**:475.

Meric, C., and Goff, S. P., 1989, Characterization of Moloney murine leukemia virus mutants with single amino acid substitutions in the Cys–His box of the nucleocapsid protein, *J. Virol.* **63**:1558.

Meric, C., and Spahr, P., 1986, Rous sarcoma virus nucleic acid binding protein p12 is necessary for viral 70S RNA dimer formation and packaging, *J. Virol.* **60**:450.

Meric, C., Darlix, J. L., and Spahr, P. F., 1984, It is Rous sarcoma virus p12 and not p19 that binds tightly to Rous sarcoma virus RNA, *J. Mol. Biol.* **173**:531.

Meric, C., Gouilloud, E., and Spahr, P., 1988, Mutations in Rous sarcoma virus nucleocapsid protein p12 (NC): Deletions of Cys–His boxes, *J. Virol.* **62**:3228.

Miksicek, R., Borgmeyer, W., and Nowock, J., 1987, Interaction of the TGGCA-binding protein with upstream sequences is required for efficient transcription of mouse mammary tumor virus, *EMBO J.* **6**:1355.

Miller, A. D., 1990a, Retrovirus packaging cells, *Hum. Gene Ther.* **1**:5.

Miller, A. D., 1990b, Progress toward human gene therapy, *Blood* **76**:271.

Miller, A. D., Jolly, D. J., Friedman, T., and Verma, I. M., 1983, A transmissible retrovirus expressing human hypoxanthine phosphoribosyltransferase (HPRT): Gene transfer into cells obtained from humans deficient in HPRT, *Proc. Natl. Acad. Sci. USA* **80**:4709.

Miller, C. K., and Temin, H. K., 1986, Insertion of several different DNAs in reticuloendotheliosis virus strain T suppresses transformation by reducing the amount of subgenomic DNA, *J. Virol.* **58**:75.

Miller, D. G., and Miller, A. D., 1992, Tunicamycin treatment of CHO cells abrogates multiple blocks to retroviral infection, one of which is due to a secreted inhibitor, *J. Virol.* **66**:78.

Miller, D. G., Adam, M. A., and Miller, A. D., 1990, Gene transfer by retrovirus vectors occurs only in cells that are actively replicating at the time of infection, *Mol. Cell. Biol.* **10**:4239.

Miller, J. T., and Stoltzfus, C. A., 1992, Regions containing *cis*-acting splicing signals facilitate 3'-end processing of avian sarcoma virus RNA, *J. Virol.*, in press.

Miller, M., Jaskolski, M., Mohana Rao, J. K., Leis, J., and Wlodawer, A., 1989, Crystal structure of a retroviral protease proves relationship to aspartic protease family, *Nature* **337**:576.

Mitchell, P. M., and Tjian, R., 1989, Transcriptional regulation in mammalian cells by sequence-specific DNA binding proteins, *Science* **245**:371.

Mitra, S., Goff, S., Gilboa, E., and Baltimore, D., 1979, Synthesis of a 600-nucleotide-long plus-strand DNA by virions of Moloney murine leukemia virus, *Proc. Natl. Acad. Sci. USA* **76**:4355.

Mitsuya, H., Yarochan, R., and Broder, S., 1990, Molecular targets for AIDS therapy, *Science* **249**:1533.

Mizuuchi, K., and Adzuma, K., 1991, Inversion of the phosphate chirality at the target site of Mu DNA strand transfer: Evidence for a one-step transesterification mechanism, *Cell* **66**:129.

Modak, M. J., and Marcus, S. L., 1977, Purification and properties of Rauscher leukemia virus DNA polymerase and selective inhibition of mammalian viral reverse transcriptase by inorganic phosphate, *J. Biol. Chem.* **252**:11.

Moelling, K., Bolognesi, D. P., Bauer, H., Busen, W., Plassmann, H. W., and Hausen, P., 1971, Association of viral reverse transcriptase with an enzyme degrading the RNA moiety of RNA–DNA hybrids, *Nature New Biol.* **234**:240.

Moore, R., Dixon, M., Smith, R., Peters, G., and Dickson, C., 1987, Complete nucleotide sequence of a milk-transmitted mouse mammary tumor virus: Two frameshift suppression events are required for translation of gag and pol, *J. Virol.* **61**:480.

Mooslehner, K., Larls, U., and Harbers, K., 1990, Retroviral integration sites in transgenic Mov mice frequently map in the vicinity of transcribed regions, *J. Virol.* **64**:3056.

Morris, D. W., 1991, Molecular biology and pathogenesis of mouse mammary tumour virus, *Rev. Med. Virol.* **1**:223.

Morris-Vasios, C., Kochan, J. P., and Skalka, A. M., 1988, Avian sarcoma-leukosis virus pol–endo proteins expressed independently in mammalian cells accumulate in the nucleus but can be directed to other cellular compartments, *J. Virol.* **62**:349.

Mount, S. M., and Rubin, G. M., 1985, Complete nucleotide sequence of the *Drosophila* transposable element copia: Homology between copia and retroviral proteins, *Mol. Cell. Biol.* **5**:1630.

Muller, M. M., Gerster, T., and Schaffner, W., 1988, Enhancer sequences and the regulation of gene transcription, *Eur. J. Biochem.* **176**:485.

Mumm, S. R., and Grandgenett, D. P., 1991, Defining nucleic acid-binding properties of avian retrovirus integrase by deletion analysis, *J. Virol.* **65**:1160.

Munroe, D., and Jacobson, A., 1990, Tales of poly(A): A review, *Gene* **91**:151.

Murphy, J. E., and Goff, S. P., 1988, Construction and analysis of deletion mutations in the U5 region of Moloney murine leukemia virus: Effects on RNA packaging and reverse transcription, *J. Virol.* **63**:319.

Nabel, G., and Baltimore, D., 1987, An inducible factor activates expression of human immunodeficiency virus in T cells, *Nature* **326**:711.

Nanduri, V. B., and Modak, M. J., 1990, Lysine-329 of murine leukemia virus reverse transcriptase: Possible involvement in the template–primer binding function, *Biochemistry* **29**:5258.

Navia, M. A., Fitzgerald, P. M. D., and McKeever, B. M., 1989, Three dimensional structure of aspartyl protease from human immunodeficiency virus HIV-1, *Nature* **337**:615.

Neil, J. C., Fulton, R., Rigby, M., and Stewart, M., 1991, Feline leukaemia virus: Generation of pathogenic and oncogenic variants, in: *Retroviral Insertion and Oncogene Activation*, (H. J. Kung and P. K. Vogt), pp. 67, Springer-Verlag, Berlin.

Nigg, E. A., Baeuerle, P. A., and Luhrmann, R., 1991, Nuclear import–export: In search of signals and mechanisms, *Cell* **66**:15.

Norton, P. A., and Coffin, J. M., 1987, Characterization of Rous sarcoma virus sequences essential for viral gene expression, *J. Virol.* **61**:1171.

Nusse, R., 1991, Insertional mutagenesis in mouse mammary tumorigenesis, in: *Retroviral Insertion and Oncogene Activation*, (H. J. Kung and P. K. Vogt), pp. 43–66, Springer-Verlag, Berlin.

Oertle, S., and Spahr, P. F., 1990, Role of the gag polyprotein precursor in packaging and maturation of Rous sarcoma virus genomic RNA, *J. Virol.* **64**:5757.

O'Hara, B., Johann, S. V., Klinger, H. P., Blair, D. G., Rubinson, H., Dunn, K. J., Sass, P., Vitek, S. M., and Robbins, T., 1990, Characterization of a human gene conferring sensitivity to infection by Gibbon ape leukemia virus, *Cell Growth Differ.* **1**:119.

Olsen, J. C., Bova-Hill, C., Grandgenett, D. P., Quinn, T. P., Manfredi, J. P., and Swanstrom, R., 1990, Rearrangements in unintegrated retroviral DNA are complex and are the result of multiple genetic determinants. *J. Virol.* **64**:5475.

Olson, E. N., and Spizz, G., 1986, Fatty acylation of cellular proteins, *J. Biol. Chem.* **261**:2458.

Omer, C. A., and Faras, A. J., 1982, Mechanism of release of the avian retrovirus RNAtrp primer molecule from viral DNA by ribonuclease H during reverse transcription, *Cell* **30**:797.

Onts, H., Kennedy, N., Skroch, P., Hynes, N. E., and Groner, B., 1985, Hormonal response region in the mouse mammary tumor virus long terminal repeat can be dissociated from the proviral promoter and has enhancer properties, *Proc. Natl. Acad. Sci. USA* **82**:1020.

Oroszlan, S., and Luftig, R. B., 1990, Retroviral proteinases, in: *Retroviruses—Strategies of Replication* (R. Swanstrom and P. K. Vogt, eds.), pp. 153–185, Springer-Verlag, Berlin.

O'Shea, E. K., Rutkowski, R., and Kim, P. S., 1989, Evidence that the leucine zipper is a coiled coil, *Science* **243**:538.

Ou, C. Y., Boone, L. R., Koh, C. K., Tennant, R. W., and Yang, W. K., 1983, Nucleotide sequence of *gag–pol* regions that determine the Fv-1 host range property of BALB/c N-tropic and B-tropic murine leukemia viruses, *J. Virol.* **48**:779.

Owens, R., Dubay, J. W., Hunter, E., and Compans, R. W., 1991, Human immunodeficiency virus envelope protein determines the site of virus release in polarized epithelial cells, *Proc. Natl. Acad. Sci. USA* **88**:3987.

Ozer, J., Faber, M., Chalkley, R., and Sealy, L., 1990, Isolation and characterization of a cDNA clone for the CCAAT transcription factor EFIa reveals a novel structural motif, *J. Biol. Chem.* **36**:22143.

Palmiter, R. D., Gagnon, J., Vogt, V. M., Ripley, S., and Eisenman, R. N., 1978, The NH2-terminal sequence of the avian oncovirus gag precursor polyprotein (Pr76gag), *Virology* **91**:423.

Panganiban, A. T., 1988, Retroviral *gag* gene amber codon suppression is caused by an intrinsic *cis*-acting component of the viral mRNA, *J. Virol.* **62**:3574.

Panganiban, A. T., and Fiore, D., 1988, Ordered interstrand and intrastrand DNA transfer during reverse transcription, *Science* **241**:1964.

Panganiban, A. T., and Temin, H. M., 1983, The terminal nucleotides of retrovirus DNA are required for integration but not virus production, *Nature* **306**:155.

Panganiban, A. T., and Temin, H. M., 1984a, Circles with two tandem LTRs are precursors to integrated retrovirus DNA, *Cell* **36**:673.

Panganiban, A. T., and Temin, H. M., 1984b, The retrovirus *pol* gene encodes a product required for DNA integration: Identification of a retrovirus *int* locus, *Proc. Natl. Acad. Sci. USA* **81**:7885.

Parslow, T. G., Blair, D. L., Murphy, W. J., and Granner, D. K., 1984, Structure of the 5' ends of immunoglobulin genes: A novel conserved sequence, *Proc. Natl. Acad. Sci. USA* **81**:2650.

Pathak, V. K., and Temin, H. M., 1990a, Broad spectrum of *in vivo* forward mutations, hypermutations, and mutational hot-spots in a retroviral shuttle vector after a single replication cycle: Substitutions, frameshifts, and hypermutations, *Proc. Natl. Acad. Sci. USA* **87**:6019.

Pathak, V. K., and Temin, H. M., 1990b, Broad spectrum of *in vivo* forward mutations, hypermutations, and mutational hot-spots in a retroviral shuttle vector after a single replication cycle: Deletions and deletions with insertions, *Proc. Natl. Acad. Sci. USA* **87**:6024.

Pato, M. L., 1989, Bacteriophage Mu, in: *Mobile DNA* (D. E. Berg and M. M. Howe, eds.), pp. 23–52, American Society for Microbiology, Washington, D.C.

Pauza, C. D., and Price, T. M., 1988, Human immunodeficiency virus infection of T cells and monocytes proceeds via receptor-mediated endocytosis, *J. Cell Biol.* **107**:959.

Payne, G. S., Courtneidge, S. A., Crittenden, L. B., Fadley, A. M., Bishop, J. M., and Varmus, H. E., 1981, Analyses of avian leukosis virus DNA and RNA in bursal tumors suggest a novel mechanism for retroviral oncogenesis, *Cell* **23**:311.

Payne, G. S., Bishop, J. M., and Varmus, H. E., 1982, Multiple arrangements of viral DNA and an activated host oncogene in bursal lymphomas, *Nature* **295**:209.

Payvar, F. D., DeFranco, D., Firestone, G. L., Edgar, B., Wrange, O., Okret, S., Gustafsson, J. A., and Yamamoto, K., 1983, Sequence-specific binding of glucocorticoid receptor to MTV DNA at sites within and upstream of the transcribed region, *Cell* **35**:381.

Pearl, L. H., and Taylor, W. R., 1987, Sequence specificity of retroviral proteases, *Nature* **328**:482.

Perez, L. G., and Hunter, E., 1987, Mutations within proteolytic cleavage site of the Rous sarcoma virus glycoprotein that block processing to gp85 and gp37, *J. Virol.* **61**:1609.

Perez, L. G., Davis, G. L., and Hunter, E., 1987, Mutants of the Rous sarcoma virus envelope glycoprotein that lack the transmembrane anchor and cytoplasmic domains: Analysis of intracellular transport and assembly into virions, *J. Virol.* **61**:2981.

Perlmann, T., and Wrange, O., 1988, Specific glucocorticoid receptor binding to DNA reconstituted in a nucleosome, *EMBO J.* **7**:3073.

Perucho, M., Hanahan, D., and Wigler, M., 1980, Genetic and physical linkage of exogenous sequences in transformed cells, *Cell* **22**:309.

Peters, G. G., and Hu, J., 1980, Reverse transcriptase as the major determinant for selective packaging of tRNAs into avian sarcoma virus particles, *J. Virol.* **36**:692.

Petropoulos, C. J., and Hughes, S. H., 1991, Replication-competent retrovirus vectors for the transfer and expression of gene cassettes in avian cells, *J. Virol.* **65**:3728.

Pettit, S. C., Simsic, J., Loeb, D. D., Everitt, L., Hutchison, C. A., and Swanstrom, R., 1991, Analysis of retroviral protease cleavage sites reveals two types of cleavage sites and the structural requirements of the P1 amino acid, *J. Biol. Chem.* **266**:14539.

Picard, D., Salser, S. J., and Yamamoto, K. R., 1988, A movable and regulable inactivation function within the steroid binding domain of the glucocorticoid receptor, *Cell* **54**:1073.

Pillemer, E. A., Kooistra, D. A., Witte, O. N., and Weissman, I. L., 1986, Monoclonal antibody to the amino-terminal L sequence of murine leukemia virus glycosylated gag polyproteins demonstrates their unusual orientation in the cell membrane, *J. Virol.* **57**:413.

Pina, B., Bruggemeier, U., and Beato, M., 1990, Nucleosome positioning modulates accessibility of regulatory proteins to the mouse mammary tumor virus promoter, *Cell* **60**:719.

Pinter, A., 1989, Functions of murine leukemia virus envelope gene products in leukemogenesis, in: *Retroviruses and Disease* (H. Hanafusa, A. Pinter, and M. E. Pullman, eds.), pp. 21–39, Academic Press, San Diego, California.

Pinter, A., and Honnen, W. J., 1988, O-linked glycosylation of retroviral envelope gene products, *J. Virol.* **62**:1016.

Pinter, A., Chen, T. E., Lowry, A., Cortez, N. G., and Silagi, S., 1986, Ecotropic murine leukemia virus-induced fusion of murine cells, *J. Virol.* **57**:1048.

Portis, J. L., Atee, F. J., and Evans, L. H., 1985, Infectious entry of murine retroviruses into mouse cells: Evidence of a post-adsorption step inhibited by acidic pH, *J. Virol.* **55**:806.

Power, M. D., Marx, P. A., Bryant, M. L., Gardner, M. B., Barr, P. J., and Luciw, P. A., 1986, Nucleotide sequence of SRV-1, a type-D simian acquired immunodeficiency syndrome retrovirus, *Science* **231**:1567.

Prats, A. C., Sarih, L., Gabus, C., Litvak, S., Keith, G., and Darlix, J., 1988, Small finger protein of avian and murine retroviruses had nucleic acid annealing activity and positions the replication primer tRNA onto genomic RNA, *EMBO J.* **7**:1136.

Prats, A. C., Billy, G. D., Wang, P., and Darlix, J., 1989, CUG initiation codon used for the synthesis of a cell surface antigen coded by the murine leukemia virus, *J. Mol. Biol.* **205**:363.

Preston, B. D., Poiez, B. J., and Loeb, L., 1988, Fidelity of HIV-1 reverse transcriptase, *Science* **242**:1168.

Proudfoot, N. J., 1989, How RNA polymerase terminates transcription in higher eucaryotes, *Trends Biochem. Sci.* **14**:105.

Proudfoot, N. J., 1991, Poly(A) signals, *Cell* **64**:671.

Pryciak, P. M., Sil, A., and Varmus, H. E., 1992, Retroviral integration into minichromosomes *in vitro*, *EMBO J.* **11**:291.

Pryciak, P. M., and Varmus, H. E., 1992, Nucleosomes, DNA-binding proteins, and DNA sequence modulate retroviral integration target site selection, *Cell* **69**:769.

Ptashne, M., and Gann, A. A. F., 1990, Activators and targets, *Nature* **346**:329.

Pulsinelli, G. A., and Temin, H. M., 1991, Characterization of large deletions occurring during a single round of retrovirus vector replication: Novel deletion mechanism involving errors in strand transfer, *J. Virol.* **65**:4786.

Putterman, D., Pepinsky, R. B., and Vogt, V. M., 1990, Ubiquitin in avian leukosis virus particles, *Virology* **176**:633.

Quinn, T. P., and Grandgenett, D. P., 1988, Genetic evidence that the avian retrovirus DNA endonuclease domain of pol is necessary for viral integration, *J. Virol.* **62**:2307.

Quintrell, N., Hughes, S. H., Varmus, H. E., and Bishop, J. M., 1980, Structure of viral DNA and RNA in mammalian cells infected with avian sarcoma virus, *J. Mol. Biol.* **143**:363.

Rasheed, S., Gardner, M. B., and Chan, E., 1976, Amphotropic host range of naturally occurring wild mouse leukemia viruses, *J. Virol.* **19**:13.

Reddy, S., DeGregori, J. V., von Melchner, H., and Ruley, H. E., 1991, Retrovirus promoter-trap vector to induce *lacZ* gene fusions in mammalian cells, *J. Virol.* **65**:1507.

Rein, A., 1982, Interference grouping of murine leukemia viruses: A distinct receptor for MCF-recombinant viruses in mouse cells, *Virology* **120**:251.

Rein, A., McClure, M. R., Rice, N. R., Luftig, R. B., and Schultz, A. M., 1986, Myristylation site in Pr65gag is essential for virus particle formation by Moloney murine leukemia virus, *Proc. Natl. Acad. Sci. USA* **83**:7246.

Renne, R., Friedl, E., Schweizer, M., Fleps, U., Turek, R., and Neumann-Haefelin, D., 1992, Genome organization and expression of simian foamy virus type 3 (SFV-3), *Virology* **186**:597.

Repaske, R., Hartley, J. W., Kavlick, M. F., O'Neill, R. R., and Austin, J. B., 1989, Inhibition of RNase H activity and viral replication by single mutations in the 3' region of Moloney murine leukemia virus reverse transcriptase, *J. Virol.* **63**:1460.

Resnick, R., Omer, C. A., and Faras, A. J., 1984, Involvement of retrovirus reverse-transcriptase-associated RNase H in the initiation of strong-stop (+) DNA synthesis and the generation of the long terminal repeat, *J. Virol.* **51**:813.

Reuss, F., and Schaller, H. C., 1991, cDNA sequence and genomic characterization of intracisternal A-particle-related retroviral elements containing an envelope gene, *J. Virol.* **65**:5702.

Rhee, S. S., and Hunter, E., 1987, Myristylation is required for intracellular transport but not for assembly of D-type retrovirus capsids, *J. Virol.* **61**:1045.

Rhee, S. S., and Hunter, E., 1990a, A single amino acid substitution within the matrix protein of a type D retrovirus converts its morphogenesis to that of a type C retrovirus, *Cell* **63**:77.

Rhee, S. S., and Hunter, E., 1990b, Structural role of the matrix protein of type D retroviruses in gag polyprotein stability and capsid assembly, *J. Virol.* **64**:4383.

Rhee, S. S., and Hunter, E., 1991, Amino acid substitutions within the matrix protein of type D retroviruses affect assembly, transport, and membrane association of a capsid, *EMBO J.* **10**:535.

Rhee, S. S., Hui, H., and Hunter, E., 1990, Preassembled capsids of type D retroviruses contain a signal sufficient for targeting specifically to the plasma membrane, *J. Virol.* **64**:3844.

Richter, A., Ozer, H. L., DesGroseillers, L., and Jolicoeur, P., 1984, An X-linked gene affecting mouse cell DNA synthesis also affects production of unintegrated linear and supercoiled DNA of murine leukemia virus, *Mol. Cell. Biol.* **4**:151.

Ridgway, A. A. G., Kung, H., and Fujita, D., 1989, Transient expression analysis of reticuloendotheliosis virus long terminal repeat, *Nucleic Acids Res.* **17**:3199.

Roberts, J. D., Bebenek, K., and Kunkel, T. A., 1988, The accuracy of reverse transcriptase from HIV-1, *Science* **242**:11171.

Roberts, J. D., Preston, B. D., Johnston, L. A., Soni, A., Loeb, L. A., and Kunkel, T., 1989, Fidelity of two retroviral reverse transcriptases during DNA-dependent DNA synthesis *in vitro*, *Mol. Cell. Biol.* **9**:469.

Roberts, M. M., and Oroszlan, S., 1989, The preparation and biochemical characterization of intact capsids of equine infectious anemia virus, *Biochem. Biophys. Res. Commun.* **160:**486.

Roberts, M. M., Copeland, T. D., and Oroszlan, S., 1991, *In situ* processing of a retroviral nucleocapsid protein by the viral proteinase, *Protein Eng.* **4:**695.

Robinson, H., 1979, Inheritance and expression of chicken genes which are related to avian-leukosis sarcoma viruses, *Curr. Top. Microbiol. Immunol.* **83:**1.

Robinson, H. L., and Gagnon, G. C., 1986, Patterns of proviral insertion in avian leukosis virus-induced lymphomas, *J. Virol.* **57:**28.

Robinson, W. S., 1990, Hepadnaviridae and their replication, in: *Virology* (B. N. Fields, D. M. Knipe, R. M., Chanock, *et al.*, eds.), pp. 2137–2169, Raven Press, New York.

Rohdewohld, H., Weiher, H., Reik, W., Jaenisch, R., and Breindl, M., 1987, Retrovirus integration and chromatin structure: Moloney murine leukemia proviral integration sites map near DNase I-hypersensitive sites, *J. Virol.* **61:**336.

Rosen, C. A., Haseltine, W. A., Lenz, J., Ruprecht, R., and Cloyd, M. W., 1985a, Tissue selectivity of murine leukemia virus infection is determined by long terminal repeat sequences, *J. Virol.* **55:**862.

Rosen, C. A., Sodroski, J. G., and Haseltine, W. A., 1985b, Location of *cis*-acting regulatory sequences in human T cell lymphotropic virus type III (HTLV-III/LAV) long terminal repeat, *Cell* **41:**813.

Rosenberg, S. A., Aebersold, P., Cornetta, K., *et al.*, 1990, Gene transfer into humans—Immunotherapy of patients with advanced melanoma, using tumor-infiltrating lymphocytes modified by retroviral gene transduction, *N. Engl. J. Med.* **323:**570.

Roth, M. G., Srinivas, R. V., and Compans, R. W., 1983, Basolateral maturation of retroviruses in polarized epithelial cells, *J. Virol.* **45:**1065.

Roth, M. J., Tanese, N., and Goff, S. P., 1985, Purification and characterization of murine retroviral reverse transcriptase expressed in *Escherichia coli, J. Biol. Chem.* **260:**9326.

Roth, M. J., Tanese, N., and Goff, S. P., 1989a, Gene product of Moloney murine leukemia virus required for proviral integration is a DNA-binding protein, *J. Mol. Biol.* **203:**131.

Roth, M. J., Schwartzberg, P. L., and Goff, S. P., 1989b, Structure of the termini of DNA intermediates in the integration of retroviral DNA: Dependence on IN function and terminal DNA sequence, *Cell* **58:**47.

Rothenberg, E., Smotkin, D., Baltimore, D., and Weinberg, R. A., 1977, *In vitro* synthesis of infectious DNA of murine leukaemia virus, *Nature* **296:**122.

Rubin, H., 1960, A virus in chick embryos which induced resistance to *in vitro* infection by Rous sarcoma virus, *Proc. Natl. Acad. Sci. USA* **46:**1105.

Ruprecht, R. M., Mullaney, S., Bernard, L. D., Gama Sosa, M. A., Hom, R. C., and Fineberg, R. W., 1990, Vaccination with a live retrovirus: The nature of the protective immune response, *Proc. Natl. Acad. Sci. USA* **87:**5558.

Ryden, T. A., and Beemon, K., 1989, Avian retroviral long terminal repeats bind CCAAT/enhancer-binding protein, *Mol. Cell. Biol.* **9:**1155.

Sali, A., Veerapandian, B., Cooper, J. B., Foundling, S. I., Hoover, D. J., and Blundell, T. L., 1989, High-resolution X-ray diffraction study of the complex between endothiapepsin and an oligopeptide inhibitor: The analysis of the inhibitor binding and description of the rigid body shift in the enzyme, *EMBO J.* **8:**2179.

Samuel, C. E., 1989, Polycistronic animal virus RNAs, *Prog. Nucleic Acid Res.* **37:**127.

Sandemeyer, S. B., Hansen, L. J., and Chalker, D. L., 1990, Integration specificity of retrotransposons and retroviruses, *Annu. Rev. Genet.* **24:**491.

Sattentau, Q. J., and Weiss, R. A., 1988, The CD4 antigen: Physiological ligand and HIV receptor, *Cell* **52:**631.

Sawadogo, M., and Sentenac, A., 1990, RNA polymerase B (II) and general transcription factors, *Annu. Rev. Biochem.* **59:**711.

Sawyer, R. C., and Hanafusa, H., 1979, Comparison of the small RNAs of polymerase-deficient and polymerase-positive Rous sarcoma virus and another species of avian retrovirus, *J. Virol.* **29**:863.

Sawyer, R. C., Harada, F., and Dahlber, J. E., 1974, Virion-associated RNA primer for Rous sarcoma virus DNA synthesis: Isolation from uninfected cells, *J. Virol.* **28**:279.

Schatz, O., Mous, J., and LeGrice, S. F. J., 1990, HIV-1 RT-associated ribonuclease H displays both endonuclease and 3′–5′ exonuclease activity, *EMBO J.* **9**:1171.

Schawaller, M., Smith, G. E., Skehel, J. J., and Wiley, D. C., 1989, Studies with crosslinking reagents on the oligomeric structure of the env glycoprotein of HIV, *Virology* **172**:367.

Scherdin, U., Rhodes, K., and Breindl, M., 1990, Transcriptionally active genome regions are preferred targets for retrovirus integration, *J. Virol.* **64**:907.

Schimmel, P., 1989, RNA pseudoknots that interact with components of the translation apparatus, *Cell* **58**:9.

Schultz, A. M., and Oroszlan, S., 1983, *In vivo* modification of retroviral *gag* gene-encoded polyproteins by myristic acid, *J. Virol.* **46**:355.

Schultz, A. M., and Rein, A., 1989, Unmyristylated Moloney murine leukemia virus Pr65gag is excluded from virus assembly and maturation events, *J. Virol.* **63**:2370.

Schultz, A. M., Henderson, L. E., and Oroszlan, S., 1988, Fatty acylation of proteins, *Annu. Rev. Cell Biol.* **4**:611.

Schwartz, D. E., Tizard, R., and Gilbert, W., 1983, Nucleotide sequence of Rous sarcoma virus, *Cell* **32**:853.

Schwartz, S., Felber, B. K., and Pavlakis, G. N., 1992, Mechanism of translation of monocistronic and multicistronic human immunodeficiency virus type 1 mRNAs, *Mol. Cell. Biol.* **12**:207.

Schwartzberg, P., Colicelli, J., and Goff, S. P., 1983, Deletion mutants of Moloney murine leukemia virus which lack glycosylated gag protein are replication competent, *J. Virol.* **46**:538.

Schwartzberg, P., Colicelli, J., and Goff, S. P., 1984a, Construction and analysis of deletion mutants in the *pol* gene of Moloney murine leukemia virus: A new viral function required for establishment of the integrated provirus, *Cell* **37**:1043.

Schwartzberg, P., Colicelli, J., Gordon, M. L., and Goff, S. P., 1984b, Mutations in the *gag* gene of Moloney murine leukemia virus: Effects on production of virions and reverse transcriptase, *J. Virol.* **49**:918.

Sealey, L., and Chalkley, R., 1987, At least two nuclear proteins bind specifically to the Rous sarcoma virus long terminal repeat enhancer, *Mol. Cell. Biol.* **7**:787.

Seeger, C., Summers, J., and Mason, W. S., 1990, Viral DNA synthesis, in: *Hepadnaviruses—Molecular Biology and Pathogenesis* (W. S. Mason and C. Seeger, eds.), pp. 41–60, Springer-Verlag, Berlin.

Shang, F., Huang, H., Revesz, K., Chen, H. C., Herz, R., and Pinter, A., 1991, Characterization of monoclonal antibodies against the human immunodeficiency virus matrix protein, p17gag: Identification of epitopes exposed at the surfaces of infected cells, *J. Virol.* **65**:4798.

Shank, P. R., Hughes, S., Kung, H. J., Majors, J., Quintrell, N., Guntaka, R. V., Bishop, J. M., and Varmus, H. E., 1978, Mapping unintegrated avian sarcoma virus DNA: Termini of linear DNA bear 300 nucleotides present once or twice in two species of circular DNA, *Cell* **15**:1383.

Shapiro, J. A., 1979, Molecular model for the transposition and replication of bacteriophage Mu and other transposable elements, *Proc. Natl. Acad. Sci. USA* **76**:1933.

Sharp, P. A., 1987, Splicing of messenger RNA precursors, *Science* **235**:766.

Sherman, P. A., and Fyfe, J. A., 1990, Human immunodeficiency virus integration protein expressed in *Escherichia coli* possesses selective DNA cleavage activity, *Proc. Natl. Acad. Sci. USA* **87**:5119.

Shih, C., Stoye, J. P., and Coffin, J. M., 1988, Highly preferred targets for retrovirus integration, *Cell* **53**:531.

Shinnick, T., Lerner, R., and Sutcliffe, J. G., 1981, Nucleotide sequence of Moloney murine leukemia virus, *Nature* **293**:543.

Shoemaker, C. S., Goff, S. P., Gilboa, E., Paskind, M., Mitra, S. W., and Baltimore, D., 1980, Structure of a cloned circular Moloney murine leukemia virus molecule containing an inverted segment: Implications for retrovirus integration, *Proc. Natl. Acad. Sci. USA* **77**:3932.

Shoemaker, C., Hoffman, J., Goff, S. P., and Baltimore, D., 1981, Intramolecular integration within Moloney murine leukemia virus DNA, *J. Virol.* **40**:164.

Short, M. K., Okenquist, S. A., and Lenz, J., 1987, Correlation of leukemogenic potential of murine retroviruses with transcriptional tissue preference of the long terminal repeats, *J. Virol.* **61**:1067.

Simpson, R. T., 1991, Nucleosome positioning: Occurrence, mechanisms, and functional consequences, *Prog. Nucleic Acid Res.* **40**:143.

Sinn, E., Muller, W., Pattengale, P., Tepler, I., Wallace, R., and Leder, P., 1987, Coexpression of MMYT/v-Ha-*ras* and MMTV/c-*myc* genes in transgenic mice: Synergistic action of oncogenes *in vivo*, *Cell* **49**:465.

Skalka, A. M., 1989, Retroviral proteases: First glimpses at the anatomy of a processing machine, *Cell* **56**:911.

Skuzeski, J. M., Nichols, L. M., Gesteland, R. F., and Atkins, J. F., 1991, The signal for a leaky UAG stop codon in several plant viruses includes the two downstream codons, *J. Mol. Biol.* **218**:365.

Smith, J. K., Cywinski, A., and Taylor, J. M., 1984, Specificity of initiation of plus-strand DNA by Rous sarcoma virus, *J. Virol.* **52**:314.

Sommerfelt, M., and Weiss, R. A., 1990, Receptor interference groups of 20 retroviruses plating on human cells, *Virology* **176**:58.

Sommerfelt, M. A., Williams, B. P., McKnight, A., Goodfellow, P. N., and Weiss, R. A., 1990, Localization of the receptor gene for type D simian retroviruses on human chromosome 19, *J. Virol.* **64**:6214.

Sorge, J., Ricci, W., and Hughes, S. H., 1983, *Cis*-acting packaging locus in the 115-nucleotide direct repeat of Rous sarcoma virus, *J. Virol.* **48**:667.

Soriano, P., Friedrich, G., and Lawinger, P., 1991, Promoter interactions in retrovirus vectors introduced into fibroblasts and embryonic stem cells, *J. Virol.* **65**:2314.

Soriano, P., Gridley, T., and Jaenisch, R., 1989, Retroviral tagging in mammalian development and genetics, in: *Mobile DNA* (D. E. Berg and M. M. Howe, eds.), pp. 927–937, American Society for Microbiology, Washington, D.C.

Speck, N. A., and Baltimore, D., 1987, Six distinct nuclear factors interact with the 75-base-pair repeat of the Moloney murine leukemia virus enhancer, *Mol. Cell. Biol.* **7**:1101.

Speck, N. A., Renjifo, B., Golemis, B., Frederickson, T. N., Hartley, J. W., and Hopkins, N., 1990a, Mutation of the core or adjacent LVb elements of the Moloney murine leukemia virus enhancer alters disease specificity, *Genes Dev.* **4**:233.

Speck, N. A., Renjifo, B., and Hopkins, N., 1990b, Point mutations in the Moloney murine leukemia virus enhancer identify a lymphoid-specific viral core motif and 1,3-phorbol myristate acetate-inducible element, *J. Virol.* **64**:543.

Speers, W. C., Gautsch, J. W., and Dixon, F. J., 1980, Silent infection of murine embryonal carcinoma cells by Moloney murine leukemia virus, *Virology* **105**:241.

Spiro, C., Li, J., Bestwick, R. K., and Kabat, D., 1988, An enhancer sequence instability that diversifies the cell repertoire for expression of a murine leukemia virus, *Virology* **164**:350.

Stein, B. S., Gowda, S. D., Lifson, J. D., Penhallow, R. C., Bensch, K. G., and Engelman, E. G., 1987, pH-independent HIV entry into CD4-positive T cells via virus envelope fusion to the plasma membrane, *Cell* **49**:659.

Stephens, E. B., and Compans, R. W., 1986, Nonpolarized expression of a secreted murine leukemia virus glycoprotein in polarized epithelial cells, *Cell* **47**:1053.

Stephenson, J. R., 1980, *Molecular Biology of RNA Tumor Viruses*, Academic Press, New York.

Stewart, L., Schatz, G., and Vogt, V. M., 1990, Properties of avian retrovirus particles defective in viral protease, *J. Virol.* **64**:5076.

Stewart, L., and Vogt, V. M., 1991, *Trans*-acting viral protease is necessary and sufficient for activation of avian leukosis virus reverse transcriptase, *J. Virol.* **65**:6218.

Stewart, R. A., Hollingshead, P. G., and Pitts, S. L., 1988, Multiple regulatory domains in the mouse mammary tumor virus long terminal repeat revealed by analysis of fusion genes in transgenic mice, *J. Virol.* **8**:473.

Stoll, E., Billeter, M. A., Palmenberg, A., and Weissmann, C., 1977, Avian myeloblastosis virus RNA is terminally redundant: Implications for the mechanism of retrovirus replication, *Cell* **12**:57.

Stoltzfus, C. M., 1988, Synthesis and processing of avian sarcoma retrovirus RNA, *Adv. Virus Res.* **35**:1.

Stoltzfus, C. M., and Fogarty, C. J., 1989, Multiple regions in the Rous sarcoma virus *src* gene intron act in *cis* to affect the accumulation of unspliced RNA, *J. Virol.* **63**:1669.

Stoltzfus, C. M., Lorenzen, S. K., and Berberich, S. L., 1987a, Noncoding region between the *env* and *src* genes of Rous sarcoma virus influences splicing efficiency of the *src* gene 3' splice site, *J. Virol.* **61**:177.

Stoltzfus, C. M., Chang, L. J., Cripe, T. P., and Turek, L. P., 1987b, Efficient transformation by Prague A Rous sarcoma virus plasmid DNA requires the presence of *cis*-acting regions within the *gag* genes, *J. Virol.* **61**:3401.

Subramani, S., and Berg, P., 1983, Homologous and nonhomologous recombination in monkey cells, *Mol. Cell. Biol.* **3**:11040.

Suguna, K., Padlan, E. A., Smith, C. W., Carlson, W. D., and Davies, D. R., 1987, Binding of a reduced peptide inhibitor to the aspartic proteinase from *Rhizopus chinensis:* Implications for a mechanism of action, *Proc. Natl. Acad. Sci. USA* **84**:7009.

Summers, J., and Mason, W. S., 1982, Replication of the genome of a hepatitis B-like virus by reverse transcription of an RNA intermediate, *Cell* **29**:403.

Swain, A. L., Miller, M. M., Green, J., Rich, D. H., Schneider, J., Kent, S. B. H., and Wlodawer, A., 1990, X-ray crystallographic structure of a complex between a synthetic protease of human immunodeficiency virus 1 and a substrate-based hydroxyethylamone inhibitor, *Proc. Natl. Acad. Sci. USA* **87**:8805.

Swanstrom, R., and Vogt, P. K., eds., 1990, *Retroviruses—Strategies of Replication*, Springer-Verlag, Berlin.

Swanstrom, R., Varmus, H. E., and Bishop, J. M., 1982, Nucleotide sequence of the 5' noncoding region and part of the *gag* gene of Rous sarcoma virus, *J. Virol.* **41**:535.

Swanstrom, R., Kaplan, A. H., and Manchester, M., 1990, The aspartic proteinase of HIV-1, in: *Retrovirus Genome Organization and Gene Expression* (I. S. Y. Chen, ed.), pp. 175–186, Saunders, Philadelphia, Pennsylvania.

Tabin, C. J., Hoffman, J. W., Goff, S. P., and Weinberg, R. A., 1982, Adaptation of a retrovirus as a eucaryotic vector in transmitting the herpes simplex virus thymidine kinase gene, *Mol. Cell. Biol.* **2**:426.

Taketo, M., and Tanaka, M., 1987, A cellular enhancer of retrovirus gene expression in embryonal carcinoma cells, *Proc. Natl. Acad. Sci. USA* **84**:3748.

Taketo, M., Gilboa, E., and Sherman, M. I., 1985, Isolation of embryonal carcinoma cell lines that express integrated recombinant genes flanked by the Moloney murine leukemia virus long terminal repeat, *Proc. Natl. Acad. Sci. USA* **82**:2422.

Takeuchi, Y., Vile, R. G., Simpson, G., O'Hara, B., Collins, M. K. L., and Weiss, R. A., 1992, Feline leukemia virus subgroup B uses the same cell surface receptor as gibbon ape leukemia virus, *J. Virol.* **66**:1219.

Takeya, T., and Hanafusa, H., 1983, Structure and sequence of the cellular gene homologous to the RSV src gene and the mechanism for generating transforming virus, *Cell* **32**:881.

Tan, G. T., Pezzuto, J. M., Kinghorn, A. D., and Hughes, S. H., 1990, Evaluation of natural products as inhibitors of human immunodeficiency virus type 1 (HIV-1) reverse transcriptase, *J. Nat. Prod.* **54**:143.

Tanese, N., and Goff, S. P., 1988, Domain structure of the Moloney murine leukemia virus reverse transcriptase: Mutational analysis and separate expression of the DNA polymerase and RNase H activities, *Proc. Natl. Acad. Sci. USA* **85**:1777.

Tanese, N., Roth, M. J., and Goff, S. P., 1986, Analysis of retroviral *pol* gene products with antisera raised against fusion proteins produced in *Escherichia coli, J. Virol.* **59**:328.

Tanese, N., Telesnitsky, A., and Goff, S. P., 1991, Abortive reverse transcription by mutants of Moloney murine leukemia virus deficient in the reverse transcriptase-associated RNase H function, *J. Virol.* **65**:4387.

Taylor, J. M., 1977, An analysis of the role of tRNA species as primers for the transcription into DNA of RNA tumor virus genomes, *Biochim. Biophys. Acta* **473**:57.

Taylor, J. M., and Illmensee, R., 1975, Site on the RNA of an avian sarcoma virus at which primer is bound, *J. Virol.* **16**:553.

Teich, N., 1982, Taxonomy of retroviruses, in: *RNA Tumor Viruses* (R. Weiss, N. Teich, H. Varmus, and J. Coffin, eds.), pp. 25–207, Cold Spring Harbor Laboratory, Cold Spring Harbor, New York.

Teich, N., 1985, Taxonomy of retroviruses, in: *RNA Tumor Viruses—Supplements and Appendixes* (R. Weiss, N. Teich, H. Varmus, and J. Coffin, eds.), pp. 1–16, Cold Spring Harbor Laboratory, Cold Spring Harbor, New York.

Teich, N. M., Weiss, R. A., Martin, G. R., and Lowy, D. R., 1977, Virus infection of murine teratocarcinoma stem cell lines, *Cell* **12**:973.

Teich, N., Wyke, J., Mak, T., Bernstein, A., and Hardy, W., 1982, Pathogenesis of retrovirus-induced disease, in: *RNA Tumor Viruses* (R. Weiss, N. Teich, H. Varmus, and J. Coffin, eds.), pp. 785–998, Cold Spring Harbor Laboratory, Cold Spring Harbor, New York.

Teich, N., Wyke, J., and Kaplan, P., 1985, Pathogenesis of retrovirus-induced disease, in: *RNA Tumor Viruses—Supplements and Appendixes* (R. Weiss, N. Teich, H. Varmus, and J. Coffin, eds.), pp. 187–248, Cold Spring Harbor Laboratory, Cold Spring Harbor, New York.

Temin, H. M., 1964, Nature of the provirus of Rous sarcoma virus, *Natl. Cancer Inst. Monogr.* **17**:557.

Temin, H. M., 1976, The DNA provirus hypothesis, *Science* **192**:1075.

Temin, H. M., 1985, Reverse transcription in the eukaryotic genome: Retroviruses, para-retroviruses, retrotransposons, and retrotranscripts, *Mol. Biol. Evol.* **6**:455.

Temin, H. M., 1988, Evolution of retroviruses and other retrotranscripts, in: *Human Retroviruses, Cancer, and AIDS: Approaches to Prevention and Therapy* (D. Bolognesi, ed.), pp. 1–28, Alan R. Liss, New York.

Temin, H. M., 1989, Retrovirus variation and evolution, *Genome* **31**:17.

Temin, H. M., and Baltimore, D., 1972, RNA-directed DNA synthesis and RNA tumor viruses, *Adv. Virus Res.* **17**:129.

Temin, H. M., and Mizutani, S., 1970, RNA-directed DNA polymerase in virions of Rous sarcoma virus, *Nature* **226**:1211.

Ten Dam, E. B., Pleij, C. W. A., and Bosch, L., 1990, RNA pseudoknots: Translational frameshifting and readthrough on viral RNAs, *Virus Genes* **4**:1211.

Thayer, R. M., Power, M. D., Bryant, M. L., Gardner, M. B., Barr, P. J., and Luciw, P. A., 1987, Sequence relationships of type D retroviruses which cause simian acquired immunodeficiency syndrome, *Virology* **157**:317.

Thornell, A., Halberg, B., and Grundstrom, T., 1988a, Differential protein binding in lymphocytes to a sequence in the enhancer of the mouse retrovirus SL3-3, *Mol. Cell. Biol.* **65**:42.

Thornell, A., Halberg, B., and Grundstrom, T., 1988b, Binding of SL3-3 enhancer factor transcriptional activators to viral and chromosome enhancer sequences, *J. Virol.* **65**:42.

Thornell, A., Hallberg, B., and Grundstrom, T., 1991, Binding of SL3-3 enhancer factor 1 transcriptional activators to viral and chromosomal enhancer sequences, *J. Virol.* **65**:42.

Tirumalai, R. S., and Modak, M. J., 1991, Photoaffinity labeling of the primer binding domain in murine leukemia virus reverse transcriptase, *Biochemistry* **30**:6436.

Toh, H., Hayashida, H., and Miyata, T., 1983, Sequence homology between retroviral reverse transcriptase and putative polymerases of hepatitis B virus and cauliflower momsaic virus, *Nature* **305**:827.

Toh, H., Kikuno, R., Hayashida, H., Miyata, T., Kugimiya, W., Inouye, S., Yuki, S., and Saigo, K., 1985, Close structural resemblance between putative polymerase of a *Drosophila* transposable genetic element 17.6 and pol gene product of Moloney murine leukemia virus, *EMBO J.* **4**:1267.

Tong-Starksen, S., and Peterlin, B. M., 1990, Mechanisms of retroviral transcriptional activation, in: *Retrovirus Genome Organization and Gene Expression* (I. S. Y. Chen, ed.), pp. 215–227, Saunders, Philadelphia, Pennsylvania.

Tong-Starksen, S., Luciw, P. A., and Peterlin, B. M., 1987, Human immunodeficiency virus long terminal repeat responds to T-cell activation signals, *Proc. Natl. Acad. Sci. USA* **84**:6845.

Tounekti, N., Mougel, M., Roy, C., Marquet, R., Darlix, J. L., Paoletti, J., Ehresmann, B., and Ehresmann, C., 1992, Effect of dimerization on the conformation of the encapsidation Psi domain of Moloney murine leukemia virus RNA, *J. Mol. Biol.* **223**:205.

Towler, D., and Glaser, L., 1986, Protein fatty acid acylation: Enzymatic synthesis of an N-myristoylglycyl peptide, *Proc. Natl. Acad. Sci. USA* **83**:2812.

Tsichlis, P. N., and Lazo, P. A., 1991, Virus–host interactions and the pathogenesis of murine and human oncogenic retroviruses, in: *Retroviral Insertion and Oncogene Activation*, (H. J. Kung and P. K. Vogt), pp. 95–172, Springer-Verlag, Berlin.

Tsukiyama, T., Niwa, O., and Yokoro, K., 1989, Mechanism of suppression of the long terminal repeat of Moloney leukemia virus in mouse embryonal carcinoma cells, *Mol. Cell. Biol.* **9**:4670.

Tucker, S. P., Srinivas, R. V., and Compans, R. W., 1991, Molecular domains involved in oligomerization of the Friend murine leukemia virus envelope glycoprotein, *Virology* **185**:710.

Valsamakis, A., Zeichner, S., Carswell, S., and Alwine, J. C., 1991, The human immunodeficiency virus type 1 polyadenylation signal: A 3′ long terminal repeat element upstream of the AAUAA necessary for efficient polyadenylation, *Proc. Natl. Acad. Sci. USA* **88**:2108.

Varmus, H. E., 1988, Retroviruses, *Science* **240**:1427.

Varmus, H., and Brown, P., 1989, Retroviruses, in: *Mobile DNA* (D. E. Berg and M. M. Howe, eds.), pp. 53–108, American Society for Microbiology, Washington, D.C.

Varmus, H., and Swanstrom, R., 1982, Replication of retroviruses, in: *RNA Tumor Viruses* (R. Weiss, N. Teich, H. Varmus, and J. Coffin, eds.), pp. 369–512, Cold Spring Harbor Laboratory, Cold Spring Harbor, New York.

Varmus, H., and Swanstrom, R., 1985, Replication of retroviruses, in: *RNA Tumor Viruses—Supplements and Appendixes* (R. Weiss, N. Teich, H. Varmus, and J. Coffin, eds.), pp. 75–134, Cold Spring Harbor Laboratory, Cold Spring Harbor, New York.

Varmus, H. E., Padgett, T., Heasley, S., Simon, G., and Bishop, J. M., 1977, Cellular functions are required for the synthesis and integration of avian sarcoma virus-specific DNA, *Cell* **11**:307.

Varmus, H. E., Quintrell, N., and Ortiz, S., 1981, Retroviruses as mutagens: Insertion and excision of a non-transforming provirus alters expression of a resident transforming provirus, *Cell* **25**:23.

Verma, I. M., 1977, The reverse transcriptase, *Biochem. Biophys. Acta* **473**:1.

Verma, I. M., 1990, Gene therapy, *Sci. Am.* **263**:68.

Vijaya, S., Steffen, D. L., and Robinson, H. L., 1986, Acceptor sites for retroviral integration map near DNase I-hypersensitive sites in chromatin, *J. Virol.* **60**:683.

Vogt, V. M., Eisenman, R. H., and Diggelmann, H., 1975, Generation of avian myeloblastosis virus structural proteins by proteolytic cleavage of a precursor polypeptide, *J. Mol. Biol.* **96**:471.

Vora, A. C., Fitzgerald, M. L., and Grandgenett, D. P., 1990, Removal of 3′-OH-terminal nucleotides from blunt-ended long terminal repeat termini by the avian retrovirus integration protein, *J. Virol.* **64**:5656.

Wang, H., Kavanaugh, M. P., North, A., and Kabat, D., 1991a, Cell-surface receptor for ecotropic murine retroviruses is a basic amino acid transporter, *Nature* **352**:729.

Wang, H., Paul, R., Burgeson, R. E., Keene, D. R., and Kabat, D., 1991b, Plasma membrane receptors for ecotropic murine retroviruses require a limiting accessory factor, *J. Virol.* **65**:6468.

Wang, J. C., 1985, DNA topoisomerases, *Annu. Rev. Biochem.* **54**:665.

Wang, S., and Speck, N. A., 1992, Purification of core-binding factor, a protein that binds the conserved core site in murine leukemia virus enhancers, *Mol. Cell. Biol.* **12**:89.

Watanabe, S., and Temin, H. M., 1979, Encapsidation sequences for spleen necrosis virus, an avian retrovirus, are between the 5′ long terminal repeat and the start of the *gag* gene, *Proc. Natl. Acad. Sci. USA* **79**:5986.

Watanabe, S., and Temin, H. M., 1982, Encapsidation sequences for spleen necrosis virus, an avian retrovirus, are between the 5′ long terminal repeat and the start of the *gag* gene, *Proc. Natl. Acad. Sci. USA* **79**:5986.

Weaver, T. A., and Panganiban, A. T., 1990, N-myristoylation of the spleen necrosis virus matrix protein is required for correct association of the gag polyprotein with intracellular membranes and for particle formation, *J. Virol.* **64**:3995.

Weber, F., and Schaffner, W., 1985, Enhancer activity correlates with the oncogenic potential of avian retroviruses, *EMBO J.* **4**:949.

Weber, H., Barklis, E., Ostertag, W., and Jaenisch, R., 1987, Two distinct sequence elements mediate retroviral gene expression in embryonal carcinoma cells, *J. Virol.* **243**:928.

Weber, I. T., 1990, Comparison of the crystal structures and intersubunit interactions of human immunodeficiency and Rous sarcoma virus proteases, *J. Biol. Chem.* **265**:10492.

Weichs an der Glon, C., Monks, J., and Proudfoot, N. J., 1991, Occlusion of the HIV poly(A) site, *Genes Dev.* **5**:244.

Weiher, H., Barklis, E., Ostertag, W., and Jaenisch, R., 1987, Two distinct sequence elements mediate retroviral gene expression in embryonal carcinoma cells, *J. Virol.* **61**:2742.

Weiner, A. M., Deininger, P. L., and Efstratiadis, A., 1986, Nonviral retroposons: Genes, pseudogenes, and transposable elements generated by the reverse flow of genetic information, *Annu. Rev. Biochem.* **55**:631.

Weiss, R., 1982, Experimental biology and assay of RNA tumor viruses, in: *RNA Tumor Viruses* (R. Weiss, N. Teich, H. Varmus, and J. Coffin, eds.), pp. 209–260, Cold Spring Harbor Laboratory, Cold Spring Harbor, New York.

Weiss, R. A., 1991, Receptors for human retroviruses, in: *The Human Retroviruses* (R. C. Gallo and G. Jay, eds.), pp. 127–139, Academic Press, New York.

Weiss, R. A., in press, Receptors and glycoproteins involved in retrovirus entry, in: *The Retroviridae, Volume 2* (J. A. Levy, ed.), Plenum Publishing Corporation, New York.

Weiss, R. A., Mason, W. S., and Vogt, P. K., 1973, Genetic recombinants and heterozygotes derived from endogenous and exogenous avian RNA tumor viruses, *Virology* **52**:535.

Weiss, R., Teich, N., Varmus, H., and Coffin, J., eds., 1982, *RNA Tumor Viruses*, Cold Spring Harbor Laboratory, Cold Spring Harbor, New York.

Weiss, R., Teich, N., Varmus, H., and Coffin, J., eds., 1985, *RNA Tumor Viruses—Supplements and Appendices*, 2nd ed., Cold Spring Harbor Laboratory, Cold Spring Harbor, New York.

White, J. M., 1990, Viral and cellular membrane fusion proteins, *Annu. Rev. Physiol.* **52**:675.

Wilcox, C., Hu, J. S., and Olson, E. N., 1987, Acylation of proteins with myristic acid occurs cotranslationally, *Science* **238**:1275.

Williams, D. A., Orkin, S. H., and Mulligan, R. C., 1986, Retrovirus-mediated transfer of human adenosine deaminase gene sequences into cells in culture and into murine hematopoietic cells *in vivo*, *Proc. Natl. Acad. Sci. USA* **83**:2566.

Wills, J. W., and Craven, R. C., 1991, Form, function, and use of retroviral gag proteins, *AIDS* **5**:639.

Wills, J. W., Shrinivas, R. V., and Hunter, E., 1984, Mutations of the Rous sarcoma virus env gene that affect the transport and subcellular location of the glycoprotein products, *J. Cell Biol.* **99**:2011.

Wills, J. W., Craven, R. C., and Achacoso, J. A., 1989, Creation and expression of myristylated forms of Rous sarcoma virus gag protein in mammalian cells, *J. Virol.* **63**:4331.

Wills, N. M., Gesteland, R. F., and Atkins, J. F., 1991, Evidence that a downstream pseudoknot is required for translational read-through of the Moloney murine leukemia virus gag stop codon, *Proc. Natl. Acad. Sci. USA* **88**:6991.

Wilson, C. A., and Eiden, M. V., 1991, Viral and cellular factors governing hamster cell infection by murine and gibbon ape leukemia viruses, *J. Virol.* **65**:5975.

Wlodawer, A., Miller, M., Jaskolski, M., Sathyanarayana, B. K., Baldwin, E., Weber, I. T., Selk, L. M., Clawson, L., Schneider, J., and Kent, S. B. H., 1989, Conserved folding in retroviral proteases: Crystal structure of a synthetic HIV-1 protease, *Science* **245**:616.

Yang, W. K., Kiggans, J. O., Yang, D. M., Ou, C. Y., Tennant, R. W., Brown, A., and Bassin, R. H., 1980, Synthesis and circularization of N- and B-tropic retroviral DNA in Fv-1 permissive and restrictive mouse cells, *Proc. Natl. Acad. Sci. USA* **77**:2994.

Yoshimoto, T., Yoshimoto, E., and Meruelo, D., 1991, Molecular cloning and characterization of a novel human gene homologous to the murine ecotropic retroviral receptor. *Virology* **185**:10.

Yoshinaka, S., Katoh, I., Copeland, T. D., and Oroszlan, S., 1985, Murine leukemia protease is encoded by the *gag-pol* gene and is synthesized through suppression of an amber termination codon, *Proc. Natl. Acad. Sci. USA* **82**:1618.

Yoshinaka, Y., and Luftig, 1977, Murine leukemia virus morphogenesis: Cleavage of P70 *in vitro* can be accompanied by a shift from a concentrically coiled internal strand ("immature") to a collapsed ("mature") form of the virus core, *Proc. Natl. Acad. Sci. USA* **74**:3446.

Young, J. A. T., Bates, P., Willert, K., and Varmus, H. E., 1990, Efficient incorporation of human CD4 protein into avian leukosis virus particles, *Science* **250**:1421.

Young, R. A., 1991, RNA polymerase II, *Annu. Rev. Biochem.* **60**:689.

Yuki, S., Ishimaru, S., Inouye, S., and Saigo, K., 1986, Identification of genes for reverse transcriptase-like enzymes in two *Drosophila* retrotransposons, 412 and gypsy; a rapid detection method of reverse transcriptase genes using YXDD box probes, *Nucleic Acids Res.* **14**:3017.

Zack, J. A., Arrigo, S. J., Weitsman, S. R., Go, A. S., Haislip, A., and Chen, I. S. Y., 1990, HIV-1 entry into quiescent primary lymphocytes: Molecular analysis reveals a labile, latent viral structure, *Cell* **61**:213.

Zavada, J., 1972, Pseudotypes of vesicular stomatitis virus with the coat of murine leukaemia and of avian myeloblastosis viruses, *J. Gen. Virol.* **15**:183.

CHAPTER 6

Biology of Avian Retroviruses

LAURENCE N. PAYNE

I. CLASSIFICATION OF AVIAN RETROVIRUSES

Avian retroviruses have, in the past, been placed taxonomically in a subgenus termed vernacularly "avian type C oncoviruses" of the genus type C oncovirus group, within the subfamily *Oncovirinae*, family Retroviridae (Matthews, 1982). The recent new classification of retroviruses place them into a separate genus (see Coffin, Chapter 2). Four distinct "species" of avian retrovirus have been described (Matthews, 1982; Porterfield, 1989):

1. Leukosis-sarcoma group viruses (also termed avian sarcoma and leukemia viruses). These are exogenous and endogenous viruses recognized mainly in the domestic fowl and which cause a variety of leukotic disorders, sarcomas, and other tumors.
2. Reticuloendotheliosis viruses. These are exogenous viruses in several species of domesticated poultry and appear to be related to mammalian retroviruses. They cause lymphomas and acute reticulum cell and other tumors.
3. Lymphoproliferative disease virus of turkeys. This is an exogenous virus of turkeys which causes a lymphoproliferative disease.
4. Pheasant type C oncoviruses. These are endogenous viruses of golden and Lady Amhurst pheasants and are apparently non-

LAURENCE N. PAYNE • AFRC Institute for Animal Health, Compton Laboratory, Compton, Newbury, Berkshire, RG16 0NN England.

The Retroviridae, Volume 1, edited by Jay A. Levy. Plenum Press, New York, 1992.

pathogenic. They are unrelated to the endogenous leukosis-sarcoma group viruses which also occur in certain species of pheasants.

II. HISTORY

A. Introduction

The study of avian retroviruses has been driven by two forces, first, a search, conducted mainly by biomedical scientists, for an infectious basis for avian leukoses and solid tumors, as part of the intellectual endeavor to understand the cause of neoplasia in humans and other animals, and second, an attempt, mainly by veterinary scientists, to control a group of disorders of great economic importance to the poultry industry. Much interchange between these two streams of work has been a continuing feature. Here an attempt will be made to summarize the main events, and mention many of the main players, that feature in the history of avian retrovirology. Other historical reviews are provided by Burmester and Purchase (1979), Gross (1983, pp. 123–252), Svoboda (1986), and Dougherty (1987).

B. Classical Studies 1908–1940

1. Transmission of Avian Leukosis

This period is characterized by demonstrations of the transmissibility of many avian neoplasms by cell-free filtrates, descriptions of the disorders caused, and preliminary characterization of the filtrable agents. The first evidence for a viral etiology of avian leukosis, and indeed for a neoplasm in any species, was provided by Ellermann and Bang (1908), a physician and a veterinarian, respectively. Working in Copenhagen, they transmitted erythroleukemia and myelogenous leukemia by inoculation of chickens with cell-free filtrates. Ellermann (1921) developed eight strains of leukosis virus, designated A–H, which he passaged up to 12 times in chickens.

Ellermann (1921) is notable also in providing a classification of the pathological forms of avian leukosis which essentially still applies. He described (1) "lymphatic leucosis," with increase of lymphoblasts, (2) "myeloic leucosis," involving myelocytes, "large mononuclear cells," and "poikilonuclear cells," and (3) "intravascular lymphoid leukosis," involving "lymphoidocytes" (he correctly went on to conclude that lymphoidocytes are erythrocytic cells and that this intravascular form must be regarded as erythroleukosis). Ellermann adopted the term leucosis, rather than leucemia, for these diseases because of the aleukemic blood picture in some tumorous cases and the occurrence of "pure anaemia" in others.

Ellermann and Bang's (1908) observations on the transmissibility of fowl leukosis were soon confirmed by others (see Engelbreth-Holm, 1942) but interest waned, partly because of uncertainty that leukemia was a neoplastic disorder. However, attention returned in the late 1920s, and signal work was done by Furth (1929) in the United States, Jármai (1929) in Hungary, Engelbreth-Holm (1931–1932) in Denmark, and Oberling and Guérin (1933a) in France. An important question arising from Ellermann's work was whether the three types of leukosis were expressions of infection by a single agent. In general, lymphoid leukosis was not obviously transmissible, whereas erythroblastosis (erythroid leukosis) and myeloblastosis (myeloid leukosis) were readily so, occurring in either pure or mixed forms within transmission experiments. Pure erythroblastosis strains included those of Jármai (1930–1931), Stubbs and Furth (1932), and Engelbreth-Holm and Rothe Meyer (1935), and mixed strains included strain 1 of Furth (1931), strain R of Engelbreth-Holm (1932), and strains of Oberling and Guérin (1934). Nyfeldt's (1933) strain produced pure myeloblastosis. During this period, strains of leukosis virus were isolated which also caused sarcomas or endotheliomas. Examples were a strain of Oberling and Guérin (1933b), strain 2 of Furth (1933), strain E-S of Rothe Meyer and Engelbreth-Holm (1933), and strain 13 of Stubbs and Furth (1935).

2. Transmission of Avian Sarcomas

At this juncture, it is appropriate to return to a second direction of work taking place during the early decades, namely, studies on the transmissibility of avian sarcomas.

In 1909, Rous, working in New York, successfully transplanted a spontaneously occurring spindle-celled sarcoma found in an adult Plymouth Rock hen into one of two other hens (Rous, 1910), and subsequently established it by chicken passage as a transplantable tumor which he termed "Chicken Tumor I." Early experiments (Rous, 1911) showed that the tumor could also be transmitted by cell-free filtrates, and the agent became designated the Rous No. 1 virus and later Rous sarcoma virus (RSV). Two other filtrable tumors were also established, Chicken Tumor VII, an osteochondrosarcoma, and Chicken Tumor XVIII, a spindle-celled sarcoma of peculiar intracanalicular pattern (Rous and Murphy, 1914). At about the same time Fujinami and Inamoto (1914) in Japan transmitted a transplantable chicken myxosarcoma with filtrates, and over the next two decades some 20 transplantable tumors of fowl were shown, in a number of laboratories, to be filtrable (Claude and Murphy, 1933).

As with avian leukosis, opinions were divided about the true nature of these tumors of fowl and their relation to malignant neoplasms in mammals. Begg (1927) summarized it thus:

> A widely entertained opinion is that these fowl neoplasms are a peculiar group of virus infections of connective tissues, comparable with, and most closely related to, the epithelial overgrowth of fowl and pigeon-pox. Their similarity to mammalian sarcomata is on this view superficial and deceptive. Those who hold the contrary opinion that, viz., these neoplasms are true sarcomata, essentially the same aetiologically and biologically as the corresponding new growths of mammals, differ among themselves, some holding that a living infectious agent (virus) need not be postulated, pinning their faith to peculiarities of enzyme action.

3. Further Early Studies

Studies on the transmissibility and relationships between the leukoses continued, including the added question of whether neurolymphomatosis (fowl paralysis, now termed Marek's disease) was etiologically related to the leukoses. This latter debate (L. N. Payne, 1985a) continued until the 1960s when Marek's disease, a herpesvirus-induced lymphomatous disorder, was clearly dissociated from the retrovirus-induced conditions. Some workers such as Furth believed that each transmissible strain represented an etiological unit because of its more or less definite, if occasionally wide, pathological range. Others adopted a unitarian viewpoint, believing that the erythroid, myeloid, and lymphoid leukoses, including the nerve and other forms, were expressions of infection by the same agent (Patterson, 1936). During this period, transmission studies brought the bone disorder osteopetrosis into the etiological ambit of the leukoses (Jungherr and Landauer, 1938). Veterinary attention was focused on natural transmission of the leukoses, particularly the occurrence of egg transmission and contact infection. The confusion between the leukoses and Marek's disease, mentioned above, ensured that no clear answers emerged.

During this period also preliminary work was done on the physical, chemical, and antigenic properties of the leukosis viruses (Engelbreth-Holm, 1942) and Rous sarcoma "agent" (R. J. C. Harris, 1953): they had yet to be purified or visualized under the electron microscope.

C. The Middle Period 1941–1960

1. Studies on Lymphoid Leukosis

This period is marked by the progress made by a number of larger groups of workers, particularly in the United States, who directed their efforts at particular transmissible tumors and their agents. Notable were the groups of B. R. Burmester (on visceral lymphomatosis, now termed lymphoid leukosis), J. W. Beard (on myeloblastosis and erythroblastosis), and W. R. Bryan (on Rous sarcoma). Significant advances were made on the morphological and biochemical characterization of the agents, on virus assay, and on virus–host interactions. Toward the end of this period, the foundations were laid by H. Rubin for studies on virus–cell interactions which were to alter the direction of subsequent studies.

In spite of progress that had been made in demonstrating the transmissibility of a number of avian leukoses and solid tumors, similar success was elusive with respect to lymphomatous disorders, the focus of greatest economic importance. In consequence, the U. S. Department of Agriculture established in 1939 the Regional Poultry Research Laboratory (RPL) at East Lansing, Michigan. Its first project was "to determine the cause of fowl paralysis and to develop measures for its prevention and control" (Anon, 1946). Initially a number of transplantable lymphoid tumor strains were developed. The best known was the RPL12 tumor strain, which had been developed originally by Olson (1941) from a hen with a lymphoid tumor. Burmester et al. (1946) were able to isolate from this tumor a filtrable agent, RPL12 virus, which became the prototype strain of lymphoid leukosis virus (LLV). This virus was shown to cause mainly erythroblastosis when given in high doses, and lymphoid leukosis ("visceral lymphomatosis") when given in low doses (Burmester and Gentry, 1956). In addition, RPL12 virus also caused osteopetrosis and hemangiomas (Gross et al., 1959). Other important host–virus interrelations were studied, including effects of virus dose, age of host, route of inoculation, source of virus, and host genetic constitution (Burmester et al., 1959a, 1960). Many similar isolates were subsequently made from field cases of avian lymphomatosis (Burmester and Fredrickson, 1964). In epizootiological studies, the occurrence of egg transmission (Cottral et al., 1954; Burmester and Waters, 1955) and contact transmission (Burmester and Gentry, 1954a) were clearly established.

2. Studies on Myeloblastosis and Erythroblastosis

During this period, Beard and his co-workers carried out extensive studies of the BAI (Bureau of Animal Industry) strain A virus, derived originally by W. J. Hall et al. (1941) from a mixture of material from two chickens with neurolymphomatosis (a herpesvirus-induced condition). This strain produced mainly myeloblastosis, but also at times lymphomatosis, nephroblastoma, osteopetrosis, sarcoma, ovarian carcinoma, endothelioma, and mesothelioma (Beard, 1963a, b). The extensive studies of Beard's group on strain A avian myeloblastosis virus (AMV) emphasized host pathological responses, and quantitative virus assays by physical, infectivity, and enzymatic means. Similar studies were conducted by Beard with Engelbreth-Holm's erythroblastosis strain R. A notable aspect of these studies was the early use of the electron microscope, to characterize the size and morphology of myeloblastosis and erythroblastosis viruses (Sharp et al., 1952; Benedetti et al., 1956; Bernard et al., 1958) and the ultrastructure of the associated neoplasms (Beard, 1963a; Dalton and Haguenau, 1962).

3. Studies on Rous Sarcoma

Meanwhile, studies had continued on the Rous sarcoma as a model for chicken tumors, although knowledge of the nature and properties of the agent was fairly rudimentary. R. J. C. Harris (1953), in his review, wrote that "it is by no means generally agreed that the actual Rous agent has ever been prepared" or "unequivocally detected microscopically." Shortly thereafter, however, Bernard et al. (1953) firmly demonstrated virus-like particles in Rous sarcoma cells, and Gaylord (1955) described their morphology in thin sections. Particularly significant also was the work of Bryan and his colleagues over a decade or more on precise quantification of the virus–host response with RSV and on virus purification (Bryan, 1957; Bryan and Moloney, 1957).

During this period, the earlier, somewhat disparate interests of medical research workers on RSV and veterinary research workers on the leukoses came closer together. The interests of Beard's group in the leukoses were an important link, and a number of collaborative studies resulted (Beard, 1957, 1963a, b).

Although virus–host studies using methodology developed during this period extended well into the 1960s, a new era of research in avian retrovirology was initiated by Rubin and his co-workers, who developed techniques for studying virus–host interactions at the cellular level. As early as 1941, Halberstaedter et al. (1941) had observed that chick embryo cells grown in tissue culture could be transformed by the Rous sarcoma agent. Manaker and Groupé (1956) described discrete foci of altered cells in RSV-infected cultures, and Temin and Rubin (1958) introduced the use of an agar overlay to provide a quantitative assay for RSV based on focus formation. This development represented a turning point: it led to the discovery of a number of virus–cell phenomena which shed light on the mechanisms of oncogenesis by avian retroviruses. These findings were complemented by detailed knowledge of the structure of retroviruses resulting from improved techniques for virus purification and the application of methods of biochemical and genetic analysis.

D. Research Since 1960

1. Rous Sarcoma Virus–Cell Interactions

The period since 1960 may be regarded as the modern era of avian retrovirology. Very generally, in terms of interest and effort, work up to the mid-1970s was concerned with recognition of a number of important phenomena involved in virus–cell interactions, and on the biochemistry of the avian retroviruses; from the mid-1970s the emphasis has been heavily on explanation of these features in terms of molecular biology.

The development of a cell culture system for studying infection and transformation by RSV, mentioned above, and the existence of a variety of different strains of RSV (Fig. 1), provided the starting point for much that followed. Much of the outstanding progress came from the research groups of H. Hanafusa, H. Rubin, H. M. Temin, and P. K. Vogt.

At this time, the concepts of defectiveness of RSV, and of helper viruses, were being elucidated (Vogt, 1965). Some foci of RSV-transformed cells, when produced by low multiplicities of virus, were found not to produce infectious RSV. Stocks of RSV contained a nontransforming "Rous associated virus" (RAV) which was able, when dually infected with RSV, to allow completion of RSV synthesis. RSV was thus regarded as a defective virus. It was capable of transforming cells, but needed RAV for its replication (H. Hanafusa et al., 1963). Nonproducer c... l viral coat protein, and the helper virus specified the antigenicity of the RSV. The Bryan high-titer strain of RSV in which RAV was detected was found to be even more heterogeneous, in that two other RAVs, RAV-2 and RAV-3, were also isolated, each determining the antigenicity of their associated RSVs. These were designated RSV(RAV-1), RSV(RAV-2) and RSV(RAV-3). RSV(RAV-1) and RSV(RAV-2), and their respective RAVs, were antigenically dissimilar. Furthermore, their interference patterns differed, with RAV-1 interfering with the ability of RSV(RAV-1) to infect and transform cells, but not with that of RSV(RAV-2), and conversely for RAV-2. These viruses also differed in ability to infect and transform genetically selectively resistant CEF (H. Hanafusa, 1965). These three helper-controlled properties of RSV, namely antigenicity, interference, and host range, defined subgroups A and B of RSV (Vogt and Ishizaki, 1966a). Subsequently, using similar criteria, it was possible to define C, D, and E subgroups (Duff and Vogt, 1969; T. Hanafusa et al., 1970). Not all strains of RSV were defective, however; some were able to transform and replicate without the need for a helper virus. The RAVs were shown to be avian leukosis viruses (ALVs), capable of causing leukosis and other tumors.

During this period, AMV was also shown to consist of a replication-

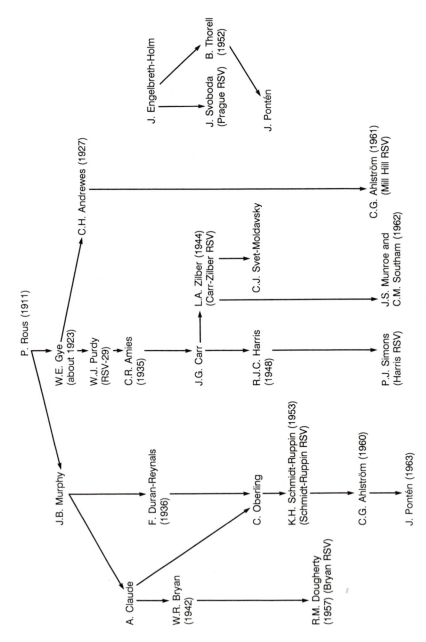

FIGURE 1. Origin of different strains of Rous sarcoma virus. Information mainly from Morgan (1964) and Gross (1983).

defective component and a replication-competent helper virus (Moscovici and Vogt, 1968; Moscovici and Zanetti, 1970). Subsequently, other acute leukemia viruses were found to be defective, giving rise to the concept of defective, acutely transforming viruses and nondefective, slowly transforming viruses (Graf and Beug, 1978; Hayman, 1983).

The oncogenicity of RSV in mammals formed a special category of research during the 1960s. In 1957, Russian workers (Zilber and Krjukova, 1957; Svet-Moldavsky, 1958) had induced cysts and sarcomas in rats with RSV. These results, of potentially far-reaching consequences, were confirmed by other laboratories, particularly in eastern Europe, but apparently not elsewhere, and they were regarded with some doubt. Subsequently, the virus strains used were exchanged and the property of transforming mammalian cells was associated particularly with subgroup D (H. Hanafusa and Hanafusa, 1966). A variety of tumors in many species, including primates, were induced by various workers (Vogt, 1965; Šimkovič, 1972). Svoboda and co-workers in particular extensively investigated the state of the RSV genome in mammalian tumors (Svoboda and Hložánek, 1970).

2. Host Response to ALV

Important to the classification of ALVs into subgroups was work on host genetic resistance to ALV. The development of strains of fowl with differing genetic susceptibilities to various avian tumor viruses had long been a central aspect of the work of the Regional Poultry Research Laboratory at East Lansing. In 1963, Crittenden et al. (1963) showed that genetic resistance or susceptibility to subgroup A RSV tumor induction in vivo was also expressed in tissue culture, with evidence of a single autosomal dominant gene controlling susceptibility (Crittenden et al., 1964). Later, other single genes were reported controlling susceptibility to subgroup B viruses (Rubin, 1965; L. N. Payne and Biggs, 1966) and subgroup C viruses (L. N. Payne and Biggs, 1970). The gene controlling susceptibility to subgroup B was found also to control response to the related subgroup D (Pani, 1975). Responses to subgroup E were more complex (L. N. Payne et al., 1971; Crittenden and Motta, 1975), with some differences in results which are still unsettled (L. N. Payne, 1985b).

3. Discovery of Endogenous Leukosis Viruses

In 1965, Dougherty and Di Stefano (1965) found that nonproducer RSV-transformed cells did release RSV-like particles, but these were apparently noninfectious. Soon, Vogt (1967a) and Weiss (1967) found that these released particles were in some instances infectious but with a restricted host range in chickens. Their occurrence was governed by the presence in some, apparently normal, chicken cells of "chick helper factor" (chf) which could complement the defective RSV genome in the same way as a helper virus. This indication of the presence of viral information in normal cells was supported by other evidence: Dougherty and Di Stefano (1966) had observed that some chicken embryos apparently free of infectious virus contained the group-specific (gs) antigen of ALV, as detected in complement-fixation or immunodiffusion tests, and L. N. Payne and Chubb (1968) showed that the gs antigen expression was inherited as a dominant autosomal gene. These antigenic expressions were shown to be due to the presence of endogenous viral (ev) genes, which in some cases were expressed as infectious virus, either spontaneously or following chemical or physical induction (H. Hanafusa et al., 1970; Weiss et al., 1971). Spontaneously produced endogenous ALV was designated RAV-0, and placed in subgroup E (Vogt and Friis, 1971). Endogenous viral DNA was detected in the host genome by hybridization techniques (Rosenthal et al., 1971), and in the following years the genetic structure and phenotypic expression of numerous ev loci were characterized (Astrin, 1978; Astrin and Robinson, 1979).

4. Epizootiology and Pathogenesis of Lymphoid Leukosis

In the course of his early studies on growth of RSV in chicken embryo fibroblasts in culture, Rubin (1960) found that some cultures which were naturally resistant to infection by RSV contained a transmissible "resistance inducing factor" (RIF), which was subsequently shown to be ALV, present as a result of congenital embryonal infection. This work led to the development of the interference assay for ALV which provided the first, much needed, rapid *in vitro* assay system for ALV, and which was widely used in many laboratories. In Rubin's laboratory and elsewhere, the RIF test was used for detailed study of the epizootiology of ALV infection in naturally affected flocks. The assay confirmed much of the earlier work on congenital and horizontal transmission of ALV by Burmester (1957) and his colleagues, and demonstrated the occurrence of immunological tolerance of congenitally infected chickens to ALV. These animals had a consequent persistent viremia and an increased propensity to develop lymphoid leukosis (Rubin et al., 1961, 1962).

The next major advance in knowledge of the epizootiology of ALV stemmed from the finding of Spencer and his colleagues (Spencer et al., 1976, 1977) that some infected hens shed large amounts of ALV and gs antigen into the albumen of their eggs and that this occurrence correlated with embryo infection. Albumen samples from eggs, or vaginal swabs from hens, provided readily available material for testing for gs antigen by means of the complement-fixation test (Sarma et al., 1964) or, later, the ELISA test (E. J. Smith *et al.*, 1979), and hence of detecting hens that shed and congenitally transmitted ALV. This test in turn led to a practical way of eradicating ALV infection in commercial breeding flocks by preventing vertical transmission (Okazaki *et al.*, 1979). During the course of these studies, it was discovered that ALV infection itself had depressive effects of economic and genetic significance on various physiological traits such as egg production (Spencer *et al.*, 1979; Gavora *et al.*, 1980).

Another milestone in research was the discovery of the bursa dependence of lymphoid leukosis. Following knowledge that thymectomy reduced leukemia in certain strains of mice, Peterson *et al.* (1964, 1966) found that avian lymphoid leukosis was prevented by removal of the bursa of Fabricius up to 5 months after virus infection; thymectomy had no influence. Subsequent studies explained the effect of bursectomy by demonstrating early neoplastic transformation of target cells in lymphoid follicles within the bursa, with later metastasis to other organs (Cooper *et al.*, 1968). Lymphoid leukosis was thus recognized as a malignancy of the B-cell lymphoid system.

5. Other Species of Avian Retroviruses

Thus far, this brief review of the history of avian retrovirology has been concerned with members of the so-called leukosis-sarcoma group of viruses. However, three other unrelated species have been identified.

a. Reticuloendotheliosis Virus

In 1958, a viral agent was isolated by M. J. Twiehaus and F. R. Robinson from a turkey with gross leukotic lesions; their early work remained unpublished for some 16 years (F. R. Robinson and Twiehaus, 1974). In other hands the isolate, designated strain T (for turkey) by Sevoian *et al.* (1964), was at first confused with other avian lymphomagenic viruses, but Theilen *et al.* (1966) provided evidence that it was distinct from the leukosis-sarcoma group, and designated the disease a reticuloendotheliosis. Subsequent extensive studies on strain T and related reticuloendotheliosis viruses (REVs) confirmed this view (Purchase and Witter, 1975).

b. Lymphoproliferative Disease Virus of Turkeys

Biggs *et al.* (1974, 1978) described a new, transmissible, lymphoproliferative disease of turkeys, apparently caused by a type C retrovirus unrelated to other avian retroviruses (McDougall *et al.*, 1978a; Yaniv *et al.*, 1979). Characterization of the virus has been hindered by the absence of a cell culture system.

c. Pheasant Type C Oncoviruses

Normal pheasant cells contain endogenous leukosis viruses which can act as helper viruses for defective RSV. The virus isolated from golden pheasant cells was placed in a new avian tumor virus subgroup, G (Fujita *et al.*, 1974). In further studies on the golden pheasant virus and on a similar isolate from Lady Amherst pheasants, T. Hanafusa *et al.* (1976) found that they differed biochemically and genetically from ALVs and REV, and apparently represented a new "class" (species) of retrovirus.

6. Biochemical and Molecular Studies

The course of developments in understanding the biochemical and molecular structure of ALVs is considered to be outside the scope of this chapter. During the 1960s the application of biochemical purification and analytical techniques, notably by P. H. Duesberg and W. S. Robinson, greatly extended knowledge of virus structure (W. S. Robinson *et al.*, 1965; Duesberg, 1968), and the discovery of reverse transcriptase and of the intermediary and proviral form of avian retrovirus by Temin and Mizutani (1970) revolutionized thinking and the direction of work on virus replication. In the 1970s the introduction of DNA cloning and sequencing shifted the emphasis of avian retrovirology onto the genetic structure of the virus and the molecular biological basis for oncogenesis. A landmark was the recognition of the *src* oncogene in RSV, responsible for sarcomatous transformation of normal cells, and of the cellular normal counterpart and origin of that gene (Stéhelin *et al.*, 1976). The detection of many more oncogenes in other acutely transforming retroviruses and of their corresponding cellular oncogenes has contributed further to our knowledge of transformation. Moreover, the second mechanism of oncogenesis by the so-called slowly transforming viruses, namely by activation of cellular oncogenes by viral promoter insertion, has added to the framework of knowledge on oncogenes which currently dominates cancer research (Glover and Hames, 1989).

III. LEUKOSIS-SARCOMA GROUP VIRUSES

A. Virology

1. Virus Morphology and Morphogenesis

In size, shape, and ultrastructural detail, the various avian leukosis-sarcoma group viruses (ALSVs) are identical, and similar to other C-type

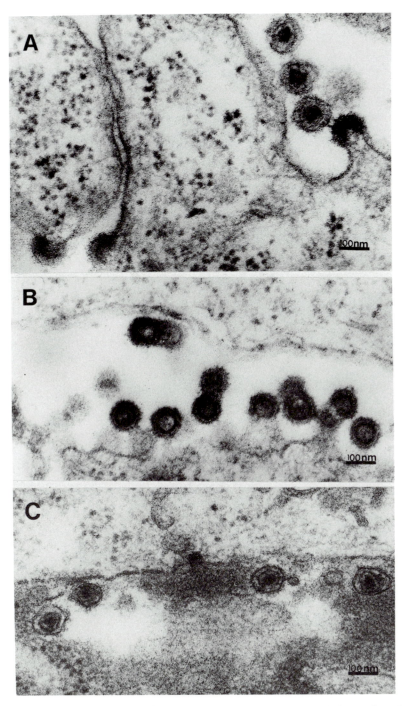

FIGURE 2. Structure of virions of (A) leukosis-sarcoma group virus, (B) reticuloendothe-liosis virus, (C) lymphoproliferative disease virus of turkeys. [Kindly provided by Dr. Judith Frazier.]

oncoviruses (Fig. 2). Overall particle diameter is 80–120 nm, with an average of 90 nm (Beard, 1973). In thin sections, the inner, centrally located, electron-dense core is about 35–45 nm in diameter; there is an inner membrane and an outer membrane bearing projections 7 nm long with knoblike ends about 6 nm in diameter (Beard, 1973; Bolognesi, 1974). With certain fixatives, ribonucleoprotein filaments can be visualized in the core; they are 3–5 nm in diameter, and of undetermined length.

As for other C-type retroviruses, the avian tumor viruses are assembled during a budding process from the cell surface or into cytoplasmic vacuoles (Haguenau and Beard, 1962; Heine et al., 1962a). A crescent-shaped structure composed of two layers appears beneath the cell plasma membrane; the inner layer collapses to form the electron-dense nucleoid and the outer layer becomes the inner membrane of the mature virion. The cell plasma membrane is modified to form the outer membrane of the virion. Because of their mode of formation, avian retroviruses contain a variety of components derived from the host cell.

In some circumstances, small cytoplasmic A-type particles which are immunologically related to the C-type virus may occur in cells infected with ALSVs (De Giuli et al., 1975).

2. Physical and Chemical Properties

The buoyant density in sucrose of the ALSVs is about 1.15–1.17 g/cm^3 (Bauer, 1974). The overall composition of AMV, which has been extensively studied, is 30–35% lipid, 60–65% protein, of which 5–7% is glycoprotein, 2.2% RNA, and small amounts of DNA thought to be of cellular origin (Beard, 1963a, b; H. Bauer, 1974; Bolognesi, 1974).

Viral RNA sediments at 60–70S, depending on virus strain, this being the viral genome, and at 4–5S, most of which is host tRNA. A tRNA is also associated with the 70S RNA, and is a primer for the DNA polymerase during transcription of viral RNA to DNA. The 60–70S RNA is a dimer which dissociates into two subunits of about 30–40S, which are believed to represent the diploid genome. The molecular weight of the 60–70S complex is between 4.5×10^6 and 7.5×10^6 (Hunter, 1980). ALVs carry three genes: gag, which codes for the internal structural proteins of the virion; pol, which codes for the RNA-dependent DNA polymerase; and env, which codes for the viral envelope. They are arranged in the order 5'gag–pol–env–poly(A)3'.

Viral lipids (mainly phospholipid) occur in the virion envelope and are of cellular origin. They have a bilayered structure similar to the outer cell membrane from which the virion envelope is derived (H. Bauer, 1974; Bolognesi, 1974).

Viral proteins fall into two categories: gs antigens and type- or subgroup-specific antigens (Bolognesi, 1974; Weiss *et al.*, 1982, 1985). The gs antigens are nonglycosylated proteins encoded by the *gag* gene: these are p27, the major gs antigen, believed to be a core shell component; p19 and p12, believed to be involved in RNA processing and packaging; and p15, a protease involved in cleavage of protein precursors. A p10 *gag* protein also occurs. The virion envelope contains two proteins encoded by the *env* genes: gp85, believed to be the knoblike structure that determines subgroup specificity, and gp37, representing the transmembrane spikes which attach the numerous knobs to the viral envelope. The two envelope proteins are linked to form a dimer. Group-specific components of viral envelope glycoproteins have been detected (Halpern and Friis, 1978). Nucleotide sequence differences in the *env* genes of different subgroups have been reported (Bova *et al.*, 1986, 1988).

Virions contain two enzymes encoded by the *pol* gene. Reverse transcriptase is present in the core, comprising α (58 kilodalton, kDa) and β (92 kDa) subunits, and has RNA- and DNA-dependent polymerase and hybrid-specific RNase H activities. Another core enzyme is an endonuclease, p32, involved in integration of viral DNA into the host chromosome. A number of other enzymic activities are also present in the virion which are believed to be cell-derived contaminants (Temin, 1974).

The foregoing aspects of the virology of ALSV, and of virus replication, which is outside the scope of this chapter, are reviewed in detail by Montelaro and Bolognesi (1980) and Weiss *et al.* (1982, 1985) (see also Chapter 5).

3. Classification and Strains

a. Pathotypes

Numerous strains of ALSVs exist, most of which were isolated from naturally occurring neoplasms 30 or more years ago. Many induce a predominant type of neoplasm, and can be designated accordingly, i.e., lymphoid leukosis virus (LLV), avian erythroblastosis virus (AEV), avian myeloblastosis virus (AMV), avian sarcoma virus (ASV), and myelocytoma/endothelioma viruses (Table I). Commonly, however, the virus strains induce other neoplasms in addition to the predominant one, and the tumor spectrum can be wide. The oncogenic spectrum tends to be characteristic of a particular virus strain, but often overlaps with that of other strains. Thus, the RPL12 strain of LLV induces lymphoid leukosis, erythroblastosis, osteopetrosis, hemangiomas, and sarcomas; the BAI A strain of AMV induces myeloblastosis, lymphoid leukosis, osteopetrosis, nephroblastomas, sarcomas, hemangiomas, thecomas, granulosa cell tumors, and epitheliomas (Beard, 1980) (Fig. 3). The spectra of neo-

TABLE I. Common Laboratory Strains of Avian Leukosis/Sarcoma Viruses of Chickens Classified According to Predominant Neoplasm Induced and Virus Envelope Subgroup[a]

Virus class according to neoplasm	Virus class according to subgroup					No subgroup (defective viruses)
	A	B	C	D	E	
Lymphoid leukosis virus (LLV)	RAV-1 RAV-3 RAV-4 RAV-5 FAV-1 RIF-1 MAV-1 RPL-12 HPRS-F42	RAV-2 RAV-6 MAV-2	RAV-7 RAV-49	RAV-50 CZAV	RAV-60	
Avian erythroblastosis virus (AEV)						AEV-ES4 AEV-R AEV-H S13 AMV-BAI-A
Avian myeloblastosis virus (AMV)						E26
Avian sarcoma virus (ASV)	SR-RSV-A PR-RSV-A EH-RSV RSV29	SR-RSV-B PR-RSV-B HA-RSV	B77 PR-RSV-C	SR-RSV-D CZ-RSV	SR-RSV-E PR-RSV-E	BH-RSV BS-RSV FuSV PRCII PRCIV ESV Y73 UR1 UR2 S1 S2 MC29
Myelocytoma and endothelioma viruses						MH2 CMII OK10
Endogenous virus (EV) (no neoplasm)					RAV-0 ILV	

[a] For original sources, see Weiss et al. (1982, 1985). Abbreviations used in the designation of virus strains: AEV-ES4, avian erythroblastosis virus-erythroblastosis-sarcoma strain 4; AEV-H, AEV Hihara strain; AEV-R, AEV R strain; BAI-A, Bureau of Animal Industry strain A; B77, Bratislava strain 77; BH, Bryan high-titer strain; BS, Bryan standard strain; CZ, Carr–Zilber; CZAV, Carr–Zilber-associated virus; CMII, strain of Loliger; E26, erythroleukemia strain 26; EH, Engelbreth-Holm strain; ESV, Esh sarcoma virus; FAV, Fujinami-associated virus; FuSV, Fujinami sarcoma virus; HA, Harris strain; HPRS-F42, Houghton Poultry Research Station field strain 42; ILV, induced leukemia virus; MAV, myeloblastosis-associated virus; MC29, myelocytomatosis strain 29; MH2, Mill Hill strain 2; OK10, Oker-Blom strain 10; PR, Prague strain; PRC II & IV, Poultry Research Centre (Edinburgh), strains II & IV; RAV, Rous associated virus; RIF, resistance-inducing factor; RPL, Regional Poultry Laboratory (East Lansing, Michigan); RSV, Rous sarcoma virus; S1 & 2, Sarcoma virus strains 1 & 2 of Hihara; S13, Stubbs strain 13; SR, Schmidt-Ruppin strain; UR 1 & 2, University of Rochester strains 1 & 2; Y73, Yamaguchi strain 73.

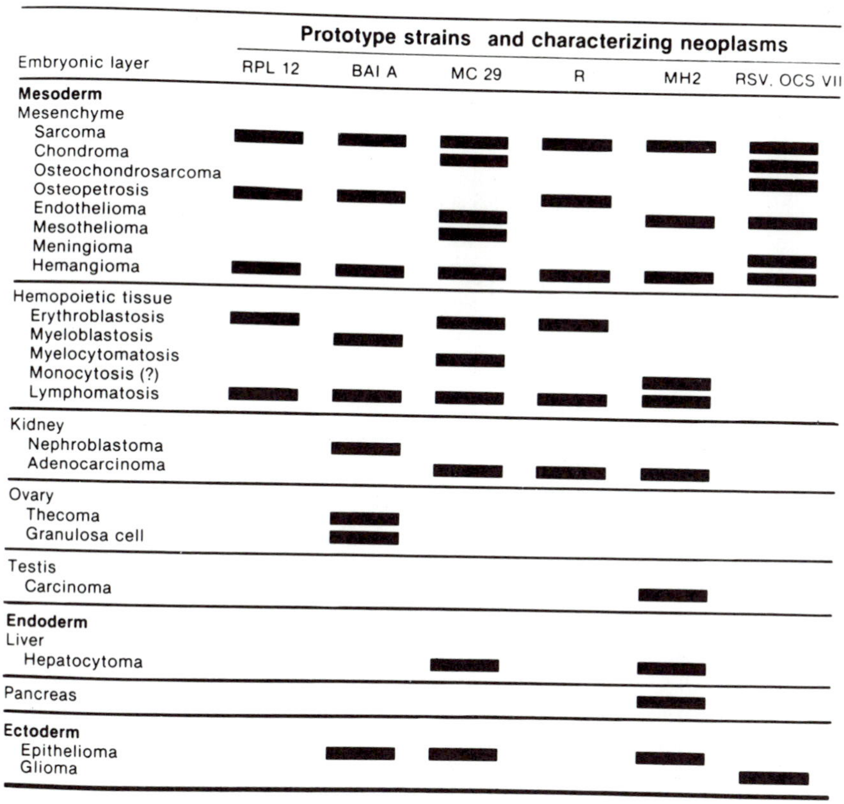

Embryonic layer	Prototype strains and characterizing neoplasms					
	RPL 12	BAI A	MC 29	R	MH2	RSV. OCS VII
Mesoderm						
Mesenchyme						
Sarcoma	X	X		X	X	X
Chondroma			X			X
Osteochondrosarcoma						X
Osteopetrosis	X	X		X		X
Endothelioma						
Mesothelioma			X		X	X
Meningioma			X			
Hemangioma	X	X	X	X	X	X
Hemopoietic tissue						
Erythroblastosis	X		X	X		
Myeloblastosis		X				
Myelocytomatosis			X			
Monocytosis (?)					X	
Lymphomatosis	X	X	X	X	X	
Kidney						
Nephroblastoma		X				
Adenocarcinoma			X	X	X	
Ovary						
Thecoma		X				
Granulosa cell		X				
Testis						
Carcinoma					X	
Endoderm						
Liver						
Hepatocytoma			X		X	
Pancreas					X	
Ectoderm						
Epithelioma	X		X		X	
Glioma						X

FIGURE 3. Oncogenic spectrum of prototype strains of avian leukosis/sarcoma virus. [From Beard (1980) and L. N. Payne and Purchase (1991), courtesy of the publishers.]

plasms induced have several explanations. Some strains of virus isolated consist of mixtures of viruses either fortuitously or because some may require coinfecting oncogenic helper viruses for replication (see below, "Acutely transforming viruses" and "Defective viruses"). However, clone-purified strains of LLV can cause a variety of neoplasms in addition to lymphoid leukosis, including erythroblastosis, osteopetrosis, and nephroblastomas (Purchase et al., 1977a), and clone-purified AEV can induce both erythroblastosis and sarcomas, irrespective of the helper virus (Graf et al., 1977). The tumor spectrum of a virus strain can also be modified in passage experiments according to the type of neoplasm used to harvest virus. Thus, selection of RPL12 strain LLV from donors with hemangiomas resulted in a higher frequency of this tumor in subsequent passage (Burmester et al., 1959a).

Virus dose is an important factor in determining tumors induced. High doses of RPL12 LLV mainly induce erythroblastosis, whereas low doses cause lymphoid leukosis (Burmester et al., 1959a) (Fig. 4). Factors

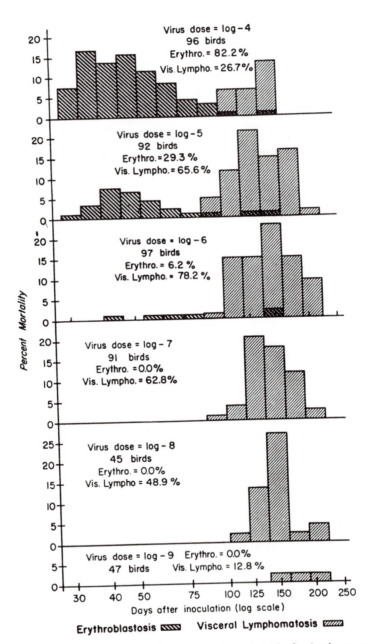

FIGURE 4. Influence of doses of RPL12 strain of lymphoid leukosis virus on mortality from erythroblastosis and lymphoid leukosis (visceral lymphomatosis). [From Burmester *et al.* (1959a) by kind permission.]

which influence effective dose, including route of inoculation, and age and genotype of host, also influence the oncogenic spectrum (Fig. 5).

In more recent years, strains of ALSV have been placed into two major classes with respect to oncogenicity:

(i) Acutely transforming viruses. These viruses are able to induce neoplastic transformation, in vivo or in vitro, within a few days or weeks. They cause either solid tumors (sarcomas and endotheliomas) or various types of acute leukemia (leukosis) (Graf and Beug, 1978; Enrietto and Wyke, 1983; Hayman, 1983; Moscovici and Gazzolo, 1987). The neoplasm induced may depend on the host species (Moscovici et al., 1981). The acutely transforming viruses contain, in their genome, oncogenes responsible for cell transformation (Table II). Some sarcoma viruses, e.g., Bryan strain RSV, are genetically defective and require a helper leukosis virus to complement them and enable virus replication; others are nondefective. In contrast, all avian acute leukemia viruses are defective: they transform cells but require a helper virus for replication. Avian sarcoma viruses containing transforming oncogenes (e.g., src) may be generated by transduction of cellular oncogenes by LLV (Hihara et al., 1984; Hagino-Yamagishi et al., 1984; Miles and Robinson, 1985).

(ii) Slowly transforming viruses. These are ALVs which do not carry oncogenes. They induce lymphoid leukosis, and sometimes erythroblastosis, by a "promoter insertion" mechanism, in which the ALV proviral genome becomes integrated adjacent to a cellular oncogene (myc or myb in the case of lymphoid leukosis, erbB in erythroblastosis) which is activated to bring about neoplastic transformation and a slow development of tumors over many weeks or months (Cooper, 1982; Enrietto and Wyke, 1983; Fung et al., 1983).

b. Subgroups

The ALSVs of the chicken are classified into five subgroups, A–E (Table I), on the basis of differences in their viral envelope glycoprotein antigens, which determine virus-serum neutralization properties, viral interference patterns, and range of infectivity in chicken and other avian cells of different phenotypes (Vogt and Ishizaki, 1966b; Duff and Vogt, 1969; T. Hanafusa et al., 1970). Endogenous viruses placed in other subgroups have been isolated from other species of birds. In general, viruses within a subgroup display interference at host-cell receptors (Table III), infect cells of the same receptor phenotype (Tables IV and V), and have related envelope antigens in serum neutralization tests. With certain exceptions, viruses in different subgroups do not share these properties. In particular, viruses of subgroups B and D interfere with each other and show partial serological cross-neutralizations; cells resistant to subgroup B are moderately resistant to subgroup D. Antisera to subgroups B and D are also reported to neutralize subgroup E virus, and subgroup B and D viruses interfere with subgroup E virus. Chicken cells resistant to subgroup B are always resistant to subgroup E but not vice versa.

A virus isolated from ring-necked pheasant belongs to subgroup F (T. Hanafusa and Hanafusa, 1973; Fujita *et al.*, 1974); an isolate (RAV-62) from Hungarian partridge is placed in subgroup H (T. Hanafusa *et al.*, 1976), and one from Gambel's quail in subgroup I (Troesch and Vogt, 1985). Viruses of these subgroups are distinct antigenically and in host range and interference patterns (Tables III and V). Other unclassified endogenous pheasant viruses exist also (Y. C. Chen and Vogt, 1977). There is some evidence that subgroup A viruses can originate in pheasants (Temin and Kassner, 1976). Recently, an exogenous ALV belonging to a new subgroup for chickens, possibly J, has been isolated from meat-type chickens (L. N. Payne *et al.*, 1991a, b).

Viruses isolated from golden and Lady Amherst pheasants constitute subgroup G, but belong to a species of avian retrovirus distinct from the ALSV group (Fujita *et al.*, 1974; T. Hanafusa *et al.*, 1976) (see Section VI).

c. Serotypes

There is evidence for varying antigenic types within subgroups, in that antiserum raised against a particular strain of virus tends to neutralize the homologous virus more strongly than heterologous viruses of the same subgroup (Ishizaki and Vogt, 1966; Chubb and Biggs, 1968; Bova *et al.*, 1988). In general, subgroup B viruses appear to be more heterogeneous than those of subgroup A.

d. Defective Viruses

A number of avian retroviruses have defective genomes; they arise either spontaneously or as a result of experimental mutagenesis (Graf and Beug, 1978; Hayman, 1983; Enrietto and Hayman, 1987). The acute leukemia viruses and many strains of avian sarcoma virus are replication-defective (*rd*) mutants, lacking in genes required for replication. The defects in the replication genes are a consequence of the insertion into them of oncogene sequences. RSV differs from other avian sarcoma viruses in possessing its oncogene, *src*, outside the genes encoding the viral structural proteins (Wang and Hanafusa, 1988). Some strains of RSV are not defective in replication, but transform rapidly due to *src*; other strains, e.g., Bryan high-titer RSV (BH-RSV), nevertheless possess defects in the structural genes, unrelated to the presence of *src*. The defective transforming viruses thus lack an envelope subgroup encoded by their own gene but acquire the subgroup of the helper virus necessary for their replication (Table I). RSV-29, a strain nearest in origin to the virus isolated by Rous (Fig. 1), was found to be defective, suggesting that the original Rous No. 1 virus was also so (Dutta *et al.*, 1985).

Certain strains of RSV spontaneously produce mutants lacking the *src* gene. They are transformation-defective (*td*) mutants and have an oncogenic potential similar to that of nondefective, slowly transforming ALV (Biggs *et al.*, 1973). They can reacquire *src* by transduction of cellular *src* (Wang and Hanafusa, 1988).

e. Endogenous Viruses

Somatic and germline cells of almost all chickens carry complete or defective DNA proviral genetic sequences of E subgroup leukosis virus integrated in their genome (H. Robinson, 1978; Crittenden, 1981, 1991; E. J. Smith, 1987). The sites of integrated viral genes are termed endogenous viral (*ev*) loci. They are detected by generating restriction fragment length polymorphisms by treatment of chicken cellular DNA with restriction enzymes, and probing with appropriate ALSV genomic sequences (Fig. 6). The viral genes are transmitted genetically in a Mendelian fashion to their progeny by both sexes (L. N. Payne and Chubb, 1968; Crittenden *et al.*, 1977; Astrin *et al.*, 1979) (Fig. 7). On average each chicken carries about five *ev* loci (Rovigatti and Astrin, 1983) and some 30 *ev* loci have been identified in different strains of chickens, although many more are likely to exist (Gudkov *et al.*, 1986; E. J. Smith, 1987; Ziemiecki *et al.*, 1988; Boulliou *et al.*, 1991) (Table VI). The chromosomal locations of a number of *ev* loci have been identified (E. J. Smith, 1987; Crittenden, 1991). Related *ev* loci also occur in red jungle fowl, ring-necked pheasants, partridge, and grouse, but not in guinea fowl, quail, peafowl, ruffed pheasants, gallo-pheasant, and turkey. This lack of a phylogenetic relationship led Frisby *et al.* (1980) to deduce that endogenous viruses in chickens had arisen postspeciation but before domestication.

Endogenous viruses are genetically related to reverse transcriptase-bearing elements termed retrotransposons present in cells from a wide variety of prokaryotic and eukaryotic life forms, and from which they may have evolved (Temin, 1985; Doolittle *et al.*, 1989) (see Chapter 1). Exogenous retroviruses are in turn believed to have evolved from endogenous viruses. The former originated after the evolution of vertebrates and have, it is suggested, a relatively short-lived evolutionary existence (Doolittle *et al.*, 1989).

The gene order of endogenous viruses is the same as that for exoge-

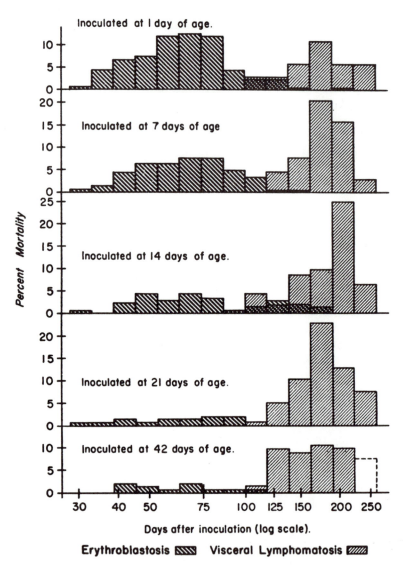

FIGURE 5. Influence of age at inoculation with RPL12 strain of lymphoid leukosis virus on mortality from erythroblastosis and lymphoid leukosis (visceral lymphomatosis). [From Burmester *et al.* (1959a) by kind permission.]

nous leukosis viruses, but the oligonucleotide sequences of endogenous and exogenous viruses differ. Heterogeneity of the p27 gs antigen in endogenous ALVs has been reported (Ignjatovic, 1988). Some *ev* loci consist of integrations of genetically defective viruses (Table VI).

The phenotypic expressions of these loci vary, depending on the

TABLE II. Transforming Properties of Acute Leukemia and Sarcoma Viruses[a]

Virus[b]	Oncogene(s) carried	Predominant neoplasm(s)	Cells transformed in vitro
RSV	src	Sarcoma	Fibroblast
B77	src	Sarcoma	Fibroblast
S1 & S2	src	Sarcoma	Fibroblast
FuSV	fps	Sarcoma	Fibroblast
PRC II & PRC IV	fps	Sarcoma	Fibroblast
UR1	fps	Sarcoma	Fibroblast
UR2	ros	Sarcoma	Fibroblast
Y73	yes	Sarcoma	Fibroblast
ESV	yes	Sarcoma	Fibroblast
AEV-ES4	erbA, erbB	Erythroblastosis, sarcoma	Erythroblast, fibroblast
AEV-H	erbB	Erythroblastosis, sarcoma	Erythroblast, fibroblast
S13	sea	Erythroblastosis, sarcoma	Erythroblast, fibroblast
E26	myb, ets	Myelobastosis, erythroblastosis	Myeloblast, erythroblast
AMV	myb	Myeloblastosis	Myeloblast
MC29	myc	Myelocytoma, endothelioma	Immature macrophage, fibroblast
CMII	myc	Myelocytoma	Immature macrophage, fibroblast
OK10	myc	Endothelioma	Immature macrophage, fibroblast
MH2	myc, mil	Endothelioma	Immature macrophage, fibroblast

[a] For original sources, see Hayman (1983), Enrietto and Wyke (1983), Enrietto and Hayman (1987), Wang and Hanafusa (1988), and Glover and Hames (1989).
[b] See Table I for abbreviations.

viral genes present and transcriptional and translational control sequences within the long terminal repeats (LTRs) (E. J. Smith, 1987; Crittenden, 1991). Four basic phenotypes have been described (Table VII). When the complete endogenous viral genome is present, subgroup E leukosis virus is produced by the cell, either spontaneously or after induction by X-irradiation or chemical mutagens and carcinogens (Weiss et al., 1971). Virus produced by an ev gene is termed an endogenous virus (EV) and numbered as for the locus. Expression of endogenous ev genes is responsible for a dominant form of genetic resistance of chicken cells to E subgroup viruses, in which virus receptors are blocked by viral envelope protein (H. L. Robinson et al., 1981). In some circumstances, expressed infectious endogenous virus may be transmitted vertically and horizontally, then behaving as an exogenous virus (E. J. Smith et al., 1986a). Subgroup E ALV, typified by RAV-0, has little or no oncogenicity in chickens (Motta et al., 1975), although it was reported to be lymphomagenic in another species, Sonnerat's jungle fowl (Weiss and

TABLE III. Interference Patterns Between ALV and RSV
of Subgroups A–G and I[a]

Subgroup of interfering ALV	Subgroup of challenge RSV:							
	A	B	C	D	E	F	G	I
A	+	−	−	−	−	−	−	−
B	−	+	−	+	+	−	−	−
C	−	−	+	−	−	−	−	−
D	−	+	−	+	+	−	−	−
E	−	−	−	−	+	−	−	?
F	−	−	−	−	−	+	−	−
G	−	−	−	−	−	−	+	−
I	−	−	−	−	−	−	−	+

[a] Susceptible avian embryo fibroblast cultures are infected with ALV of each subgroup, and challenged several days later with RSV of each subgroup. Reduction in RSV focus formation in infected cultures, compared with uninfected controls, is indicative of viral interference. +, Interference. −, No interference. ?, Not known. Information about subgroup H is incomplete. Subgroup G pheasant viruses belong to a species that differs from that of the other subgroups. Data from Vogt and Ishizaki (1966b), Duff and Vogt (1969), Weiss (1969), Chen and Vogt (1977), and Troesch and Vogt (1985).

Frisby, 1981). Similarly, spontaneously produced endogenous virus did not appear to be oncogenic (Crittenden *et al.*, 1979a). The lack of oncogenicity of endogenous subgroup E ALV is not due to its envelope antigens but probably to LTR promoter sequences (H. L. Robinson *et al.*, 1980).

The biological value of *ev* loci appears variable, although their persistence suggests that birds carrying them are at no great disadvantage. Crittenden *et al.* (1982, 1984a) found that the presence of *ev2* or *ev3* protected birds from a nonneoplastic syndrome caused by infection with

TABLE IV. Examples of Host Range of Subgroup A–E
Avian Leukosis/Sarcoma Viruses of Chickens
in Chicken Embryo Cells of Different Phenotypes[a]

Phenotype of cells	Subgroup of virus:				
	A	B	C	D	E
C/O	S	S	S	S	S
C/AE	R	S	S	S	R
C/BE	S	R	S	(R)	R
C/C	S	S	R	S	S
C/E	S	S	S	S	R

[a] S, susceptible; R, resistant; (R), partially resistant. The cell phenotype designation denotes chicken (C) cells resistant to (/) the specified subgroup (0: no subgroup; AE: A and E subgroups; etc.)

TABLE V. Host Range of Different Subgroups of RSV in Embryo Fibroblasts from Various Avian Species[a]

Avian species	Subgroup of RSV:							
	A	B	C	D	E	F	G	I
Chicken	R/S	R/S	R/S	R/S	R/S	S	S	S
Golden pheasant	S	R/S	S	R/S	S	S	S	?
Ghighi pheasant	S	R	S	S	S	?	?	?
Silver pheasant	S	R	S	R	R	S	S	?
Swinhoe pheasant	S	R	S	S	R	S	S	?
Mongolian pheasant	S	R	S	R	S	S	S	?
Ring-necked pheasant	S	R	SR	R	S	S	S	?
Green pheasant	S	R	SR	R	S	S	S	?
Reeve's pheasant	S	R	S	R	R	?	?	?
Japanese quail	S	R	SR	SR	S	S	S	S
Chinese quail	R	R	R	R	R	S	S	?
Bobwhite quail	R	R	R	SR	R	?	?	?
Chukar	S	R	R	R	R/S	S	S	?
Guinea fowl	S	S	S	S	S	?	?	?
Turkey	S	R	S	SR	S	?	?	?
Peking duck	R	R	S	R	R	S	R	S
Moscovy duck	R	R	S	R	R	S	R	?
Goose	R	R	S	R	R	R	R	S
Pigeon	R	R	R	SR	R	?	?	?
Parakeet	R	R	R	R	R	?	?	?

[a] Embryo fibroblast cultures from avian species are challenged with RSV of each subgroup and susceptibility to RSV focus formation determined. R/S, individuals segregate for susceptibility or resistance; S, susceptible; R, resistant; SR, partially resistant; ?, not known. Information about host range of subgroup H is unavailable. Data from Vogt (1977), Chen and Vogt (1977), Meyers et al., (1978), and Troesch and Vogt (1985)

subgroup A ALV. Yet, embryonic infection with RAV-0 caused a more persistent viremia and more neoplasms following infection with exogenous ALV, apparently due to tolerant depression of humoral immunity (Crittenden et al., 1987). Similarly, the presence of ev21 increased susceptibility to infection by exogenous ALV (Bacon et al., 1988; E. J. Smith and Fadly, 1988). Endogenous viral gp85 glycoprotein may be present in serum of chickens, but its significance is unclear (Bosch et al., 1983; Hadwiger and Bosch, 1985); probably it is related to tolerance induction. Evidence exists also that ev genes that code for complete EV have a direct and detrimental influence on production traits (Kuhnlein et al., 1988, 1989; Gavora et al., 1991) (see also Section III.F.4c). Iraqi et al. (1991) have suggested that ev genes are generally deleterious but that under certain circumstances some may have favorable effects.

    ev loci are not essential to chickens, as it has been possible to produce and breed chickens (line 0) free of ev genes (Astrin et al., 1979; Crittenden and Fadly, 1985). Also, Gudkov et al., (1981) found no ev

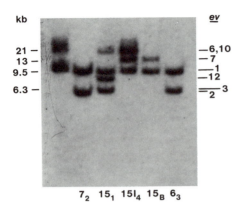

FIGURE 6. *ev* loci in six lines of chickens detected by generation of restriction fragment length polymorphisms by digestion of red blood cell DNA with Sac-1 endonuclease and probing with pRAV-2. [From E. J. Smith (1987) by kind permission of author and publisher.]

loci in Italian partridge-colored chickens, and neither did Gavora *et al.* (1989) in an inbred line of White Leghorns.

Evidence exists also for two other families of endogenous avian retrovirus-like sequences distinct from the RAV-0 family. One consists of the numerous chicken repeat 1 (CR1) retrotransposon elements (Silva

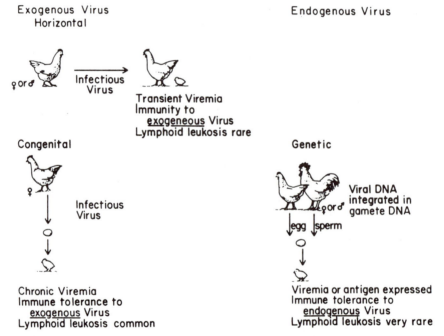

FIGURE 7. Horizontal and congenital transmission of exogenous avian leukosis virus and endogenous transmission of endogenous virus. [From Crittenden (1981) and by kind permission of author and publisher.]

TABLE VI. Phenotypes of Endogenous
Retroviral (ev) Genes in Inbred and
Commerical Lines of White
Leghorn Chickens[a]

ev	Phenotype	Line or source[b]
1	$gs^- chf^-$	Most lines
2	$V-E^+$	$RPRL7_2$
3	$gs^+ chf^+$	$RPRL6_3$
4	$gs^- chf^-$	SPAFAS
5	$gs^- chf^-$	SPAFAS
6	$gs^- chf^+$	RPRL15
7	$V-E^+$	RPRL15B
8	$gs^- chf^-$	K18
9	$gs^- chf^+$	K18
10	$V-E^+$	$RPRL\ 15I_4$
11	$V-E^+$	$RPRL\ 15I_4$
12	$V-E^+$	$RPRL\ 15_1$
14	$V-E^+$	H & N
15(C)	None	K28 × K16
16(D)	None	K28 × K16
17	$gs^- chf^-$	RC-P
18	$V-E^+$	RI
19	$V-E^+$ (?)[c]	RW
20	$V-E^+$ (?)[c]	RW
21	$V-E^+$	Hyline FP

[a] ev13 is associated with the $gs^- chf^-$ phenotype, but restriction fragments have not been characterized. Table modified from E. J. Smith (1987), by kind permission of author and publisher.

[b] Not exclusive to line or source. K, Kimber; R, Reaseheath; H & N, Heisdorf and Nelson; for references see E. J. Smith (1987).

[c] The presence of five ev loci in Reaseheath line W birds precludes definitive assignment with the $V-E^+$ phenotype. Definitive association requires further segregation of ev genes. Hyline FP birds also carry ev1, ev3, and ev6.

and Burch, 1989) and the other consists of moderately repetitive ALV-like elements called the endogenous avian retrovirus (EAV) family detected in line 0 chickens (free of ev loci) by low-stringency hybridization (Boyce-Jacino et al., 1989). In evolutionary terms, the CR1 elements appear to be most ancient, and ev genes the most recent (Crittenden, 1991). The functions of the CR1 and EAV sequences are unknown.

f. Nomenclature

A variety of conventions are used in designating ALSVs, many of which are illustrated in Table I. They are given a full and an abbreviated

TABLE VII. Phenotypic Expression of Representative Endogenous Viral (ev) Genes in Normal Chicken Cells[a]

Phenotype	Symbol	ev locus
No detectable viral product	gs$^-$ chf$^-$	1
		4, 5
Expression of subgroup E envelope antigen	gs$^-$ chf$^+$	9
Coordinate expression of group-specific and envelope antigens	gs$^+$ chf$^+$	3
Spontaneous production of subgroup E virus	V-E$^+$	2

[a] Table modified from E. J. Smith (1987) by kind permission of author and publisher.

designation on the basis of the predominant neoplasm they induce, e.g., lymphoid leukosis virus (LLV) or avian erythroblastosis virus (AEV). Representative strains may be named from their association with an individual, e.g., Rous sarcoma virus (RSV), or place, e.g., the Regional Poultry Research Laboratory isolate 12 (RPL12) of LLV. Strains of RSV are designated according to individuals who worked with them, e.g., Schmidt–Ruppin strain (SR-RSV) or place, e.g., Prague (PR-RSV).

Nondefective viruses may also be designated according to subgroup, e.g., subgroup A strain of SR-RSV (SR-RSV-A). Helper viruses isolated from stocks of defective viruses are named, for example, as Rous-associated virus (RAV) or myeloblastosis-associated virus (MAV), and isolates are numbered (RAV-1, MAV-1, etc.).

Defective virus replicated with the aid of a helper virus are also designated. Thus, Bryan high-titer RSV with RAV-1 as a helper is designated BH-RSV(RAV-1). Other abbreviations used are those for resistance-inducing factor (RIF), induced leukemia virus (ILV), and endogenous leukosis virus (EV).

Since its isolation in 1911, RSV has been studied in many laboratories, and strains with varying properties have arisen. Historical relationships between these strains where known are given in Fig. 1. The factors involved in bringing about these strain variations are generally not known, but are likely to include mutation, involvement of various helper viruses, and genetic recombination.

4. Sources for Virus Isolation

When present, tumors are the first choice of material for virus isolation. However, although the nondefective LLVs are widespread in commercial flocks, only a small proportion of infected birds develop tumors. Sources of virus from such nontumorous birds are whole blood, buffy coat cells, parenchymatous tissues such as liver, and vaginal and cloacal swabs. Plasma or serum is a good source in tolerant viremic birds. LLV may also be isolated from albumen from newly-laid eggs, from 10-day-old embryos, or from meconium from day-old chicks, when present as a consequence of vertical transmission (Purchase and Fadly, 1980). Because ALSVs are thermolabile, samples should be collected from live or freshly killed birds as appropriate, and be stored and shipped at −70°C if there is delay between collection and testing. Practical procedures are described by Spencer (1987) and Fadly (1989).

5. Propagation and Assay

a. Chicken Inoculation

For *in vivo* propagation, RSV and other sarcoma viruses are usually inoculated subcutaneously into the wingweb of young chickens, and visible tumors develop within a few days and can be harvested for virus extraction. Semipurified virus stocks can be prepared by the method of Moloney (1956). Using the criteria of tumor size, weight, or latent period, dose–response assays of potency of virus stocks are available (Bryan, 1956; Dougherty, 1964). Intramuscular, intraperitoneal, and intracerebral routes of inoculation have also been used. For tumor induction by, and propagation of, LLV, day-old infection-free chickens of a susceptible strain, such as RPL line 15I, are inoculated intraabdominally with virus. An LL response is obtained within 270 days (Burmester and Gentry, 1956). Other tumors may also be induced, depending partly on the strain of virus, including erythroblastosis, sarcomas, hemangiomas, and osteopetrosis. Other factors affecting the responses include strain of chickens, host age, virus dose, and route of inoculation (Burmester *et al.*, 1959a, 1960) (Figs. 4 and 5). Quantitative assays of some strains of LLV may be obtained within 63 days of inoculation of day-old chickens, using the less sensitive erythroblastosis response (Burmester, 1956b). AMV also may be titrated by intravenous inoculation of young chickens (Eckert *et al.*, 1954).

b. Chicken Embryo Inoculation

Inoculation of RSV and other avian sarcoma viruses onto the chorioallantoic membrane of 11-day-old chicken embryos is followed by the development of tumor pocks which can be counted 8 days later (Fig. 8). The number of pocks is linearly related to virus dose, which is expressed in pock-forming units (Dougherty *et al.*, 1960). The pocks may also be used as a source of virus. In very early chicken embryos RSV is not tumorigenic (Dolberg and Bissell, 1984).

Dose–response assays for leukosis viruses are available by inoculat-

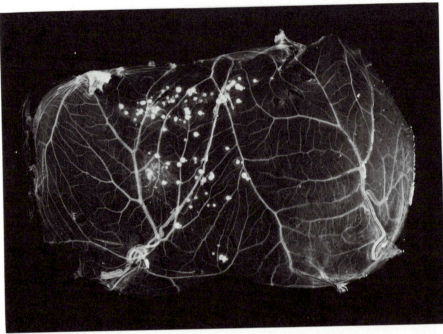

FIGURE 8. Rous sarcoma virus-induced tumor pocks on the chorioallantoic membrane of a chicken embryo.

ing virus intravenously into 11-day-old chicken embryos (via a chorioallantoic vein) and recording the erythroblastosis and other tumor responses which develop after hatching. By recording erythroblastosis and other tumors to 46 days postinoculation, virus titers were obtained (0.7–2) $\log_{10}$ higher than following chicken inoculation (Piraino *et al.*, 1963). Embryonic inoculation has also been used for assay of AMV (Baluda and Jamieson, 1961).

c. Cell Culture

RSV and other avian sarcoma viruses induce neoplastic transformation of chicken embryo fibroblasts (CEF) growing in cell culture. Cells from other avian species can be infected and transformed similarly. RSV of subgroup D and B77 virus (subgroup C), transform mammalian fibroblasts in culture (Vogt, 1965; Šimkovič, 1972). Transformed cells are rounded or fusiform, depending on the transforming virus, lose contact inhibition, and pile up to form discrete foci. Foci are usually visible in 3 or 4 days microscopically and grossly from 7 days (Fig. 9). When the infected cultures are grown under an agar overlay, secondary focus formation is prevented, and a linear virus dose–response curve is obtained,

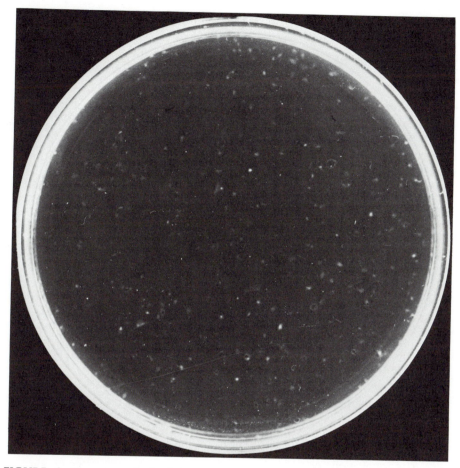

FIGURE 9. Numerous foci of Rous sarcoma virus-transformed chicken embryo fibroblasts.

from which the infectivity of virus stocks may be assayed in focus-forming units (Temin and Rubin, 1958).

Infectivity and focus formation by RSV of subgroups B, C, D, and E, but not A, are enhanced by inclusion in the culture medium of the polycations diethylaminoethyldextran and polybrene (Vogt, 1967b; Toyoshima and Vogt, 1969). Heparin, a polyanion, inhibits infectivity of subgroup B RSV, but not subgroups A and C (Toyoshima and Vogt, 1969). A factor released by cells infected by subgroup A LLV also enhanced infectivity of subgroups B and C RSV (T. Hanafusa and Hanafusa, 1967; Kass and Levinson, 1974). Several other factors which influence infectivity and focus formation have been reported (Spencer, 1987).

RSV will also transform other cell types, including chondroblasts, macrophages, myoblasts, pigmented and neural retinal cells, and erythroid cells (Weiss *et al.*, 1982; Palmieri, 1985).

In contrast to RSV, most nontransforming ALVs grow in CEF cultures without inducing any obvious morphological changes. Their presence can therefore be detected only by indirect means (see Section III.B). Under some circumstances ALVs may induce disorganized growth of CEFs or morphological changes (Rubin, 1960; Calnek, 1964). ALV of subgroups B, D, and F may induce plaques which can be used for virus assay (Graf, 1972; Moscovici *et al.*, 1976).

The acute, defective, leukemia viruses will infect and transform hematopoietic cells *in vitro*. AMV will transform blastoderm, yolk sac, and bone marrow cells into neoplastic myeloblasts, and focus assays have been developed (Baluda and Goetz, 1961; Moscovici *et al.*, 1975, 1983; Boettiger and Olsen, 1989). AEV will transform erythroblasts in blastoderm and bone marrow cultures (Graf, 1975; Moscovici *et al.*, 1983) and transformation of hematopoietic cells by MH2, MC29, and OK10 viruses has also been reported (Graf and Beug, 1978; Hayman, 1983; Moscovici and Gazzolo, 1987).

B. Diagnostic Tests

1. Resistance-Inducing Factor (RIF) Test

The RIF test was the first rapid tissue culture test for detecting ALV. It was widely used in the 1960s, but has tended to be replaced by more recently developed tests.

As mentioned in Section III.A.5c, nontransforming ALVs usually infect and grow in susceptible CEF without producing obvious morphological changes, but such cells become resistant to superinfection and transformation by RSV of the same subgroup (Rubin, 1960; Vogt and Rubin, 1963; Calnek, 1965; Steck and Rubin, 1966). This interference, which depends on the ability of ALV to block cellular virus receptors for RSV, forms the basis of the RIF test for detection of ALV in test material. Procedures for the test are described by Solomon *et al.* (1971) and Fadly (1989). In general, interference occurs only between viruses of the same subgroup (Vogt and Ishizaki, 1966b). However, reciprocal interference is seen between B and D subgroup viruses (Duff and Vogt, 1969), and ALVs of these subgroups also interfere with E subgroup RSV (T. Hanafusa *et al.*, 1970) (Table III). Conventionally, the test for the presence of ALV in a culture is considered positive when the number of foci produced by RSV of the same subgroup is reduced tenfold or more compared with the number in RSV-inoculated control cultures. Because of its specificity, the RIF test is useful for determining the subgroup of a virus.

2. Nonproducer (NP) Cell Activation Test

The NP test (Rispens et al., 1970) depends on the ability of ALV in test material to mix phenotypically with envelope-defective RSV in transformed NP cells. Solitary infections of CEF by envelope-defective BH-RSV, without concomitant infection by the helper virus, results in RSV-transformed cells which are nonproducers of RSV of subgroups A–D. Infection of a culture of NP cells by ALV in the test material results in activation of the NP cells and the production of infectious RSV of the subgroup of the ALV, which is assayed on susceptible cells. NP cells produced from CEF with endogenous subgroup E virus genome may spontaneously produce subgroup E RSV, which must be differentiated in the NP test from RSV induced by ALV of other subgroups by assaying the supernatants on C/E cells, which exclude subgroup E RSV (cell phenotype designations are described in Section III.C.2).

The R(−)Q cell test is a useful modification of the NP test in which use is made of nonproducing Japanese quail cells which have been transformed with envelope-defective BH-RSV (Crittenden et al., 1979b). These R(−)Q cells can be activated to produce infectious RSV by cocultivating them with C/E CEF infected with the ALV under test.

R(−)Q cells can also be used to detect "chick helper factor" (chf), namely endogenous viral env gene envelope glycoproteins able to complement defective RSV to produce infectious subgroup E RSV, and to detect infectious subgroup E ALV. The test involves cocultivation of R(−)Q cells with test cells (for chf detection) or test cell supernatants (for infectious ALV), and assay of supernatants on quail embryo fibroblasts for production of subgroup E RSV (Crittenden et al., 1979b).

3. Phenotypic Mixing (PM) Test

The PM test (Okazaki et al., 1975) is a variant of the NP test. Cultures of C/0 CEF are heavily infected with BH-RSV(RAV-0), the E subgroup RSV pseudotype, to produce RSV-transformed cells. These cultures are superinfected with test material suspected of containing ALV. After further incubation, culture fluid is harvested and assayed for infectious RSV on C/E CEF. Any RSV foci that develop are indicative of the presence of RSV of a subgroup other than E, and thus of the presence of ALV in the test material. Procedures for the PM test are given by Spencer (1987) and Fadly (1989).

4. Complement-Fixation Test for Avian Leukosis Virus (COFAL)

In the COFAL test, antiserum raised against the major gs antigen, p27, present in the core of ALSVs, is used in a complement-fixation test to detect gs antigen in susceptible CEF inoculated with ALV (Sarma *et al.*, 1964). Cells must be cultured for a period sufficient to allow ALV to replicate to detectable levels, usually 14 days, and these are frozen and thawed to release antigen before testing. Among the controls used are uninoculated CEF, to allow gs antigen derived from endogenous leukosis virus to be recognized. Titers of gs antigen from exogenous viruses are much higher than those from endogenous virus. Procedures are given by Fadly (1989).

Complement-fixing antisera can be obtained from hamsters bearing sarcomas induced by SR-RSV (Sarma *et al.*, 1964), from rabbits or other mammals inoculated with gs antigens derived from AMV (Stephenson *et al.*, 1973, 1975; E. J. Smith, 1977), or from pigeons bearing Rous sarcomas (Sazawa *et al.*, 1966; Sarma *et al.*, 1969).

The COFAL test as originally described includes growth of ALV in CEF, but the reagents can be used also in direct complement-fixation tests to detect gs antigen in, for example, tissue extracts or egg albumen; however difficulties arise in differentiating antigen from exogenous and endogenous viruses.

5. Enzyme-Linked Immunosorbent Assay (ELISA) for Leukosis Virus

Sensitive ELISA tests for detecting gs antigens using antisera raised in hamsters or rabbits, as described above, have been developed by E. J. Smith *et al.* (1979) and Clark and Dougherty (1980). They may be used for direct tests on materials such as egg albumen or vaginal swabs, but again with the problem of distinguishing between gs antigen derived from exogenous and endogenous ALV, or in indirect tests in which ALV is first cultivated in CEF.

ELISA tests have to a considerable degree replaced the previously mentioned tests for detecting presence of ALV and related viruses, and test kits or reagents are available commercially. Several groups of workers have developed monoclonal antibodies against p19 and p27 gs antigens which can be used in the ELISA (De Boer and Osterhaus, 1985; Lee *et al.*, 1986). Protocols for the ELISA are given by Spencer (1987) and Fadly (1989).

6. Serological Tests for Antibody

Subgroup-specific antibody to ALSV may be detected in plasma, serum, or yolk by an RSV focus reduction neutralization assay (Rubin *et al.*, 1961; Purchase, 1965), microneutralization-ELISA assay (Fadly *et al.*, 1989), or antibody ELISA (Mizuno and Hatakeyama, 1983; E. J. Smith *et al.*, 1986b; Tsukamoto *et al.*, 1985). Practical procedures are given by Spencer (1987) and Fadly (1989).

7. Other *in vitro* Tests

Other tests available for detecting ALSV include radioimmunoassays for gs antigen (Estola *et al.*, 1974; Sandelin *et al.*, 1974), reverse transcriptase assays (Kelloff *et al.*, 1972; Panet *et al.*, 1975; Tereba and Murti, 1977), and immunohistochemical tests for gs antigen or subgroup- or type-specific antigens (Kelloff and Vogt, 1966; F. E. Payne *et al.*, 1966; Dougherty *et al.*, 1974; Spencer, 1987; Stoker and Bissell, 1987). AMV can be assayed by its adenosine triphosphatase activity (Beaudreau and Becker, 1958).

C. Host Range

1. Species Affected

The natural hosts of ALSVs of subgroups A–E are domestic chickens. Cells from other avian species, including jungle fowl, Bobwhite quail, Japanese quail, goose, duck, turkey, guinea fowl, chukar and pigeon, and various species of pheasant, are susceptible to one or more of these subgroups (Vogt, 1977) (Table V). Host-range studies on viruses of subgroups F, G (but which belongs to another species of avian retrovirus), H, and I have been reported (T. Hanafusa *et al.*, 1976; Chen and Vogt, 1977; Troesch and Vogt, 1985). A virus of a novel subgroup, possibly J, isolated from chickens had a wide host range in chickens, but Japanese quail were resistant (L. N. Payne *et al.*, 1991a). This virus causes myelocytomatosis, renal adenomas and other tumors (Payne *et al.*, 1991b).

Some strains of RSV and other ALSVs induce tumors in various mammals and transform mammalian fibroblasts in culture: these properties are associated particularly with subgroup D viruses, and with the subgroup C virus B77 (Vogt, 1965; Šimkovič, 1972). Ability to infect mammalian cells may be related to viral possession of an unstable temperature-sensitive and fusogenic envelope glycoprotein (Bova-Hill *et al.*,

1991). Phenotypic mixing between RSV and murine leukemia virus can give rise to RSV able to infect mouse and other mammalian cells (Levy, 1977).

2. Genetic Susceptibility to Infection

Two types of genetic resistance to ALSV are described: cellular resistance to virus infection and resistance to tumor development (Crittenden, 1975; Weiss, 1981; L. N. Payne, 1985b; Bacon, 1987).

Resistance to infection depends upon the lack of specific virus receptors in the cell membrane which interact with viral envelope glycoproteins. The specificity of the interaction between virus and receptor appears to be at the stage of virus penetration or uncoating, because both resistant and susceptible cells adsorb virus with similar facility (Crittenden, 1968). Little is known of these processes. Morphologically distinctive virus attachment sites on the surface of CEF and entry of virus in phagocytic vacuoles were described ultrastructurally by Dales and Hanafusa (1972), but viral dissolution was not visualized. The molecular nature of the virus receptors is unknown. If virus is introduced into a resistant cell by other means, the cell can support virus replication (Piraino, 1967; Crittenden, 1968).

Inheritance of the virus receptors is of a simple Mendelian type. Three independent autosomal loci, designated Tv-A (i.e., tumor virus A subgroup), Tv-B, and Tv-C, control responses to infection by ALSV of subgroups A, B, and C, respectively (Table VIII). The Tv-B locus also controls responses to subgroup D virus (Pani, 1975), and linkage occurs between Tv-A and Tv-C loci (L. N. Payne and Pani, 1971). At each locus, alleles for susceptibility and resistance exist, which are designated Tv-A^s, tv-A^r, Tv-B^s, tv-B^r, and Tv-C^s, tv-C^r; these genes are usually abbreviated A^s, A^r, etc. Susceptibility genes are dominant over resistance

TABLE VIII. Genes Controlling Cellular Susceptibility
to Subgroups A–E of Avian Leukosis/Sarcoma Viruses[a]

Virus subgroup	Locus	Alleles	Dominant trait
A	Tv-A	Tv-A^s, tv-A^r	Susceptibility
B and D	Tv-B	Tv-$B^{s1,s2,s3}$, tv-B^r	Susceptibility
C	Tv-C	Tv-C^s, tv-C^r	Susceptibility
E	Tv-E	Tv-E^s, tv-E^r	Susceptibility
	I-E	I-E, i-E	Resistance

[a] Gene nomenclature of Somes (1980) is used. The existence of independent Tv-B and Tv-E loci is not settled. The I-E allele represents expression of an ev locus which inhibits cell penetration by subgroup E virus

genes; the resistant state may simply represent the absence of the susceptibility gene. Different degrees of susceptibility or resistance are encountered, suggesting that multiple alleles occur at each locus, but this has not been studied in detail.

Inheritance of resistance to subgroup E virus is more complex and the subject of unresolved dispute. L. N. Payne et al. (1971) reported that genetic resistance to subgroup E virus involved two autosomal loci, Inhibitor-E and Tv-E, with pairs of alleles I-E and i-E, and Tv-E^s and tv-E^r. The dominant I-E gene apparently masked the expression of the dominant Tv-E^s gene. For this reason embryos of genotype I-E I-E E^{s}E^s, for example, were resistant to infection by subgroup E virus, even though they carried the E^s susceptibility gene. The occurrence of a dominant resistant gene, I-E, was confirmed by Crittenden et al. (1973), but these workers questioned the existence of a Tv-E locus and presented evidence that multiple allelism at the Tv-B locus explained their observations (Crittenden and Motta, 1975). This claim was in turn disputed by Pani (1976, 1977) and the argument is not settled. The action of the I-E gene was ascribed to control of endogenous virus production, which blocked subgroup E virus receptors. Subsequent studies have suggested strongly, however, that the I-E locus was in fact an ev locus, with subgroup E envelope glycoproteins blocking the receptor (H. L. Robinson et al., 1981).

Susceptibility or resistance to infection arising from the expression of the genes at these tumor virus loci is expressed in CEF in culture using the RSV focus assay and in chicken embryos using the RSV pock assay on chorioallantoic membranes. These parameters of infectivity are also observed in hatched chickens using tumor incidence or mortality as measures of response following subcutaneous, intramuscular, or intracerebral inoculation of RSV (L. N. Payne, 1985b). Chicken assays have the disadvantage that responses may be quantified only by quantal methods, and genetic factors not operative in cultures or embryos, e.g., immune responses, may influence tumor development. Consequently, the in vitro or in ovo enumeration assays are preferred. Macrophages from embryos of susceptible phenotypes show a specific type of resistance to certain subgroups of ALSV (Gazzolo et al., 1975).

The response phenotypes of cells, embryos, or chickens from different genetic sources to the five subgroups of ALSV are designated by a convention that indicates chicken (C) cells and the subgroups to which they are resistant (bar, /). For example, C/AE cells are resistant to subgroup A and E, and susceptible to subgroups B, C, and D; C/0 cells are resistant to no subgroup, that is, susceptible to all (see also Table IV). It should be noted that resistant cells can be more than 6 log$_{10}$ units more resistant than susceptible cells.

Although RSV mainly is used to determine the response phenotype as a consequence of the method of assay, resistant cells are of course also resistant to infection by nontransforming ALV of the same subgroup (Vogt and Ishizaki, 1965). Pseudotypes between vesicular stomatitis virus (VSV) and ALV can also be used, in which VSV particles have the envelope antigens, and hence the host range, of the ALV, while retaining the ability of VSV to form rapidly cytocidal plaques in cultured cells (Love and Weiss, 1974).

Recently, Salter and Crittenden (1989) have produced transgenic chickens containing an ALV proviral DNA insert of the subgroup A *env* gene, which showed dominant resistance to infection by subgroup A RSV and to lymphoid leukosis induction by subgroup A ALV.

3. Genetic Susceptibility to Disease

Genetic factors can influence the development of tumors in birds susceptible to virus infection. The best-studied genes are those that influence the growth or regression of Rous sarcomas. A dominant gene *R-Rs-1* within the major histocompatibility complex (MHC) is responsible for tumor regression (Collins *et al.*, 1977; Schierman *et al.*, 1977). The responsible gene was more closely linked to the *Ir-GAT* locus than to the *Ea-B* locus in the MHC, suggesting that it is located in the immune response (B-L) region (Gebriel *et al.*, 1979). A number of B haplotypes have been associated with Rous sarcoma regression (Nordskog and Gebriel, 1983; Collins *et al.*, 1985; Schierman and Collins, 1987) and metastasis (Collins *et al.*, 1986). The chicken MHC (*Ea-B* locus) also influences the incidence of erythroblastosis and to a lesser extent lymphoid leukosis (Bacon *et al.*, 1981) and an influence of B haplotype on ALV shedding has been reported (Bacon, 1987).

The mechanism of resistance to Rous sarcomas is specific and thymus- and bursa-dependent (McBride *et al.*, 1978). Evidence suggests that tumor progression takes place in birds unable to respond to tumor antigens resembling their particular MHC antigens (Heinzelmann *et al.*, 1981a, b). They showed that regressor chickens made tolerant to lysed white blood cells from progressor birds became more susceptible to tumors, and demonstrated a serological cross-reaction between tumor cells and an MHC antigen. Ability to regress sarcomas could be transferred by blood lymphocytes and macrophages from regressor chickens to histocompatible progressor chickens (Whitfill *et al.*, 1986). For a fuller discussion of the role of MHC in tumor development, see Collins and Zsigray (1984), Bumstead (1985), Schierman and Collins (1987), and Gavora (1990).

Intrinsic resistance of the target B cell to neoplastic transformation has been identified as of major importance in one strain of fowl, line 6_1, which is susceptible to infection but resistant to lymphoid leukosis

(Purchase *et al.*, 1977b). The genetic basis for this resistance is not known. No obvious difference in the pattern of bursal infection by ALV in susceptible and resistant lines was detected (Fung *et al.*, 1982; Baba and Humphries, 1984).

Some influence of the lymphocyte antigen *Bu-1* locus on Rous sarcoma regression and of the *Th-1* locus on LL has also been reported (Bacon *et al.*, 1985).

D. Pathology and Pathogenesis

As discussed in Section III.A.3a, ALSVs induce a wide variety of neoplasms. A particular virus often produces a predominant type of tumor, but it may also induce other neoplasms (Fig. 3). The tumor spectrum can be wide and often overlaps with that of other virus strains. In this section the pathology and pathogenesis of different neoplasms and nonneoplastic conditions will be discussed, usually without regard to the inducing agent (see also Purchase, 1987; L. N. Payne and Purchase, 1991).

1. Nonneoplastic Effects

Experimental infection with many strains of ALV, particularly with large doses of virus inoculated into young chickens or chicken embryos, often causes depressed growth rate and stunting of chickens. The virus multiplies in most tissues and organs of the body (Dougherty and Di Stefano, 1967; Di Stefano and Dougherty, 1968; Welt *et al.*, 1977), although different strains vary in tissue tropism (Brown and Robinson, 1988). Transitory nonneoplastic lymphoid hyperplasias occur after ALV infection in chickens (Calnek, 1968a) and turkeys (Elmubarak *et al.*, 1983).

Some leukosis viruses, reputedly those of subgroups B and D (R. E. Smith and Schmidt, 1982), cause anemia, due to failure of erythrocytes to incorporate iron into hemoglobin (Cummins and Smith, 1988). Many viruses cause immunosuppression (see Section III.E.4).

RAV-7 causes obesity, high triglyceride and cholesterol levels, reduced thyroxine levels, and increased insulin levels, together with stunting and lymphoid atrophy (Carter and Smith, 1984). Neurological signs resulting from a meningoencephalomyelitis have also been associated with infection with this virus strain (Whalen *et al.*, 1988).

An affinity of ALV for myocardial cells, with viral matrix inclusion body formation and myocarditis, was reported by Gilka and Spencer (1985, 1990) and this infection may lead to a chronic cardiovascular disease and ascites.

Pathophysiological effects of ALV infection must be responsible for a variety of changes in important production traits which have been observed in naturally infected chickens. These effects were recognized during investigations into procedures to eradicate exogenous ALV from commercial breeding stock as a means of disease control (see Section III.G.1). Gavora *et al.* (1980) observed that hens that shed ALV into their eggs, when compared to nonshedder hens, matured later sexually and produced 20–35 fewer eggs during the production period. The eggs were smaller, with thinner shells, and the fertility rate was 2.4% lower and hatchability 12.4% lower. These viruses have similar effects on broiler breeders and also caused a consistent but small reduction in broiler growth rate (Gavora *et al.*, 1982; Crittenden *et al.*, 1983). Recently, evidence for deleterious effects of ALV infection on semen quality and fertility has been reported (Segura *et al.*, 1988). These and other studies on productivity in ALV-infected chickens, and on the important genetic consequences, have been reviewed by Gavora (1987, 1990). The physiological bases for these effects remain to be studied.

2. Lymphoid Leukosis

Fully developed lymphoid leukosis occurs in chickens of about 4 months of age and older. Exceptionally, certain laboratory recombinant viruses cause lymphoid leukosis within 5–7 weeks (Kanter *et al.*, 1988). Clinical signs are nonspecific and include inappetance, emaciation, and weakness; the comb may be pale, shriveled, or occasionally cyanotic, and abdominal enlargement may be apparent.

Lymphoid leukosis is characterized by great enlargement of the liver caused by infiltrating lymphoblasts; the pattern of involvement is usually diffuse or miliary, but large nodular tumors, and mixed forms, sometimes occur (Figs. 10 and 11). Other organs are usually also tumorous, including the spleen, kidneys, gonads, lungs, thymus, and bone marrow (Gross *et al.*, 1959). Nodular tumorous involvement of the bursa of Fabricius is present in nearly all cases (Cooper *et al.*, 1968). Occasionally a terminal leukemia may be present.

Microscopically the lesions in all affected organs consist of diffuse areas or coalescing foci of extravascular immature lymphoid cells (Fig. 12). In the bursa, a follicular pattern of tumor growth can often be seen. The tumor cells vary slightly in size, but invariably have the morphology of large lymphocytes or lymphoblasts. They have B-cell antigen markers (L. N. Payne and Rennie, 1975) and produce and carry IgM on their surface (Cooper *et al.*, 1974; Neumann and Witter, 1979; Ishii and Oki, 1981). Transplantable lymphoid leukosis tumors can be established (Olson, 1941) and leukotic cell lines derived from transplantable or primary

FIGURE 10. Lymphoid leukosis. Enlarged liver with miliary tumor foci.

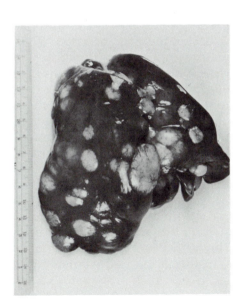

FIGURE 11. Lymphoid leukosis. Nodular lymphomas in the liver.

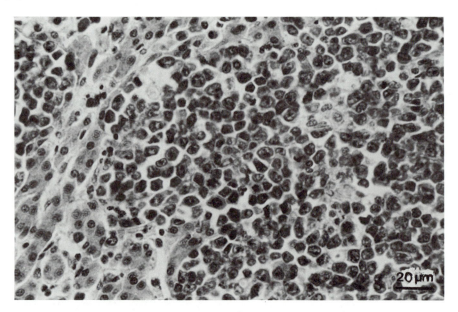

FIGURE 12. Lymphoid leukosis. Focus of neoplastic lymphoblasts in the liver.

lymphoid leukosis tumors (Okazaki *et al.*, 1980; Hihara *et al.*, 1983a; Baba *et al.*, 1985).

The target cells for neoplastic transformation in lymphoid leukosis reside in the bursa of Fabricius and the bursa itself is a target organ for virus (Suni *et al.*, 1978). Medullary macrophages appear to be the principal bursal host cells for virus replication (Gilka and Spencer, 1987). At a variable time after infection, which can be as short as 4 weeks in experimental studies, a proliferation of lymphoblasts can be observed in one or more lymphoid follicles in the bursa. The affected follicle becomes filled and enlarged by proliferating lymphoblasts (Fig. 13). Sometimes many follicles are transformed, but the majority of these appear to regress and only a few continue to grow to give rise to nodular tumors in the bursa which are visible grossly from about 14 weeks of age (Cooper *et al.*, 1968; Neiman *et al.*, 1980). Each of these tumors is a clonal growth. Lymphomagenesis appears to be a multistage process, involving more than one transformational event and activation of at least two genes, c-*myc* and *Blym-1* (Goubin *et al.*, 1983).

Another putative oncogene, c-*bic*, has also been implicated in a late stage of tumor progression or metastasis (Hayward, 1989; Clurman and Hayward, 1989). Humphries and Baba (1984, 1986) regarded the initial transformation in the bursal follicles as a focal preneoplastic hyperplasia, from which progression to frank neoplasia occurs in some follicles. The latency period during the development of the lymphoma is a prop-

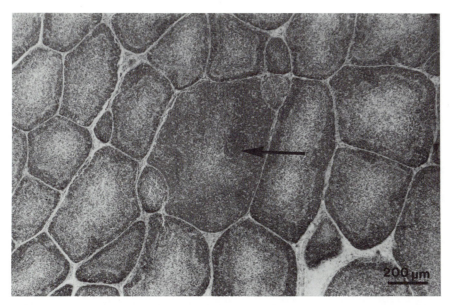

FIGURE 13. Lymphoid leukosis. A transformed lymphoid follicle (arrow) in the bursa of
Fabricius.

erty of the transformed B cell rather than the physiological environment
of the host (Fadly *et al.*, 1981c). Arrest of maturation of the transformed
B cell results in interference of the normal intraclonal switch of immunoglobulin production from IgM to IgG, hence the surface IgM that charac-
terizes lymphoid leukosis cells. Molecular events occurring during lymphomagenesis are reviewed elsewhere (Kung and Maihle, 1987). From
about 12 weeks of age, cells in the clonal tumors metastasize to other
organs and tissues, and result in the florid disease. Metastatic tumors in
the viscera usually have the same DNA fragments as bursal tumors from
the same birds, supporting their bursal origin (Crittenden and Kung,
1984), but multiple bursal tumors can give rise to polyclonal metastatic
disease (Smith *et al.*, 1980).

Kanter *et al.* (1988) and Pizer and Humphries (1989) also induced
B-cell lymphomas by c-*myb* activation following embryonic infection
with ALV: the tumors were unusual in that metastatic disease developed
within 7 weeks of infection, and preneoplastic and primary bursal neoplasms were not detected. Interestingly, early transformed bursal follicles and bursal lymphomas can also be induced by the v-*myc* gene present in MC29 or HB-1 virus (Hayward *et al.*, 1983; Neiman *et al.*, 1985).
Cases of apparent lymphoid leukosis have been observed also in a line of
chickens free from exogenous avian tumor viruses (Crittenden
et al., 1979c).

A consequence of the bursal origin and dependence of lymphoid leukosis is that treatments that destroy or remove the target cell before transformation or metastasis prevent the development of the disease. These treatments include surgical bursectomy between 1 day and 5 months of age (Peterson et al., 1966), treatment of embryos or hatched chickens with androgens or androgen analogues (Burmester, 1969; Kakuk et al., 1977), chemical bursectomy with cyclophosphamide (Purchase and Gilmour, 1975), and infection with infectious bursal disease virus (Purchase and Cheville, 1975). Susceptibility to lymphoma development can be restored in chemically bursectomized birds by transplantation with bursal cells from genetically susceptible chickens, but not with cells from chickens genetically resistant to tumor development although susceptible to infection (Purchase et al., 1977b).

3. Erythroblastosis

Natural cases of erythroblastosis (erythroid leukosis) usually occur in birds between 3 and 6 months of age. Signs are similar to those in lymphoid leukosis. After experimental inoculation of day-old chickens with slowly transforming strains of ALV, such as RPL12 virus, the incubation period varies from 21 to over 100 days (Burmester et al., 1959a). After inoculation of such strains into 11-day-old embryos, erythroblastosis occurs from the first week of age (Piraino et al., 1963). Induction of erythroblastosis by slowly transforming ALV involves activation of the cellular oncogene c-erbB by LTR insertion (Fung et al., 1983) and new AEV strains with transduced c-erbB genes may arise (Miles and Robinson, 1985). Acutely transforming AEV strains, such as ES4 and R, cause mortality from erythroblastosis 7–14 days after inoculation (Graf and Beug, 1978). ES4 carries the v-erbA and v-erbB oncogenes (Enrietto and Hayman, 1987). Hihara et al. (1983b) isolated an acutely transforming AEV, strain H, from erythroblastosis induced by a slowly transforming ALV.

The liver of birds affected by erythroblastosis is moderately swollen, and the spleen often greatly enlarged, from a diffuse intravascular infiltration of neoplastic erythroblasts, and these organs are soft and cherry red in color. The bone marrow is largely replaced by proliferating erythroblasts and is semiliquid and red in color. Petechial hemorrhages occur in muscles, subdermis, and viscera, and there may be ascites and hydropericardium.

Microscopically, the liver sinusoids, splenic red pulp, bone marrow, and sinusoids in other organs are filled by proliferating erythroblasts (Pontén and Thorell, 1957) (Fig. 14). There is a marked leukemia, anemia, and thrombocytopenia; blood smears show many erythroblasts and various immature stages of the erythrocyte series.

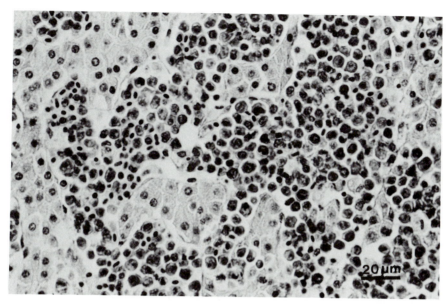

FIGURE 14. Erythroblastosis. Intrasinusoidal accumulations of neoplastic erythroblasts in the liver.

4. Myeloblastosis

Natural cases of myeloblastosis (myeloid leukosis) are uncommon and usually occur in adult birds. Signs are similar to those in lymphoid leukosis. The BAI-A strain of AMV inoculated into day-old chickens induces leukemia and death in about 14 days and mortality continues for several weeks (Eckert *et al.*, 1954; Burmester *et al.*, 1959b; Lagerlöf and Sundelin, 1963a, b).

The liver in myeloblastosis is greatly enlarged and firm with diffuse grayish tumor infiltrates which give a mottled or granular ("Morocco leather") appearance. The spleen and kidneys are also diffusely infiltrated and moderately enlarged, and the bone marrow is replaced by a solid, yellowish-gray tumor cell infiltration. There is a severe leukemia, with myeloblasts comprising up to 75% of peripheral blood cells and forming a thick buffy coat, and usually an anemia and thrombocytopenia.

Microscopically, the parenchymatous organs show a severe intravascular and extravascular infiltration and proliferation by myeloblasts and promyelocytes (Fig. 15).

The v-*myb* gene of AMV is responsible for neoplastic transformation of the target myeloblasts (Enrietto and Hayman, 1987). Experimental infection is followed within a few days by the appearance of multiple foci of proliferating myeloblasts in the extrasinusoidal areas of the bone

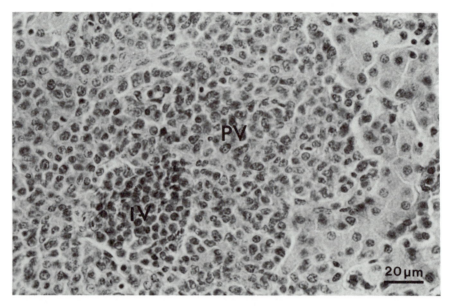

FIGURE 15. Myeloblastosis. Intravascular (IV) and perivascular (PV) accumulations of neoplastic myeloblasts in the liver.

marrow, followed rapidly by leukemia and infiltration of the liver, spleen, and other organs.

5. Myelocytomatosis

Most naturally occurring cases occur in immature chickens. The MC29 strain of virus induces myelocytomatosis in 3–11 weeks (Mladenov *et al.*, 1967). The recently isolated HPRS-103 strain of ALV induced myelocytomatosis, following embryonal infection, from 9 weeks of age (Payne *et al.*, 1991b).

Myelocytomas have a characteristic gross appearance. They usually occur on the surface of bones and are often found at the costochondral junctions of the ribs, and on the sternum, pelvis, mandible, and skull (Fig. 16). Occasionally tumors occur in the viscera. The tumors are often multiple and nodular, with a soft, friable consistency and of a yellowish-white color; they consist of solid masses of well-differentiated myelocytes (Fig. 17). The experimental disease may be leukemic. Ultrastructural features of myelocytoma cells are described by Mladenov *et al.* (1967). The v-*myc* gene of MC29 virus is the element responsible for neoplastic transformation of myelocytes (Enrietto and Hayman, 1987).

After experimental infection, intersinusoidal spaces in the bone marrow become filled by two types of cell, myelocytes and primitive

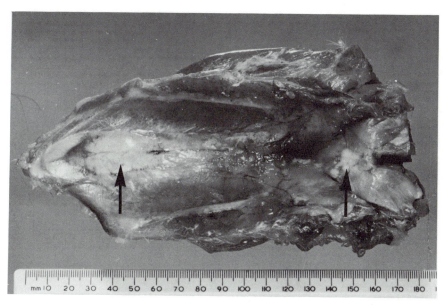

FIGURE 16. Myelocytoma. Tumor (arrows) on the ventral sternum.

hemocytoblast-like cells from which they are derived, and which may be the stem cell of the myelocyte–macrophage series. The myelocytomas extend through the bone and periosteum and some may arise by metastasis.

6. Fibrosarcoma and Other Connective Tissue Tumors

A variety of benign and malignant connective tissue tumors occur naturally, usually sporadically, in young and mature chickens, and transmission of many of these by cell-free filtrates has been demonstrated. These tumors include fibromas and fibrosarcomas, myxomas and myxosarcomas, histiocytic sarcomas, osteomas and osteosarcomas, and chondromas and chondrosarcomas. The benign tumors grow slowly, are localized, and are noninfiltrative. In contrast, the malignant counterparts grow more rapidly, infiltrating surrounding tissue, and they may metastasize. Gross and histological features of these tumors are similar to those in other species and have been described by Campbell (1969), Fredrickson and Helmboldt (1991), and L. N. Payne and Purchase (1991).

The incubation period between infection and the appearance of tumors varies greatly, from a few days with viruses such as RSV to several months for less acutely-transforming strains. The acute transforming sarcoma viruses carry one of a variety of viral oncogenes (Table II)

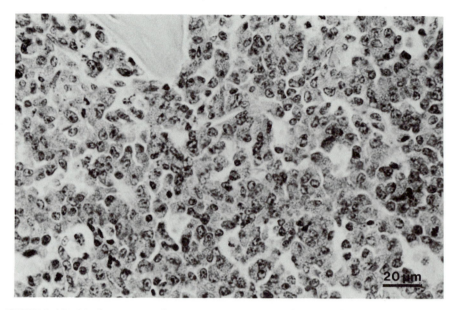

FIGURE 17. Myelocytoma. The tumor is made up of well-differentiated neoplastic myelocytes.

(Enrietto and Hayman, 1987; Wang and Hanafusa, 1988). Sarcoma viruses apparently with unidentified oncogenes also arise (Cavalieri *et al.*, 1985). Slowly transforming viruses also may induce sarcomas, erythroblastosis, and other tumors, presumably by a promoter insertion mechanism or as a result of genetic recombination with a cellular oncogene (Kung and Maihle, 1987; Wang and Hanafusa, 1988). Most virus strains that induce tumors of connective tissue are multipotent, inducing a variety of tumors. Sarcomas can also be induced by inoculation of v-*src* DNA (Halpern *et al.*, 1990).

Recently, the cocarcinogenic effect of wounding on RSV tumorigenesis has been studied (Sieweke *et al.*, 1989, 1990). It appears that cells latently infected with RSV may become activated by wounding, probably as a result of release of the growth factor TGF-β.

7. Hemangioma

This tumor is found sporadically in the skin or in visceral organs in chickens of various ages and may give rise to hemorrhages. Hemangiomas formed the largest group (26.5%) among naturally occurring non-leukotic tumors in broilers (Campbell and Appleby, 1966). They appear as blood-filled cystic or more solid tumor masses, and consist of distended blood-filled spaces lined by endothelium or as more cellular, proliferative, lesions (Campbell, 1969) (Fig. 18). They are often multiple and may rupture, causing fatal hemorrhage. Most field isolates or virus strains of ALV cause hemangiomas, which appear from 3 weeks to 4 months after infection of young chickens (Fredrickson *et al.*, 1964, 1965). Burstein *et al.* (1984) described mortality from outbreaks of hemangiosarcomas in the field in birds of 6–9 months. The virus isolated induced hemangiomas, was cytocidal, and had an affinity for endothelial cells (Resnick-Roguel *et al.*, 1989, 1990; Soffer *et al.*, 1990). An avian angiosarcoma virus has been generated by transduction of c-*erbB* into RAV-1 (Tracy *et al.*, 1985) and subgroup F ring-necked pheasant ALV induced lung angiosarcomas in chickens with a 2 week latency (Carter *et al.*, 1983; Simon *et al.*, 1987).

8. Renal Tumors

Two types of renal tumor occur: nephroblastomas (Wilms' tumor) and adenomas and carcinomas. The nephroblastoma is a common, naturally occurring tumor and accounted for 16.4% of broiler tumors in the series of Campbell and Appleby (1966). Experimentally, nephroblastomas have been induced by BAI-A AMV MAV-2 (N), MAV-2-0 (Burmester *et al.*, 1959b; Baluda and Jamieson, 1961; Walter *et al.*, 1962; Beard, 1980; Watts and Smith, 1980; Böni-Schnetzler *et al.*, 1985), most occur-

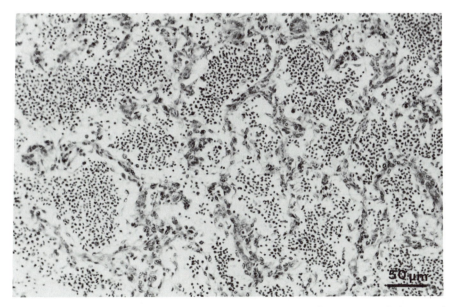

FIGURE 18. Hemangioma. Blood-filled spaces lined by proliferating endothelium in the liver.

ring between 2 and 6 months of age. They vary from small nodules embedded in the kidney parenchyma to large, lobulated masses which replace most of the kidney tissue. Histologically there is much variation between different tumors or areas of the same tumor, with both epithelial and mesenchymal elements, including malformed glomeruli, tubules, and sarcomatous areas (Ishiguro *et al.*, 1962; Helmboldt and Jortner, 1966). Ultrastructural features were described by Heine *et al.* (1962b). They originate from embryonic rests or nephrogenic buds in the kidneys. A target oncogene for ALV-induced nephroblastomas has not been consistently identified (Collart *et al.*, 1990).

Adenomas, adenocarcinomas, and solid carcinomas arise only from the epithelial part of embryonic rests (Carr, 1956, 1960; Chouroulinkov and Rivière, 1959) (Fig. 19). They are induced by the MC29, ES4, and MH2 virus strains and various field isolates. These tumors may be solitary or multiple.

9. Osteopetrosis

Osteopetrosis is characterized by a diffuse, usually symmetrical, overgrowth of the diaphysis of the long bones (especially the tibiotarsus, femur, tarsometatarsus, and humerus) and in the bones of the pelvis, shoulder girdle, and ribs (Figs. 20 and 21). Clinically the shanks are characteristically thickened, and affected birds are usually stunted and

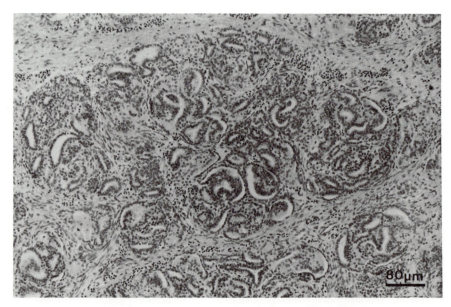

FIGURE 19. Adenomatous nephroma in the kidney.

pale and have a stilted gait. Excessive periosteal bone is deposited which at first is soft and spongy and later very hard (Sanger and Holt, 1982). The marrow cavity becomes obliterated by endosteal bone formation, leading to anemia. Osteopetrosis is most commonly seen in birds 8–12 weeks of age, and in older birds may be accompanied by lymphoid leukosis. A detailed review of osteopetrosis is provided by R. E. Smith (1982).

Experimentally, palpable osteopetrosis can be induced by 7 days after hatching following inoculation of 11- or 12-day-old embryos with MAV-2-0 virus and by 17 days of embryogenesis following infection of the blastoderm (R. M. Franklin and Martin, 1980). Other strains induce the disease from 1 month of age following inoculation of day-old chickens (Holmes, 1964; Sanger et al., 1966; Price and Smith, 1981). Guinea fowl develop multiple bone tumors following inoculation of MAV-2-0

FIGURE 20. Osteopetrosis, showing greatly thickened shanks.

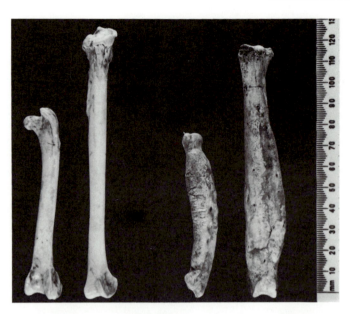

FIGURE 21. Osteopetrosis. Greatly thickened osteopetrotic (right) and normal (left) femur and tibiotarsus.

virus (Kirev, 1988). The propensity for certain strains of ALV to cause osteopetrosis depends on sequences in the *gag–pol* region of the viral genome (Shank *et al.*, 1985). The disease is associated with persistent synthesis of viral DNA, with no evidence of proviral insertions or oncogene transduction (H. L. Robinson and Miles, 1985).

The osteopetrotic lesion is proliferative or hyperplastic and may be neoplastic (Boyde *et al.*, 1978; Schmidt *et al.*, 1981). The periosteum over the lesion is greatly thickened by an increase in the number and size of basophilic osteoblasts (Fig. 22). The number of osteoclasts per tibia increases, but their density decreases (Schmidt *et al.*, 1981). Spongy bone converges centripetally toward the center of the bone shaft, and there is an increase in size and irregularity of the haversian canals and lacunae.

10. Other Tumors

Other rare tumors associated with ALSV infection include mesothelioma, hepatocarcinoma, thecoma, granulosa cell tumor, adenocarcinoma of the pancreas, and squamous cell carcinoma (Beard, 1980). The nature of the so-called endothelioma induced by MH2 (Begg, 1927; Murray and Begg, 1930) and MC29 viruses is debatable: it may be a histiocytic sarcoma (Enrietto *et al.*, 1983) (Fig. 23). The Pts-56 strain of

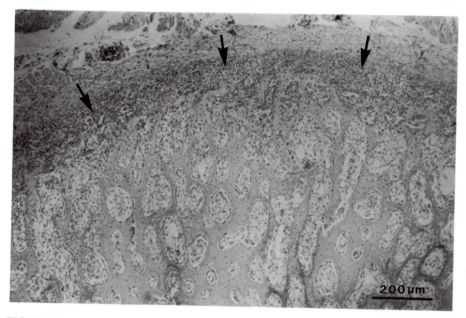

FIGURE 22. Osteopetrosis. Early experimental lesion showing marked proliferation of periosteal osteoblasts (arrows) with excessive formation of bone with enlarged and irregular Haversian canals.

osteopetrosis virus produced bone tumors, pancreatic adenomas and adenocarcinomas, and duodenal papillomas in guinea fowl (Kirev, 1984; Kirev et al., 1986, 1987). Rhabdomyosarcomas of the heart and skin were induced by inoculation of MC29 virus into early chicken embryos (Saule et al., 1987).

E. Immunology

1. Humoral Immunity

Chickens infected horizontally with ALV develop, after a transient viremia, subgroup-specific virus-neutralizing antibodies directed against virus envelope antigens that rise to a high titer and persist throughout the life of the bird (Rubin et al., 1962; Solomon et al., 1966). After inoculation of birds with ALV at 4 weeks of age or older, transient viremia was detectable at 1 week and was followed by antibodies at 3 weeks and later (Maas et al., 1982). In the study of Rubin et al. (1962), birds naturally infected after hatching first developed antibodies at 9 weeks of age, with a marked increase in the proportion with antibodies

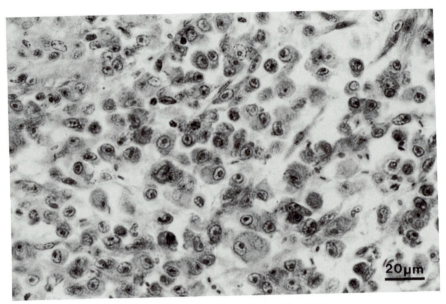

FIGURE 23. "Endothelioma" induced by MH2 virus.

between 14 and 18 weeks of age, when 80% were positive. A similar slow increase in the proportion of antibody-positive birds after contact infection was observed by L. N. Payne *et al.* (1982). Weyl and Dougherty (1977) noted that the younger the bird at infection, the longer the duration of viremia and the greater the delay in antibody production. Birds genetically resistant to infection do not develop antibodies (Crittenden and Okazaki, 1966).

Serum antibodies, which are mainly in the IgG fraction (Meyers and Dougherty, 1972), are passed on by the hen to her progeny via the egg yolk and provide a passive immunity to infection which lasts 3–4 weeks, and which delays infection by ALV (Witter *et al.*, 1966) and reduces the incidence of tumors (Burmester, 1955) and of viremia and ALV shedding (Fadly, 1988). Passively administered antibody decreased or eliminated virus replication in some tissues, and virus multiplication was prolonged in bursectomized birds (Welt *et al.*, 1979). In studies with RSV, Dougherty *et al.* (1960) found that passive antibodies had a small effect on the time of appearance of sarcomas, depending on doses of antibody and virus, but concluded that there was little, if any, effect on established tumors. Subsequent studies on the role of antiviral antibodies have supported this view (Schat, 1987; Wainberg and Halpern, 1987).

Antibodies against gs antigens also occur in ALV-infected birds

(Roth et al., 1971), but these apparently have no influence on tumor growth. Antibodies against reverse transcriptase have also been detected in virus-infected and virus-free chickens (Graevskaya et al., 1982).

An early rise in serum IgG levels, unrelated to neutralizing antibodies was observed following ALV infection (E. J. Smith et al., 1980). IgM was elevated during the leukotic stage, probably shed from lymphoma cells.

2. Cell-Mediated Immunity

Most studies on the role of cell-mediated immunity to avian retrovirus-induced tumors have been conducted using Rous sarcoma, and little has been done on lymphoid leukosis. Several authors have shown that neonatal thymectomy of chickens and quails increased the incidence of Rous sarcomas and prevents tumor regression, thus implicating T-cell-mediated mechanisms in tumor immunity (Radzichovskaja, 1967; Yamanouchi et al., 1971; Cotter et al., 1976a). In quail, antilymphocyte serum prevented regression of Rous sarcomas and increased metastasis (Yamanouchi and Hayami, 1970). Peripheral blood leukocytes from birds with regressing Rous sarcomas showed migration inhibition in vitro in the presence of tumor extracts (Cotter et al., 1976b).

Tests based on the release of lymphokines or on blastogenic transformation also implicate cell-mediated immunity in regression of Rous sarcomas (McArthur et al., 1972; Cotter et al., 1976b; Israel and Wainberg, 1977; Wainberg et al., 1977a, b). In these studies, the target for the immunological response was frequently the tumor cell, and no distinction was made between viral structural antigens and possible neoantigens arising on tumor cells. However, Bauer et al. (1976) and Kurth and Bauer (1972) demonstrated the presence of cytotoxic lymphocytes against viral envelope antigens in birds immunized with ALV or RSV.

Using a leukocyte blastogenesis assay, M. R. Hall et al. (1979) found that cellular immunity to viral envelope components developed 4–5 weeks after inoculation of chickens with RSV, during the period of tumor growth and regression. Cellular responses to putative non-virion transformation-specific cell-surface antigens were weak during this period, but strong following challenge with RSV. Since then, the involvement of non-virion virus-induced antigen has been called into question, and the importance of viral proteins expressed at the tumor cell surface has been emphasized (Wainberg and Halpern, 1987). For detailed review of this complex topic the reader is referred to Wainberg and Phillips (1976), H. Bauer and Fleischer (1981), Wainberg and Halpern (1987), and Schat (1987).

3. Immune Tolerance

Chickens congenitally infected with ALV become immunologically tolerant to the infecting virus, developing a persistent viremia in the absence of neutralizing antibodies (Rubin et al., 1962; Meyers, 1976). Tolerant infection may also be induced by inoculating chickens with ALV at up to 2 weeks of age (Weyl and Dougherty, 1977). Tolerant viremic birds have a hypergammaglobulinemia apparently unrelated to immunity (Qualtiere and Meyers, 1976). Tolerant birds are more likely to develop lymphoid leukosis than are infected immune birds.

Meyers (1976) showed that tolerance to viral envelope antigens of A subgroup ALV which follows congenital infection could be abrogated or circumvented by challenge with other A subgroup viruses, as measured by the virus-neutralizing antibody response. This response was possibly due to multiple viral envelope antigen determinants.

Roth et al. (1971) found that a high proportion of young chickens produced gs antibody after inoculation with A or B subgroup ALV, but that birds congenitally infected with A subgroup ALV were immunologically tolerant and did not respond. The responders made antibody in spite of presumably expressing E subgroup ALV as embryos; this finding is explicable on the basis that the major endogenous ALV gs antigen (p19) differed from the exogenous ALV gs antigen (p27). E. J. Smith et al. (1979) found evidence that A subgroup ALV infection could abrogate tolerance to the p19 of endogenous virus.

Interestingly, and contrary to expectations on the basis of Rubin's (1962) work, the persistent tolerant viremia which follows embryonic infection with ALV may not occur in all circumstances. Nehyba et al. (1990), using ducks as hosts for a td mutant of PR-RSV-C, found that embryo infection led to a transient viremia which was followed by the development of virus-neutralizing antibodies together with persistent infection. The basis for the loss of tolerance requires study.

4. Immunosuppression

Infection by ALV can depress primary and secondary antibody responses and cell-mediated immunity (Peterson et al., 1966; Dent et al., 1968; Purchase et al., 1968; Meyers and Dougherty, 1971), although these effects have been variable in the different studies. Atrophy of the bursa and thymus occurs in chickens bearing Rous sarcomas (Ionescu and Simu, 1972; Fadly and Bacon, 1979). More recent studies have confirmed the variability of some effects and underlined the need for more systematic studies.

Fadly et al. (1982), in a study of congenital infection with an A subgroup virus, RAV-1, failed to detect effects on B- and T-cell functions during the early and late stages of infection, and they reported no histological damage to the bursa, thymus, or spleen. In contrast, B subgroup viruses have been reported to induce a marked suppression of the humoral immune response. MAV-2-0, a helper virus for AMV, caused marked decrease in the IgM, IgG, and IgA responses against sheep red blood cells, *Brucella abortus* antigens, and human γ-globulins, following *in ovo* infection (R. E. Smith and Van Eldik, 1978; Hirota *et al.*, 1980). Subgroup B viruses also caused a marked decrease in responsiveness to several mitogens (R. E. Smith and Van Eldik, 1978; Hirota *et al.*, 1980; Rup *et al.*, 1982; Rao *et al.*, 1990). Price and Smith (1982) were able to restore responsiveness to concanavalin A by addition of macrophages from uninfected birds, suggesting that the immunosuppressive effects of ALV might be mediated by an effect on macrophage function. Rao *et al.* (1990) identified transient T-suppressor cell activity and thymic atrophy in avian erythroblastosis virus-infected chickens. For recent reviews, see R. E. Smith (1987), Wainberg and Halpern (1987), and Storms and Bose (1989).

F. Epizootiology

1. Prevalence of Infection and Disease

Infection of commercial flocks by exogenous ALV is almost certain to be present unless the stock is genetically resistant to infection or efforts have been made to eliminate the infection. Subgroup A viruses occur frequently and subgroup B viruses more rarely (Calnek, 1968b; De Boer *et al.*, 1981). Subgroup C and D viruses have not been widely recognized in the field, but were reported in Finland by Sandelin and Estola (1974). Endogenous viral (*ev*) genes are present in virtually all domestic fowl and in certain other species within the Phasianidae (H. Robinson, 1978; E. J. Smith, 1987). A virus of a new subgroup has been reported recently in meat-type chickens (L. N. Payne *et al.*, 1991).

Although infection by exogenous ALV is very common, the incidence of clinical neoplastic disease is usually low. Mortality from LL is commonly up to about 2% (Randall *et al.*, 1977; De Boer *et al.*, 1981), although occasionally losses can be much higher, e.g., 23% in certain flocks (Purchase *et al.*, 1972). The other leukoses, erythroblastosis, myeloblastosis, and myelocytomastosis, occur sporadically. Rare epizootics have occurred of erythroblastosis (Hamilton and Sawyer, 1939), histiocytic sarcomas (Perek, 1960), and hemangiosarcomas (Burstein *et al.*, 1984). Hemangiomas and nephroblastomas are the most frequently observed nonleukotic tumors, usually occurring sporadically (Campbell and Appleby, 1966; Purchase *et al.*, 1972).

2. Spread of Infection

a. Modes of Transmission

Three modes of natural transmission of ALV occur: (1) Horizontal transmission, in which exogenous virus spreads from bird to bird by direct contact, or indirectly by contact between uninfected birds and fomites. This mode is of particular importance for the high incidence of infection in flocks. (2) Congenital or egg transmission, the form of vertical transmission in which exogenous infectious virus is passed from hen to offspring. Although usually only a small minority of chickens becomes infected in this way, the route is important for maintaining infection from generation to generation, and as a source of virus for horizontal transmission. (3) Genetic transmission, a form of vertical transmission in which endogenous viral genomes, sometimes capable of encoding infectious ALV, but often genetically defective, are transmitted in a Mendelian fashion from parents to offspring (Fig. 7).

b. Horizontal Transmission

Natural sources of virus from infected birds include feces and saliva (Burmester and Gentry, 1954b; Burmester, 1956a) and skin (Spencer and Gilka, 1983). Congenitally infected chickens are an important source of contact infection in the hatchery and during the brooding period (Burmester and Waters, 1955) and meconium from day-old congenitally infected chickens contains high concentrations of ALV (Spencer *et al.*, 1977). Various routes of infection have been identified, notably tracheal, nasal, oral, conjunctival, and cloacal (Burmester and Gentry, 1954c; Weyl and Dougherty, 1977).

c. Congenital Transmission

Direct evidence for the occurrence of congenital transmission was reported by Cottral *et al.* (1949, 1954) in which ALV was detected in livers of congenitally infected embryos. The propensity of infected dams to transmit ALV to embryos was higher in lymphoid leukosis-susceptible infected lines of fowl than in a leukosis-resistant infected line (Burmester *et al.*, 1955). Rubin *et al.* (1961) also demonstrated the presence of ALV in some chicken embryos. Most embryo infection was attributable to ALV viremic hens that lacked antibody; congenital transmission by nonviremic, antibody-positive hens was much more erratic. Viremic roosters, even when their testicular cells produced ALV in cell culture, were shown not to infect their progeny, and later Spencer et al. (1980) showed that insemination with semen containing ALV did not result in infected embryos. Although most workers believe that infected roosters do not give rise to infected embryos, there is some circumstantial evidence to the contrary (Krasselt et al., 1986).

Embryo infection is believed to result mainly if not entirely from virus that is excreted from the oviduct into the egg albumen and thence to the embryo. Spencer et al. (1976) discovered the presence of virus in the albumen of unincubated eggs, consistent with earlier evidence of Di Stefano and Dougherty (1966) for abundant virus replication throughout the magnum of the oviduct. Subsequently Spencer et al. (1977) observed strong associations between virus in vaginal swabs, egg albumen, and embryos, and these associations have been amply confirmed by others (L. N. Payne, 1987).

In contrast, although ALV can also be detected in the yolk of eggs from viremic hens (Rubin et al., 1961; Spencer et al., 1977), there has been no firm evidence that a transovarial route of infection is important. Nevertheless, De Boer et al. (1980a) believed embryo infection to be more closely related to viremia than to virus in the albumen, and Ignjatovic (1990) found embryo transmission in the absence of detectable virus in the albumen, possibly pointing to a nonoviductal, perhaps ovarial, route of infection.

d. Genetic Transmission

The term "genetic transmission" refers to a form of vertical transmission in which viral genes are transmitted from one generation to the next in a Mendelian fashion along with other host genes (Weiss, 1973, 1975). Genetic transmission requires the viral genes to be present in the genome of germ cells. In the fowl this phenomenon occurs in both sexes, so that either the hen or the rooster can pass retroviral genome to their progeny. This form of transmission applies only to ALV of subgroup E, and has not been recognized to occur naturally for subgroups A, B, C, or D. However, subgroup A and other ALV gene insertions into the chicken germ line have been achieved experimentally (Salter et al., 1986, 1987; Crittenden et al., 1989); for detailed discussion, see Crittenden (1991).

The locations of the subgroup E viral genes in the host cell genome are termed "endogenous viral (ev)" loci, and some 30 such loci have been identified in different strains of fowl (see Section III.A.3e). Some ev loci contain all the proviral genes necessary for the production of infectious ALV particles: such virus, when produced spontaneously or after certain stimulations, may then be transmitted additionally congenitally or horizontally as for exogenous ALV (E. J. Smith *et al.*, 1986a). The ability of endogenous viruses to undergo congenital transmission may be determined by their p27 capsid proteins (H. L. Robinson and Eisenman, 1984). Other *ev* loci are genetically defective and unable to produce virus although viral antigens may be produced.

3. Development of Flock Infection

Patterns of infection within a flock were first described by Rubin *et al.* (1962). Four serological classes of birds are recognized: (1) Viremia, no antibody (V+A−), (2) no viremia, with antibody (V−A+), (3) viremia, with antibody (V+A+), and (4) no viremia, no antibody (V−A−). Birds in the V+A− class come from congenitally infected chickens. They are immunologically tolerant to ALV and consequently lack virus-neutralizing antibodies, but have high levels of viremia and of virus in other tissues. Birds in this class are normally in a minority: 20% of hens in the flock studied by Rubin *et al.* (1962) and 7% and 16% in flocks studied by L. N. Payne *et al.* (1979, 1982). Similarly, the proportion of infected embryos or viremic chickens from infected flocks is low: 5% in the study of Solomon *et al.* (1966), and 6% and 10% in the studies of L. N. Payne *et al.* (1979, 1982). According to Rubin *et al.* (1962), V+A− hens transmit ALV to most (94%) of their progeny, but others have found congenital transmission by such hens to be more variable (L. N. Payne *et al.*, 1979, 1982).

Birds in the V−A+ class acquire their infection by contact after hatching, commonly from congenitally-infected hatchmates. In a flock genetically susceptible to infection, the majority of birds are in this class: 78% of hens in the flock studied by Rubin *et al.* (1962) and 64% and 56% in the flocks studied by L. N. Payne *et al.* (1979, 1982). V−A+ hens transmit virus to their progeny much more erratically than do V+A− hens. Rubin *et al.* (1962) found that 14% of nonviremic (i.e., V−A+) hens were congenital transmitters, with 6% of their embryos being infected. In the study of L. N. Payne *et al.* (1979), 24% of V−A+ hens transmitted virus to their progeny, with 15% of their embryos being infected.

The V+A+ class is small. Rubin *et al.* (1962) found viremia levels to be low and apparently infectious antibody-bound virus could have been responsible. Possibly the class represents a transient phase in birds destined to become V−A+. The fourth class, V−A−, occurs in a susceptible population when horizontal spread of virus is still incomplete, and in populations or individuals genetically resistant to infection. The classes are relevant only in the context of a virus subgroup, and status for one subgroup is independent, as far as is known, of that of another subgroup.

A second way of classifying hens derives from the work of Spencer *et al.* (1976, 1977); some infected hens shed infectious ALV and gs antigen into the albumen of their eggs and this occurrence is correlated with embryo infection. The term "shedding" denotes release of virus into the albumen and environment, and "congenital transmission" denotes infection of the embryo. Accordingly, hens may be classified as shedders or nonshedders. Figure 24 shows the epizootiological relationships between the different classes of hens, based on published information.

4. Factors Affecting Epizootiology

a. Viral Factors

Virus isolates or strains vary in the variety and incidence of the different types of tumor they induce under given conditions. These differences appear to be a consequence of genetic variation between viruses and because most strains are a mixture of different viral entities. No influence of virus subgroup per se on the incidence of lymphoid leukosis or other neoplasms has been observed, although the ability of ALV to cause anemia does appear to depend on subgroups, particularly subgroups B and D (see Section III.D.1). Strains of ALV also vary in the persistence of the viremia they induce, in the antibody response, in shedding rates, and in tolerance production (Crittenden *et al.*, 1984a; Fadly *et al.*, 1987). In experimental studies, increasing the dose of ALV increases the incidence of lymphoid leukosis, erythroblastosis, and other tumors (Burmester *et al.*, 1959a; Fredrickson *et al.*, 1964). However, in the field, erythroblastosis and nonleukotic tumors are not common, even though some birds will have been infected with high doses of ALV as a result of congenital infection. Route of infection has a marked effect on tumor incidence (Burmester and Gentry, 1954a; Burmester *et al.*, 1959a), presumably by an influence on the dose of virus received.

b. Host Factors

The importance of host genetic factors on infection and disease has been discussed in Sections III.C.2 and III.C.3. Sex of the bird can affect disease incidence but not infection. Burmester (1945) found that lymphoid leukosis (lymphomatosis) was twice as common in females than in males when the overall incidence of disease was low. Castration abolished the resistance seen in males, and this was restored in capons by treatment with testosterone (Burmester and Nelson, 1945). The sex difference is believed to be a consequence of the earlier natural regression of the bursa in males, under the influence of testosterone (Cooper *et al.*, 1968). A sex-maternal interaction on overall incidence of lymphoid leu-

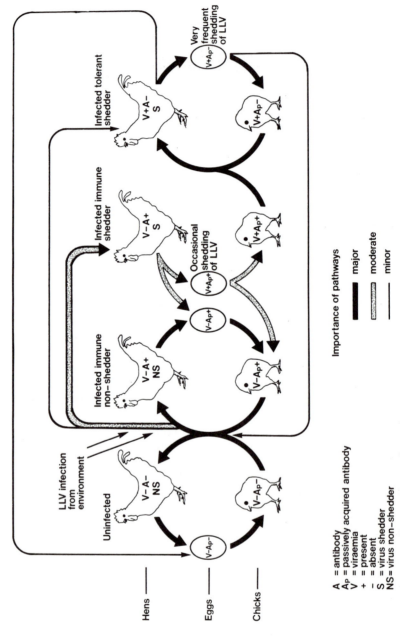

FIGURE 24. Different classes of uninfected and lymphoid leukosis virus-infected chickens in relation to congenital and horizontal transmission of the virus.

kosis and erythroblastosis, suggestive of an influence of sex-linked genes, was discussed by Crittenden *et al.* (1972) in diallel crosses exposed to ALV.

Age at infection markedly influences the course of infection and disease incidence. Congenital infection, leading to immunological tolerance and high viremia, is associated with a high incidence of disease (Rubin, 1962). Burmester *et al.* (1959a, 1960) clearly showed that with increasing age at inoculation from 1 day to 168 days of age, overall neoplastic mortality from erythroblastosis or lymphoid leukosis (visceral lymphomatosis) declined, and early high mortality from erythroblastosis gave way to later mortality from lymphoid leukosis (Fig. 5). As discussed in Section III.E.1, passively acquired antibodies delay ALV infection and its consequent effects.

c. Presence of Endogenous Leukosis Virus

Endogenous viruses and *ev* loci influence the response of the bird to infection by exogenous ALV. Birds lacking *ev* gene expression, compared with their positive counterparts, showed, following exogenous ALV infection, and depending on viral strain, a high incidence of an acute runting syndrome (nonneoplastic syndrome). They also had higher virus-neutralizing antibody responses, lower viremia, and probably lower virus shedding from the cloaca (Crittenden *et al.*, 1982, 1984a; Crittenden and Fadly, 1985; E. J. Smith and Fadly, 1988). Mortality from neoplasms in these chickens can be increased. These effects are believed to be due to the absence of partial tolerance to exogenous ALV induced by expression of the endogenous virus. Infection of susceptible chicken embryos with the endogenous virus RAV-0 prolonged the viremia and increased the frequency of neoplasms following infection posthatching with exogenous ALV (Crittenden *et al.*, 1987).

An association between the presence of the sex-linked slow-feathering gene *K*, important commercially in autosexing of chickens, and increased congenital and horizontal transmission of exogenous ALV has been noted. It is apparently due to the close linkage on the Z chromosome of the *K* gene and the *ev21* locus, which codes for the complete endogenous virus, EV21, and has a consequent tolerizing effect on response to exogenous ALV (D. L. Harris *et al.*, 1984; Bacon *et al.*, 1988). Strategies to abrogate this effect have been studied by E. J. Smith *et al.* (1990a, b). For detailed discussion, see Crittenden (1991).

d. Other Intercurrent Infections

Infection by various immunosuppressive viruses can influence ALV infection. Infectious bursal disease virus, the cause of Gumboro disease,

reduced markedly the incidence of lymphoid leukosis when administered to ALV-infected chickens, presumably as a consequence of its de-
structive effect on target cells for leukotic transformation in the bursa of
Fabricius (Purchase and Cheville, 1975). A moderately virulent strain of
infectious bursal disease virus used to vaccinate against Gumboro disease similarly reduced leukosis, whereas an avirulent vaccinal strain did
not (Cheville et al., 1978). Infection by bursal disease virus also increased cloacal shedding of ALV (Fadly et al., 1985).

Infection by Marek's disease virus or reticuloendotheliosis virus
also increased cloacal shedding of ALV and/or ALV viremia (Fadly et al.,
1985). Bacon et al. (1989) reported that the use of SB-1 (serotype 2)
Marek's disease virus vaccines increased the incidence of lymphoid leukosis in the field and experimentally. The mechanism for this phenomenon is not known.

e. Environmental Factors

Influences of certain management practices and dietary factors on
ALV infection and disease have been reported. These include practices
which increase contact infection, such as manual vent-sexing, rearing on
solid floors, and rearing with infected chicks (Fadly et al., 1981b). In
certain circumstances ALV can be transferred from chick to chick by the
needle during vaccination against Marek's disease (De Boer et al.,
1980b). The infection can also be transferred to uninfected hens by artificial insemination with infected semen (Spencer et al., 1980). Several
dietary factors have been reported to increase the incidence of lymphoid
leukosis (L. N. Payne, 1987).

Attempts to influence ALV infection and shedding by various stressors were not successful (Fadly et al., 1989).

G. Prevention and Control

1. Eradication

Eradication of exogenous ALV from a flock depends on breaking the
vertical transmission of virus from dam to progeny, and prevention of
reinfection. Various eradication procedures have been used for about 30
years in the development of specific pathogen-free flocks for research
purposes or vaccine production (Payne and Purchase, 1991). More recently, however, techniques developed by Spencer et al. (1977) and
others have allowed eradication procedures to be used on a commercial
scale by poultry breeding companies (Spencer, 1984; De Boer, 1987).
These procedures depend on the close associations between virus infec-

tions in hens, egg albumen, embryos, and chicks (Section III.F.2c). Hens with a low probability of producing infected embryos are selected for providing replacement progeny; these hens are those negative for ALV or gs antigen in egg albumen or in vaginal swabs.

A procedure for eradication of ALV involves: (1) selection of fertile eggs from hens negative for ALV in tests on egg albumen or vaginal swabs, usually using the ELISA for gs antigen; (2) hatching of chicks in isolation in small groups in wire-floored cages and avoiding manual vent-sexing and vaccination with a common needle to prevent mechanical spread of any residual infection; (3) testing of newly-hatched chicks for gs antigen or ALV in blood or meconium, discarding residual reactors and contact chicks; and (4) rearing ALV-free groups in isolation (Okazaki et al., 1979; Fadly et al., 1981a, b; L. N. Payne et al., 1982; Crittenden et al., 1984b; L. N. Payne and Howes, 1991). Cocks may be selected for freedom from ALV infection by tests on cloacal swabs or semen samples. Selection of hens with a low shedding rate is simpler than the subsequent chick testing and isolation rearing needed to achieve complete eradication, and consequently some commercial breeders concentrate on reduction of infection rate by hen testing only.

In some instances shedding in the albumen of gs antigen originating from endogenous ALV can interfere with ALV eradication programs by giving "false-positive" test reactions. Titers of endogenous gs antigen are usually lower than those of exogenous virus and can be discriminated by testing dilutions of albumen (Crittenden and Smith, 1984; Ignjatovic, 1986). Monoclonal antibodies which discriminate between endogenous and exogenous gs antigen may also become available (Lee et al., 1986).

2. Selection for Genetic Resistance

As an alternative or adjunct to eradication procedures, some poultry breeders select for genetic resistance to infection by exogenous ALV. The frequencies of the alleles which encode cellular susceptibility or resistance to infection vary greatly among commercial lines of chickens (Crittenden and Motta, 1969; Motta et al., 1973) (see Section III.C.1). In some lines high frequencies of a resistant allele may be found naturally; in others, frequencies of the resistance alleles can be increased by artificial selection. Usually emphasis is placed on the A and B subgroups.

In artificial selection, genotypes of unknown parents may be determined in a progeny test by mating them to recessive tester birds for the subgroup in question [e.g., $A^r A^r$ for subgroup A virus (Pani and Biggs, 1973)]. Depending on the segregation of susceptible and resistant progeny in a particular mating, the genotype of the unknown parent may be deduced. Progeny phenotypes may be determined by inoculation of RSV

onto the chorioallantoic membrane, the embryos being scored as susceptible or resistant on the basis of the pock count (L. N. Payne, 1985b) (Fig.
25). The phenotypes of hatched birds may also be determined directly by
RSV susceptibility tests on fibroblasts cultured from pulp of plucked pin
feathers (Crittenden *et al.*, 1971; Payne *et al.*, 1985; Spencer *et al.*, 1987).
In the future, transgenic techniques are likely to be used to produce
genetically resistant stock (Salter and Crittenden, 1989; Federspiel
et al., 1991).

3. Vaccination

Although use of antiviral vaccines to increase resistance is attrac-
tive, attempts to produce killed or attenuated vaccines had limited or no
success (Burmester *et al.*, 1957; Okazaki *et al.*, 1982). More recently,
Heider and Klaczinski (1986) and Bennett and Wright (1987) obtained
protection against AMV and RSV tumors with envelope glycoprotein
preparations, and a recombinant ALV with characteristics of RAV-0 yet
expressing subgroup A envelope glycoproteins has also been produced
which could have potential as a vaccine (McBride and Shuman, 1988;

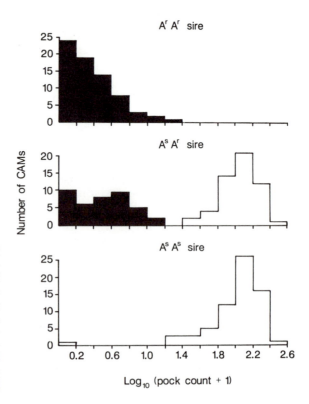

FIGURE 25. Distribution of
pock counts on chorioallantoic membranes (CAMs) of
embryos from crosses be-
tween commercial A^rA^r tester
dams and putative A^rA^r, A^sA^r,
and A^sA^s sires after inoculation of subgroup A Rous sarcoma virus. [From Pani and
Biggs (1973) by kind permission of authors and publisher.]

Shuman and McBride, 1988). The value of any such vaccine would lie in providing active and passive immunity to horizontal infection, for use as an adjunct to reduction of congenital infection.

IV. RETICULOENDOTHELIOSIS VIRUS

A. Virology

1. Virus Morphology and Morphogenesis

In morphology and size, various isolates of reticuloendotheliosis virus (REV) are identical and the virions are about 75–100 nm in diameter (Zeigel et al., 1966; Padgett et al., 1971) (Fig. 2). The virion consists of a central shell-like core, a dense, homogeneous intermediate layer, and an outer envelope bearing surface projections about 6 nm long and 10 nm in diameter. REV virions can be distinguished from those of ALSV in that immature forms of the latter lack a distinct intermediate membrane and the nucleoid core of mature ALSV is smaller and denser than that of REV (Zeigel et al., 1966; Moelling et al., 1975). REV virions resemble murine leukemia virus more closely, and other evidence supports a closer relationship with mammalian retroviruses than with ALSVs. REV virions proliferate by budding from the cytoplasmic membrane of connective tissue and reticular and endothelial cells. The nucleoid forms as an electron-opaque crescent, surrounded by the intermediate layer and the envelope membrane.

2. Physical and Chemical Properties

The buoyant density of REV virions in sucrose is 1.16–1.18 g/cm³ (Baxter-Gabbard et al., 1971). Genomic single-stranded RNA consists of a 60–70S dimer containing two 30–40S subunits each of about 3.9×10^6 daltons in size (Moldonado and Bose, 1975; Beemon et al., 1976). Nondefective REV has a genome of about 9.0 kilobases (kb) and consists of gag, pol, and env genes. The replication-defective strain T REV has a smaller genome of about 5.7 kb due mainly to deletions in the gag–pol and env regions, but it also carries a substitution of 0.8–1.5 kb in the env region identified as the transforming gene v-rel most probably derived from the cellular oncogene c-rel present in normal turkey cells (Weiss et al., 1982; Moore and Bose, 1988; Witter, 1991).

The gag gene encodes five structural proteins, p10, p12, pp18, pp20, and p30, of which p30 (30 kD) is the major REV gs antigen (Tsai et al., 1985). The env gene encodes two envelope glycoproteins, gp90 and gp120 (Tsai et al., 1986). The pol gene encodes a reverse transcriptase which differs from that of the ALSV but is related to the enzyme of mammalian type retroviruses (Moelling *et al.*, 1975; G. Bauer and Temin, 1980).

3. Classification and Strains

The REV group was first characterized by Purchase *et al.* (1973), and included initially REV isolated from turkeys, duck infectious anemia virus (DIAV) and spleen necrosis virus (SNV) isolated from ducks, and chicken syncytial virus (CSV) isolated from chickens. Subsequently many other isolates from chickens, ducks, geese, pheasants, and turkeys have been included, as listed by P.-Y. Chen *et al.* (1987). With the exception of REV strain T, all REV isolates are nondefective. Chen *et al.* compared 26 isolates by neutralization assays and concluded that they probably constituted a single serotype. However, three antigenic subtypes were suggested by minor differences in neutralization titers and by monoclonal antibodies, which recognized at least three epitopes (Cui *et al.*, 1986). REV strain T was included in subtype 1, DIAV and SNV in subtype 2, and CSV in subtype 3 (P.-Y. Chen *et al.*, 1987). Viruses of subtypes 1 and 2 could not be differentiated by receptor interference tests, indicating the absence of major subgroup differences (Federspiel *et al.*, 1989).

The nondefective REVs differ in pathogenicity in ducks and chickens as determined by occurrence of reticuloendotheliosis, anemia, and nerve lesions (Purchase *et al.*, 1973; Purchase and Witter, 1975). DIAV and CSV were less pathogenic than SNV and some strains of REV. The replication-defective strain T REV, which carries the oncogene v-*rel* and requires a nondefective REV as a helper virus for replication, differs from the other REVs by being rapidly and highly oncogenic (Sevoian *et al.*, 1964; Theilen *et al.*, 1966; Mussman and Twiehaus, 1971; F. R. Robinson and Twiehaus, 1974).

4. Sources for Virus Isolation

Tumor tissue when available is the best source of REV, but virus may also be isolated from spleen or other tissues, and whole blood (Witter, 1989, 1991). Cellular inocula are generally better than cell-free inocula. In tolerant viremic, congenitally infected birds, plasma or serum is a good source. In congenital infections, virus may also be isolated sometimes from embryos and rarely from albumen.

5. Propagation and Assay

A variety of birds, including chickens, ducks, geese, quail, and turkeys, are susceptible to infection by REV and can be used for virus propagation and assay (Purchase and Witter, 1975; Witter, 1989, 1991). Chicken embryos are also susceptible and can be used for bioassay (Sevoian et al., 1964; Theilen et al., 1966).

Nondefective REV replicates in cultured fibroblasts from several avian species, including chicken, duck, turkey, and quail (Witter, 1989), and also in QT35 quail sarcoma cells and D17 dog osteosarcoma cells (Barbacid et al., 1979; Cowen and Braune, 1988). Initially REV is cytocidal, but then a chronic infection is established (Temin and Kassner, 1975). It is suggested that the cytocidal effect is a result of superinfection of cells already infected by REV, leading to an accumulation of unintegrated DNA which in some way is cytocidal (Keshet and Temin, 1979; Weller and Temin, 1981). The defective strain T REV, with its nondefective REV helper virus, loses its oncogenicity during passage in fibroblast cultures, due to loss of the defective viral component (Campbell et al., 1971; Breitman et al., 1980). Oncogenicity is maintained during culture of hematopoietic cells (Hoelzer et al., 1979) and in transformed chicken embryo fibroblasts (R. B. Franklin et al., 1977).

B. Diagnostic Tests

Virus may be detected by the presence of viral antigens, cytopathic effects with some strains, and reverse transcriptase. Under agar, foci of cells containing fluorescent antigen can be used to quantify infectious virus (Purchase et al., 1973), and a plaque assay has been described (Temin and Kassner, 1974). Viral antigens may also be detected by complement-fixation tests (E. J. Smith et al., 1977) or the more sensitive ELISA (Ignjatovic et al., 1987; Cui et al., 1988). Antibodies to REV may be detected by the indirect fluorescent test, virus neutralization, agar gel precipitin test, or ELISA test (Witter, 1989). Nondefective REV forms pseudotypes with defective RSV (Vogt et al., 1977) and vesicular stomatitis virus (Kang and Lambright, 1977), which can be used in neutralization tests to detect REV antibodies.

C. Host Range

Natural hosts for REV infection and associated disease include chickens, ducks, geese, Japanese quail, and turkeys, the last being the species most frequently associated with the infection. Cells from these species are also susceptible to infection in vitro.

REV has not been recognized in nonavian species in vivo, but viral replication does occur in certain mammalian cells in culture (Witter, 1991).

D. Pathology and Pathogenesis

1. Runting Disease Syndrome

Nondefective strains of REV can cause a variety of nonneoplastic lesions in chickens and ducks, including anemia (Ludford et al., 1972; Kawamura et al., 1976), abnormal feather growth ("nakanuke") (Koyama et al., 1980), atrophy of the bursa and thymus (Mussman and Twiehaus, 1971), enlarged peripheral nerves due to lymphoid cell infiltration (Witter et al., 1970), enteritis (McDougall et al., 1978b), proventriculitis (Jackson et al., 1977), necrosis of liver and spleen (Purchase and Witter, 1975), and runting (Theilen et al., 1966; Witter et al., 1970; Mussman and Twiehaus, 1971).

2. Acute Reticulum Cell Neoplasia

This form of the disease, also termed reticuloendotheliosis, is induced by the replication-defective strain T virus. It causes death commonly 1–3 weeks after inoculation, and is characterized by enlargement of the liver and spleen with diffuse or focal infiltrative lesion, and commonly lesions also in the gonads, heart, kidney, and pancreas, and there may be a terminal leukemia (Sevoian et al., 1964; Theilen et al., 1966; Olson, 1967; Mussman and Twiehaus, 1971; F. R. Robinson and Twiehaus, 1974). There are few clinical signs until a few hours before death, when the birds become lethargic (Purchase and Witter, 1975).

Microscopically the neoplastic infiltrations are made up of large vesicular cells, considered to be primitive mesenchymal cells (Sevoian et al., 1964; F. R. Robinson and Twiehaus, 1974) or reticuloendothelial cells (Theilen et al., 1966). In some instances more differentiated lymphocytes are present in the lesion. A number of cell lines have been established from these tumors (R. B. Franklin et al., 1974; Keller et al., 1979; Koyama et al., 1981).

The nature of the target cell in the disease is unsettled. Various cell lines transformed by strain T virus expressed markers of immature B lymphocytes, in some instances with IgM markers (Beug et al., 1981; Lewis et al., 1981; Shibuya et al., 1982; Benatar et al., 1991), or in addition markers of immature T cells (Beug et al., 1981). Other evidence supports the idea that there is more than one type of target cell (Barth and Humphries, 1988). This form of the disease is not prevented by neonatal thymectomy or bursectomy (Thompson and Linna, 1973).

3. Chronic Lymphoid Neoplasia

In chickens, two types of chronic lymphomatous disease have been induced experimentally with nondefective REV. One form, described by

Witter and Crittenden (1979), had a long latent period and was similar pathologically to lymphoid leukosis induced by ALVs (see Section III.D.2), with bursal and liver involvement by IgM-bearing lymphoma cells, and prevention by surgical or chemical bursectomy (Witter *et al.*, 1981; Fadly and Witter, 1983). Furthermore, lymphoma induction was associated with integration of REV proviral genome adjacent to c-*myc* (Noori-Daloii *et al.*, 1981; Swift *et al.*, 1985). An IgM-producing B-lymphoblastoid cell line was established from lymphomas by Nazerian *et al.* (1982).

The second type of chronic lymphoma in chickens occurs more rapidly, by 6 weeks, and involves the thymus, liver, spleen, heart, and peripheral nerves but not the bursa of Fabricius (Witter *et al.*, 1986). It is a T-cell neoplasm associated with activation of c-*myc* by REV insertional mutagenesis (Isfort *et al.*, 1987).

In turkeys, REV causes a lymphomatous disease involving the liver and other organs, and occurs about 8–12 weeks after experimental inoculation (Paul *et al.*, 1977; McDougall *et al.*, 1978b) or between 15 and 20 weeks following natural infection (Paul *et al.*, 1976; McDougall *et al.*, 1978b). Chronic lymphomas caused by REV have also been described in ducks (Perk *et al.*, 1982; Li *et al.*, 1983), geese (Dren *et al.*, 1988), pheasants (Dren *et al.*, 1983), and quail (Schat *et al.*, 1976). Histiocytic and spindle cell sarcomas may also be induced (Li *et al.*, 1983).

E. Immunology

1. Humoral Immunity

Infection by REV after hatching results, after several weeks, in the development of antibodies which persist and which may be detected in serum using a variety of methods, including virus neutralization (Purchase *et al.*, 1973; Crittenden *et al.*, 1982), agar gel precipitin (Ianconescu, 1977), immunofluorescence (Aulisio and Shelakov, 1969), and enzyme immunoassay (E. J. Smith and Witter, 1983). These antibodies curtail the initial viremia which follows infection, and apparently control REV-induced tumors. Neonatal bursectomy causes increased mortality after inoculation of strain T REV (Linna *et al.*, 1974), and administration of immune serum increases tumor resistance (Hu and Linna, 1976).

2. Cell-Mediated Immunity

Thymectomy increased mortality from strain T REV, suggesting that cell-mediated immunity was important in protection (Linna *et al.*,

1974), and these responses appeared to be directed against viral antigens and to be MHC-restricted (Maccubbin and Schierman, 1986).

3. Immune Tolerance

Inoculation of REV into chicken embryos (Ianconescu and Aharonovici, 1978; Witter et al., 1981) results in a persistent viremia without the development of neutralizing antibodies. Tolerant infection can also be produced, but more variably, by infection of chickens or turkeys at hatching (Bagust and Grimes, 1979; McDougall et al., 1980; Witter et al., 1981).

4. Immunosuppression

REV can have immunosuppressive effects on both humoral and cell-mediated immunity. T strain REV in chickens (Witter et al., 1979) and RU-1 strain in ducks (Li et al., 1983) caused decreased humoral responses against several antigens, varying with challenge dose and virus strain. Primary responses were more severely affected than secondary.

Infection with T strain REV caused depressed responses to several mitogens, mediated via a population of suppressor cells (Carpenter et al., 1977; Scofield and Bose, 1978; Walker et al., 1983). Purified virus or REV-transformed cells themselves were not able to suppress the blastogenic response of normal lymphocytes (Carpenter et al., 1978). Suppression of cytotoxic activity against T strain REV tumor cells was believed to be important in allowing the development of the acute leukemic disease (Walker et al., 1983). For recent review, see Storms and Bose (1989).

The immunosuppressive properties of REV have attracted practical attention because of its ability to exacerbate other infections such as those of Eimera tenella, Salmonella typhimurium, and fowl pox virus (Motha and Egerton, 1983, 1984, 1987), and to impair vaccinal immunity, such as to Newcastle disease (Ianconescu and Aharonovici, 1978) and Marek's disease (Bülow, 1977; Witter et al., 1979). Furthermore, contamination of Marek's disease vaccines by REV has caused deleterious immunosuppressive and other effects (Kawamura et al., 1976; Koyama et al., 1976; Jackson et al., 1977).

F. Epizootiology

1. Prevalence of Infection and Disease

REV infection occurs in a variety of avian species and the infection is more prevalent than the clinical disease. Antibodies are present, with geographical variations, in a significant proportion of commercial layer and meat-type chickens and turkey flocks in the United States, but no disease or deleterious effect on performance was evident (Witter *et al.*, 1982; Witter and Johnson, 1985).

Chronic lymphomatous disease associated with REV has been observed in various countries in ducks (Grimes and Purchase, 1973; Paul and Werdin, 1978; Perk *et al.*, 1982), geese (Dren *et al.*, 1988), pheasants (Dren *et al.*, 1983), quail (Carlson *et al.*, 1974; Schat *et al.*, 1976), and turkeys (F. R. Robinson and Twiehaus, 1974; Paul *et al.*, 1976; Solomon *et al.*, 1976; McDougall *et al.*, 1978b; Witter and Glass, 1984; Witter and Salter, 1989).

2. Spread of Infection

REV is transmitted both horizontally and vertically. Horizontal infection occurs by contact with infected chickens and turkeys and the virus is present in oral-nasal washings, feces, and litter from infected flocks (Peterson and Levine, 1971; Witter *et al.*, 1981). There is some evidence that mosquitoes may play a role in transmission in particular circumstances (Motha *et al.*, 1984).

Vertical transmission occurs through the egg from tolerantly infected chicken and turkey dams, usually at a low rate (McDougall *et al.*, 1980; Bagust *et al.*, 1981; Witter *et al.*, 1981). Such transmission from nontolerantly infected chickens and turkeys is uncommon (Witter and Salter, 1989). Evidence exists that vertical transmission can also occur from infected male chickens and turkeys (McDougall *et al.*, 1980; Salter *et al.*, 1986). There is no evidence for natural genetic transmission of REV; however, there is evidence for experimental insertion of REV genes into chicken germ line (Salter *et al.*, 1986).

G. Prevention and Control

Because of the usually sporadic and subclinical nature of REV infections, no control procedures have been necessary commercially. However, it seems probable that eradication could be achieved by prevention of vertical transmission by testing albumen samples for REV gs antigen, testing of males, and rearing of progeny in isolation (Witter and Salter, 1989), essentially as for the eradication of ALV (see Section III.G.1).

V. LYMPHOPROLIFERATIVE DISEASE VIRUS OF TURKEYS

A. Virology and Diagnostic Tests

Studies on the virology of lymphoproliferative disease virus (LPDV) of turkeys have been handicapped by the lack of an *in vitro* cell culture system (McDougall et al., 1978a). Consequently, virus characterization has depended on use of virus prepared from tissues or plasma from turkeys affected with LPD, with which the disease can be reproduced experimentally.

The LPD-associated virus is a C-type retrovirus measuring 90–120 nm in diameter (Biggs et al., 1978; McDougall et al., 1978a; Perk et al., 1979). (Fig. 2). It has an electron-dense core, with a less dense intermediate layer and an outer envelope, and replicates by budding from the membrane of cells in the lymphoproliferative lesions.

The buoyant density of the virions is 1.16–1.18 g/cm^3 in sucrose gradients (McDougall et al., 1978a; Yaniv et al., 1979). Genomic RNA has a sedimentation coefficient of about 70S. The virus has a reverse transcriptase which has a preference for Mg^{2+} over Mn^{2+} in both endogenous and exogenous reactions, differing in this respect from REV (Yaniv et al., 1979). There is no nucleic acid sequence homology between LPDV and ALSV or REV (Yaniv et al., 1979) or serological relationship (McDougall et al., 1978a; Patel and Shilleto, 1987a). More recently however, Gak et al. (1989, 1991) have suggested an evolutionary relationship with the ALSV group based on relatedness of the pol genes. No evidence of any known or new oncogene was found.

Five structural polypeptides with molecular weights of 15, 20, 28, 31, and 76 kDa were described by Gazit et al. (1986). The 76-kDa polypeptide was glycosylated and considered to be the major component of the virion envelope; p28 and p31 were also virus structural proteins. A p42 host protein, identified as actin, was incorporated with the virion. Patel and Shilleto (1987b) identified gp76 as probably a virion envelope protein, p13.5/13 (as probably a ribonucleoprotein), p26 and p32 in the viral core, p22/21 as a probably an intramembranous protein, and two other minor structural proteins, p12 and p41. Further studies are needed to equate the findings of these two groups of workers.

LPDV-specific sequences were not present in normal turkey cells, indicating that the virus is not endogenous in turkeys (Gazit et al., 1979; Gak et al., 1989).

Detection of LPDV has proved difficult because of the lack of tissue culture systems and absence of antibodies in infected turkeys. Recently, an ELISA using antibodies raised in rabbits has been used to detect LPDV in plasma pellets (Patel and Shilleto, 1987a), and hyperimmunized rabbit sera have also been used to detect the virus in buffy coat smears and spleen sections by indirect immunofluorescence (Patel and Shilleto, 1987c).

B. Host Range, Pathology, and Pathogenesis

LPD occurs naturally in turkeys (Biggs et al., 1978; Ianconescu et al., 1979). It can be reproduced experimentally in turkeys (McDougall et al.,

1978a) and chickens but not in ducks or geese (Ianconescu *et al.*, 1983). The natural disease occurs in turkeys mainly between 7 and 18 weeks of age, and males may be more susceptible than females. Clinical signs are few and nonspecific and the course is rapid, with death of up to 25% of the flock. The disease is characterized by marked splenomegaly, affected spleens being pale in color with a marbled appearance. Livers are moderately enlarged, with miliary grayish-white foci, and similar infiltrative lesions also occur in the kidney, gonad, intestine, pancreas, lungs, and myocardium. The thymus is enlarged and congested. In some cases peripheral nerves may be enlarged. Microscopically in all affected organs there is diffuse or focal proliferation of lymphoid cells, including lymphocytes, lymphoblasts, reticulum cells, and plasma cells. Ultrastructurally, mainly intercellular C-type virus particles are seen.

Experimentally, these lymphoproliferative lesions were first seen in the red pulp of the spleen and in the thymus 14 days after inoculation of 4-week-old turkey poults. In the thymus, there was cortical atrophy with lymphoproliferative lesions in the medulla. By 21 days the normal splenic architecture was lost and similar proliferative lesions were observed in other tissues. Focal lesions made up mainly of mature lymphocytes and germinal centers were also seen to increase later in the development of the disease, possibly representing a regressing lesion.

In a study of the organotropism of LPDV using molecular hybridization techniques, Gazit *et al.* (1982) showed the bone marrow to be first infected, with virus extending to the thymus, then to the spleen and bursa of Fabricius, and eventually to other organs. Their findings suggested that LPDV is lymphotropic.

C. Immunology

LPDV infection has some unusual features. For reasons that are not known, poults infected at 4 weeks of age develop a higher incidence of disease than those infected at 1 day of age (McDougall *et al.*, 1978a). Furthermore, infection of 3- or 4-week-old turkeys is followed by a persistent viremia, raised serum IgG levels, but no viral antibodies in the serum (Gazit *et al.*, 1982; Zimber *et al.*, 1983; Patel and Shilleto, 1987a). The development of LPD in turkey poults was not influenced by chemical or surgical bursectomy, suggesting that B cells are not the targets for neoplastic transformation (Zimber *et al.*, 1984). However, the hypergammaglobulinemia of LPD was reduced by chemical bursectomy, and viremia was increased. Infection of intact birds with LPDV did not impair the humoral antibody response to *Brucella abortus* or sheep red blood cells, although some reduction due to infection was observed in the agglutinin responses of surgically bursectomized birds, indicating that LPDV has immunosuppressive activity. Furthermore, turkeys infected with LPDV had marked depression of lymphocyte blastogenic responses to phytohemagglutinin and concanavalin A (Zimber *et al.*, 1983).

D. Epizootiology

LPD was first recognized in turkeys in the United Kingdom in 1973 (Biggs *et al.*, 1974, 1978), causing mortality of up to 20% in a number of rearing farms over a 6-month period in turkeys from 10 to 18 weeks of age. The disease continued in successive crops of turkeys, in spite of complete depopulation between crops, and ceased only when the strain of turkeys was changed. The disease has also been recognized in Israel since 1976, mainly in one of two strains of turkeys (Ianconescu *et al.*, 1979). These two strains did not differ in susceptibility to experimental infection (Ianconescu *et al.*, 1981).

Almost nothing is known of the natural epizootiology of the infection. LPDV is apparently an exogenous virus of turkeys, and spread by contact has been observed experimentally (McDougall *et al.*, 1978a). These workers observed differences in susceptibility to experimental infection among four strains of turkeys. No information is available about the possible occurrence of vertical transmission.

E. Prevention and Control

No methods of prevention or control have been reported. Change of strain of turkey and attention to general hygiene procedures are recommended (Biggs, 1991).

VI. PHEASANT TYPE C ONCOVIRUSES

Fujita *et al.* (1974) isolated an endogenous leukosis-like virus from golden pheasants (*Chrysolophus pictus*) which differed in host range, interference, and envelope antigens from viruses of subgroups A–F, and allocated the virus to a new subgroup G. Similar viruses were isolated from golden pheasant and Amherst pheasant (*Chrysolophus amherstiae*) by T. Hanafusa *et al.* (1976). They showed that the gs antigen of these viruses differed from that of the ALSV, that their viral proteins were quite different from those of the ALSV group and REV, that no homologies could be found among the nucleic acids of the three types of virus, nor differences between the DNA polymerases. On this basis, the subgroup G pheasant virus was placed in a new class (species) of ribodeoxy-virus, termed pheasant virus.

Morphologically the pheasant virus differed from ALSV. The density of pheasant virus in sucrose was greater than for ALSV; its RNA sedimented as a 70–75S complex.

On the basis of the above properties, pheasant virus apparently differs also from LPDV. Since the initial descriptions, no further information on pheasant virus appears to have been published.

VII. CONCLUSIONS

We have attempted in this chapter to provide an outline of the basic facts of the biology of avian retroviruses. Molecular aspects such as the mechanisms of viral replication and oncogenesis do not fall within the remit of the review, but nevertheless they have been mentioned briefly where necessary to illuminate other aspects of retroviral biology. Further details are provided in Chapter 5. The bibliography of avian retrovirology is enormous. The references given here are selective and can but provide entry points to the literature.

One of the attractions and impressive features of avian retrovirology is the coherent thread of understanding of virus–cell interactions which started some 30 years ago and has led through studies on cellular, subcellular, and molecular phenomena to the current and dominating emphasis on the nature of neoplastic transformation at the molecular level. This emphasis is likely to continue as attempts are made to discover the basis for the pathogenesis of the many avian retrovirus-induced neoplasms. It is remarkable how biological phenomena observed many years ago are being explained by molecular mechanisms. For example, the roles of cell oncogene activation and transduction are appreciated in understanding the multipotency of avian retroviruses and the ability to change tumor spectra of viruses by selection.

Studies on avian retroviruses have provided a rich seam of discoveries with far-reaching implications, such as the recognition of endogenous viruses and the subsequent investigation of their nature and biological and evolutionary significance, and the discovery of cellular and viral oncogenes and of their normal and abnormal functions. Retroviruses are now also being used as tools to study questions of stem cell differentiation and of the role of cell growth factors in cocarcinogenesis.

Emphasis on these biomedical aspects of avian retrovirology has tended to overshadow developments of particular interest to agricultural scientists. Effects of subclinical exogenous ALV infection on traits such as egg production and growth rate have been described and are important economically, but they now need to be explained physiologically. These effects also have practical and theoretical significance to poultry geneticists and breeders. Recently the question has arisen of whether *ev* genes have similar effects either directly or by interaction with exogenous ALV infection. Economic consequences have encouraged breeders to eradicate exogenous ALV from their stock; will there be a rationale for eradicating endogenous loci and viruses?

The poultry world continues to spawn new retrovirus-related disease problems which not only require practical solutions but also provide new material for basic studies: recent field outbreaks of hemangiosarcomas discussed in this chapter and the novel thrombogenic properties of the virus responsible provide an example.

Retroviral technology is also providing new approaches to the control of ALV infection. For example, the *env* gene of an exogenous virus has been introduced into the germ line of poultry, providing a form of dominant genetic resistance to virus infection by receptor blocking, in mimicry of a mechanism which occurs naturally with endogenous virus. More generally, avian retroviral constructs are being developed as vectors for other beneficial genes for producing transgenic chickens. This technique will be used also as a way of manipulating the chicken genome in studies on cell differentiation and embryogenesis. Molecular studies in progress on the chicken MHC are likely to be linked with transgenic techniques to produce disease-resistant stock, and recombinant ALV vaccines may also have a role in disease control.

ACKNOWLEDGMENTS. I am grateful to Helen Tiddy for secretarial assistance and to S. Hodgson for photography.

VIII. REFERENCES

Anon, 1946, Poultry Disease Investigations at the U. S. Regional Poultry Research Laboratory, Miscellaneous Publication No. 609, U. S. Department of Agriculture, Washington, D. C.

Astrin, S. M., 1978, Endogenous viral genes of the White Leghorn chicken: Common site of residence and sites associated with specific phenotypes of viral gene expression, *Proc. Natl. Acad. Sci. USA* **75**:5941.

Astrin, S. M., and Robinson, H. L., 1979, *Gs*, an allele of chicken for endogenous avian leukosis viral antigen segregates with *ev3*, a genetic locus that contains structural genes for virus, *J. Virol.* **31**:420.

Astrin, W. M., Robinson, H. L., Crittenden, L. B., Buss, E. G., Wyban, J., and Hayward, W. S., 1979, Ten genetic loci in the chicken that contain structural genes for endogenous avian leukosis viruses, *Cold Spring Harbor Symp. Quant. Biol.* **44**:1105.

Aulisio, C. G., and Shelokov, A., 1969, Prevalence of reticuloendotheliosis in chickens: Immunofluorescence studies, *Proc. Soc. Exp. Biol. Med.* **130**:178.

Baba, T. W., and Humphries, E. H., 1984, Avian leukosis virus infection: Analysis of viremia and DNA integration in susceptible and resistant chicken lines, *J. Virol.* **51**:123.

Baba, T. W., Giroir, B. P., and Humphries, E. H., 1985, Cell lines derived from avian lymphomas exhibit two distinct phenotypes, *Virology* **144**:139.

Bacon, L. D., 1987, Influence of the major histocompatability complex on disease resistance and productivity, *Poultry Sci.* **66**:802.

Bacon, L. D., Witter, R. L., Crittenden, L. B., Fadly, A., and Motta, J., 1981, B haplotype influence on Marek's disease, Rous sarcoma, and lymphoid leukosis virus-induced tumors in chickens, *Poultry Sci.* **60**:1132.

Bacon, L. D., Fredrickson, T. L., Gilmour, D. G., Fadly, A. M., and Crittenden, L. B., 1985, Tests of association of lymphocyte alloantigen genotypes with resistance to viral oncogenesis in chickens. 2. Rous sarcoma and lymphoid leukosis in progeny derived from $6_3 \times 15_1$ and $100 \times 6_3$ crosses, *Poultry Sci.* **64**:39.

Bacon, L. D., Smith, E., Crittenden, L. B., and Havenstein, G. B., 1988, Association of the slow feathering (K) and an endogenous viral (ev21) gene on the Z chromosome of chickens, *Poultry Sci.* **67**:191.

Bacon, L. D., Witter, R. L., and Fadly, A. M., 1989, Augmentation of retrovirus-induced lymphoid leukosis by Marek's disease herpesvirus in White leghorn chickens, *J. Virol.* **63**:504.

Bagust, T. J., and Grimes, T. M., 1979, Experimental infection of chickens with an Australian strain of reticuloendotheliosis virus. 2. Serological responses and pathogenesis, *Avian Pathol.* **8**:375.

Bagust, T. J., Grimes, T. M., and Ratnamohan, N., 1981, Experimental infection of chickens with an Australian strain of reticuloendotheliosis virus. 3. Persistent infection and transmission by the adult hen, *Avian Pathol.* **10**:375.

Baluda, M. A., and Goetz, I. E., 1961, Morphological conversion of cell cultures by avian myeloblastosis virus, *Virology* **15**:185.

Baluda, M. A., and Jamieson, P. P., 1961, *In vivo* infectivity studies with avian myeloblastosis virus, *Virology* **14**:33.

Barbacid, M., Hunter, E., and Aaronson, S. A., 1979, Avian reticuloendotheliosis viruses: Evolutionary linkages with mammalian type C retroviruses, *J. Virol.* **30**:508.

Barth, C. F., and Humphries, E. H., 1988, A nonimmunosuppressive helper virus allows high efficiency induction of B cell lymphomas by reticuloendotheliosis virus strain T, *J. Exp. Med.* **167**:89.

Bauer, G., and Temin, H. M., 1980, Specific antigenic relationships between the RNA-dependent DNA polymerases of avian reticuloendotheliosis viruses and mammalian type C retroviruses, *J. Virol.* **34**:168.

Bauer, H., 1974, Virion and tumor cell antigens of C-type RNA tumor viruses, *Adv. Cancer Res.* **20**:275.

Bauer, H., and Fleischer, B., 1981, The immunobiology of avian RNA tumor virus induced cell surface antigens, in: *Mechanisms of Immunity to Virus-Induced Tumors* (J. W. Blasecki, ed.), pp. 69–118, Dekker, New York.

Bauer, H., Kirth, R., and Gelderblum, H., 1976, Immune response to oncornaviruses and tumor-associated antigens in the chicken, *Cancer Res.* **36**:598.

Baxter-Gabbard, K. L., Campbell, W. F., Padgett, F., Raitano-Fenton, A., and Levine, A. S., 1971, Avian reticuloendotheliosis virus (strain T). II. Biochemical and biophysical properties, *Avian Dis.* **15**:850.

Beard, J. W., 1957, Etiology of avian leukosis, *Ann. N. Y. Acad. Sci.* **68**:473.

Beard, J. W., 1963a, Avian virus growths and their etiological agents, *Adv. Cancer Res.* **7**:1.

Beard, J. W., 1963b, Viral tumors of chickens with particular reference to the leukosis complex, *Ann. N. Y. Acad. Sci.* **108**:1057.

Beard, J. W., 1973, Oncornaviruses. 1. The avian tumor viruses, in: *Ultrastructure in Biological Systems*, Vol. 5, *Ultrastructure of Animal Viruses and Bacteriophages: An Atlas* (A. J. Dalton and F. Haguenau, eds.), pp. 261–281, Academic Press, New York.

Beard, J. W., 1980, Biology of avian oncornaviruses, in: *Viral Oncology* (G. Klein, ed.), pp. 55–87, Raven Press, New York.

Beaudreau, G. S., and Becker, C., 1958, Virus of avian myeloblastosis X. Photometric microdetermination of adenosine triphosphatase activity, *J. Natl. Cancer Inst.* **20**:339.

Beemon, K. L., Faras, A. J., Haase, A. T., Duesberg, P. H., and Maisel, J. E., 1976, Genomic complexities of murine leukemia and sarcoma, reticuloendotheliosis and visna viruses, *J. Virol.* **17**:525.

Begg, A. M., 1927, A filterable endothelioma of the fowl, *Lancet* **1927**(i):912.

Benatar, T., Iacampo, S., Tkalec, L., and Ratcliffe, M. J. H., 1991, Expression of immunoglobulin genes in the avian embryo bone marrow revealed by retroviral transformation, *Europ. J. Immunol.* **21**:2529.

Benedetti, E. L., Bernhard, W., and Oberling, C., 1956, Présence de corpuscules d'aspect virusal dans des cellules spléniques et médullaires de poissins leucémiques et normaux, *C. R. Acad. Sci. Paris* **242**:2891.

Bennett, D. D., and Wright, S. E., 1987, Immunization with envelope glycoprotein of an avian RNA tumor virus protects against sarcoma virus tumor induction: Role of subgroup, *Virus Res.* **8**:73.

Bernard, W., Dontcheff, A., Oberling, C., and Vigier, P., 1953, Corpuscules-d'aspect virusal dans les cellules du sarcome de Rous, *Bull. Cancer (Paris)* **40**:311.

Bernard, W., Bonar, R. A., Beard, D., and Beard, J. W., 1958, Ultrastructure of viruses of myeloblastosis and erythroblastosis isolated from plasma of leukemic chickens, *Proc. Soc. Exp. Biol. Med.* **97**:48.

Beug, H., Muller, H., Grieser, S., Doederlein, G., and Graf, T., 1981, Hematopoietic cells transformed *in vitro* by REV-T avian reticuloendotheliosis virus express characteristics of very immature lymphoid cells, *Virology* **115**:295.

Biggs, P. M., 1991, Lymphoproliferative disease of turkeys, in: *Diseases of Poultry*, 9th ed. (B. W. Calnek, ed.), pp. 456–459, Iowa State University Press, Ames, Iowa.

Biggs, P. M., Milne, B. S., Graf, T., and Bauer, H., 1973, Oncogenicity of nontransforming mutants of avian sarcoma viruses, *J. Gen. Virol.* **18**:399.

Biggs, P. M., Milne, B. S., Frazier, J. A., McDougall, J. S., and Stuart, J. C., 1974, Lymphoproliferative disease in turkeys, in: *Proceedings 15th World Poultry Congress New Orleans*, pp. 55–56.

Biggs, P. M., McDougall, J. S., Frazier, J. A., and Milne, B. S., 1978, Lymphoproliferative disease of turkeys. 1. Clinical aspects, *Avian Pathol.* **7**:131.

Boettiger, D., and Olsen, M., 1989, Induction of leukemia by avian myeloblastosis virus: A mechanistic hypothesis, *Curr. Top. Microbiol. Immunol.* **149**:157.

Bolognesi, D. P., 1974, Structural components of RNA tumor viruses, *Adv. Virus Res.* **19**:315.

Böni-Schnetzler, M., Böni, J., Ferdinand, F.-J., and Franklin, R. M., 1985, Developmental and molecular aspects of nephroblastomas induced by avian myeloblastosis-associated virus 2-0, *J. Virol.* **55**:213.

Bosch, V., Gebhardt, A., Friis, R. R., and Vielitz, E., 1983, Differential expression of endogenous virus glycoproteins in fibroblasts and sera of some adult chickens, *J. Gen. Virol.* **64**:225.

Boulliou, A., Le Pennec, J. P., Hubert, G., Donal, R., and Smiley, M., 1991, Restriction fragment length polymorphism analysis of endogenous avian leukosis viral loci: determination of frequencies in commercial broiler lines, *Poult. Sci.* **70**:1287.

Bova, C. A., Manfredi, J. P., and Swanstrom, R., 1986, *env* genes of avian retroviruses: Nucleotide sequence and molecular recombinants define host range determinants, *Virology* **152**:343.

Bova, C. A., Olsen, J. C., and Swanstrom, R., 1988, The avian retrovirus *env* gene family: Molecular analysis of host range and antigenic variants, *J. Virol.* **62**:75.

Bova-Hill, C., Olsen, J. C., and Swanstrom, R., 1991, Genetic analysis of the Rous sarcoma virus subgroup D env gene: Mammal tropism correlates with temperature sensitivity of gp 85, J. Virol. 65:2073.

Boyce-Jacino, M. T., Resnick, R., and Faras, A. J., 1989, Structural and functional characterization of the unusually short long terminal repeats and their adjacent regions of a novel endogenous avian retrovirus, Virology 173:157.

Boyde, A., Banes, A. J., Dillaman, R. M., and Mechanic, G. L., 1978, Morphological study of an avian bone disorder caused by myeloblastosis-associated virus, Metab. Bone Dis. Relat. Res. 1:235.

Breitman, M. L., Lai, M. M. C., and Vogt, P. K., 1980, Attenuation of avian reticuloendotheliosis virus: Loss of the defective transforming component during serial passage of oncogenic virus in fibroblasts, Virology 101:304.

Brown, D. W., and Robinson, H. L., 1988, Influence of env and long terminal repeat sequences on the tissue tropism of avian leukosis viruses, J. Virol. 62:4828.

Bryan, W. R., 1956, Biological studies on the Rous sarcoma virus. IV. Interpretation of tumor response data involving one inoculation site per chicken, J. Natl. Cancer Inst. 16:843.

Bryan, W. R., 1957, Interpretation of host response in quantitative studies on animal viruses, Ann. N. Y. Acad. Sci. 69:698.

Bryan, W. R., and Moloney, J. B., 1957, Rous sarcoma virus: The purification problem, Ann. N. Y. Acad. Sci. 68:441.

Bülow, V. von, 1977, Immunological effects of reticuloendotheliosis virus as potential contaminant of Marek's disease vaccines, Avian Pathol. 6:383.

Bumstead, N., 1985, Genetics of the major histocompatibility complex in chickens, in: Poultry Genetics and Breeding (W. G. Hill, J. M. Manson, and D. Hewitt, eds.), pp. 25–35, British Poultry Science, Harlow.

Burmester, B. R., 1945, The incidence of lymphomatosis among male and female chickens, Poultry Sci. 24:469.

Burmester, B. R., 1955, Immunity to visceral lymphomatosis in chicks following injection of virus into dams, Proc. Soc. Exp. Biol. Med. 88:153.

Burmester, B. R., 1956a, The shedding of the virus of visceral lymphomatosis in the saliva and feces of individual normal and lymphomatous chickens, Poultry Sci. 35:1089.

Burmester, B. R., 1956b, Bioassay of the virus of visceral lymphomatosis. I. Use of short experimental period, J. Natl. Cancer Inst. 16:1121.

Burmester, B. R., 1957, Transmission of tumors inducing avian viruses under natural conditions, Tex. Rep. Biol. Med. 15:540.

Burmester, B. R., 1969, The prevention of lymphoid leukosis with androgens, Poultry Sci. 48:401.

Burmester, B. R., and Fredrickson, T. N., 1964, Transmission of virus from field cases of avian lymphomatosis. I. Isolation of virus in line 15I chickens, J. Natl. Cancer Inst. 32:37.

Burmester, B. R., and Gentry, R. F., 1954a, The transmission of avian visceral lymphomatosis by contact, Cancer Res. 14:34.

Burmester, B. R., and Gentry, R. F., 1954b, The presence of the virus causing visceral lymphomatosis in the secretions and excretions of chickens, Poultry Sci. 33:836.

Burmester, B. R., and Gentry, R. F., 1954c, A study of possible avenues of infection with the virus of avian visceral lymphomatosis, in: Proceedings 91st Annual Meeting American Veterinary Medical Association, p. 311.

Burmester, B. R., and Gentry, R. F., 1956, The response of susceptible chickens to graded doses of the virus of visceral lymphomatosis, Poultry Sci. 35:17.

Burmester, B. R., and Nelson, N. M., 1945, The effect of castration and sex hormones upon the incidence of lymphomatosis in chickens, Poultry Sci. 24:509.

Burmester, B. R., and Purchase, H. G., 1979, The history of avian medicine in the United States. V. Insights into avian tumor virus research, *Avian Dis.* **23**:1.

Burmester, B. R., and Waters, N. F., 1955, The role of the infected egg in the transmission of visceral lymphomatosis, *Poultry Sci.* **34**:1415.

Burmester, B. R., Prickett, C. O., and Belding, T. C., 1946, A filtrable agent producing lymphoid tumors and osteopetrosis in chickens, *Cancer Res.* **6**:189.

Burmester, B. R., Gentry, R. F., and Waters, N. F., 1955, The presence of the virus of visceral lymphomatosis in embryonated eggs of normal appearing hens, *Poultry Sci.* **34**:609.

Burmester, B. R., Walter, W. G., and Fontes, A. K., 1957, The immunological response of chickens after treatment with several vaccines of visceral lymphomatosis, *Poultry Sci.* **36**:79.

Burmester, B. R., Gross, M. A., Walter, W. G., and Fontes, A. K., 1959a, Pathogenicity of a viral strain (RPL12) causing avian visceral lymphomatosis and related neoplasms. II. Host–virus interrelations affecting response, *J. Natl. Cancer Inst.* **22**:103.

Burmester, B. R., Walter, W. G., Gross, M. A., and Fontes, A. K., 1959b, The oncogenic spectrum of two 'pure' strains of avian leukosis, *J. Natl. Cancer Inst.* **23**:277.

Burmester, B. R., Fontes, A. K., and Walter, W. G., 1960, Pathogenicity of a viral strain (RPL12) causing avian visceral lymphomatosis and related neoplasms. III. Influence of host age and route of inoculation. *J. Natl. Cancer Inst.* **24**:1423.

Burstein, H., Gilead, M., Bendheim, U., and Kotler, M., 1984, Viral aetiology of haemangiosarcoma outbreaks among layer hens, *Avian Pathol.* **13**:715.

Calnek, B. W., 1964, Morphological alteration of RIF-infected chick embryo fibroblasts, *Natl. Cancer Inst. Monogr.* **17**:425.

Calnek, B. W., 1965, Studies on the RIF test for the detection of an avian leukosis virus, *Avian Dis.* **9**:545.

Calnek, B. W., 1968a, Lesions in young chickens induced by lymphoid leukosis virus, *Avian Dis.* **12**:111.

Calnek, B. W., 1968b, Lymphoid leukosis virus: A survey of commercial breeding flocks for genetic resistance and incidence of embryo infection, *Avian Dis.* **12**:104.

Campbell, J. G., 1969, *Tumours of the Fowl*, William Heinemann Medical Books, London.

Campbell, J. G., and Appleby, E. C., 1966, Tumours in young chickens bred for rapid body growth (broiler chickens): A study of 351 cases, *J. Pathol. Bacteriol.* **92**:77.

Campbell, W. F., Baxter-Gabbard, K. L., and Levine, A. S., 1971, Avian reticuloendotheliosis virus (strain T). I. Virological characterization, *Avian Dis.* **15**:837.

Carlson, H. C., Seawright, G. L., and Pettit, J. R., 1974, Reticuloendotheliosis in Japanese quail, *Avian Pathol.* **3**:169.

Carpenter, C. R., Bose, H. R., and Rubin, A. S., 1977, Contact-mediated suppression of mitogen-induced responsiveness by spleen cells in reticuloendotheliosis virus-induced tumorigenesis, *Cell. Immunol.* **33**:392.

Carpenter, C. R., Kempf, K. E., Bose, H. R., and Rubin, A. S., 1978, Characterization of the interaction of reticuloendotheliosis virus with the avian lymphoid system, *Cell. Immunol.* **39**:307.

Carr, J. G., 1956, Renal adenocarcinoma induced by fowl leukemia virus, *Br. J. Cancer* **10**:379.

Carr, J. G., 1960, Kidney carcinomas of the fowl induced by the MH2 reticuloendothelioma virus, *Br. J. Cancer* **14**:77.

Carter, J. K., and Smith, R. E., 1984, Specificity of avian leukosis virus-induced hyperlipidemia, *J. Virol.* **50**:301.

Carter, J. K., Proctor, S. J., and Smith, R. E., 1983, Induction of angiosarcomas by ring-necked pheasant virus, *Infect. Immun.* **40**:310.

Cavallieri, F., Ruscio, T., Tinoco, R., Benedict, S., Davis, C., and Vogt, P. K., 1985, Isolation of three new avian sarcoma viruses: ASV9, ASV17, and ASV25, *Virology* **143**:680.

Chen, P.-Y., Cui, Z., Lee, L. F., and Witter, R. L., 1987, Serological differences among non-defective reticuloendotheliosis viruses, *Arch. Virol.* **93**:233.

Chen, Y. C., and Vogt, P. K., 1977, Endogenous leukosis viruses in the avian family Phasianidae, *Virology* **76**:740.

Cheville, N. F., Okazaki, W., Lukert, P. D., and Purchase, H. G., 1978, Prevention of avian lymphoid leukosis by induction of bursal atrophy with infectious bursal disease viruses, *Vet. Pathol.* **15**:376.

Chouroulinkov, I., and Rivière, M. R., 1959, Tumeurs rénales à virus de la poule. 1. Etude morphologique, *Bull. Cancer* **46**:722.

Chubb, R. C., and Biggs, P. M., 1968, The neutralisation of Rous sarcoma virus, *J. Gen. Virol.* **3**:87.

Clark, D. P., and Dougherty, R. M., 1980, Detection of avian oncovirus group-specific antigens by the enzyme-linked immunosorbent assay, *J. Gen. Virol.* **47**:283.

Claude, A., and Murphy, J. G., 1933, Transmissible tumors of the fowl, *Physiol. Rev.* **13**:246.

Clurman, B. E., and Hayward, W. S., 1989, Multiple proto-oncogene activations in avian leukosis virus-induced lymphomas: Evidence for stage-specific events, *Mol. Cell. Biol.* **9**:2657.

Collart, K. L., Aurigemma, R., Smith, R. E., Kawai, S., and Robinson, H. L., 1990, Infrequent involvement of c-*fos* in avian leukosis virus-induced nephroblastoma, *J. Virol.* **64**:3541.

Collins, W. M., and Zsigray, R. M., 1984, Genetics of the response to Rous sarcoma virus-induced tumours in chickens, *Anim. Blood Groups Biochem. Genet.* **15**:159.

Collins, W. M., Briles, W. E., Zsigray, R. M., Dunlop, W. R., Corbett, A. C., Clark, K. K., Marks, J. L., and McGrail, T. P., 1977, The B locus (MHC) in the chicken: Association with the fate of RSV-induced tumors, *Immunogenetics* **5**:333.

Collins, W. M., Zervas, N. P., Urban, W. E., Jr., Briles, W. E., and Aeed, P. A., 1985, Response of B complex haplotypes B^{22}, B^{24}, and B^{26} to Rous sarcomas, *Poultry Sci.* **64**:2017.

Collins, W. M., Dunlop, W. R., Zsigray, R. M., Briles, R. W., and Fite, R. W., 1986, Metastasis of Rous sarcoma tumors in chickens is influenced by the major histocompatibility (B) complex and sex, *Poultry Sci.* **65**:1642.

Cooper, G. M., 1982, Cellular transforming genes, *Science*, **217**:801.

Cooper, M. D., Payne, L. N., Dent, P. B., Burmester, B. R., and Good, R. A., 1968, Pathogenesis of avian lymphoid leukosis. 1. Histogenesis. *J. Natl. Cancer Inst.* **41**:373.

Cooper, M. D., Purchase, H. G., Bockman, D. E., and Gathings, W. E., 1974, Studies on the nature of the abnormality of B cell differentiation in avian lymphoid leukosis: Production of heterogeneous IgM by tumor cells, *J. Immunol.* **113**:1210.

Cotter, P. F., Collins, W. M., Dunlop, W. R., and Corbett, A. C., 1976a, The influence of thymectomy on Rous sarcoma regression, *Avian Dis.* **20**:75.

Cotter, P. F., Collins, W. M., Dunlop, W. R., and Corbett, A. C., 1976b, Detection of cellular immunity to Rous tumors of chickens by the leukocyte migration inhibition reaction, *Poultry Sci.* **55**:1008.

Cottral, G. E., Burmester, B. R., and Waters, N. F., 1949, The transmission of visceral lymphomatosis with tissues from embryonated eggs and chicks of "normal" parents, *Poultry Sci.* **28**:761 (abstract).

Cottral, G. E., Burmester, B. R., and Waters, N. F., 1954, Egg transmission of avian lymphomatosis, *Poultry Sci.* **33**:1174.

Cowen, B. S., and Braune, M. O., 1988, The propagation of avian viruses in a continuous cell line (QT35) of Japanese quail origin, *Avian Dis.* **32**:282.

Crittenden, L. B., 1968, Observations on the nature of a genetic cellular resistance to avian tumor viruses, *J. Natl. Cancer Inst.* **41**:145.

Crittenden, L. B., 1975, Two levels of genetic resistance to lymphoid leukosis, *Avian Dis.* **19**:281.

Crittenden, L. B., 1981, Exogenous and endogenous leukosis virus genes—A review, *Avian Pathol.* **10**:101.

Crittenden, L. B., 1991, Retroviral elements in the genome of the chicken: Implications for poultry genetics and breeding, in: *Critical Reviews of Poultry Biology*, Vol. 3, Part 2, pp. 73–109, Elsevier, Amsterdam.

Crittenden, L. B., and Fadly, A. M., 1985, Responses of chickens lacking or expressing endogenous avian leukosis virus genes to infection with exogenous virus, *Poultry Sci.* **64**:454.

Crittenden, L. B., and Kung, H.-J., 1984, Mechanism of induction of lymphoid leukosis and related neoplasms by avian leukosis viruses, in: *Mechanisms of Viral Leukaemogenesis* (J. M. Goldman and O. Jarrett, eds.), pp. 64–88, Churchill Livingstone, Edinburgh.

Crittenden, L. B., and Motta, J. V., 1969, A survey of genetic resistance to leukosis sarcoma viruses in commercial stocks of chickens, *Poultry Sci.* **48**:1751.

Crittenden, L. B., and Motta, J. V., 1975, The role of the tvb locus in genetic resistance to RSV(RAV-0), *Virology* **67**:327.

Crittenden, L. B., and Okazaki, W., 1966, Genetic influence of the Rs locus on susceptibility to avian tumor viruses. II. Rous sarcoma virus antibody production after strain RPL12 virus inoculation, *J. Natl. Cancer Inst.* **36**:299.

Crittenden, L. B., and Smith, E. J., 1984, A comparison of test materials for differentiating avian leukosis virus group-specific antigens of exogenous and endogenous origin, *Avian Dis.* **28**:1057.

Crittenden, L. B., Okazaki, W., and Reamer, R., 1963, Genetic resistance to Rous sarcoma virus in embryo cell cultures and embryos, *Virology* **20**:541.

Crittenden, L. B., Okazaki, W., and Reamer, R. H., 1964, Genetic control of responses to Rous sarcoma and strain RPL12 viruses in the cells, embryos, and chickens of two inbred lines, *Natl. Cancer Inst. Monogr.* **17**:161.

Crittenden, L. B., Wendel, E. J., and Ratzsch, D., 1971, Genetic resistance to the avian leukosis-sarcoma virus group: Determining the phenotype of adult birds, *Avian Dis.* **15**:503.

Crittenden, L. B., Purchase, H. G., Solomon, J. J., Okazaki, W., and Burmester, B. R., 1972, Genetic control of susceptibility to the avian leukosis complex. 1. The leukosis-sarcoma virus group, *Poultry Sci.* **51**:242.

Crittenden, L. B., Wendel, E. J., and Motta, J. V., 1973, Interaction of genes controlling resistance to RSV(RAV-0), *Virology* **52**:378.

Crittenden, L. B., Motta, J. V., and Smith, E. J., 1977, Genetic control of RAV-0 production in chickens, *Virology* **76**:90.

Crittenden, L. B., Witter, R. L., and Fadly, A. M., 1979a, Low incidence of lymphoid tumors in chickens continuously producing endogenous virus, *Avian Dis.* **23**:646.

Crittenden, L. B., Eagen, D. A., and Gulvas, F. A., 1979b, Assays for endogenous and exogenous lymphoid leukosis viruses and chick helper factor with RSV(−) cell lines, *Infect. Immun.* **24**:379.

Crittenden, L. B., Witter, R. L., Okazaki, W., and Neiman, P. E., 1979c, Lymphoid neoplasms in chicken flocks free of infection with exogenous avian tumor viruses, *J. Natl. Cancer Inst.* **63**:191.

Crittenden, L. B., Fadly, A. M., and Smith, E. J., 1982, Effect of endogenous leukosis virus genes on response to infection with avian leukosis and reticuloendotheliosis viruses, *Avian Dis.* **26**:279.

Crittenden, L. B., Okazaki, W., and Smith, E. J., 1983, Incidence of avian leukosis virus infection in broiler stocks and its effect on early growth, *Poultry Sci.* **62**:2383.

Crittenden, L. B., Smith, E. J., and Fadly, A. M., 1984a, Influence of endogenous viral (ev) gene expression and strain of exogenous avian leukosis virus (ALV) on mortality and ALV infection and shedding in chickens, *Avian Dis.* **28**:1037.

Crittenden, L. B., Smith, E. J., and Okazaki, W., 1984b, Identification of broiler breeders congenitally transmitting avian leukosis virus by enzyme-linked immunosorbent assay, *Poultry Sci.* **63**:492.

Crittenden, L. B., McMahon, S., Halpern, M. S., and Fadly, A. M., 1987, Embryonic infection with the endogenous avian leukosis virus Rous-associated virus-0 alters responses to exogenous avian leukosis virus infection, *J. Virol.* **61**:722.

Crittenden, L. B., Salter, D. W., and Federspiel, M. J., 1989, Segregation, viral phenotype, and proviral structure of 23 avian leukosis virus inserts in the germ line of chickens, *Theor. Appl. Genet.* **77**:505.

Cui, Z.-Z., Lee, L. F., Silva, R. F., and Witter, R. L., 1986, Monoclonal antibodies against avian reticuloendotheliosis virus: Identification of strain-specific and strain-common epitopes, *J. Immunol.* **136**:4237.

Cui, Z.-Z., Lee, L. F., Smith, E. J., Witter, R. L., and Chang, T. S., 1988, Monoclonal-antibody-mediated enzyme-linked immunosorbent assay for detection of reticuloendotheliosis viruses, *Avian Dis.* **32**:32.

Cummins, T. J., and Smith, R. E., 1988, Analysis of hematopoietic and lymphopoietic tissue during a regenerative aplastic crisis induced by avian retrovirus MAV-2(0), *Virology* **163**:452.

Dales, S., and Hanafusa, H., 1972, Penetration and intracellular release of the genomes of avian RNA tumor viruses, *Virology* **50**:440.

Dalton, A. J., and Haguenau, F., eds., 1962, *Tumors Induced by Viruses: Ultrastructural Studies*, Academic Press, New York.

De Boer, G. F., 1987, Approaches to control avian lymphoid leukosis, in: *Avian Leukosis* (G. F. De Boer, ed.), pp. 261–286, Martinus Nijhoff, Boston.

De Boer, G. F., and Osterhaus, A. D. M. E., 1985, Application of monoclonal antibodies in the avian leukosis virus gs-antigen ELISA, *Avian Pathol.* **14**:39.

De Boer, G. F., Van Vloten, J., and Hartog, L., 1980a, Comparison of complement fixation and phenotypic mixing tests for the detection of lymphoid leukosis virus in egg albumen and embryos of individual eggs, *Avian Pathol.* **9**:207.

De Boer, G. F., van Vloten, J., and van Zaane, D., 1980b, Possible horizontal spread of lymphoid leukosis virus during vaccination against Marek's disease, in: *Resistance and Immunity to Marek's Disease* (P. M. Biggs, ed.), pp. 552–565, CEC Luxembourg.

De Boer, G. F., Devos, O. J. H., and Maas, H. J. L., 1981, The incidence of lymphoid leukosis in chickens in the Netherlands, *Zootec. Int.* **1981**(10):32.

De Giuli, C., Hanafusa, H., Kawai, S., Dales, S., Chen, J. H., and Hsu, K. C., 1975, Relationship between A-type and C-type particles in cells infected by Rous sarcoma virus, *Proc. Natl. Acad. Sci. USA* **72**:3706.

Dent, P. B., Cooper, M. D., Payne, L. N., Solomon, J. J., Burmester, B. R., and Good, R. A., 1968, Pathogenesis of avian lymphoid leukosis. II. Immunologic reactivity during lymphomagenesis, *J. Natl. Cancer Inst.* **41**:391.

Di Stefano, H. S., and Dougherty, R. M., 1966, Mechanisms for congenital transmission of avian leukosis virus, *J. Natl. Cancer Inst.* **37**:869.

Di Stefano, H. S., and Dougherty, R. M., 1968, Multiplication of avian leukosis virus in the reproductive system of the rooster, *J. Natl. Cancer Inst.* **41**:451.

Dolberg, D. S., and Bissell, M. J., 1984, Inability of Rous sarcoma virus to cause sarcomas in the avian embryo, *Nature* **309**:552.

Doolittle, R. F., Feng, D.-F., Johnson, M. S., and McClure, M. A., 1989, Origins and evolutionary relationship of retroviruses, *Q. Rev. Biol.* **64**:1.

Dougherty, R. M., 1964, Animal virus titration techniques, in: *Techniques in Experimental Virology* (R. J. C. Harris, ed.), pp. 169–223, Academic Press, New York.

Dougherty, R. M., 1987, A historical review of avian retrovirus research, in: *Avian Leukosis* (G. F. De Boer, ed.), pp. 1–27, Martinus Nijhoff, Boston.

Dougherty, R. M., and Di Stefano, H. S., 1965, Virus particles associated with "nonproducer" Rous sarcoma cells, *Virology* **27**:351.

Dougherty, R. M., and Di Stefano, H. S., 1966, Lack of relationship between infection with avian leukosis virus and the presence of COFAL antigen in chick embryos, *Virology* **29**:586.

Dougherty, R. M., and Di Stefano, H. S., 1967, Sites of avian leukosis virus multiplication in congenitally infected chickens, *Cancer Res.* **27**:322.

Dougherty, R. M., Stewart, J. A., and Morgan, H. R., 1960, Quantitation studies of the relationships between infecting dose of Rous sarcoma virus, antiviral immune response, and tumor growth in chickens, *Virology* **11**:349.

Dougherty, R. M., Di Stefano, H. S., and Marucci, A. A., 1974, Application of soluble antigen–antibody complexes to the immune histochemical study of avian leukosis virus antigen, in: *Viral Immunodiagnosis* (E. Kurstak and R. Morisset, eds.), pp. 88–89, Academic Press, New York.

Dren, C. N., Saghy, E., Glavits, R., Ratz, F., Ping, J., and Sztojkov, V., 1983, Lymphoreticular tumour in pen-raised pheasants associated with a RE-like virus infection, *Avian Pathol.* **12**:55.

Dren, C. N., Nemeth, I., Sari, I., Ratz, F., Glavits, R., and Somogyi, P., 1988, Isolation of a reticuloendotheliosis-like virus from naturally occurring lymphoreticular tumours of domestic goose, *Avian Pathol.* **17**:259.

Duesberg, P. H., 1968, On the structure of RNA tumor viruses, *Curr. Top. Microbiol. Immunol.* **51**:79.

Duff, R. G., and Vogt, P. K., 1969, Characteristics of two new avian tumor virus subgroups, *Virology* **39**:18.

Dutta, A., Wang, L.-H., Hanafusa, T., and Hanafusa, H., 1985, Partial nucleotide sequence of Rous sarcoma virus-29 provides evidence that the original Rous sarcoma virus was replication-defective, *J. Virol.* **55**:728.

Eckert, E. A., Beard, D., and Beard, J. W., 1954, Dose–response relations in experimental transmission of avian erythromyeloblastic leukemia. III. Titration of the virus, *J. Natl. Cancer Inst.* **14**:1055.

Ellermann, V., 1921, *The Leucosis of Fowls and Leucemia Problems*, Gyldendal, London.

Ellermann, V., and Bang, O., 1908, Experimentelle Leukämie bei Hühnern, *Zentralbl. Bakteriol. Parasitenkd. Infectionskr. Hyg. Abt. Orig.* **46**:595.

Elmubarak, A. K., Sharma, J. M., Witter, R. L., Crittenden, L. B., and Sanger, V. L., 1983, Comparative response of turkeys and chickens to avian lymphoid leukosis virus, *Avian Pathol.* **12**:235.

Engelbreth-Holm, J., 1931–1932, Bericht über einen neuen Stamm Hühnerleukose, *Z. Immunitätsforsch.* **73**:126.

Engelbreth-Holm, J., 1932, Untersuchungen über die Sogerannte Erythroleukose bei Hühnern, *Z. Immunitätsforsch.* **75**:425.

Engelbreth-Holm, J., 1942, *Spontaneous and Experimental Leukaemia in Animals*, Oliver and Boyd, Edinburgh.

Engelbreth-Holm, J., and Rothe Meyer, A., 1935, On the connection between erythroblastosis (haemocytoblastosis), myelosis and sarcoma in chickens, *Acta Pathol. Scand.* **12**:352.

Enrietto, P. J., and Hayman, M. J., 1987, Structure and virus-associated oncogenes of avian sarcoma and leukemia viruses, in: *Avian Leukosis* (G. F. De Boer, ed.), pp. 29–46, Martinus Nijhoff, Boston.

Enrietto, P. J., and Wyke, J. A., 1983, The pathogenesis of oncogenic avian retroviruses, *Adv. Cancer Res.* **39**:269.

Enrietto, P. J., Hayman, M. J., Ramsay, G. M., Wyke, J. A., and Payne, L. N., 1983, Altered pathogenicity of avian myelocytomatosis (MC29) viruses with mutations in the v-*myc* gene, *Virology* **124**:164.

Estola, T., Sandelin, K., Vaheri, A., Ruoslahti, E., and Suni, J., 1974, Radioimmunoassay for detecting group-specific avian RNA tumor virus antigens and antibodies, *Dev. Biol. Stand.* **25**:115.

Fadly, A. M., 1988, Avian leukosis virus (ALV) infection, shedding, and tumors in maternal ALV antibody-positive and -negative chickens exposed to virus at hatching, *Avian Dis.* **32**:89.

Fadly, A. M., 1989, Leukosis and sarcomas, in: *A Laboratory Manual for the Isolation and Identification of Avian Pathogens* (H. G. Purchase, L. H. Arp, C. H. Domermuth, and J. E. Pearson, eds.), pp. 135–142, American Association of Avian Pathologists, Kendall/Hunt Publishing Company, Dubuque, Iowa.

Fadly, A. M., and Bacon, L. D., 1979, Bursal and thymic lesions in chickens bearing progressive Rous sarcomas, *Avian Dis.* **23**:529.

Fadly, A. M., and Witter, R. L., 1983, Studies of reticuloendotheliosis virus-induced lymphomagenesis in chickens, *Avian Dis.* **27**:271.

Fadly, A. M., Okazaki, W., Smith, E. J., and Crittenden, L. B., 1981a, Relative efficiency of test procedures to detect lymphoid leukosis virus infection, *Poultry Sci.* **60**:2037.

Fadly, A. M., Okazaki, W., and Witter, R. L., 1981b, Hatchery-related contact transmission and short-term small-group-rearing as related to lymphoid-leukosis-virus-eradication programs, *Avian Dis.* **25**:667.

Fadly, A. M., Purchase, H. G., and Gilmour, D. G., 1981c, Tumor latency in avian lymphoid leukosis, *J. Natl. Cancer Inst.* **66**:549.

Fadly, A. M., Lee, L. F., and Bacon, L. D., 1982, Immunocompetence of chickens during early and tumorigenic stages of Rous-associated virus-1 infection, *Infect. Immun.* **37**:1156.

Fadly, A. M., Witter, R. L., and Lee, L. F., 1985, Effects of chemically or virus-induced immunodepression on response of chickens to avian leukosis virus, *Avian Dis.* **29**:12.

Fadly, A. M., Crittenden, L. B., and Smith, E. J., 1987, Variation in tolerance induction and oncogenicity due to strain of avian leukosis virus, *Avian Pathol.* **16**:665.

Fadly, A. M., Davison, T. F., Payne, L. N., and Howes, K., 1989, Avian leukosis virus infection and shedding in Brown Leghorn chickens treated with corticosterone or exposed to various stressors, *Avian Pathol.* **18**:283.

Federspiel, M. J., Crittenden, L. B., and Hughes, S. H., 1989, Expression of avian reticuloendotheliosis virus envelope confers host resistance, *Virology* **173**:167.

Federspiel, M. J., Crittenden, L. B., Provencher, L. P., and Hughes, S. H., 1991, Experimentally introduced defective endogenous proviruses are highly expressed in chickens, *J. Virol.* **65**:313.

Franklin, R. B., Maldonado, R. L., and Bose, Jr., H. R., 1974, Isolation and characterization of reticuloendotheliosis virus transformed bone marrow cells, *Intervirology* **3**:342.

Franklin, R. B., Kang, C. Y., Wan, K. M. M., and Bose, H. R., 1977, Transformation of chick embryo fibroblasts by reticuloendotheliosis virus, *Virology* **83**:313.

Franklin, R. M., and Martin, M.-T., 1980, In ovo tumorigenesis induced by avian osteope-
    trosis virus, Virology 105:245.
Fredrickson, T. N., and Helmboldt, C. F., 1991, Tumors of unknown etiology, in: Dis-
    eases of Poultry, 9th ed. (B. W. Calnek ed.), pp. 459–470, Iowa State University Press,
    Ames, Iowa.
Fredrickson, T. N., Purchase, H. G., and Burmester, B. R., 1964, Transmission of virus
    from field cases of avian lymphomatosis. III. Variation in the oncogenic spectra of
    passaged virus isolates, Natl. Cancer Inst. Monogr. 17:1.
Fredrickson, T. N., Burmester, B. R., and Okazaki, W., 1965, Transmission of virus from
    field cases of avian lymphomatosis. II. Development of strains by serial passage in line
    15I chickens, Avian Dis. 9:82.
Frisby, D., MacCormick, R., and Weiss, R., 1980, Origin of RAV-0. The endogenous
    retrovirus of chickens, in: Cold Spring Harbor Conference on Cell Proliferation, Vol.
    7, p. 509, Cold Spring Harbor Laboratory, Cold Spring Harbor, New York.
Fujinami, A., and Inamoto, K., 1914, Ueber Geschwülste bei japanischen Haushühnern
    insbesondere über einen transplantablen tumor, Z. Krebsforsch. 14:94.
Fujita, D. J., Chen, Y. C., Friis, R. R., and Vogt, P. K., 1974, RNA tumor viruses of phea-
    sants: Characterization of avian leukosis subgroups F and G, Virology 60:558.
Fung, Y.-K. T., Fadly, A. M., Crittenden, L. B., and Kung, H.-J., 1982, Avian lymphoid
    leukosis virus infection and DNA integration in the preleukotic bursal tissues: A
    comparative study of susceptible and resistant lines, Virology 119:411.
Fung, Y.-K. T., Lewis, W. G., Crittenden, L. B., and Kung, H.-J., 1983, Activation of the
    cellular oncogene c-erbB by LTR insertion: Molecular basis for induction of erythro-
    blastosis by avian leukosis virus, Cell 33:357.
Furth, J., 1929, On the transmissibility of the leucosis of fowls, Proc. Soc. Exp. Biol. Med.
    27:155.
Furth, J., 1931, Nature of the agent transmitting leucosis of the fowl, Proc. Soc. Exp. Biol.
    Med. 28:449.
Furth, J., 1933, Lymphomatosis, myelomatosis and endothelioma of chickens caused by a
    filterable agent. 1. Transmission experiments, J. Exp. Med. 58:253.
Gak, E., Yaniv, A., Chajut, A., Ianconescu, M., Tronick, S. R., and Gazit, A., 1989, Molecu-
    lar cloning of an oncogenic replication-competent virus that causes lymphoprolifera-
    tive disease in turkeys, J. Virol. 63:2877.
Gak, E., Yaniv, A., Sherman, L., Ianconescu, M., Tronick, S. R., and Gazit, A., 1991,
    Lymphoproliferative disease virus of turkeys: Sequence analysis and transcriptional
    activity of the long terminal repeat, Gene 99:157.
Gavora, J. S., 1987, Influences of avian leukosis virus infection on production and mortal-
    ity and the role of genetic selection in the control of lymphoid leukosis, in: Avian
    Leukosis (G. F. De Boer, ed.), pp. 241–260, Martinus Nijhoff, Boston.
Gavora, J. S., 1990, Disease genetics, in: Poultry Breeding and Genetics (R. D. Crawford,
    ed.), pp. 805–846, Elsevier, Amsterdam.
Gavora, J. S., Spencer, J. L., Gowe, R. S., and Harris, D. L., 1980, Lymphoid leukosis virus
    infection: Effects on production and mortality and consequences in selection for high
    egg production, Poultry Sci. 59:2165.
Gavora, J. S., Spencer, J. L., and Chambers, J. R., 1982, Performance of meat-type chickens
    test-positive and -negative for lymphoid leukosis virus infection, Avian Pathol. 11:29.
Gavora, J. S., Kuhnlein, U., and Spencer, J. L., 1989, Absence of endogenous viral genes
    from an inbred line of leghorn chickens selected for high egg production and Marek's
    disease resistance, J. Anim. Breed. and Genet. 106:217.
Gavora, J. S., Kuhnlein, U., Crittenden, L. B., Spencer, J. L., and Sabour, M. P., 1991,
    Endogenous viral genes: Association with reduced egg production rate, and egg size in
    White Leghorns, Poultry Sci. 70:618.

Gaylord, W. H., 1955, Virus-like particles associated with the Rous sarcoma as seen in sections of the tumor, *Cancer Res.* **15**:80.

Gazit, A., Yaniv, M., Ianconescu, M., Perk, K., Aizenberg, B., and Zimber, Z., 1979, Molecular evidence for a type C retrovirus etiology of lymphoproliferative disease of turkeys, *J. Virol.* **31**:639.

Gazit, A., Schwarzbard, Z., Yaniv, A., Ianconescu, M., Perk, K., and Zimber, A., 1982, Organotropism of the lymphoproliferative disease virus (LPDV) of turkeys, *Int. J. Cancer* **29**:599.

Gazit, A., Basri, R., Ianconescu, M., Perk, K., Zimber, A., and Yaniv, A., 1986, Analysis of structural polypeptides of the lymphoproliferative disease virus (LPDV) of turkeys, *Int. J. Cancer* **37**:241.

Gazzolo, L., Moscovici, M. G., and Moscovici, C., 1975, Susceptibility and resistance of chicken macrophages to avian RNA tumor viruses, *Virology* **67**:553.

Gebriel, G. M., Pevzner, I. Y., and Nordskog, A. W., 1979, Genetic linkage between immune response to GAT and the fate of RSV-induced tumors in chickens, *Immunogenetics* **9**:327.

Gilka, F., and Spencer, J. L., 1985, Viral matrix inclusion bodies in myocardium of lymphoid leukosis virus-infected chickens, *Am. J. Vet. Res.* **46**:1953.

Gilka, F., and Spencer, J. L., 1987, Importance of the medullary macrophage in the replication of lymphoid leukosis virus in the bursa of Fabricius in chickens, *Am. J. Vet. Res.* **48**:613.

Gilka, F., and Spencer, J. L., 1990, Chronic myocarditis and circulatory syndrome in a White leghorn strain induced by an avian leukosis virus: Light and electron microscopic study, *Avian Dis.* **34**:174.

Glover, D. M., and Hames, B. D., eds., 1989, *Oncogenes*, IRL Press, Oxford.

Goubin, G., Goldman, D. S., Luce, J., Neiman, P. E., and Cooper, G. M., 1983, Molecular cloning and nucleotide sequence of a transforming gene detected by transfection of chicken B-cell lymphoma DNA, *Nature* **302**:114.

Graevskaya, N. A., Heider, G., Dementieva, S. P., and Ebner, D., 1982, Antibodies to reverse transcriptase of avian oncoviruses in sera of specific-pathogen-free chickens, *Acta Virol.* **26**:333.

Graf, T., 1972, A plaque assay for avian RNA tumor viruses, *Virology* **50**:567.

Graf, T., 1975, *In vitro* transformation of chicken bone marrow cells with avian myeloblastosis virus, *Z. Naturforsch.* **30c**:847.

Graf, T., and Beug, H., 1978, Avian leukemia viruses. Interaction with their target cells *in vivo* and *in vitro*, *Biochim. Biophys. Acta* **516**:269.

Graf, T., Fink, D., Beug, H., and Royer-Pokora, B., 1977, Oncornavirus-induced sarcoma formation observed by rapid development of lethal leukemia, *Cancer Res.* **37**:59.

Grimes, T. M., and Purchase, H. G., 1973, Reticuloendotheliosis in a duck, *Aust. Vet. J.* **49**:466.

Gross, L., 1983, *Oncogenic Viruses*, 3rd ed., Pergamon Press, Oxford.

Gross, M. A., Burmester, B. R., and Walter, W. G., 1959, Pathogenicity of a viral strain (RPL12) causing avian visceral lymphomatosis and related neoplasms. 1. Nature of the lesions, *J. Natl. Cancer Inst.* **22**:83.

Gudkov, A. V., Obukh, I. B., Serov, S. M., and Naroditsky, B. S., 1981, Variety of endogenous proviruses in the genomes of chickens of different breeds, *J. Gen. Virol.* **57**:85.

Gudkov, A. V., Korec, E., Chernov, M. V., Tikhonenko, A. T., Obukh, I. B., and Hlozanek, I., 1986, Genetic structure of the endogenous proviruses and expression of the *gag* gene in Brown Leghorn chickens, *Folia Biol. (Praha)* **32**:65.

Hadwiger, A., and Bosch, V., 1985, Characterization of the endogenous retroviral envelope glycoproteins found in the sera of *ev3* and *ev6* chickens, *J. Gen. Virol.* **66**:2051.

Hagino-Yamagishi, K., Ikawa, S., Kawai, S., Hihara, H., Yamamoto, T., and Toyoshima,

K., 1984, Characterization of two strains of avian sarcoma virus isolated from avian lymphatic leukosis virus-induced sarcomas, *Virology* **137**:266.

Haguenau, F., and Beard, J. W., 1962, The avian sarcoma-leukosis complex: Its biology and ultrastructure, in: *Tumors Induced by Viruses: Ultrastructural Studies* (A. J. Dalton and F. Haguenau, eds.), pp. 1–59, Academic Press, New York.

Halberstaedter, L., Doljanski, L., and Tenenbaum, E., 1941, Experiments on the cancerization of cells *in vitro* by means of Rous sarcoma agent, *Br. J. Exp. Pathol.* **22**:179.

Hall, M. R., Qualtiere, L. F., and Meyers, P., 1979, Cellular and humoral immune reactivity to tumor-associated antigens in chickens infected with Rous sarcoma virus, *J. Immunol.* **123**:1097.

Hall, W. J., Bean, C. W., and Pollard, M., 1941, Transmission of fowl leukosis through chick embryos and young chicks, *Am. J. Vet. Res.* **2**:272.

Halpern, M. S., and Friis, R. R., 1978, Immunogenicity of the envelope glycoprotein of avian sarcoma virus, *Proc. Natl. Acad. Sci. USA* **75**:1962.

Halpern, M. S., Ewert, D. L., and England, J. M., 1990, Wing web or intravenous inoculation of chickens with v-*src* DNA induces visceral sarcomas, *Virology* **175**:328.

Hamilton, C. M., and Sawyer, C. E., 1939, Transmission of erythroleukosis in young chickens, *Poultry Sci.* **18**:388.

Hanafusa, H., 1965, Analysis of the defectiveness of Rous sarcoma virus. III. Determining influence of a new helper virus on the host range and susceptibility to interference of RSV, *Virology* **25**:248.

Hanafusa, H., and Hanafusa, T., 1966, Determining factor in the capacity of Rous sarcoma virus to induce tumors in mammals, *Proc. Natl. Acad. Sci. USA* **55**:532.

Hanafusa, H., Hanafusa, T., and Rubin, H., 1963, The defectiveness of Rous sarcoma virus, *Proc. Natl. Acad. Sci. USA* **49**:572.

Hanafusa, H., Miyamoto, T., and Hanafusa, T., 1970, A cell-associated factor essential for formation of an infectious form of Rous sarcoma virus, *Proc. Natl. Acad. Sci. USA* **66**:314.

Hanafusa, T., and Hanafusa, H., 1967, Interaction among avian tumor viruses giving enhanced infectivity, *Proc. Natl. Acad. Sci. USA* **58**:818.

Hanafusa, T., and Hanafusa, H., 1973, Isolation of leukosis-type virus from pheasant embryo cells: Possible presence of viral genes in cells, *Virology* **51**:247.

Hanafusa, T., Hanafusa, H., and Miyamoto, T., 1970, Recovery of a new virus from apparently normal chick cells by infection with avian tumor viruses, *Proc. Natl. Acad. Sci. USA* **67**:1797.

Hanafusa, T., Hanafusa, H., Metroka, C. E., Hayward, W. S., Rettenmier, C. W., Sawyer, R. C., Dougherty, R. M., and Di Stefano, H. S., 1976, Pheasant virus: New class of ribodeoxyvirus, *Proc. Natl. Acad. Sci. USA* **73**:1333.

Harris, D. L., Garwood, V. A., Lowe, P. C., Hester, P. Y., Crittenden, L. B., and Fadly, A. M., 1984, Influence of sex-linked feathering phenotypes of parents and progeny upon lymphoid leukosis virus infection status and egg production, *Poultry Sci.* **63**:401.

Harris, R. J. C., 1953, Properties of the agent of Rous No. 1 sarcoma, *Adv. Cancer Res.* **1**:233.

Hayman, M. J., 1983, Avian acute leukemia viruses, in: Retroviruses 1 (P. K. Vogt and H. Koprowski, eds.), *Curr. Top. Microbiol. Immunol.* **103**:109.

Hayward, W. S., 1989, Multiple stages in avian leukosis virus-induced B cell lymphoma, in: *Retroviruses and Disease* (H. Hanafusa, A. Pinter, and M. E. Pullman, eds.), pp. 57–65, Academic Press, San Diego, California.

Hayward, W. S., Shi, C.-K., and Moscovici, C., 1983, Induction of bursal lymphoma by myelocytomatosis virus 29 (MC29), in: *Tumor Viruses and Differentiation* (E. M. Scolnick and A. J. Levine, eds.), pp. 279–287, Liss, New York.

Heider, G., and Klaczinski, K., 1986, Immunisierung gegen aviäre Leukosen mit dem Hüllprotein gp85 des aviären Myeloblastosevirus, *Monatsh. Veterinärmed.* **41**:454.

Heine, U., de Thé, G., Ishiguro, H., and Beard, J. W., 1962a, Morphologic aspects of Rous sarcoma virus elaboration, *J. Natl. Cancer Inst.* **29**:211.

Heine, U., de Thé, G., Ishiguro, H., Sommer, J. R., Beard, D., and Beard, J. W., 1962b, Multiplicity of cell response to the BAI strain A (myeloblastosis) avian tumor virus. II. Nephroblastoma (Wilms' tumor): Ultrastructure, *J. Natl. Cancer Inst.* **29**:41.

Heinzelmann, E. W., Zsigray, R. M., and Collins, W. M., 1981a, Increased growth of RSV-induced tumors in chickens partially tolerant to MHC alloantigens, *Immunogenetics* **12**:275.

Heinzelmann, E. W., Zsigray, R. M., and Collins, W. M., 1981b, Cross-reactivity between RSV-induced tumor antigen and B₅ MHC alloantigen in the chicken, *Immunogenetics* **13**:29.

Helmboldt, C. F., and Jortner, B. S., 1966, Histological patterns of the avian embryonal nephroma, *Avian Dis.* **10**:452.

Hihara, H., Shimuzu, T., and Yamamoto, H., 1983a, Establishment of tumor cell lines cultured from chickens with avian lymphoid leukosis, *Jpn. J. Vet. Sci.* **45**:519.

Hihara, H., Yamamoto, H., Shimohira, H., Arai, K., and Shimizu, T., 1983b, Avian erythroblastosis virus isolated from chick erythroblastosis induced by lymphatic leukemia virus subgroup A, *J. Natl. Cancer Inst.* **70**:891.

Hihara, H., Shimizu, T., Yamamoto, H., and Yoshino, T., 1984, Two strains of avian sarcoma virus newly isolated from chick fibrosarcomas induced by lymphatic leukemia virus subgroup A in two lines of chickens, *J. Natl. Cancer Inst.* **72**:631.

Hirota, Y., Martin, M. T., Viljanen, M., Toivanen, P., and Franklin, R. M., 1980, Immunopathology of chickens infected *in ovo* and at hatching with the avian osteopetrosis virus MAV2-0, *Eur. J. Immunol.* **10**:929.

Hoelzer, J. D., Franklin, R. B., and Bose, H. R., 1979, Transformation by reticuloendotheliosis virus: Development of a focus assay and isolation of a non-transforming virus, *Virology* **93**:20.

Holmes, J. R., 1964, Avian osteopetrosis, *Natl. Cancer Inst. Monogr.* **17**:63.

Hu, C.-P., and Linna, T. J., 1976, Serotherapy of avian reticuloendotheliosis virus-induced tumors, *Ann. N. Y. Acad. Sci.* **277**:634.

Humphries, E. H., and Baba, T. W., 1984, Follicular hyperplasia in the prelymphomatous avian bursa: Relationship to the incidence of B-cell lymphomas, *Curr. Top. Microbiol. Immunol.* **113**:47.

Humphries, E. H., and Baba, T. W., 1986, Restrictions that influence avian leukosis virus-induced lymphoid leukosis, *Curr. Top. Microbiol. Immunol.* **132**:215.

Hunter, E., 1980, Avian oncoviruses: Genetics, in: *Viral Oncology* (G. Klein, ed.), pp. 1–38, Raven Press, New York.

Ianconescu, M., 1977, Reticuloendotheliosis antigen for the agar gel precipitation test, *Avian Pathol.* **6**:259.

Ianconescu, M., and Aharonovici, A., 1978, Persistent viraemia in chickens, subsequent to *in ovo* inoculation of reticuloendotheliosis virus, *Avian Pathol.* **7**:237.

Ianconescu, M., Perk, K., Zimber, A., and Yaniv, A., 1979, Reticuloendotheliosis and lymphoproliferative disease of turkeys, *Ref. Vet.* **36**:2.

Ianconescu, M., Gazit, A., Yaniv, A., Perk, K., and Zimber, A., 1981, Comparative susceptibility of two turkey strains to lymphoproliferative disease virus, *Avian Pathol.* **10**:131.

Ianconescu, M., Yaniv, A., Gazit, A., Perk, K., and Zimber, A., 1983, Susceptibility of domestic birds to lymphoproliferative disease virus (LPDV) of turkeys, *Avian Pathol.* **12**:291.

Ignjatovic, J., 1986, Replication-competent endogenous avian leukosis virus in commercial lines of meat chickens, *Avian Dis.* **30**:264.

Ignjatovic, J., 1988, Isolation of a variant endogenous avian leukosis virus: Non-productive exogenous infection with endogenous viruses containing p27 and p27⁰, *J. Gen. Virol.* **69**:641.

Ignjatovic, J., 1990, Congenital transmission of avian leukosis virus in the absence of detectable shedding of group specific antigen, *Aust. Vet. J.* **67**:299.

Ignjatovic, J., Fahey, K. J., and Bagust, T. J., 1987, An enzyme-linked immunosorbent assay for detection of reticuloendotheliosis virus infection in chickens, *Avian Pathol.* **16**:609.

Ionescu, D., and Simu, G., 1972, Thymus and other lymphoid structures in chickens with Rous sarcoma, *Folia Biol.* **18**:444.

Iraqi, F., Soller, M., and Beckmann, J. S., 1991, Distribution of endogenous viruses in some commercial chicken layer populations, *Poultry Sci.* **70**:665.

Isfort, R., Witter, R. L., and Kung, H.-J., 1987, C-*myc* activation in an unusual retrovirus-induced avian T-lymphoma resembling Marek's disease: Proviral insertion 5' of exon one enhances the expression of an intron promoter, *Oncogene Res.* **2**:81.

Ishiguro, H., Beard, D., Sommer, J. R., Heine, U., de Thé, G., and Beard, J. W., 1962, Multiplicity of cell response to the BAI strain A (myeloblastosis) avian tumor virus. 1. Nephroblastoma (Wilms' tumor): Gross and microscopic pathology, *J. Natl. Cancer Inst.* **29**:1.

Ishii, H., and Oki, Y., 1981, Purification, immunochemical characteristics and quantification of low molecular weight IgM appearing in serum of chickens with avian lymphoid leukosis, *Jpn. J. Vet. Sci.* **43**:369.

Ishizaki, R., and Vogt, P. K., 1966, Immunological relationships among envelope antigens of avian tumor viruses, *Virology* **30**:375.

Israel, E., and Wainberg, M. A., 1977, Development of cellular anti-tumor immunity in chickens bearing tumors induced by Rous sarcoma virus, *J. Immunol.* **118**:2237.

Jackson, C. A. W., Dunn, S. E., Smith, D. I., Gilchrist, P. T., and MacQueen, P. A., 1977, Proventriculitis, "Nakanuke" and reticuloendotheliosis in chickens following vaccination with herpesvirus of turkeys (HVT), *Aust. Vet. J.* **53**:457.

Jármai, K., 1929, Ueber die Hühnerleukose. 1. Mitt, *Allatorvosi Lapok (Hung.)* **52**:229.

Jármai, K., 1930–1931, Beiträge zur Kenntris der Hühnerleukose, *Arch. Tierheilk.* **62**:113.

Jungherr, E., and Landauer, W., 1938, Studies on Fowl Paralysis. III. A Condition Resembling Osteopetrosis (Marble Bone) in Common Fowl, Storrs Agricultural Experiment Station Bulletin, 222.

Kakuk, T. J., Frank, F. R., Weddon, T. E., Burmester, B. R., Purchase, H. G., and Romero, C. H., 1977, Avian lymphoid leukosis prophylaxis with miboleron, *Avian Dis.* **21**:280.

Kang, C.-Y., and Lambright, P., 1977, Pseudotypes of vesicular stomatitis virus with the mixed coat of reticuloendotheliosis virus and vesicular stomatitis virus, *J. Virol.* **21**:1252.

Kanter, M. R., Smith, R. E., and Hayward, W. S., 1988, Rapid induction of B-cell lymphomas: Insertional activation of c-*myb* by avian leukosis virus, *J. Virol.* **62**:1423.

Kass, S. J., and Levinson, W., 1974, Enhancing factor for Rous sarcoma virus: *In vitro* assay and action on both cells and virus, *Virology* **57**:291.

Kawamura, H., Wakabayashi, T., Yamaguchi, S., Taniguchi, T., Takayanagi, N., Sato, S., Sekiya, S., and Horiuchi, T., 1976, Inoculation experiment of Marek's disease vaccine contaminated with reticuloendotheliosis virus, *Natl. Inst. Anim. Health Q.* **16**:135.

Keller, L. H., Rufner, R., and Sevoian, M., 1979, Isolation and development of a reticuloendotheliosis virus-transformed lymphoblastoid cell line from chicken spleen cells, *Infect. Immun.* **25**:694.

Kelloff, G., and Vogt, P. K., 1966, Localisation of avian tumor virus group-specific antigen in cell and virus, *Virology* **29**:377.

Kelloff, G., Hatanaka, M., and Gilden, R. V., 1972, Assay of C-type virus infectivity by measurement of RNA-dependent DNA polymerase activity, *Virology* **48**:266.

Keshet, E., and Temin, H. M., 1979, Cell killing by spleen necrosis virus is correlated with a transient accumulation of spleen necrosis virus DNA, *J. Virol.* **31**:376.

Kirev, T. T., 1984, Characterisation of osteopetrosis induced by viral strain Pts 56 in guinea fowl, *Avian Pathol.* **13**:647.

Kirev, T. T., 1988, Neoplastic response of guinea fowl to osteopetrosis virus strain MAV-2(0), *Avian Pathol.* **17**:101.

Kirev, T. T., Toshkov, I. A., and Mladenov, Z. M., 1987, Virus-induced duodenal adenomas in guinea fowl, *J. Natl. Cancer Inst.* **79**:1117.

Kirev, T. T., Toshkov, I. A., and Mladenov, Z. M., 1986, Virus-induced pancreatic cancer in guinea fowl: A morphological study, *J. Natl. Cancer Inst.* **77**:713.

Koyama, H., Susuki, Y., Ohwada, Y., and Saito, Y., 1976, Reticuloendotheliosis group virus pathogenic to chickens isolated from material infected with turkey herpesvirus (HVT), *Avian Dis.* **20**:429.

Koyama, H., Sasaki, T., Ohwada, Y., and Saito, Y., 1980, The relationship between feathering abnormalities ("Nakanuke") and tumour production in chickens inoculated with reticuloendotheliosis virus, *Avian Pathol.* **9**:331.

Koyama, H., Hodatsu, T., Sasaki, T., Ohwada, Y., Saito, Y., and Saito, H., 1981, Continuous cell culture from chicken embryos inoculated with REV strain T, *Avian Pathol.* **10**:151.

Krasselt, M., Simons, P. C. M., Maas, H. J. L., Elzinga, N., and Jansonius, F. A. T., 1986, The role of viraemic male breeder birds in congenital infection with lymphoid leukosis virus, *Arch. Geflügelk.* **50**:198.

Kuhnlein, U., Gavora, J. S., Spencer, J. L., Bernon, D. E., and Sabour, M., 1988, Incidence of endogenous viral genes in two strains of White Leghorn chickens selected for egg production and susceptibility or resistance to Marek's disease, *Theor. Appl. Genet.* **77**:26.

Kuhnlein, U., Sabour, M., Gavora, J. S., Fairfull, R. W., and Bernon, D. E., 1989, Influence of selection for egg production and Marek's disease resistance on the incidence of endogenous viral genes in White Leghorns, *Poultry Sci.* **68**:1161.

Kung, H.-J., and Maihle, N. J., 1987, Molecular basis of oncogenesis by nonacute avian retroviruses, in: *Avian Leukosis* (G. F. De Boer, ed.), pp. 77–99, Martinus Nijhoff, Boston.

Kurth, R., and Bauer, H., 1972, Cell-surface antigens induced by avian RNA tumor viruses: Detection by a cytotoxic microassay, *Virology* **47**:426.

Lagerlöf, B., and Sundelin, P., 1963a, The histogenesis and haematology of virus-induced myeloid leukaemia in the fowl, *Acta Haematol.* **30**:111.

Lagerlöf, B., and Sundelin, P., 1963b, Variations in the pathogenic effect of myeloid fowl leukaemia virus, *Acta Pathol. Microbiol. Scand.* **59**(2):129.

Lee, L. F., Silva, R. F., Cheng, Y.-Q., Smith, E. J., and Crittenden, L. B., 1986, Characterization of monoclonal antibodies to avian leukosis viruses, *Avian Dis.* **30**:132.

Levy, J. A., 1977, Murine xenotropic type C viruses. III. Phenotypic mixing with avian leukosis and sarcoma viruses, *Virology* **77**:811.

Lewis, R. B., McClure, J., Rup, B., Niesel, D. W., Garry, R. F., Hoelzer, J. D., Nazerian, K., and Bose, H. R., 1981, Avian reticuloendotheliosis virus: Identification of the hematopoietic target cell for transformation, *Cell* **25**:421.

Li, J., Calnek, B. W., Schat, K. A., and Graham, D. L., 1983, Pathogenesis of reticuloen-dotheliosis virus infection in ducks, *Avian Dis.* **27**:1090.

Linna, T. J., Hu, C., and Thompson, K. D., 1974, Development of systemic and local tumors induced by avian reticuloendotheliosis virus after thymectomy or bursec-tomy, *J. Natl. Cancer Inst.* **53**:847.

Love, D. N., and Weiss, R. A., 1974, Pseudotypes of vesicular stomatitis virus determined by exogenous and endogenous avian RNA tumor viruses, *Virology* **57**:271.

Ludford, C. G., Purchase, H. G., and Cox, H. W., 1972, Duck infectious anemia virus associated with *Plasmodium lophurae, Exp. Parasitol.* **31**:29.

Maas, H. J. L., de Boer, G. F., and Groenendal, J. E., 1982, Age related resistance to avian leukosis virus. III. Infectious virus, neutralising antibody and tumours in chickens inoculated at various ages, *Avian Pathol.* **11**:309.

Maccubbin, D., and Schierman, L., 1986, MHC-restricted cytotoxic response of chicken T cells: Expression, augmentation, and clonal characterization, *J. Immunol.* **136**:12.

Manaker, R. A., and Groupé, V., 1956, Discrete foci of altered chicken embryo cells associated with Rous sarcoma virus in tissue culture, *Virology* **2**:838.

Matthews, R. E. F., 1982, *Classification and Nomenclature of Viruses,* Karger, Basel.

McArthur, W. P., Carswell, E. A., and Thorbecke, G. J., 1972, Brief communication: Growth of Rous sarcomas in bursectomized chickens, *J. Natl. Cancer Inst.* **49**:907.

McBride, M. A. T., and Shuman, R. M., 1988, Immune response of chickens inoculated with a recombinant avian leukosis virus, *Avian Dis.* **32**:96.

McBride, R. A., Watanabe, D. H., and Schierman, L. W., 1978, Role of B cells in the expression of genetic resistance to growth of Rous sarcoma in the chicken, *Eur. J. Immunol.* **8**:147.

McDougall, J. S., Biggs, P. M., Shilleto, R. W., and Milne, B. S., 1978a, Lymphoprolifera-tive disease of turkeys. II. Experimental transmission and aetiology, *Avian Pathol.* **7**:141.

McDougall, J. S., Biggs, P. M., and Shilleto, R. W., 1978b, A leukosis in turkeys associated with infection with reticuloendotheliosis virus, *Avian Pathol.* **7**:557.

McDougall, J. S., Shilleto, R. W., and Biggs, P. M., 1980, Experimental infection and vertical transmission of reticuloendotheliosis virus in the turkey, *Avian Pathol.* **9**:445.

Meyers, P., 1976, Antibody response to related leukosis viruses induced in chickens toler-ant to an avian leukosis virus, *J. Natl. Cancer Inst.* **56**:381.

Meyers, P., and Dougherty, R. M., 1971, Immunologic reactivity to viral antigens in chickens infected with avian leukosis viruses, *J. Natl. Cancer Inst.* **46**:701.

Meyers, P., and Dougherty, R. M., 1972, Analysis of immunoglobulins in chicken anti-body to avian leucosis viruses, *Immunology* **23**:1.

Meyers, P., Ritts, G. D., Heise, J. M., and Qualtiere, L. F., 1978, Limited host range of avian tumor viruses on pigeon cells, *Virology* **90**:162.

Miles, B. D., and Robinson, H. L., 1985, High-frequency transduction of c-*erb*B in avian leukosis virus-induced erythroblastosis, *J. Virol.* **54**:295.

Mizuno, Y., and Hatakeyama, H., 1983, Detection of antibodies against avian leukosis viruses with indirect immunoperoxidase absorbance test, *Jpn. J. Vet. Sci.* **45**:31.

Mladenov, Z., Heine, U., Beard, D., and Beard, J. W., 1967, Strain MC29 avian leukosis virus. Myelocytoma, endothelioma, and renal growths: Pathomorphological and ultra-structural aspects, *J. Natl. Cancer Inst.* **38**:251.

Moelling, K., Gelderblom, H., Pauli, G., Friis, R., and Bauer, H., 1975, A comparative study of the avian reticuloendotheliosis virus: Relationship to murine leukemia virus and viruses of the avian sarcoma–leukosis complex, *Virology* **65**:546.

Moldonado, R. L., and Bose, H. R., 1975, Polypeptide and RNA composition of the reticu-loendotheliosis viruses, *Intervirology* **5**:194.

Moloney, J. B., 1956, Biological studies on the Rous sarcoma virus. V. Preparation of improved standard lots of the virus for use in quantitative investigations, *J. Natl. Cancer Inst.* **16**:877.

Montelaro, R. C., and Bolognesi, D. P., 1980, Retroviruses, in: *Cell Membranes and Viral Envelopes* (H. A. Blough and J. M. Tiffany, eds.), Vol. 2, pp. 683–707, Academic Press, New York.

Moore, B. E., and Bose, H. R., 1988, Transformation of avian lymphoid cells by reticuloendotheliosis virus, *Mutat. Res.* **195**:79.

Morgan, H. R., 1964, Origin of Rous sarcoma strains, *J. Natl. Cancer Inst. Monogr.* **17**:392.

Moscovici, C., and Gazzolo, L., 1987, Virus–cell interactions of avian sarcoma and defective leukemia viruses, in: *Avian Leukosis* (G. F. De Boer, ed.), pp. 151–169, Martinus Nijhoff, Boston.

Moscovici, C., and Vogt, P. K., 1968, Effects of genetic cellular resistance on cell transformation and virus replication in chicken hematopoietic cell cultures infected with avian myeloblastosis virus (BAI-A), *Virology* **35**:487.

Moscovici, C., and Zanetti, M., 1970, Studies on single foci of hematopoietic cells transformed by avian myeloblastosis virus, *Virology* **42**:61.

Moscovici, C., Gazzolo, L., and Moscovici, M. G., 1975, Focus assay and defectiveness of avian myeloblastosis virus, *Virology* **68**:173.

Moscovici, C., Chi, D., Gazzolo, L., and Moscovici, M. G., 1976, A study of plaque formation with avian RNA tumor viruses, *Virology* **73**:181.

Moscovici, C., Samarut, J., Gazzolo, L., and Moscovici, M. G., 1981, Myeloid and erythroid neoplastic responses to avian defective leukemia viruses in chickens and quail, *Virology* **113**:765.

Moscovici, M. G., Samarut, J., Jurdic, P., Gazzolo, L., and Moscovici, C., 1983, Transforming ability of avian defective leukemia viruses in early embryogenesis, *Virology* **124**:185.

Motha, M. X. J., and Egerton, J. R., 1983, Effect of reticuloendotheliosis virus on the response of chickens to *Salmonella typhimurium* infection, *Res. Vet. Sci.* **34**:188.

Motha, M. X. J., and Egerton, J. R., 1984, Influence of reticuloendotheliosis on the severity of *Eimeria tenella* infection in broiler chickens, *Vet. Microbiol.* **9**:121.

Motha, M. X. J., and Egerton, J. R., 1987, Outbreak of atypical fowlpox in chickens with persistent reticuloendotheliosis viraemia, *Avian Pathol.* **16**:177.

Motha, M. X. J., Egerton, J. R., and Sweeney, A. W., 1984, Some evidence of mechanical transmission of reticuloendotheliosis virus by mosquitoes, *Avian Dis.* **28**:858.

Motta, J. V., Crittenden, L. B., and Pollard, W. O., 1973, The inheritance of resistance to subgroup C leukosis-sarcoma viruses in New Hampshire chickens, *Poultry Sci.* **52**:578.

Motta, J. V., Crittenden, L. B., Purchase, H. G., Stone, H. A., and Witter, R. L., 1975, Low oncogenic potential of avian endogenous RNA tumor virus infection or expression, *J. Natl. Cancer Inst.* **55**:685.

Murray, J. A., and Begg, A. M., 1930, Histology and histogenesis of a filterable endothelioma of the fowl, in: *Ninth Scientific Report Imperial Cancer Research Fund*, pp. 1–13.

Mussman, H. C., and Twiehaus, M. J., 1971, Pathogenesis of reticuloendothelial virus disease in chicks—An acute runting syndrome, *Avian Dis.* **15**:483.

Nazerian, K., Witter, R. L., Crittenden, L. B., Noori-Dalloii, M. R., and Kung, H. J., 1982, An IgM-producing B lymphoblastoid cell line established from lymphomas induced by a non-defective reticuloendotheliosis virus, *J. Gen. Virol.* **58**:351.

Nehyba, J., Svoboda, J., Karakoz, I., Geryk, J., and Hejnar, J., 1990, Ducks: A new experimental host system for studying persistent infection with avian leukaemia retroviruses, *J. Gen. Virol.* **71**:1937.

Neiman, P. E., Jordan, L., Weiss, R. A., and Payne, L. N., 1980, Malignant lymphoma of the bursa of Fabricius: Analysis of early transformation, in: *Cold Spring Harbor Conference on Cell Proliferation,* Vol. 7, p. 519, Cold Spring Harbor Laboratory, Cold Spring Harbor, New York.

Neiman, P., Wolf, C., Enrietto, P. J., and Cooper, G. M., 1985, A retroviral *myc* gene induces preneoplastic transformation of lymphocytes in a bursal transplantation assay, *Proc. Natl. Acad. Sci. USA* **82**:222.

Neumann, U., and Witter, R. L., 1979, Differential diagnosis of lymphoid leukosis and Marek's disease by tumor-associated criteria. 1. Studies on experimentally infected chickens, *Avian Dis.* **23**:417.

Noori-Daloii, M. R., Swift, R. A., Kung, H. J., Crittenden, L. B., and Witter, R. L., 1981, Specific integration of REV proviruses in avian bursal lymphomas, *Nature* **294**:574.

Nordskog, A. W., and Gebriel, G. M., 1983, Genetic aspects of Rous sarcoma-induced tumor expression in chickens, *Poultry Sci.* **62**:725.

Nyfeldt, A., 1933, Studier over Hønseleukoser. 1. En reen Myeloblastosestamme, *Hospitalstidende* **76**:29.

Oberling, C., and Guérin, M., 1933a, Lesions tumorales en rapport avec la leucémie transmissible des poules, *Bull. Cancer* **22**:180.

Oberling, C., and Guérin, M., 1933b, Nouvelles recherches sur la production de tumeurs malignes avec le virus de la leucémie transmissible des poules, *Bull. Cancer* **22**:326.

Oberling, C., and Guérin, M., 1934, La leucémie erythroblastique ou érythroblastose transmissible des poules, *Bull. Cancer* **23**:38.

Okazaki, W., Purchase, H. G., and Burmester, B. R., 1975, Phenotypic mixing test to detect and assay avian leukosis viruses, *Avian Dis.* **19**:311.

Okazaki, W., Burmester, B. R., Fadly, A., and Chase, W. B., 1979, An evaluation of methods for eradication of avian leukosis virus from a commercial breeder flock, *Avian Dis.* **23**:688.

Okazaki, W., Witter, R. L., Romero, C., Nazerian, K., Sharma, J. M., Fadly, A., and Ewert, D., 1980, Induction of lymphoid leukosis transplantable tumors and the establishment of lymphoblastoid cell lines, *Avian Pathol.* **9**:311.

Okazaki, W., Purchase, H. G., and Crittenden, L. B., 1982, Pathogenicity of avian leukosis viruses, *Avian Dis.* **26**:553.

Olson, C., 1941, A transmissible lymphoid tumor of the chicken, *Cancer Res.* **1**:384.

Olson, L. D., 1967, Histopathologic and hematologic changes in moribund stages of chicks infected with T-virus, *Am. J. Vet. Res.* **28**:1501.

Padgett, F., Baxter-Gabbard, K. L., Raitano-Fenton, A., and Levine, A. S., 1971, Avian reticuloendotheliosis virus. III. Ultrastructural studies, *Avian Dis.* **15**:963.

Palmieri, S., 1985, Transformation of erythroid cells by Rous sarcoma virus (RSV), *Virology* **140**:269.

Panet, A., Baltimore, D., and Hanafusa, T., 1975, Quantitation of avian RNA tumor virus reverse transcriptase by radioimmunoassay, *J. Virol.* **16**:146.

Pani, P. K., 1975, Genetic control of resistance of chick embryo cultures to RSV (RAV-50), *J. Gen. Virol.* **27**:163.

Pani, P. K., 1976, Further studies in genetic resistance of fowl to RSV(RAV-0): Evidence for interaction between independently segregating tumour virus *b* and tumour virus *e* genes, *J. Gen. Virol.* **32**:441.

Pani, P. K., 1977, Evidence for complementary action of *tvb* and *tve* genes that control susceptibility to subgroup E RNA tumour virus in chickens, *J. Gen. Virol.* **37**:639.

Pani, P. K., and Biggs, P. M., 1973, Genetic control of susceptibility to an A subgroup sarcoma virus in commercial chickens, *Avian Pathol.* **2**:27.

Patel, J. R., and Shilleto, R. W., 1987a, Detection of lymphoproliferative disease virus by an enzyme-linked immunosorbent assay, *Epidemiol. Infect.* **99**:711.

Patel, J. R., and Shilleto, R. W., 1987b, Characterisation of lymphoproliferative disease virus of turkeys. Structural polypeptides of the C-type particles, *Arch. Virol.* **95**:159.

Patel, J. R., and Shilleto, R. W., 1987c, Diagnosis of lymphoproliferative disease virus infection of turkeys by an indirect immunofluorescent test, *Avian Pathol.* **16**:367.

Patterson, F. D., 1936, Fowl leukosis, *J. Am. Vet. Med. Assoc.* **88**:32.

Paul, P. S., and Werdin, R. W., 1978, Spontaneously occurring lymphoproliferative disease in ducks (case reports), *Avian Dis.* **22**:191.

Paul, P. S., Pomeroy, K. A., Sarma, P. S., Johnson, K. H., Barnes, D. M., Kumar, M. C., and Pomeroy, B. S., 1976, Brief communication: Naturally occurring reticuloendotheliosis in turkeys: Transmission, *J. Natl. Cancer Inst.* **56**:419.

Paul, P. S., Johnson, K. H., Pomeroy, K. A., Pomeroy, B. S., and Sarma, P. S., 1977, Experimental transmission of reticuloendotheliosis in turkeys with the cell-culture-propagated reticuloendotheliosis viruses of turkey origin, *J. Natl. Cancer Inst.* **58**:1819.

Payne, F. E., Solomon, J. J., and Purchase, H. G., 1966, Immunofluorescent studies of group-specific antigen of the avian sarcoma-leukosis viruses, *Proc. Natl. Acad. Sci. USA* **55**:341.

Payne, L. N., 1985a, Historical review, in: *Marek's Disease* (L. N. Payne, ed.), pp. 1–15, Martinus Nijhoff, Boston.

Payne, L. N., 1985b, Genetics of cell receptors for avian retroviruses, in: *Poultry Genetics and Breeding* (W. G. Hill, J. M. Manson, and D. Hewitt, eds.), pp. 1–16, British Poultry Science, Harlow.

Payne, L. N., 1987, Epizootiology of avian leukosis virus infections, in: *Avian Leukosis* (G. F. De Boer, ed.), pp. 47–75, Martinus Nijhoff, Boston.

Payne, L. N., and Biggs, P. M., 1966, Genetic basis of cellular susceptibility to the Schmidt-Ruppin and Harris strains of Rous sarcoma virus, *Virology* **29**:190.

Payne, L. N., and Biggs, P. M., 1970, Genetic resistance of fowl to MH2 reticuloendothelioma virus, *J. Gen. Virol.* **7**:177.

Payne, L. N., and Chubb, R. C., 1968, Studies on the nature and genetic control of an antigen in normal chick embryos which reacts in the COFAL test, *J. Gen. Virol.* **3**:379.

Payne, L. N., and Howes, K., 1991, Eradication of exogenous avian leukosis virus from commercial layer breeder lines, *Vet. Rec.* **128**:8.

Payne, L. N., and Pani, P. K., 1971, Evidence of linkage between genetic loci controlling response of fowl to subgroup A and subgroup C sarcoma viruses, *J. Gen. Virol.* **13**:253.

Payne, L. N., and Purchase, H. G., 1991, Leukosis/sarcoma group, in: *Diseases of Poultry*, 9th ed. (B. W. Calnek, ed.), pp. 386–439, Iowa State University Press, Ames, Iowa.

Payne, L. N., and Rennie, M., 1975, B cell antigen markers on avian lymphoid leukosis tumour cells, *Vet. Rec.* **96**:454.

Payne, L. N., Pani, P. K., and Weiss, R. A., 1971, A dominant epistatic gene which influences cellular susceptibility to RSV(RAV-0), *J. Gen. Virol.* **13**:455.

Payne, L. N., Holmes, A., Howes, K., Pattison, M., and Walters, D. F., 1979, Studies on the associations between natural infections of hens, cocks and their progeny with lymphoid leukosis virus, *Avian Pathol.* **8**:411.

Payne, L. N., Holmes, A. E., Howes, K., Pattison, M., Pollock, D. L., and Waters, D. E., 1982, Further studies on the eradication and epizootiology of lymphoid leukosis virus infection in a commercial strain of chickens, *Avian Pathol.* **11**:145.

Payne, L. N., Howes, K., and Adene, D. F., 1985, A modified feather pulp culture method for determining the genetic susceptibility of adult chickens to leukosis-sarcoma viruses, *Avian Pathol.* **14**:261.

Payne, L. N., Brown, S. R., Bumstead, N., Howes, K., Frazier, J. F., and Thouless, M. E.,

1991a, A novel subgroup of exogenous avian leukosis virus in chickens, *J. Gen. Virol.* **72**:801.

Payne, L. N., Gillespie, A. M., and Howes, K., 1991b, Induction of myeloid leukosis and other tumours with the HPRS-103 strain of ALV, *Vet. Rec.* **129**:447.

Perek, M., 1960, An epizootic of histiocytic sarcomas in chickens induced by a cell-free agent, *Avian Dis.* **4**:85.

Perk, K., Ianconescu, M., Yaniv, A., and Zimber, A., 1979, Morphological characterization of proliferative cells and virus particles in turkeys with lymphoproliferative disease, *J. Natl. Cancer Inst.* **62**:1483.

Perk, K., Malkinson, M., Gazit, A., Yaniv, A., and Zimber, A., 1982, Reappearance of an acute undifferentiated leukemia in a flock of Muscovy ducks, in: *Proceedings of the 10th International Symposium for Comparative Research on Leukemia and Related Diseases* (D. S. Yohn and J. R. Blakeslee, eds.), pp. 99–100, Elsevier Biomedical, New York.

Peterson, D. A., and Levine, A. S., 1971, Avian reticuloendotheliosis virus (strain T). IV. Infectivity and transmissibility in day-old cockerels, *Avian Dis.* **15**:874.

Peterson, R. D. A., Burmester, B. R., Fredrickson, T. N., Purchase, H. G., and Good, R. A., 1964, Effect of bursectomy and thymectomy on the development of visceral lymphomatosis in the chicken, *J. Natl. Cancer Inst.* **32**:1343.

Peterson, R. D. A., Purchase, H. G., Burmester, B. R., Cooper, M. D., and Good, R. A., 1966, Relationships among visceral lymphomatosis, bursa of Fabricius, and bursa-dependent lymphoid tissue of the chicken, *J. Natl. Cancer Inst.* **36**:585.

Piraino, F., 1967, The mechanism of genetic resistance of chicken embryo cells to infection by Rous sarcoma virus-Bryan strain (BS-RSV), *Virology* **32**:700.

Piraino, F., Okazaki, W., Burmester, B. R., and Fredrickson, T. N., 1963, Bioassay of fowl leukosis virus in chickens by the inoculation of 11-day-old embryos, *Virology* **21**:396.

Pizer, E., and Humphries, E. H., 1989, RAV-1 insertional mutagenesis: Disruption of the c-*myb* locus and development of avian B-cell lymphomas, *J. Virol.* **63**:1630.

Pontén, J., and Thorell, B., 1957, The histogenesis of virus-induced chicken leukemia, *J. Natl. Cancer Inst.* **18**:443.

Porterfield, J. S., 1989, *Andrewes' Viruses of Vertebrates*, 5th ed., Bailliere Tindall, London.

Price, J. A., and Smith, R. E., 1981, Influence of bursectomy on bone growth and anemia induced by avian osteopetrosis viruses, *Cancer Res.* **41**:752.

Price, J. A., and Smith, R. E., 1982, Inhibition of concanavalin A response during osteopetrosis virus infection, *Cancer Res.* **42**:3617.

Purchase, H. G., 1965, Rous sarcoma and its helper viruses (A review), *Avian Dis.* **9**:127.

Purchase, H. G., 1987, The pathogenesis and pathology of neoplasms caused by avian leukosis viruses, in: *Avian Leukosis* (G. F. De Boer, ed.), pp. 171–196, Martinus Nijhoff, Boston.

Purchase, H. G., and Cheville, N. F., 1975, Infectious bursal agent of chickens reduces the incidence of lymphoid leukosis, *Avian Pathol.* **4**:239.

Purchase, H. G., and Fadly, A. M., 1980, Leukosis and sarcomas, in: *Isolation and Identification of Avian Pathogens* (S. B. Hitchner, C. H. Domermuth, H. G. Purchase, and J. E. Williams, eds.), pp. 54–58, American Association of Avian Pathologists, College Station, Texas.

Purchase, H. G., and Gilmour, D. G., 1975, Lymphoid leukosis in chickens chemically bursectomized and subsequently inoculated with bursa cells, *J. Natl. Cancer Inst.* **55**:851.

Purchase, H. G., and Witter, R. L., 1975, The reticuloendotheliosis viruses, *Curr. Top. Microbiol. Immunol.* **71**:103.

Purchase, H. G., Chubb, R. C., and Biggs, P. M., 1968, Effect of lymphoid leukosis and

Marek's disease on the immunological responsiveness of the chicken, *J. Natl. Cancer Inst.* **40**:583.

Purchase, H. G., Okazaki, W., and Burmester, B. R., 1972, Long-term field trials with the herpesvirus of turkeys vaccine against Marek's disease, *Avian Dis.* **16**:57.

Purchase, H. G., Ludford, C., Nazerian, K., and Cox, H. W., 1973, A new group of oncogenic viruses: Reticuloendotheliosis, chick syncytial, duck infectious anemia, and spleen necrosis viruses, *J. Natl. Cancer Inst.* **51**:489.

Purchase, H. G., Okazaki, W., Vogt, P. K., Hanafusa, H., Burmester, B. R., and Crittenden, L. B., 1977a, Oncogenicity of avian leukosis viruses of different subgroups and mutants of sarcoma viruses, *Infect. Immun.* **15**:423.

Purchase, H. G., Gilmour, D. G., Romero, C. H., and Okazaki, W., 1977b, Post infection genetic resistance to avian lymphoid leukosis resides in a B target cell, *Nature* **270**:61.

Qualtiere, L. F., and Meyers, P., 1976, Hypergammaglobulinemia in chickens congenitally infected with an avian leukosis virus, *J. Immunol.* **117**:1127.

Radzichovskaja, R., 1967, Effect of thymectomy on Rous virus tumor growth induced in chickens, *Proc. Soc. Exp. Biol. Med.* **126**:13.

Randall, C. J., Blandford, T. B., Borland, E. D., Brooksbank, N. H., and Hall, S. A., 1977, A survey of mortality in 51 caged laying flocks, *Avian Pathol.* **6**:149.

Rao, A., Kline, K., and Sanders, B. G., 1990, Immune abnormalities in avian erythroblastosis virus-infected chickens, *Cancer Res.* **50**:4764.

Resnick-Roguel, N., Burstein, H., Hamburger, J., Panet, A., Eldor, A., Vlodavsky, I., and Kotler, M., 1989, Cytocidal effect caused by the envelope glycoprotein of a newly isolated avian hemangioma-inducing retrovirus, *J. Virol.* **63**:4325.

Resnick-Roguel, N., Eldor, A., Burstein, H., Hy-Am, E., Vlodavsky, I., Panet, A., Blajckman, M. A., and Kotler, M., 1990, Envelope glycoprotein of avian hemangioma retrovirus induces a thrombogenic surface in human and bovine endothelial cells, *J. Virol.* **64**:4029.

Rispens, B. H., Long, P. A., Okazaki, W., and Burmester, B. R., 1970, The NP activation test for assay of avian leukosis/sarcoma viruses, *Avian Dis.* **14**:738.

Robinson, F. R., and Twiehaus, M. J., 1974, Isolation of the avian reticuloendothelial virus (strain T), *Avian Dis.* **18**:278.

Robinson, H., 1978, Inheritance and expression of chicken genes that are related to avian leukosis sarcoma virus genes, *Curr. Top. Microbiol. Immunol.* **83**:1.

Robinson, H. L., and Eisenman, R. N., 1984, New findings on the congenital transmission of avian leukosis viruses, *Science* **225**:417.

Robinson, H. L., and Miles, B. D., 1985, Avian leukosis virus-induced osteopetrosis is associated with the persistent synthesis of viral DNA, *Virology* **141**:130.

Robinson, H. L., Pearson, M. N., DeSimone, D. W., Tsichlis, P. N., and Coffin, J. M., 1980, Subgroup-E avian-leukosis-virus-associated disease in chickens, *Cold Spring Harbor Symp. Quant. Biology* **44**:1133.

Robinson, H. L., Astrin, S. M., Senior, A. M., and Salazar, F. H., 1981, Host susceptibility to endogenous viruses: Defective glycoprotein-expressing proviruses interfere with infections, *J. Virol.* **40**:745.

Robinson, W. S., Pitkauen, A., and Rubin, H., 1965, The nucleic acid of Rous sarcoma virus: Purification of the virus and isolation of nucleic acid, *Proc. Natl. Acad. Sci. USA* **54**:137.

Rosenthal, P. N., Robinson, H. L., Robinson, W. S., Hanafusa, T., and Hanafusa, H., 1971, RNA in uninfected and virus infected cells complementary to avian tumor virus RNA, *Proc. Natl. Acad. Sci. USA* **68**:2336.

Roth, F. K., Meyers, P., and Dougherty, R. M., 1971, The presence of avian leukosis virus group-specific antibodies in chicken sera, *Virology* **45**:265.

Rothe Meyer, A., and Engelbreth-Holm, J., 1933, Exerimentalle Studien über die Bezie-

hungen zwischen Hühnerleukose und Sarkom an der Hand eines Stammes von über-
tragbarer Leukose-Sarkom-Kombination, *Acta Pathol. Scand.* **10**:380.

Rous, P., 1910, A transmissible avian neoplasm. (Sarcoma of the common fowl), *J. Exp. Med.* **12**:696.

Rous, P., 1911, Transmission of a malignant new growth by means of a cell-free filtrate, *J. Am. Med. Assoc.* **56**:198.

Rous, P., and Murphy, J. B., 1914, On the causation by filterable agents of three distinct chicken tumors, *J. Exp. Med.* **19**:52.

Rovigatti, V. G., and Astrin, S. M., 1983, Avian endogenous viral genes, *Curr. Top. Microbiol. Immunol.* **103**:1.

Rubin, H., 1960, A virus in chick embryos which induces resistance *in vitro* to infection with Rous sarcoma virus, *Proc. Natl. Acad. Sci. USA* **46**:1105.

Rubin, H., 1962, Conditions for establishing immunological tolerance to a tumour virus, *Nature* **195**:342.

Rubin, H., 1965, Genetic control of cellular susceptibility to pseudotypes of Rous sarcoma virus, *Virology* **26**:270.

Rubin, H., Cornelius, A., and Fanshier, L., 1961, The pattern of congenital transmission of an avian leukosis virus, *Proc. Natl. Acad. Sci. USA* **47**:1058.

Rubin, H., Fanshier, L., Cornelius, A., and Hughes, W. F., 1962, Tolerance and immunity in chickens after congenital and contact infection with an avian leukosis virus, *Virology* **17**:143.

Rup, B. J., Hoelzer, J. D., and Bose, H. R., Jr., 1982, Helper viruses associated with avian acute leukemia viruses inhibit the cellular immune response, *Virology* **116**:61.

Salter, D. W., and Crittenden, L. B., 1989, Artificial insertion of a dominant gene for resistance to avian leukosis virus into the germ line of the chicken, *Theor. Appl. Genet.* **77**:457.

Salter, D. W., Smith, E. J., Hughes, S. H., Wright, S. E., Fadly, A. M., Witter, R. L., and Crittenden, L. B., 1986, Gene insertion into the chicken germ line by retroviruses, *Poultry Sci.* **65**:1445.

Salter, D. W., Smith, E. J., Hughes, S. H., Wright, S. E., and Crittenden, L. B., 1987, Transgenic chickens: Insertion of retroviral genes into the chicken germ line, *Virology* **157**:236.

Sandelin, K., and Estola, T., 1974, Occurrence of different subgroups of avian leukosis virus in Finnish poultry, *Avian Pathol.* **3**:159.

Sandelin, K., Estola, T., Ristimäki, S., Ruoslahti, E., and Vaheri, A., 1974, Radioimmunoassays of the group-specific antigen in detection of avian leukosis virus infection, *J. Gen. Virol.* **25**:415.

Sanger, V. L., and Holt, J. A., 1982, Bone density and ash studies in avian osteopetrosis, *Avian Dis.* **26**:177.

Sanger, V. L., Fredrickson, T. N., Morrill, C. C., and Burmester, B. R., 1966, Pathogenesis of osteopetrosis in chickens, *Am. J. Vet. Res.* **27**:1735.

Sarma, P. S., Turner, H. C., and Huebner, R. J., 1964, An avian leucosis group-specific complement fixation reaction. Application for the detection and assay of non-cytopathogenic leucosis viruses, *Virology* **23**:313.

Sarma, P. S., Log, T. S., Huebner, R. J., and Turner, H. C., 1969, Studies of avian leukosis group-specific complement-fixing serum antibodies in pigeons, *Virology* **37**:480.

Saule, S., Mérigaud, J. P., Al-Moustafa, A.-E. M., Ferré, F., Rong, P. M., Amonyel, P., Quatannens, B., Stéhelin, D., and Dieterlen-Lièvre, F., 1987, Heart tumors specifically induced in young avian embryos by the v-*myc* oncogene, *Proc. Natl. Acad. Sci. USA* **84**:7982.

Sazawa, H., Sugimori, T., Miura, Y., and Shimizu, T., 1966, Specific complement fixation test of Rous sarcoma with pigeon serum, *Natl. Inst. Anim. Health Q.* **6**:208.

Schat, K. A., 1987, Immunity in Marek's disease and other tumors, in: *Avian Immunology: Basis and Practice* (A. Toivanen and P. Toivanen, eds.), Vol. II, pp. 101–128, CRC Press, Boca Raton, Florida.

Schat, K. A., Gonzales, J., Solorzano, A., Avila, E., and Witter, R. L., 1976, A lymphoproliferative disease in Japanese quail, *Avian Dis.* **20**:153.

Schierman, L. W., and Collins, W. M., 1987, Influence of the major histocompatibility complex on tumor regression and immunity in chickens, *Poultry Sci.* **60**:812.

Schierman, L. W., Watanabe, D. H., and McBride, R. A., 1977, Genetic control of Rous sarcoma regression in chickens: Linkage with the major histocompatibility complex, *Immunogenetics* **5**:325.

Schmidt, E. V., Crapo, J. D., Harrelson, J. R., and Smith, R. L., 1981, A quantitative histological study of avian osteopetrotic bone demonstrating normal osteoclast numbers and osteoblastic activity, *Lab. Invest.* **44**:164.

Scofield, V. L., and Bose, H. R., 1978, Depression of mitogen response in spleen cells from reticuloendotheliosis virus-infected chickens and their suppressive effect on normal lymphocyte response, *J. Immunol.* **120**:1321.

Segura, J. C., Gavora, J. S., Spencer, J. L., Fairfull, R. W., Gowe, R. J., and Buckland, R. B., 1988, Semen traits and fertility of White Leghorn males shown to be positive or negative for lymphoid leukosis virus in semen and feather pulp, *Br. Poultry Sci.* **29**:545.

Sevoian, M., Larose, R. N., and Chamberlain, D. M., 1964, Avian lymphomatosis VI. A virus of unusual potency and pathogenicity, *Avian Dis.* **8**:336.

Shank, P. R., Schatz, P. J., Jensen, L. M., Tsichlis, P. N., Coffin, J. M., and Robinson, J. L., 1985, Sequences in the *gag–pol–5' env* region of avian leukosis viruses confer the ability to induce osteopetrosis, *Virology* **145**:94.

Sharp, D. G., Eckert, E. A., Beard, D., and Beard, J. W., 1952, Morphology of the virus of avian erythromyeloblastic leucosis and a comparison with the agent of Newcastle disease, *J. Bacteriol.* **63**:151.

Shibuya, T., Chen, I., Howatson, A., and Mak, T. W., 1982, Morphological, immunological and biochemical analyses of chicken spleen cells transformed *in vitro* by reticuloendotheliosis virus strain T, *Cancer Res.* **42**:2722.

Shuman, R. M., and McBride, M. A. T., 1988, Resistance of chickens to Rous sarcoma virus challenge following immunization with a recombinant avian leukosis virus, *Avian Dis.* **32**:410.

Sieweke, M. H., Stoker, A. W., and Bissell, M. J., 1989, Evaluation of the cocarcinogenic effect of wounding in Rous sarcoma virus tumorigenesis, *Cancer Res.* **49**:6419.

Sieweke, M. H., Thompson, N. L., Sporn, M. B., and Bissell, M. J., 1990, Mediation of wound-related Rous sarcoma virus tumorigenesis by TGF-β, *Science* **248**:1656.

Silva, R., and Burch, J. B. E., 1989, Evidence that chicken CRI elements represent a novel family of retroposons, *Mol. Cell. Biol.* **9**:3563.

Šimkovič, D., 1972, Characteristics of tumors induced in mammals, especially rodents, by viruses of the avian leukosis sarcoma group, *Adv. Virus Res.* **17**:95.

Simon, M. C., Neckameyer, W. S., Hayward, W. S., and Smith, R. E., 1987, Genetic determinants of neoplastic diseases induced by a subgroup F avian leukosis virus, *J. Virol.* **61**:1203.

Smith, E. J., 1977, Preparation of antisera to group-specific antigens of avian leukosis-sarcoma viruses: An alternative approach, *Avian Dis.* **21**:290.

Smith, E. J., 1987, Endogenous avian leukemia viruses, in: *Avian Leukosis* (G. F. De Boer, ed.), pp. 101–120, Martinus Nijhoff, Boston.

Smith, E. J., and Fadly, A. M., 1988, Influence of congenital transmission of endogenous virus-21 on the immune response to avian leukosis virus infection and the incidence of tumors in chickens, *Poultry Sci.* **67**:1674.

Smith, E. J., and Witter, R. L., 1983, Detection of antibodies against reticuloendotheliosis viruses by an enzyme-linked immunosorbent assay, *Avian Dis.* **27**:225.

Smith, E. J., Solomon, J. J., and Witter, R. L., 1977, Complement-fixation test for reticuloendotheliosis viruses. Limits of sensitivity in infected avian cells, *Avian Dis.* **21**:612.

Smith, E. J., Fadly, A., and Okazaki, W., 1979, An enzyme-linked immunosorbent assay for detecting avian leukosis-sarcoma viruses, *Avian Dis.* **23**:698.

Smith, E. J., Neumann, U., and Okazaki, W., 1980, Immune response to avian leukosis virus infection in chickens: Sequential expression of serum immunoglobulins and viral antibodies, *Comp. Immunol. Microbiol. Infect. Dis.* **2**:519.

Smith, E. J., Salter, D. W., Silva, R. F., and Crittenden, L. B., 1986a, Selective shedding and congenital transmission of endogenous avian leukosis viruses, *J. Virol.* **60**:1050.

Smith, E. J., Fadly, A. M., and Crittenden, L. B., 1986b, Observations on an enzyme-linked immunosorbent assay for the detection of antibodies against avian leukosis-sarcoma viruses, *Avian Dis.* **30**:488.

Smith, E. J., Fadly, A. M., and Crittenden, L. B., 1990a, Interactions between endogenous virus loci *ev*6 and *ev*21. 1. Immune response to exogenous avian leukosis virus infection, *Poultry Sci.* **69**:1244.

Smith, E. J., Fadly, A. M., and Crittenden, L. B., 1990b, Interactions between endogenous virus loci *ev*6 and *ev*21. 2. Congenital transmission of EV21 viral product to female progeny from slow-feathering dams, *Poultry Sci.* **69**:1251.

Smith, R. E., 1982, Avian osteopetrosis, *Curr. Top. Microbiol. Immunol.* **101**:75.

Smith, R. E., 1987, Immunology of avian leukosis virus infections, in: *Avian Leukosis* (G. F. De Boer, ed.), pp. 121–129, Martinus Nijhoff, Boston.

Smith, R. E., and Schmidt, E. V., 1982, Induction of anemia by avian leukosis viruses of five subgroups, *Virology* **117**:516.

Smith, R. E., and Van Eldik, L. J., 1978, Characterization of the immunosuppression accompanying virus-induced avian osteopetrosis, *Infect. Immun.* **22**:452.

Soffer, D., Resnick-Roguel, N., Eldor, A., and Kotler, M., 1990, Multifocal vascular tumors in fowl induced by a newly isolated retrovirus, *Cancer Res.* **50**:4787.

Solomon, J. J., Burmester, B. R., and Fredrickson, R. N., 1966, Investigations of lymphoid leukosis infection in genetically similar chicken populations, *Avian Dis.* **10**:477.

Solomon, J. J., Long, P. A., and Okazaki, W., 1971, Procedures for the *In vitro* Assay of Viruses and Antibody of Avian Lymphoid Leukosis and Marek's Disease, Agriculture Handbook No. 404, Agricultural Research Service, U. S. Department of Agriculture, U. S. Government Printers Office, Washington D. C.

Solomon, J. J., Witter, R. L., and Nazerian, K., 1976, Studies on the etiology of lymphomas in turkeys: Isolation of reticuloendotheliosis virus, *Avian Dis.* **20**:735.

Somes, R. G., 1980, Alphabetical list of the genes of domestic fowl, *J. Hered.* **71**:168.

Spencer, J. L., 1984, Progress towards eradication of lymphoid leukosis viruses—A review, *Avian Pathol.* **13**:599.

Spencer, J. L., 1987, Laboratory diagnostic procedures for detecting avian leukosis virus infections, in: *Avian Leukosis* (G. F. de Boer, ed.), pp. 213–235, Martinus Nijhoff, Boston.

Spencer, J. L., and Gilka, F., 1983, Role of the skin and gut in horizontal transmission of avian leukosis virus, *J. Am. Vet. Med. Assoc.* **183**:355 (abstract).

Spencer, J. L., Crittenden, L. B., Burmester, B. R., Romero, C., and Witter, R. L., 1976, Lymphoid leukosis viruses and gs antigen in unincubated chicken eggs, *Avian Pathol.* **5**:221.

Spencer, J. L., Crittenden, L. B., Burmester, B. R., Okazaki, W., and Witter, R. L., 1977, Lymphoid leukosis: Interrelations among virus infections in hens, eggs, embryos and chicks, *Avian Dis.* **21**:331.

Spencer, J. L., Gavora, J. S., and Gowe, R. S., 1979, Effect of selection for high egg production in chickens on shedding of lymphoid leukosis virus and gs antigen into eggs, *Poultry Sci.* **58**:279.

Spencer, J. L., Gavora, J. S., and Gowe, R. S., 1980, Lymphoid leukosis virus: Natural transmission and nonneoplastic effects, in: *Cold Spring Harbor Conference on Cell Proliferation*, Vol. 7, p. 553.

Spencer, J. L., Gavora, J. S., and Gilka, F., 1987, Feather pulp organ cultures for assessing host resistance to infection with avian leukosis-sarcoma viruses, *Avian Pathol.* **16**:425.

Steck, F. T., and Rubin, H., 1966, The mechanism of interference between an avian leukosis virus and Rous sarcoma virus. 1. Establishment of interference, *Virology* **29**:628.

Stéhelin, D., Varmus, H. E., Bishop, J. M., and Vogt, P. K., 1976, DNA related to the transforming gene(s) of avian sarcoma viruses is present in normal avian DNA, *Nature* **260**:170.

Stephenson, J. R., Wilsnack, R. E., and Aaronson, S. A., 1973, Radioimmunoassay for avian C-type virus group-specific antigen: Detection in normal and virus-transformed cells, *J. Virol.* **11**:893.

Stephenson, J. R., Smith, E. J., Crittenden, L. B., and Aaronson, S. A., 1975, Analysis of antigenic determinants of structural polypeptides of avian type C tumor viruses, *J. Virol.* **16**:27.

Stoker, A. W., and Bissell, M. J., 1987, Quantitative immunocytochemical assay for infectious avian retroviruses, *J. Gen. Virol.* **68**:2481.

Storms, R. W., and Bose, H. R., Jr., 1989, Avian retroviruses, in: *Virus-Induced Immunosuppression* (S. Specter, M. Bendinelli, and H. Friedman, eds.), pp. 375–393, Plenum Press, New York.

Stubbs, E. L., and Furth, J., 1932, Anemia and erythroleucosis occurring spontaneously in the common fowl, *J. Am. Vet. Med. Assoc.* **81**:209.

Stubbs, E. L., and Furth, J., 1935, The relation of leukosis to sarcoma of chickens. 1. Sarcoma and erythroleukosis (strain 13), *J. Exp. Med.* **61**:593.

Suni, J., Hortling, L., and Vaheri, A., 1978, Selective expression of endogenous and exogenous C-type viruses in chicken lymphoid tissues, *Clin. Lab. Immunol.* **1**:37.

Svet-Moldavsky, G. J., 1958, Sarcoma in albino rats treated during the embryonic stage with Rous virus, *Nature* **182**:1452.

Svoboda, J., 1986, Rous sarcoma virus, *Intervirology* **26**:1.

Svoboda, J., and Hložánek, I., 1970, Role of cell association in virus infection and virus rescue, *Adv. Cancer Res.* **13**:217.

Swift, R. A., Shaller, E., Witter, R. L., and Kung, H.-J., 1985, Insertional activation of c-*myc* by reticuloendotheliosis virus in chicken B lymphoma: Nonrandom distribution and orientation of the provirus, *J. Virol.* **54**:869.

Temin, H. M., 1974, The Bertner Foundation Memorial Award Lecture—From proviruses to protoviruses: RNA-directed DNA synthesis by RNA tumor viruses and cells, in: *Molecular Studies in Viral Neoplasia*, pp. 7–38, Williams and Wilkins, Baltimore.

Temin, H. M., 1985, Reverse transcription in the eukaryotic genome: Retroviruses, pararetroviruses, retrotransposons, and retrotranscripts, *Mol. Biol. Evol.* **6**:455.

Temin, H. M., and Kassner, V. K., 1974, Replication of reticuloendotheliosis viruses in cell cultures: Acute infection, *J. Virol.* **13**:291.

Temin, H. M., and Kassner, V. K., 1975, Replication of reticuloendotheliosis viruses in cell culture: Chronic infection, *J. Gen. Virol.* **27**:267.

Temin, H. M., and Kassner, V. K., 1976, Avian leukosis viruses of different subgroups and types isolated after passage of Rous sarcoma virus—Rous associated virus-0 in cells from different ring-necked pheasant embryos, *J. Virol.* **19**:302.

Temin, H. M., and Mizutani, S., 1970, RNA-dependent DNA polymerase in virions of Rous sarcoma virus, *Nature* **226**:1211.

Temin, H. M., and Rubin, H., 1958, Characteristics of an assay for Rous sarcoma virus and Rous sarcoma cells in tissue culture, *Virology* **6**:669.

Tereba, A., and Murti, K. G., 1977, A very sensitive biochemical assay for detecting and quantitating avian oncornaviruses, *Virology* **80**:166.

Theilen, G. H., Zeigel, R. F., and Twiehaus, M. J., 1966, Biological studies with RE virus (strain T) that induces reticuloendotheliosis in turkeys, chickens and Japanese quail, *J. Natl. Cancer Inst.* **37**:731.

Thompson, K. D., and Linna, T. J., 1973, Bursa-dependent and thymus-dependent "surveillance" of a virus-induced tumor in the chicken, *Nature New Biol.* **245**:10.

Toyoshima, K., and Vogt, P. K., 1969, Enhancement and inhibition of avian sarcoma viruses by polycations and polyanions, *Virology* **38**:414.

Tracy, S. E., Woda, B. A., and Robinson, H. L., 1985, Induction of angiosarcoma by a c-*erbB* transducing virus, *J. Virol.* **54**:304.

Troesch, C. D., and Vogt, P. K., 1985, An endogenous virus from Lophortyx quail is the prototype for envelope subgroup I of avian retroviruses, *Virology* **143**:595.

Tsai, W.-P., Copeland, T. D., and Oroszlan, S., 1985, Purification and chemical and immunological characterization of avian reticuloendotheliosis virus *gag*-gene-encoded structural proteins, *Virology* **140**:289.

Tsai, W.-P., Copeland, T. D., and Oroszlan, S., 1986, Biosynthesis and chemical and immunological characterization of avian reticuloendotheliosis virus *env* gene-encoded proteins, *Virology* **155**:567.

Tsukamoto, K., Kono, Y., Arai, K., Kitahara, H., and Takahashi, K., 1985, An enzyme-linked immunosorbent assay for detection of antibodies to exogenous avian leukosis virus, *Avian Dis.* **29**:1118.

Vogt, P. K., 1965, Avian tumor viruses, *Adv. Virus Res.* **11**:293.

Vogt, P. K., 1967a, A virus released by "non-producing" Rous sarcoma cells, *Proc. Natl. Acad. Sci. USA* **58**:801.

Vogt, P. K., 1967b, DEAE-Dextran: Enhancement of cellular transformation induced by avian sarcoma viruses, *Virology* **33**:175.

Vogt, P. K., 1977, Genetics of RNA tumor viruses, in: *Genetics of Animal Viruses* (H. Fraenkel-Conrat and R. R. Wagner, eds.), pp. 341–455, Plenum Press, New York.

Vogt, P. K., and Friis, R. R., 1971, An avian leukosis virus related to RSV (0): Properties and evidence for helper activity, *Virology* **43**:223.

Vogt, P. K., and Ishizaki, R., 1965, Reciprocal patterns of genetic resistance to avian tumor viruses in two lines of chickens, *Virology* **26**:664.

Vogt, P. K., and Ishizaki, R., 1966a, Criteria for the classification of avian tumor viruses, in: *Viruses Inducing Cancer: Implications for Therapy* (W. J. Burdette, ed.), pp. 71–90, University of Utah Press, Salt Lake City, Utah.

Vogt, P. K., and Ishizaki, R., 1966b, Patterns of viral interference in the avian leukosis and sarcoma complex, *Virology* **30**:368.

Vogt, P. K., and Rubin, H., 1963, Studies on the assay and multiplication of avian myeloblastosis virus, *Virology* **19**:92.

Vogt, P. K., Spencer, J. L., Okazaki, W., Witter, R. L., and Crittenden, L. B., 1977, Phenotypic mixing between reticuloendotheliosis virus and avian sarcoma viruses, *Virology* **80**:127.

Wainberg, M. A., and Halpern, M. S., 1987, Avian sarcomas: Immune responsiveness and pathology, in: *Avian Leukosis* (G. F. De Boer, ed.), pp. 131–152, Martinus Nijhoff, Boston.

Wainberg, M. A., and Phillips, E. R., 1976, Immunity against avian sarcomas, *Isr. J. Med. Sci.* **12**:388.

Wainberg, M. A., Yu, M., Schwartz-Luft, E., and Israel, E., 1977a, Cellular and humoral anti-tumor immune responsiveness in chickens bearing tumors induced by avian sarcoma virus, *Int. J. Cancer* **19**:680.

Wainberg, M. A., Israel, E., Schwartz-Luft, E., and Yu, E., 1977b, Differential expression of relevant Rous sarcoma-associated antigens in cultured cells, *Cancer Res.* **37**:3026.

Walker, M. H., Rup, B. J., Rubin, A. S., and Bose, H. R., Jr., 1983, Specificity of the immunosuppression induced by avian reticuloendotheliosis, *Infect. Immun.* **40**:225.

Walter, W. G., Burmester, B. R., and Cunningham, C. H., 1962, Studies on the transmission and pathology of a viral-induced avian nephroblastoma (embryonal nephroma), *Avian Dis.* **6**:455.

Wang, L.-H., and Hanafusa, H., 1988, Avian sarcoma viruses, *Virus Res.* **9**:159.

Watts, S. L., and Smith, R. E., 1980, Pathology of chickens infected with avian nephroblastoma virus MAV-2(N), *Infect. Immun.* **27**:501.

Weiss, R. A., 1967, Spontaneous virus production from "non-virus producing" Rous sarcoma cells, *Virology* **32**:719.

Weiss, R., 1969, Interference and neutralization studies with Bryan strain Rous sarcoma virus synthesised in the absence of helper virus, *J. Gen. Virol.* **5**:529.

Weiss, R. A., 1973, Ecological genetics of RNA tumor viruses and their hosts, in: *Analytic and Experimental Epidemiology* (W. Nakahara, T. Hirayama, K. Nishioka, and H. Sugano, eds.), pp. 201–233, University of Tokyo Press, Tokyo.

Weiss, R. A., 1975, Genetic transmission of RNA tumor viruses, in: *Antiviral Mechanisms* (M. Pollard, ed.), pp. 165–203, Academic Press, New York.

Weiss, R. A., 1981, Retrovirus receptors and their genetics, in: *Virus Receptors Part 2* (K. Lonberg-Holm and L. Philipson, eds.), pp. 187–202, Chapman and Hall, London.

Weiss, R. A., and Frisby, D. P., 1982, Are avian endogenous viruses pathogenic? in: *Proceedings of the 10th International Symposium for Comparative Research on Leukemia and Related Diseases* (D. S. Yohn and J. R. Blakeslee, eds.), pp. 303–308, Elsevier Biomedical, New York.

Weiss, R. A., Friis, R. R., Katz, E., and Vogt, P. K., 1971, Induction of avian tumor viruses in normal cells by physical and chemical carcinogens, *Virology* **46**:920.

Weiss, R., Teich, N., Varmus, H., and Coffin, J., eds., 1982, *RNA Tumor Viruses*, Cold Spring Harbor Laboratory, Cold Spring Harbor, New York.

Weiss, R. A., Teich, N., Varmus, H., and Coffin, J., eds., 1985, *RNA Tumor Viruses 2. Supplements and Appendixes*, Cold Spring Harbor Laboratory, Cold Spring Harbor, New York.

Weller, S. K., and Temin, H. M., 1981, Cell killing by avian leukosis viruses, *J. Virol.* **39**:713.

Welt, S., Purchase, H. G., and Thorbecke, G. J., 1977, Viral protein synthesis by tissues from avian leukosis virus-infected chickens. 1. Susceptible chickens infected after hatching, *J. Immunol.* **119**:1800.

Welt, S., Okazaki, W., Purchase, H. G., and Thorbecke, G. J., 1979, Viral protein synthesis by tissue from avian leukosis virus-infected chickens. II. Effect of passive neutralizing antibody in normal and agammaglobulinaemic chickens, *Immunology* **37**:587.

Weyl, K. G., and Dougherty, R. M., 1977, Contact transmission of avian leukosis virus, *J. Natl. Cancer Inst.* **58**:1019.

Whalen, L. R., Wheeler, D. W., Gould, D. H., Fiscus, S. A., Boggie, L. C., and Smith, R. E., 1988, Functional and structural alterations of the nervous system induced by avian retrovirus RAV-7, *Microb. Pathog.* **4**:401.

Whitfill, C. E., Akbar, W. J., Gyles, N. R., and Thoma, J. A., 1986, Transfer of blood lymphocytes and macrophages between histocompatible progressor and regressor chickens infected with Rous sarcoma virus, *J. Natl. Cancer Inst.* **76**:1185.

Witter, R. L., 1989, Reticuloendotheliosis, in: *A Laboratory Manual for the Isolation and*

Identification of Avian Pathogens, 3rd ed. (H. G. Purchase, L. H. Arp, C. H. Domermuth, and J. E. Pearson, eds.), pp. 143–148, American Association of Avian Pathologists, Kandall/Hunt Publishing Company, Dubuque, Iowa.

Witter, R. L., 1991, Reticuloendotheliosis, in: *Diseases of Poultry*, 9th ed. (B. W. Calnek, ed.), pp. 439–456, Iowa State University Press, Ames, Iowa.

Witter, R. L., and Crittenden, L. B., 1979, Lymphomas resembling lymphoid leukosis in chickens inoculated with reticuloendotheliosis virus, *Int. J. Cancer* 23:673.

Witter, R. L., and Glass, S. W., 1984, Reticuloendotheliosis in breeder turkeys, *Avian Dis.* 28:742.

Witter, R. L., and Johnson, D. C., 1985, Epidemiology of reticuloendotheliosis virus in broiler breeder flocks, *Avian Dis.* 29:1140.

Witter, R. L., and Salter, D. W., 1989, Vertical transmission of reticuloendotheliosis virus in breeder turkeys, *Avian Dis.* 33:226.

Witter, R. L., Calnek, B. W., and Levine, P. P., 1966, Influence of naturally occurring parental antibody on visceral lymphomatosis virus infection in chickens, *Avian Dis.* 10:43.

Witter, R. L., Purchase, H. G., and Burgoyne, G. H., 1970, Peripheral nerve lesions similar to those of Marek's disease in chickens inoculated with reticuloendotheliosis virus, *J. Natl. Cancer Inst.* 45:567.

Witter, R. L., Lee, L. F., Bacon, L. D., and Smith, E. J., 1979, Depression of vaccinal immunity to Marek's disease by infection with reticuloendotheliosis virus, *Infect. Immun.* 26:90.

Witter, R. L., Smith, E. J., and Crittenden, L. B., 1981, Tolerance, viral shedding, and neoplasia in chickens infected with non-defective reticuloendotheliosis viruses, *Avian Dis.* 25:374.

Witter, R. L., Peterson, I. L., Smith, E. J., and Johnson, D. C., 1982, Serological evidence in commercial chicken and turkey flocks of infection with reticuloendotheliosis virus, *Avian Dis.* 26:753.

Witter, R. L., Sharma, J. M., and Fadly, A. M., 1986, Nonbursal lymphomas by nondefective reticuloendotheliosis virus, *Avian Pathol.* 15:467.

Yamanouchi, K., and Hayami, M., 1970, Cellular immunity induced by Rous sarcoma virus in Japanese quail. I. Effect of anti-lymphocyte serum on oncogenesis of Rous sarcoma virus, *Jpn. J. Med. Sci. Biol.* 23:395.

Yamanouchi, K., Hayami, M., Miyakura, S., Fukuda, A., and Kobune, F., 1971, Cellular immunity induced by Rous sarcoma virus in Japanese quail. II. Effect of thymectomy and bursectomy on oncogenesis of Rous sarcoma virus, *Jpn. J. Med. Sci. Biol.* 24:1.

Yaniv, A., Gazit, A., Ianconescu, I., Perk, K., Aizenberg, B., and Zimber, A., 1979, Biochemical characterization of the type C retrovirus associated with lymphoproliferative disease of turkeys, *J. Virol.* 30:351.

Zeigel, R. F., Theilen, G. H., and Twiehaus, M. J., 1966, Electron microscopic observation on RE virus (strain T) that induces reticuloendotheliosis in turkeys, chickens, and Japanese quail, *J. Natl. Cancer Inst.* 37:709.

Ziemiecki, A., Krömer, G., Mueller, R. G., Hàla, K., and Wick, G., 1988, ev22, a new endogenous avian leukosis virus locus found in chickens with spontaneous autoimmune thyroiditis, *Arch. Virol.* 100:267.

Zilber, L. A., and Krjukova, I. N., 1957, Haemorrhagic disease in rats caused by Rous sarcoma virus, *Vopr. Virusol.* 4:239.

Zimber, A., Heller, E. D., Perk, K., Ianconescu, M., and Yaniv, A., 1983, Effect of lymphoproliferative disease virus and of niridazole on the *in vitro* blastogenic response of peripheral blood lymphocytes of turkeys, *Avian Dis.* 27:1012.

Zimber, A., Perk, K., Ianconescu, M., Schwarzbard, Z., and Yaniv, A., 1984, Lymphoproliferative disease of turkeys: Effect of chemical and surgical bursectomy on viraemia, pathogenesis and on the humoral immune response, *Avian Pathol.* 13:277.

CHAPTER 7

Retroviruses in Rodents

CHRISTINE A. KOZAK AND SANDRA RUSCETTI

I. INTRODUCTION

Rodents have long been used as model systems for the analysis of onco-genesis. The early recognition that rodent retroviruses played an etiolog-ical role in neoplastic diseases has resulted in the extensive characteriza-tion of these viruses and the virus–host interaction. The first evidence for a tumorigenic mammalian retrovirus was, in fact, provided in 1936 by Bittner (1936), who identified an agent responsible for mammary carcinomas in the milk of nursing female C3H mice. Since that time, numerous replication-competent as well as replication-defective viruses have been described, and the mechanisms by which these viruses infect and transform cells have been under intensive study.

Retrovirus infection can have a number of serious consequences for its natural host, and many of these phenomena have been studied exten-sively in rodents, particularly the mouse. First, DNA copies of retroviral genomes integrate into host chromosomes and persist as stable genetic elements. Although most virus integrations are not overtly detrimental, provirus integrations occurring at or near specific coding regions can alter normal gene activity. Integrations into somatic tissue can result in altered growth regulation leading to neoplastic disease, and specific in-

CHRISTINE A. KOZAK • Laboratory of Molecular Microbiology, National Institute of Allergy and Infectious Diseases, National Institutes of Health, Bethesda, Maryland 20892. SANDRA RUSCETTI • Laboratory of Molecular Oncology, National Cancer Institute, Frederick Cancer Research and Development Center, Frederick, Mary-land 21702-1201.

The Retroviridae, Volume 1, edited by Jay A. Levy. Plenum Press, New York, 1992.

tegrations into germline tissue have been associated with specific developmental aberrations or lethal mutations. Second, replication-competent retroviruses can recombine with endogenous retroviral or host sequences to acquire novel segments which significantly alter the biological properties of the virus. Finally, rodent retroviruses have been associated with neurological disorders and nonneoplastic immune system defects, including autoimmunity and immune deficiency disorders. Several retroviruses are also thought to function as specific mitogens in hematopoietic cells. This chapter describes the various retroviruses which have been described in rodents, the distribution of related sequences within the rodent genome, the mechanisms by which these viruses transform cells or alter expression of normal cellular genes, and the host genes which restrict their replication and disease induction.

II. NATURALLY OCCURRING RETROVIRUSES

Many different retroviruses have been isolated from rodents. Some of these represent unique recombinant viruses isolated from neoplastic tissue or following serial animal inoculations or long-term passage in cultured cells. Others, however, are the products of germline proviruses or are congenitally transmitted in specific breeding populations. These naturally occurring retroviruses can be routinely isolated as infectious viruses or have been identified as germline-associated retroviral related sequences in the members of several rodent families (Table I).

TABLE I. Retroviruses and Virus-Related Sequences in the Order Rodentia

Family[a]	Genus	Common name	Retrovirus type[b]				
			C-type	B-type	VL30	IAP	Other
Muridae	Mus	Mouse	V, S	V, S	S^v	P, S	S^B, S^G
	Praomys	Mastomys	V	?	?	?	?
	Rattus	Rat	V, S	?	S^v	P, S	S^B, S^G
	Vandeleuria	Tree mouse	V	?	?	P	?
Cricetidae	Cricetulus	Chinese hamster	P, S	?	?	P, S	S^B
	Mesocricetulus	Syrian hamster	?	?	?	P	S^B
	Gerbillus	Gerbil	P	?	?	P, S	S^B
Sciuridae	Sciurus	Squirrel	?	?	?	?	S^B
	Marmota	Woodchuck	?	?	?	?	S^B
Caviidae	Cavia	Guinea pig	?	?	?	?	P, S^B, S^G
Agoutidae	Agouti	Agouti	V	?	?	?	?

[a] Taxonomy from Honacki et al. (1982).
[b] V, Infectious virus; P, particles; S, germline sequences related to VL30s (S^v), IAPs (S^A) B-26 endogenous retrovirus (Obata and Khan, 1988) (S^B), or the GLN-3 retrovirus element (Itin and Keslet, 1986) (S^G).

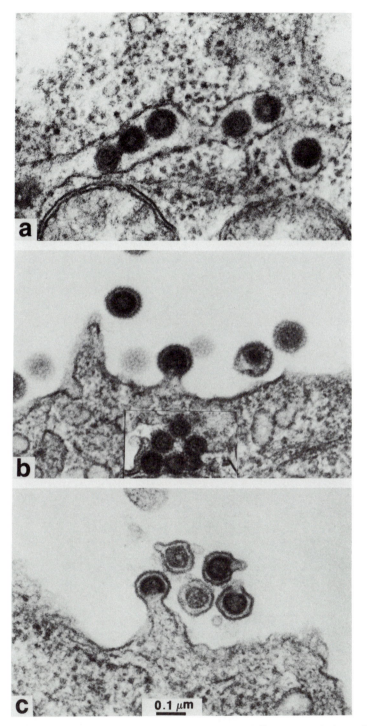

FIGURE 1. Electron micrographs representing the three different morphological classifications of rodent retroviruses: (A) intracytoplasmic A-type particles; (B) B-type particles; and (C) C-type particles. [Kindly provided by Dr. M. Gonda.]

A. Mice

Mouse retroviruses can be classified into four groups based on differences in virion morphology (Fig. 1) and sequence homology (Chapter 2). Only two classes, the type C murine leukemia viruses (MuLVs or MLVs) and the type B mammary tumor viruses (MMTVs), are replication-competent. Type A particles occur only intracellularly. The type A intracisternal particles (IAPs) are found only in association with endoplasmic reticulum and have no extracellular phase. Particles resembling IAPs have also been detected in the cytoplasm, but generally represent precursors of B-type viruses, although intracytoplasmic particles which have been observed in some wild mouse species, notably *Mus cervicolor*, have an extracellular phase which most closely resembles C-type virions (see Section II.A.3). VL30s are endogenous retrovirus-like sequences which are not known to produce any virion structural components, but can be efficiently packaged and transmitted as pseudotypes of type C viruses. Mice contain no known retroviruses analogous to the primate type D retroviruses.

The genomes of the replication-competent viruses contain sequences that code for virion components as well as sequences that function in virus replication and transcription. The proviral copies integrated into host DNA are generally colinear with the viral RNA genome, although they can be disrupted by major deletions or insertions. The number of proviral copies for each subgroup found in the mouse germ line varies from a few to many thousand, and it has been estimated that 0.04% of the mouse genome is composed of these sequences (Callahan and Todaro, 1978). Most proviruses share a common genomic organization in which the regions which code for the core, polymerase, and envelope virion proteins, termed *gag*, *pol*, and *env*, are flanked by regulatory sequences in the long terminal repeats (LTRs). The MMTV genome contains an additional short coding region, *orf* (open reading frame) (Dickson *et al.*, 1981; Donehower *et al.*, 1981). Although the function of the *orf* product is not known, it has been suggested that it may act as a transcriptional transactivator (Van Klaveren and Bentvelzen, 1988) or a negative-acting regulator (Salmon *et al.*, 1990). The *gag*, *pol*, and *env* viral genes can either be expressed as components of the virion or, alternatively, *gag* or *env* gene products have been detected as cell surface components or serum proteins (Elder *et al.*, 1977; Morse *et al.*, 1979; Obata *et al.*, 1975; Tung *et al.*, 1975). There is substantial heterogeneity within as well as between the basic mouse retrovirus subgroups, in part because these viruses can acquire novel sequences by recombination with viral or host cellular sequences. Such recombinants are usually identified and isolated by virtue of their unique biological properties, which most notably include altered host range or transforming potential.

1. C-Type Murine Leukemia Viruses

The type C mouse leukemia viruses are the best characterized of the mouse retroviruses. These viruses can be divided into several different host range or interference classes (Table II) (Levy, 1973, 1975; Fischinger et al., 1975; Hartley et al., 1977; Hartley and Rowe, 1976; Rasheed et al., 1976b; Rein, 1982). These host-range classes differ predominantly in their env-coded major viral glycoprotein, SUenv (formerly gp70). The ecotropic, amphotropic, and polytropic host range groups interact with cell surface receptors encoded by different mouse genes to infect mouse cells (Gazdar et al., 1977; Kozak, 1983). The receptor for ecotropic viruses has been molecularly cloned (Albritton et al., 1989) and shown to be a basic amino-acid transporter (Kim et al., 1991; Wang et al., 1991). Recent evidence suggests that the receptor for ectropic virus may require a limiting accessory factor (Wang et al., 1991).

TABLE II. Distribution of the Four MuLV Host Range Groups Among Inbred Laboratory Mouse Strains and *Mus* Species[a]

| Interference group | Host range | | Prototype virus | Source of infectious virus | Mice with SUenv-related germline copies |
	Rodent	Nonrodent			
Ecotropic	+	−	Akv MuLV	VIS,[a] M. m. molossinus	VIS, M. m. molossinus
			C-II MuLV	Southeast Asian mice	?
			Ho MuLV	M. hortulanus	None
			Cas-Br-E	Lake Casitas mice, M. m. castaneus	Lake Casitas mice, M. m. molossinus, M. m. castaneus
Xenotropic	−	+	BALB-IU	VIS, M. m. molossinus, M. m. castaneus	All inbred strains, M. m. molossinus, M. m. castaneus, M. m. musculus
			C-I MuLV	All Mus species	?
Polytropic	+	+	Akv MCF	Inbred strains with high ecotropic MuLV	All inbred strains, M. m. domesticus
Amphotropic	+	+	1504A	Lake Casitas mice	None

[a] VIS, Various inbred strains.

A variety of methods can be used to detect and titrate infectious MuLVs. As is the case for retroviruses in general, these methods include molecular techniques based on the recognition of specific sequences, biochemical techniques such as the direct assay for particle-associated reverse transcriptase, and immunological assays using MuLV-specific antisera. For the MuLVs, there are also a number of unique focus or plaque assays useful for specific MuLV subgroups. For example, in the presence of an appropriate replication-competent helper virus, the acute, replication-defective sarcoma viruses have transforming activity *in vitro* as well as *in vivo* and can induce foci on monolayers of suscepti- ble cells (Fig. 2). In addition, a number of nontransformed cell lines have been developed which contain a sarcoma virus genome that can trans- form the cell following addition of a competent helper virus (Fig. 3A). Such S$^+$L$^-$ cell lines have been developed using mouse, mink, and cat cells (Bassin *et al.*, 1970; Peebles, 1975; Haapala *et al.*, 1985). Some of the leukemia viruses can also cause direct cytopathic changes in cultured cells, like the polytropic mink cell focus-forming (MCF) viruses on mink lung cells (Hartley *et al.*, 1977) (Fig. 3B). Finally, the ecotropic MuLVs can produce syncytial plaques on rat XC cells, which were de- rived from a rat tumor induced by Rous sarcoma virus (Klement *et al.*, 1969) (Fig. 3C). This ability has been used to develop a very effective plaque assay for ecotropic MuLVs (Rowe *et al.*, 1970), although not all ecotropic viruses can produce these foci.

a. Ecotropic MuLVs

The first infectious MuLVs isolated were ecotropic in host range, that is, infectious only for cells of their natural host (Levy, 1974). The prototype for ecotropic MuLVs, Akv MuLV, can be readily isolated from the high-virus-expression mouse strain AKR, which is character- ized by a lifelong viremia. Other inbred strains cannot produce virus or, like BALB/c, produce high titers of virus only late in life (Table III). Evidence that genetic elements inherited in mice are responsible for this virus production came from the demonstration that cultured nonpro- ducer cells can produce virus after chemical induction (Lowy *et al.*, 1971; Aaronson *et al.*, 1971). Subsequent studies identified specific Mendelian loci responsible for virus production (Rowe, 1972) and showed that viral sequences are inherited with the induction phenotype (Chattopadhyay *et al.*, 1975). Classical genetic crosses and Southern blotting using ecotropic *env*-specific segments as hybridization probes have shown that some, but not all, mice contain chromosomally inte- grated copies of proviral loci which differ in number and chromosomal location in the different inbred strains (Chan *et al.*, 1980; Chattopad- hyay *et al.*, 1980a; Jenkins *et al.*, 1982; Kozak and Rowe, 1982; Taylor

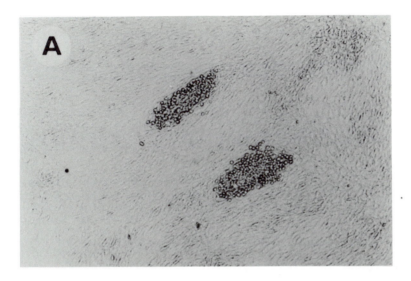

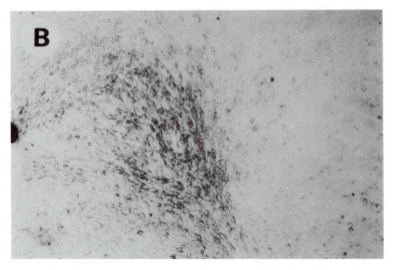

FIGURE 2. (A) A Kirsten murine sarcoma virus (Ki-MSV)-induced focus in normal rat kidney (NRK) cells. ×25. (B) A Moloney murine sarcoma virus (M-MSV)-induced focus in mouse embryo fibroblasts. ×25. [Photos provided by Dr. J. Levy.]

and Rowe, 1989). The variation in virus expression seen among inbred strains can be attributed to several factors, most notably small differences in the proviral genomes they carry and host restriction genes which retard virus replication.

Infectious ecotropic MuLVs, as well as other mouse-tropic MuLVs, can be subclassed into several groups based on their ability to infect cells

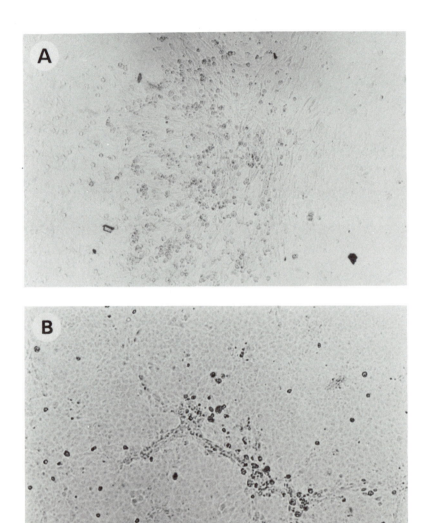

FIGURE 3. (A) Focus induced in mouse S^+L^- cells by ecotropic MuLV. ×25. [Provided by Dr. R. Bassin.] (B) Focus induced in mink lung cells by polytropic MuLV. ×25. [Provided by Dr. J. Hartley.] (C) Plaque induced in rat XC cells by ecotropic MuLV. ×15. [Provided by Dr. J. Hartley and Dr. T. Fredrickson.]

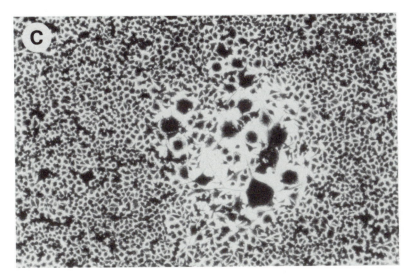

FIGURE 3. (Continued)

of different inbred strains (Hartley et al., 1970). Thus, viruses which replicate more efficiently in cells of BALB/c mice are designated B-tropic, and viruses which replicate more efficiently in cells of NIH Swiss mice are N-tropic. Some viruses (NB-tropic) are not restricted in either cell type. A more recently described fourth subgroup is, like N-tropic viruses, restricted in B-type cells but is also restricted in some, but not all, N-type mouse cells (Rowe and Hartley, 1983; Kozak, 1985a). The underlying molecular mechanism for host restriction is unknown, although resistance has been attributed to allelic variation at a single genetic locus in mice, Fv-1 (see Section III.C), which apparently affects virus replication at a point occurring after penetration but before chromosomal integration of the viral DNA copy (Jolicoeur and Baltimore, 1976; Sveda and Soeiro, 1976; W. K. Yang et al., 1980). Viral sequences responsible for virus restriction by Fv-1 were shown to be in CAgag, formerly p30gag (Gautsch et al., 1978). Sequence analysis of representative N- and B-tropic MuLVs has identified two adjacent amino acids in CAgag which differ in these viruses and which are responsible for their biological differences (DesGrosseillers and Jolicoeur, 1983; Ou et al., 1983a). Comparative analysis of NB-tropic MuLVs has identified substitutions at numerous additional positions in this same gag region. Most of the endogenous ecotropic MuLV copies in inbred mice produce N-tropic virus, although a few endogenous B-tropic MuLVs have been identified (Moll et al., 1979; Datta et al., 1978; Langdon et al., 1984). Also, B-type mice, such as BALB/c, can produce B-tropic virus late in life, which is presumed to arise by recombination between the endoge-

TABLE III. Ecotropic Proviruses Present in Common Laboratory Mice[a]

Strain	Proviral locus[b]	Chromosome	PvuII restriction fragment size (kb)	Expression pattern[c]
AKR/N	Akv-1 (Emv-11)	7	4.8	High
	Akv-2 (Emv-12)	16	5.0	High
AKR/J	Akv-1 (Emv-11)	7	4.8	High
	Akv-3 (Emv-13)	2	3.9	Not expressed
	Akv-4 (Emv-14)	11	8.3	High
C58/J	Emv-20	1	8.2	Unknown
	Emv-21	18	6.6	Unknown
	Emv-22	19	6.6	Unknown
	Emv-23 (C58v-4)	7	3.7	Unknown
	Emv-24	5	3.5	Unknown
	Emv-25	10	3.4	High
	Emv-26 (C58v-1)	8	2.9	High
	Emv-27	3	9.4	Unknown
BALB/c, CBA, C3H/He, A/He	Cv (Emv-1)	5	4.3	Low
DBA/2	Emv-3	9	5.4	Low
C57BL, C57BR	Emv-2	8	5.2	Low
NIH Swiss	None	—	—	—

[a] Taken from Jenkins et al. (1982) and Taylor and Rowe (1989).
[b] Other names for the same locus given in parentheses.
[c] High refers to high virus titers in lymphoid tissue early in life and spontaneous or easily induced virus production in cultured cells; low refers to production of infectious virus late in life and virus easily induced in cultured cells.

nous N-ecotropic MuLV and other germline retroviral sequences (Gautsch et al., 1980).

Numerous studies show that while viruses of inbred mouse strains closely resemble Akv MuLV, there are some structural and sequence differences among proviral copies and virus isolates which are often reflected in biological differences (Table III). Southern blot analysis of DNAs from certain mouse strains, such as SJL and C58, has identified ecotropic MuLV-related fragments of unusual size, suggesting the presence of either different proviral genomes or, as is likely in the case of the SJL mouse, defective or deleted proviruses (Yetter et al., 1985). Small mutations which affect biological phenotypes have also been described. Thus, the single ecotropic proviral copy of DBA mice (Emv-3) is very inefficiently expressed to produce infectious virus, and sequence analysis of this provirus has identified a single base substitution in MAgag (formerly p15gag) which affects virus assembly (Mercer et al., 1990). The proviral copy present in BALB/c and C3H mouse cells (Emv-1) is efficiently induced in vitro, but differs from most ecotropic viruses in its inability to produce syncytia on the rat XC cell line (Bikel, 1976; Rapp and Nowinski, 1976; Hopkins and Jolicoeur, 1975). This viral phenotype has been attributed to a single amino acid difference at the end of SU*env* (King *et al.*, 1987). For both DBA and BALB/c mice, the amino acid differences present in the proviral copies are not found in infectious viruses isolated from these mice, which generally resemble Akv MuLVs. These alterations have been attributed to a simple mutation or recombination with other endogenous MuLVs.

MuLVs with ecotropic host range have also been isolated from a variety of wild-trapped mice or laboratory breeding colonies of wild-derived mice (Table II). While few of these viruses have been molecularly cloned and sequenced, the data indicate that wild-mouse viruses are more diverse than those carried by the inbred laboratory mice. Only viruses isolated from the Japanese mouse, *Mus musculus molossinus*, closely resemble the Akv-type MuLVs of the common laboratory mouse strains (Chattopadhyay *et al.*, 1980b; Steffen *et al.*, 1980). Viruses isolated from animals trapped in southern California (Cas-Br-E) and eastern Europe (Ho MuLV) have now been molecularly cloned and sequenced (Rassart *et al.*, 1986; Voytek and Kozak, 1989). While these viruses are similar to each other and to Akv MuLV in genomic structure and sequence, the *env* genes of these three types of ecotropic MuLV diverge by approximately 30%. Cloned *env* segments of Akv MuLV, Cas-Br-E, and Ho MuLV have been used to confirm that viruses related to Akv MuLV and Cas-Br-E are found as germline integrations in specific wild mouse populations (Table II) (Kozak and O'Neill, 1987; Inaguma *et al.*, 1991), whereas Ho MuLV is apparently transmitted only as an exogenous agent (Voytek and Kozak, 1989). The observation that *M. m. molossinus* is the only known wild mouse known to contain the laboratory-mouse Akv-type MuLV is consistent with the suggestion that these Japanese mice contributed to the gene pool from which the common inbred strains of mice were developed.

Other viruses with ecotropic host range have been isolated from wild mouse populations which are more distantly related to the inbred mouse. Benveniste *et al.* (1977) first described several distinct subclasses of type C viruses from the Southeast Asian mouse, *M. cervicolor*, one of which, termed C-II, is classed as ecotropic because of its ability to replicate in cells of all *Mus* species with the exception of *M. cervicolor* (Table II). C-II viruses differ from the laboratory-mouse MuLVs on the basis of their antigenic properties and nucleic acid homologies defined by liquid hybridization and by the fact that they interact with a different cell surface receptor to infect mouse cells (Rapp and Marshall, 1980). The C-II viruses have not been molecularly cloned and sequenced, precluding any more precise description of their relationship with other MuLVs.

Few wild mouse populations indigenous to Africa have been typed for retroviruses. Southern blot analysis of the pygmy mouse, *Mus minutoides*, failed to identify sequences cross-reactive with either ecotropic or nonecotropic MuLV *env*-specific segments as hybridization probes (Kozak, 1985b; Kozak and O'Neill, 1987). Analysis of *Praomys natalensis*, however, detected viral antigens related to mouse C-type viruses and to the feline leukemia virus. Infectious virus was obtained from these animals which replicated in some but not all of the mouse cells tested for susceptibility, suggesting ecotropic host range (van Pelt *et al.*, 1976).

b. Xenotropic MuLVs

Viruses with xenotropic host range have been isolated from a variety of inbred and wild mouse species (Levy, 1973, 1975; Callahan and Todaro, 1978) [reviewed by Levy (1978)] (Table II). Unlike ecotropic MuLVs, xenotropic viruses are unable to productively infect cells of their natural host, but easily infect cells of other nonrodent species. Although direct nucleotide sequencing of two infectious xenotropic MuLVs, CWM-S-5S and NZB-IU-6, show them to be 98% homologous (O'Neill *et al.*, 1985; Massey *et al.*, 1990), examination of various xenotropic MuLVs by fingerprint analysis, by hybridization, by restriction mapping and by sequencing show that substantial differences among these viruses do exist (Elder *et al.*, 1977; Callahan *et al.*, 1975; Chattopadhyay *et al.*, 1982a, b; Lamont *et al.*, 1991). These molecular differences can be reflected in their biological properties. Most notably, differences have been detected in the ability of xenotropic viruses to infect cells of different nonrodent species. For some isolates, the ability to infect rabbit, rat, human, and bat cells can vary dramatically, even among isolates obtained from the same mouse (Callahan *et al.*, 1975; Varnier *et al.*, 1984; Cloyd *et al.*, 1985).

As is the case for ecotropic MuLVs, proviral copies of xenotropic MuLVs which are capable of producing infectious virus exist in inbred and wild mouse populations, although many fewer xenotropic proviruses have this capability (Table II). Many of the common inbred strains share the same proviral locus, *Bxv-1*, first described in BALB/c mice (Aaronson and Stephenson, 1973; Kozak and Rowe, 1980), and cells from these mice can be readily induced to produce infectious virus with halogenated pyrimidines such as 5-iododeoxyuridine (IUdR) (in fibroblasts) or B-cell mitogens such as lipopolysaccharide (in spleen cells) (Aaronson and Stephenson, 1973; S. M. Phillips *et al.*, 1976). The high-xenotropic-virus strain NZB produces high titers of virus *in vivo* and *in vitro* (Levy and Pincus, 1970), whereas strains such as NFS are not inducible for virus *in vitro* and yield virus only rarely *in vivo*. Genetic studies have shown that the NZB mouse contains two independently assorting loci for production of xenotropic MuLV (Datta and Schwartz, 1977) and have identified additional loci for xenotropic MuLV production in MA/My inbred mice and the Japanese mouse, *M. m. molossinus* (Kozak *et al.*, 1984). Other proviral sequences are capable of producing viral proteins in the absence of infectious virus, and some of the serologically-detected products of nonecotropic SU*env* genes, such as G_{IX} and XenCSA, have been extensively studied (Stockert *et al.*, 1971; Morse *et al.*, 1979). Finally, although the pattern of infectious virus expression is generally a property of the integrated provirus, proviral gene expression can also be under independent genetic control. A recessive locus termed *cxv-1* controls high constitutive expression of an unlinked xenotropic MuLV provirus in F/St mice (Yetter *et al.*, 1983), and the *Gv-1* locus regulates the transcriptional activity of numerous different nonecotropic proviruses (Levy *et al.*, 1985).

Use of *env*-derived hybridization probes has shown that the laboratory mouse genome contains more than 15 xenotropic MuLV-related copies (O'Neill *et al.*, 1986; Stoye and Coffin, 1988). The genomic structure and transcriptional activity of most of these individual copies are unknown, but the specific proviral sequences associated with *Bxv-1* and with the very poorly expressed NFS/N virus have been identified on Southern blots (Hoggan *et al.*, 1986). Genetic studies have mapped the expressed provirus *Bxv-1* to a position on mouse chromosome 1 (Kozak and Rowe, 1980), and most of the xenotropic MuLV *env*-related copies in the common inbred strains have now been localized to specific positions in the mouse linkage map (Frankel *et al.*, 1989a).

MuLVs with xenotropic host range have also been isolated from wild mice. Viruses related to laboratory mouse viruses by their physical maps and by hybridization have been isolated from the Asian mice *M. m. molossinus* and *Mus musculus castaneus* (Lieber *et al.*, 1975; Chattopadhyay *et al.*, 1981) (Table II). A distinct subgroup of xenotropic viruses, termed C-I, was first isolated from Southeast Asian mice and has now been isolated from every mouse species tested except *M. m. musculus*. C-I viruses are unrelated to other mouse type C viruses, but are antigenically and by hybridization related to infectious primate type C viruses (Benveniste and Todaro, 1973). This resemblance to the gibbon ape leukemia virus provides an argument for interspecies transmission. A similar type C virus has been isolated from *Vandeleuria oleracea* (Callahan *et al.*, 1979), a long-tailed tree mouse indigenous to south central Asia which is morphologically and karyotypically distinct from the *Mus* species. Cultured kidney cells produced an infectious virus with xenotropic host range, and molecular hybridization and immunological criteria suggest that this virus is closely related to simian sarcoma associated virus and to the C-I xenotropic MuLVs.

env-specific probes derived from laboratory-mouse xenotropic

MuLVs have been used to screen DNAs from wild mice. Few contain related sequences, although the mice which do, the *M. musculus-molossinus-castaneus* lineage, contain multiple copies (14–30) (Kozak and O'Neill, 1987) (Table II). This suggests that xenotropic MuLVs were acquired and amplified in the wild mice of Asia and eastern Europe and that these mice were among the progenitors of the laboratory strains. cDNAs as well as a molecular clone of a xenotropic MuLVs of the C-I class have also been used as hybridization probes to identify multiple retrovirus-related sequences in various laboratory and wild mouse species, a number of which are on the Y chromosome (Benveniste *et al.*, 1974; S. J. Phillips *et al.*, 1982). Eicher *et al.* (1989) have now shown that the Y chromosome contains a repeated segment containing C-I virus-related as well as nonretroviral sequences. The presence of multiple xenotropic MuLV-related sequences in the wild mouse germ line may be related to the fact that, unlike laboratory mice, wild mice can be infected with exogenous xenotropic MuLVs (Hartley and Rowe, 1975; Kozak, 1985b). Wild mouse species closely related to laboratory mice such as *Mus musculus domesticus* and *M. m. musculus* are generally resistant to infection by xenotropic MuLVs, whereas more evolutionarily divergent species such as *Mus caroli*, *Mus cookii*, *Mus spretus*, and *Mus pahari* can be productively infected although virus replication is inefficient (Kozak, 1985b). Susceptibility has been attributed to a single dominant gene (Kozak, 1985b), and this phenomenon provides a mechanism which may account for the germline acquisition of these sequences in *Mus*.

c. Polytropic MuLVs

The polytropic MuLVs, also termed MCF MuLVs (for mink cell focus-forming MuLVs) are replication-competent MuLVs which differ from ecotropic and xenotropic viruses in host range and interference pattern (Fischinger *et al.*, 1975; Hartley *et al.*, 1977; Rein, 1982). These viruses have a broader host range than ecotropic or xenotropic MuLVs in that they are capable of infecting cells of both rodent and nonrodent species, although they generally infect a more restricted group of nonrodent species than do xenotropic viruses (Cloyd *et al.*, 1985). Their frequent designation as MCF MuLVs is based on the ability of many isolates to induce characteristic foci on mink lung cells (Hartley *et al.*, 1977) (Fig. 3B). Unlike ecotropic and xenotropic MuLVs, polytropic MuLV proviruses capable of producing infectious virus do not preexist in the mouse germ line. Rather, polytropic viruses arise *de novo* as recombinants between infectious ecotropic viruses and endogenous polytropic MuLV-related sequences (Fig. 4). They are typically isolated from high leukemic strains of mice which chronically express high levels of

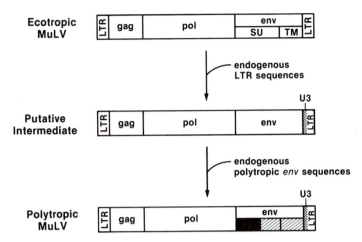

FIGURE 4. Recombinational events leading to the generation of AKR polytropic viruses. Dotted area, endogenous U3 LTR sequences; black area, polytropic *env* sequences; hatched area, *env* sequences that can be either ecotropic or polytropic in origin.

ecotropic virus, such as AKR mice, or from mice inoculated with various exogenous ecotropic MuLVs.

Since polytropic viruses are generated *de novo* in the mouse, individual isolates can vary significantly in pathogenicity, structure, and host range. Many, but not all, of these viruses have leukemogenic potential defined as the ability to accelerate the onset of thymomas after inoculation into newborn AKR mice (Cloyd *et al.*, 1980). Exogenous infection of mice with most polytropic viruses appears to be most efficient in the presence of ecotropic virus, allowing for the polytropic viral genome to be packaged in an ecotropic coat. These polytropic viral pseudotypes, unlike polytropic genomes packaged in their own envelope coats, have been shown to be resistant to inactivating factors in the blood (Haas and Patch, 1980).

Structurally, all polytropic MuLVs contain polytropic-specific envelope gene sequences that encode at least the amino half of SU*env*. Sequences encoding the carboxyl terminus of TM*env* (formerly p15E) through the U3 region of the LTR of pathogenic AKR viruses are nonecotropic in origin (Holland *et al.*, 1983; Kelly *et al.*, 1983). It has been suggested that the leukemogenic polytropic viruses isolated from spontaneous lymphomas of AKR and CWD mice are produced following successive recombinational events resulting in either 3' TM*env*/U3 LTR or 5' SU*env* gene substitutions (Thomas and Coffin, 1982; Khan *et al.*, 1982; Stoye *et al.*, 1991) (Fig. 4). The increased pathogenicity of polytropic MuLVs compared with their ecotropic MuLV parents is due, in large part, to their unique envelope genes, but the LTR sequences are also important both for pathogenicity and target cell specificity (Lung et al., 1983; DesGrosseillers et al., 1983; Lenz and Haseltine, 1983).

The env gene substitutions in polytropic MuLVs are responsible for the host range characteristics of this class of viruses, which use unique cell surface receptors to enter mouse cells (Rein, 1982). Although most polytropic MuLVs can infect nonrodent species, there is considerable variability with respect to the particular nonrodent species individual isolates can infect. For example, one type of recombinant MuLV, termed erv (for ecotropic recombinant viruses), lacks the ability to infect any nonrodent cells, although they resemble polytropic MuLVs in genome structure, serological cross-reactivity, and interference patterns (Cloyd and Chattopadhyay, 1986). Analysis of a variety of different polytropic isolates by restriction mapping, fingerprinting, and sequencing confirms that there is a great deal of heterogeneity in the env and LTR sequences among polytropic MuLVs (Chattopadhyay et al., 1982a, b), although interference tests confirm that they all bind the same cell surface receptor to infect mouse cells (Rein, 1982).

It is not known what regulates the recombinational events that generate leukemogenic polytropic viruses, where these events occur in the mouse, or what specific endogenous sequences participate. The substituted LTR sequences in leukemogenic AKR polytropic viruses appear to be derived from the xenotropic provirus Bxv-1 (Quint et al., 1984; Hoggan et al., 1986; and Stoye et al., 1991), although compared to the LTR of Bxv-1, they contain duplicated enhancer regions (Stoye et al., 1991). Multiple copies of polytropic envelope genes exist in the mouse germ line (Hoggan et al., 1983; O'Neill et al., 1986; Stoye and Coffin, 1988), but the number and identity of sequences that can recombine with ecotropic MuLVs to generate polytropic viruses is unknown, although it has been suggested that polytropic viruses derived from different ecotropic MuLVs are generated by recombination with different endogenous sequences (Evans and Cloyd, 1985). Endogenous polytropic env gene sequences can be distinguished by the presence or absence of specific restriction enzyme sites and the presence or absence of specific deletions or insertions, including a 28-base pair insertion in env and a 190-base pair insertion in the LTR (Stoye and Coffin, 1988; Ch'ang et al., 1989; Policastro et al., 1989; Khan and Martin, 1983; Ou et al., 1983b). Genetic studies using oligonucleotide probes which span the 28-base pair env gene deletion have been used to distinguish two subgroups of endogenous env genes termed polytropic (Pmv) and modified polytropic (Mpmv). Multiple copies of both subtypes are present in the mouse genome, and Southern blot analysis of the progeny of genetic crosses has been used to map these genes to almost 100 positions throughout the laboratory mouse genome (Frankel et al., 1989b, 1990).

Hybridization probes which distinguish polytropic env sequences from other MuLVs have been used to examine wild mouse populations for endogenous copies. Unlike laboratory strains, all of which contain multiple copies of this sequence, polytropic MuLV *env*-related sequences are absent from most wild mouse species (Kozak and O'Neill, 1987). Multiple copies are, however, detected in *M. m. domesticus*, mice which are found in North America, western Europe, and North Africa and which are recognized as progenitors of the common laboratory mouse strains. Thus, analysis of inbred and wild mice with probes that distinguish ecotropic, xenotropic, and polytropic MuLVs has shown that related sequences are present in the wild mouse germ line, but that the different MuLV subgroups are specific to different species or subspecies (Table II).

Typical infectious polytropic MuLVs have not been isolated from wild mouse populations, including those spontaneously expressing high levels of ecotropic MuLVs. This may be due to the fact that most wild mice which express high levels of ecotropic MuLVs, such as *M. m. molossinus* and *M. m. castaneus*, contain endogenous xenotropic but not polytropic MuLV-related sequences (Kozak and O'Neill, 1987). One unusual isolate has, however, been obtained from embryo fibroblasts of a California wild mouse. This virus, designated Cas E No. 1, has a xenotropic host range, but more closely resembles polytropic MuLVs in its interference pattern (Cloyd *et al.*, 1985). More typical polytropic MuLVs, as well as lymphomas, have been induced in North African wild mice following inoculation of these mice with ecotropic MuLV, indicating that the sequences necessary for production of infectious polytropic virus are present in these animals (Villar *et al.*, 1988).

d. Amphotropic MuLVs

Wild mice trapped in the Lake Casitas (LC) and La Puente regions of southern California contain a novel host range class of MuLVs (Rasheed *et al.*, 1976b; Hartley and Rowe, 1976). These viruses have not been found in inbred mice, nor in wild mice trapped in other regions of California or in other parts of the world (Table II). They have an amphotropic host range, that is, they replicate in mouse cells as well as cells of nonrodent species. The amphotropic viruses are, however, distinct from other MuLVs in their antigenic and interference properties (Rein, 1982). Analysis of these viruses by tryptic peptide mapping and restriction enzyme analysis suggests that there is little variation among the different isolates (Chattopadhyay *et al.*, 1981; Byrant *et al.*, 1978), although host range studies show that the two most commonly studied isolates, 1504A and 4070A, differ in their ability to infect Chinese hamster cells (Cloyd *et al.*, 1985).

Amphotropic MuLVs, like the ecotropic MuLVs also isolated from

LC mice, are capable of inducing lymphomas, although the incidence is lower and the latency period longer for amphotropic MuLVs (Gardner, 1978) (see Section III.A.2d). Introduction of amphotropic MuLVs into inbred mice has resulted in the production of recombinant viruses, often with altered host range and oncogenic potential (Rasheed et al., 1982, 1983). One type of recombinant appears to have an ecotropic host range and is nonpathogenic in other mice (Rasheed et al., 1983). These recombinants differ from other ecotropic MuLVs serologically and by oligonucleotide fingerprinting, and their resemblance to nonecotropic MuLVs suggests that they have acquired a novel type of env gene. These recombinants have not been further characterized. Another recombinant, 10A1, is more pathogenic than its amphotropic progenitor (Rasheed et al., 1982) and appears to have acquired novel env gene sequences (Lai et al., 1982; Ott et al., 1990). Its one-directional pattern of interference with amphotropic viruses is unique among the MuLVs and has been used to identify a retroviral receptor on mouse cells distinct from the receptors for ecotropic, polytropic, and amphotropic MuLVs (Rein and Schultz, 1984; Ott et al., 1990).

An amphotropic MuLV type-specific probe has been used to analyze DNAs of inbred and wild mouse populations to show that none of these mice contain amphotropic viral sequences in their genomic DNA (O'Neill et al., 1987) (Table II). The mice tested included recently trapped LC mice which were viremic with amphotropic virus, suggesting that this virus is transmitted by congenital or horizontal infection in this population, as previous studies had suggested (Gardner et al., 1979). The origin of amphotropic MuLVs is unknown. Analysis of other mammalian and avian species failed to identify amphotropic MuLV-related sequences in any other species, arguing against interspecies transmission (O'Neill et al., 1987). There is evidence, however, to suggest that amphotropicvirus-positive LC mice were derived from the interbreeding of North American mice with mice, probably M. castaneus, introduced by humans into southern California from Asia (Kozak and O'Neill, 1987). While it is thus possible that amphotropic MuLVs were introduced along with these Asian mice, this cannot be determined until mice endemic to the regions of Asia populated by M. castaneus have been systematically typed for retroviral sequences.

e. Laboratory Strains of Murine Leukemia Viruses

In addition to the naturally occurring MuLVs which can be reproducibly isolated from individual mice of the same genetic background, a variety of laboratory MuLV variants have also been described which differ from the naturally occurring mouse viruses and from each other. Many of these laboratory strains, usually named after the investigator who first isolated them, were initially identified in conjunction with efforts to find etiological agents of specific neoplastic diseases. Most were, therefore, isolated from tumor tissue, in many cases after extensive serial passage in mice of cells or cell-free filtrates (Tables IV and V). Some of these viral stocks, such as Friend MuLV, were subsequently found to contain mixtures of defective and nondefective viruses, and some, such as Friend and Moloney MuLVs, contained mixtures of different host range types. Unlike the naturally occurring MuLVs, which are usually poorly oncogenic, if at all, these laboratory-derived isolates are often capable of inducing a variety of neoplastic diseases and therefore have been important in the study of pathogenic processes (Section III). Sequence analysis of the most commonly studied ecotropic strains, Friend and Moloney, show that these isolates differ in env and LTR sequences from their naturally occurring ecotropic MuLV counterparts (Herr, 1984; Shinnick et al., 1981; Koch et al., 1983, 1984). Inoculation of these ecotropic strains into inbred mice often results in the formation of polytropic recombinants. These polytropic viruses differ from those recovered from the inbred strains that express high titers of endogenous ecotropic MuLVs, in that the SUenv sequences are derived from the host but the TMenv and LTR sequences are derived from the ecotropic progenitors (Koch et al., 1984; Adachi et al., 1984; Bosselman et al., 1982).

2. B-Type Mouse Mammary Tumor Viruses

The MMTVs represent a class of murine retroviruses distinct from the MuLVs by virion morphology, by the absence of sequence homology, and by their sensitivity to induction by glucocorticoid hormones (Ringold, 1983; Chapter 2). Like the MuLVs, these viruses were first identified by their association with neoplastic disease (Nandi and McGrath, 1973). Early studies established that infectious MMTVs as well as MMTV-associated mammary carcinomas could be inherited vertically in certain inbred strains or horizontally through the milk of infected females (Bentvelzen and Daams, 1969; Muhlbock and Bentvelzen, 1968; Nandi and McGrath, 1973). Genetic studies have identified specific proviral integrations by their distinct pattern of virus expression (van Nie and Verstraeten, 1975; Verstraeten and van Nie, 1978; van Nie et al., 1977; Nusse et al., 1980) or by use of specific hybridization probes (Majors and Varmus, 1981). Nearly 30 unique proviral integrations have now been identified in the various inbred mouse strains, many of which have been chromosomally mapped (Kozak et al., 1987; Eicher and Lee, 1990; Lee and Eicher, 1990). Two of these proviruses, Mtv-1 and Mtv-2, can be expressed as infectious virus, and this expression is associated with induction of mammary carcinomas (van Nie and Verstraeten, 1975; Verstraeten and van Nie, 1978; van Nie et al., 1977).

Use of the inbred mouse MMTV genome as a hybridization probe to analyze wild mouse species demonstrated that not all wild mice contain sequences related to this retrovirus (Cohen and Varmus, 1979; Callahan et al., 1982). A recent study analyzed an interspecies backcross for *M. spretus* MMTV-related proviral sequences (Siracusa et al., 1991). Ten proviral integrations distinct from those found in laboratory mice were identified and chromosomally localized. Wild mouse species have not been systemically screened for infectious MMTVs, although a virus morphologically and antigenically related to the MMTV viruses has been observed in the milk of *M. cervicolor* (Schlom et al., 1978). This virus, however, has not been characterized further.

3. Intracisternal A-Type Particles

Intracisternal A-type particles (IAPs) are retrovirus-like structures that are commonly found in embryonic tissue and certain tumors (Wivel and Smith, 1971; Kuff et al., 1972; Calarco and Szollosi, 1973). These particles resemble immature type B virions but are generally found within the endoplasmic reticulum. IAPs are not found extracellularly and are not infectious, although the virions do contain viral RNA, have reverse transcriptase activity, and contain coding information for the structural elements of the viral particle (Paterson et al., 1978). IAPs show some homology with murine type B and primate type D retroviruses (Chiu et al., 1984), and one infectious mouse virus (M432) isolated from the Southeast Asian species *M. cervicolor* and *M. caroli* is thought to represent a recombinant C-type virus which has acquired IAP sequences (Kuff et al., 1978; Callahan et al., 1981).

Southern blot analysis of mouse genomic DNA with IAP-specific probes indicates that there are approximately 1000 IAP-related chromosomal genes which are scattered throughout the genome (Lueders and Kuff, 1980). Many of these endogenous copies contain terminal repeats with typical U3 and U5 structures (Kuff et al., 1981; Cole et al., 1981). While the great majority of IAPs contain transcriptionally inactive *env* genes (Meitz et al., 1987; Kuff and Lueders, 1988), the mouse does contain about 200 copies of an element termed IAPE (IAP envelope) that contains a retroviral-like open reading frame (Reuss and Schaller, 1991). Several different families of envelope-defective IAP genes have been defined by sequence homology and the presence of specific deletions or insertions (Ono et al., 1980; Shen-Ong and Cole, 1982; Lueders and Meitz, 1986). IAP genes are expressed at enhanced levels in early embryos and specific tumors, and analysis of the transcripts indicates that not all families are actively transcribed in all of these cell types and their expression may be developmentally regulated (Paterson et al., 1978; Shen-Ong and Cole, 1982, 1984; Pikó et al., 1984; Takayama et al.,

1991). Individual IAP-associated LTRs differ in their promoter activity (Christy and Huang, 1988).

Analysis of wild mouse genomes has identified IAP-related sequences, and relatively high copy numbers are present in taxonomic groups which are more closely related to inbred strains. Particles have been observed in the early embryos of LC mice, *M. cervicolor*, and *M. pahari* (Calarco *et al.*, 1980; Yotsuyanagi and Szöllösi, 1984). More distantly related families contain fewer copies, and *M. m. molossinus* lacks at least one IAP variant (Kuff *et al.*, 1981; Shen-Ong and Cole, 1982).

4. VL30s

Mice contain DNA sequences which produce 30S RNA species that can be packaged into type C virions and horizontally transmitted to other cells. Although proviral copies are produced (Howk *et al.*, 1978; Sherwin *et al.*, 1978; Besmer *et al.*, 1979; Scolnick *et al.*, 1979a), protein products of these RNAs have not been described. Endogenous DNA sequences that produce VL30s are reiterated at least 100–200 times in the laboratory mouse genome (Keshet *et al.*, 1980; Keshet and Itin, 1982). The VL30 genes are generally smaller than the proviral copies of MuLVs or MMTVs, but, like these retroviruses, VL30s can have terminal repeats which are 400 base pairs in length (Itin and Keshet, 1983a). Analysis of genomic DNA as well as cloned VL30 sequences indicates that VL30s are not all identical (Keshet *et al.*, 1980; Keshet and Shaul, 1981).

VL30 sequences can recombine with other retroviral sequences. Recombinant structures containing sequences from VL30s and type C viruses have been identified both as exogenous viral genomes and as endogenous proviral genes. Thus, both the Harvey and Kirsten sarcoma viruses contain rat VL30 sequences and the rat c-*ras* oncogene flanked by MuLV-related sequences (see Section III.A.1a) (Shih *et al.*, 1978; Chien *et al.*, 1979; Young *et al.*, 1980; Ellis *et al.*, 1980). VL30 sequences have also been found flanking other proviral elements in the mouse genome (Itin and Keshet, 1983b; Horowitz and Risser, 1985).

VL30 sequences are present in all inbred mice and various *Mus* species, although the number of copies varies. Some Asian species only distantly related to laboratory mice contain fewer than ten copies (Sherwin *et al.*, 1978; Courtney *et al.*, 1982; Itin *et al.*, 1983).

5. Other Retroviral Elements

Analysis of the mouse genome has identified families of DNA sequences which resemble retroviral elements or which represent subgenomic fragments of related sequences. Such families include solitary long terminal repeats (Wirth *et al.*, 1983) and deleted MuLV-related genomes (Khan *et al.*, 1982; Schmidt *et al.*, 1985; Ch'ang *et al.*, 1989; Policastro *et al.*, 1989). In addition to such retroviral sequences which are clearly related to MuLVs, MMTVs, IAPs, or VL30s, other retroviruses or retroviral sequences have been identified which are not closely related to members of any of these well-defined families (Chapter 4). Itin and Keshet (1986) cloned a retrovirus-like element, termed GLN-3, by its limited homology to VL30 probes; however, VL30-related sequences are restricted to a 0.9-kilobase (kb) segment of the 8-kb GLN-3 element. No sequence homology was detected with other known retroviruses, and Southern blot analysis revealed that the mouse genome contained 20–50 copies related to the internal region along with 1000–1500 copies of the LTR. Obata and Khan (1988) subsequently identified a GLN-3-like LTR in association with a different retroviral element (termed B-26) which showed some homology to VL30s and MuLVs. Southern blot analysis of genomic DNAs with a *pol–env* fragment of this element showed it to be more broadly dispersed in the rodent germ line than MuLVs and MMTVs, and it is present in DNAs of the rodent families Cricetidae, Sciuridae, and Caviidae (Table I). Because of their widespread distribution among rodent genera, GLN-3 and B-26 must represent retroviral families established early in the evolution of the family Rodentia.

B. Other Rodents

Infectious retroviruses, retroviral particles, and retrovirus-related sequences have all been detected in rodents other than mice, but few are as well characterized as the mouse retroviruses. Many of the endogenous proviral genes identified in these species were, in fact, identified using mouse retroviral probes, and most of the retroviral sequences identified in these other rodent genera are related to replication-defective mouse IAPs, VL30s, and GLN-3 (Table I).

1. Rats

In the rat, there is evidence for the presence and expression of endogenous sequences corresponding to C-type viruses, IAPs, and VL30s.

a. C-Type Viruses

C-type viruses have been isolated in the rat from a variety of tissues and cell lines (Bergs *et al.*, 1972; Chopra *et al.*, 1970). Such viruses have also been induced from rat cells by halogenated pyrimidines (Klement *et* al., 1973; Verwoerd and Sarma, 1973). These viruses appear to be eco-tropic in host range (Rasheed et al., 1976a) and, although most replicate inefficiently in cultured cells, one isolate has been reported which is capable of transforming cells in culture (Bergs et al., 1972). Although none of these viruses has been cloned and sequenced, solution hybridiza-tion indicates that the rat viruses are not identical, are distinct from C-type viruses isolated from other species, and are present in multiple copies in the rat genome (Tsuchida et al., 1975; Rasheed et al., 1976a).

Replication-defective sarcoma viruses have also been isolated from rats following inoculation with murine leukemia viruses. Thus, Harvey murine sarcoma virus was isolated from a leukemic Chester-Beatty rat inoculated with Moloney MuLV (Harvey, 1964) and Kirsten sarcoma virus was isolated after passage of cell-free extracts from a C3H mouse thymoma (containing Kirsten MuLV) into Wistar-Furth rats (Kirsten and Mayer, 1967). Both Harvey and Kirsten sarcoma viruses were subse-quently shown to be recombinants containing rat endogenous sequences responsible for their acute transforming ability (Scolnick and Parks, 1974). Finally, a rat sarcoma virus was isolated by in vitro cocultivation of chemically-transformed rat cells with Sprague-Dawley rat embryo cells. This isolate contains the same rat transforming gene as Harvey sarcoma virus (Rasheed et al., 1978) (see Section III.A.1a).

b. VL30s

Sequences related to retrovirus-like 30S RNAs (VL30s) were first identified by the analysis of the Harvey and Kirsten murine sarcoma viruses. In addition to the rat sequences responsible for their transform-ing potential, these viruses were also found to contain additional rat-derived sequences present in multiple copies in the rat genome (Scolnick et al., 1976; Tsuchida et al., 1974) (Fig. 5). As with the mouse VL30s, C-type viruses grown in rat cells can package these 30S RNAs, although the VL30s of rats are not closely related to the VL30s of mice (Scolnick et al., 1979a).

c. Intracisternal A-Type Particles

Sequences related to those in IAPs have been isolated from rats by virtue of their homology to mouse IAP sequences. The rat genome was estimated to contain about 500 mouse IAP-related genes (Lueders and Kuff, 1983); however, heteroduplex analysis showed that these rat IAPs contain only short regions of homology with mouse IAPs interspersed with regions of nonhomology (Lueders and Kuff, 1983). Heteroduplex analysis using pairs of rat clones also showed that the rat contains heterogeneous IAP genomes. Subcloned fragments used as hybridization probes confirmed this heterogeneity.

Southern blot analysis of five different inbred rat lines revealed comparable restriction patterns, although the hybridization patterns in the species *Rattus norwegicus* and *Rattus rattus* were somewhat different (Lueders and Kuff, 1983). Both species, however, contained similar numbers of copies, suggesting that IAP sequences were acquired and amplified prior to the divergence of these rat species.

2. Hamsters

Retroviral particles, as well as endogenous retroviral-related genes, have been identified in both Syrian hamsters (*Mesocricetus auratus*) and Chinese hamsters (*Cricetulus griseus*) (Table I). Both hamster genera contain genetic sequences for IAPs and C-type viruses, some of which have been cloned and sequenced. These genes are transcribed and, in some cases, translated into retroviral proteins and particles.

a. C-Type Viruses

Cultured Chinese hamster ovary (CHO) cells produce particle-associated reverse transcriptase activity and C-type virions detectable by electron microscopy either spontaneously (Lieber *et al.*, 1973) or following chemical induction (Tihon and Green, 1973; Manly *et al.*, 1978). In all cases, however, none of these viruses was able to infect other mammalian cell lines, suggesting that they are replication-defective (Hojman *et al.*, 1988).

Replication-defective sarcoma viruses have also been isolated from hamster tumors induced by murine sarcoma viruses (Bassin *et al.*, 1968). These sarcoma viruses transform hamster cells and induce tumors in hamsters but do not infect mice. These sarcomogenic viral stocks have been reported to contain a hamster-tropic helper virus (HaLV), but infectious helper virus stocks have not been prepared for further analysis (Kelloff *et al.*, 1970). cDNAs prepared from HaLV were used to show that this virus is unrelated to mouse or rat viruses and to identify related sequences in hamster DNA by solution hybridization (Okabe *et al.*, 1974). One study has also described the isolation of a replication-competent hamster retrovirus capable of transforming mouse, rat, and hamster cells and inducing solid tumors in hamsters. It appears to have a full-length genome, but has not been the subject of subsequent analysis (Russell *et al.*, 1979).

More recent studies have analyzed several cDNA clones from particle-associated RNA (K. P. Anderson *et al.*, 1990a). Significant homology (73%) was observed to the endonuclease gene of the Moloney strain of

MuLV, although none of the clones contained an open reading frame. Thus, although some retroviral genes in hamsters are clearly capable of directing production of proteins which can be assembled into particles, the transcriptionally active copies have not yet been cloned.

Multiple copies of genes related to murine C-type retroviral genes are present in Chinese hamster liver and CHO cells. Southern blot analysis using various MuLV genes as hybridization probes identified multiple fragments in hamster DNA, although reactivity with *env*-derived probes was generally poor (Hojman *et al.*, 1988). Use of cDNA clones isolated from hamster C-type particles identified about 100 copies/haploid genome in CHO cells and Chinese hamster liver. Cultured cells from related species, particularly gerbils, are also known to produce virions that are morphologically similar to type C particles (Yelle and Berthiaune, 1982).

b. Intracisternal A-Type Particles

Both CHO and Syrian hamster cells have been shown to contain intracytoplasmic particles which morphologically and biochemically resemble mouse IAPs (Heine *et al.*, 1979), and both hamster genera contain multiple IAP-related genes (Lueders and Kuff, 1981). A number of the Syrian hamster IAP genes have been cloned (Lueders and Kuff, 1983; A. Suzuki *et al.*, 1982) and show regions of strong homology with mouse IAPs. One clone which represents an entire Syrian hamster IAP genome was sequenced (Ono *et al.*, 1985). The coding sequences of this gene are flanked by LTRs, and the *gag* and *pol* regions show limited sequence homology to the avian Rous sarcoma virus, to the B-type MMTV, and to the D-type squirrel monkey retrovirus.

Southern blotting using Syrian hamster and mouse IAP genes has identified related sequences in Chinese hamster DNA (K. P. Anderson *et al.*, 1990b). Several of these hamster IAPs have been cloned from genomic or cDNA libraries and sequenced (Anderson *et al.*, 1990b; Servenay *et al.*, 1988; Dorner *et al.*, 1991). There are probably at least three distinct families of IAP-related genes in the hamster genome. Family I is characterized by a specific internal deletion and shows some sequence homology with the *gag* and *env* regions of the Syrian hamster IAP. Family II contains no obvious major deletions, appears to contain LTR sequences, and is related to MMTVs and D-type retroviruses as well as Syrian hamster IAPs (K. P. Anderson *et al.*, 1990b). The third family shows some homology to the LTR and *gag* regions of the Syrian hamster IAP (Servenay *et al.*, 1988). These observations also suggest that IAP proviruses may be more heterogeneous in Chinese hamster than Syrian hamster, since Syrian hamster IAPs appear by restriction mapping and electron microscopy to be fairly homogeneous (Lueders and Kuff, 1983).

Among other rodents in this same family (Cricetidae), Southern blot analysis of gerbils has identified a moderate number of fragments reactive with an IAP probe (Lueders and Kuff, 1981). A-type particles have also been observed in cultured gerbil cells (Tumilowicz and Cholon, 1971).

3. Guinea Pigs and Agoutis

Guinea pigs and agoutis represent New World rodents and are, therefore, more evolutionarily divergent members of Rodentia. Retroviruses have been isolated from both species. An unusual retrovirus has been isolated from guinea pig neoplasms (Opler, 1967; Dunkel, 1974) as well as normal tissues treated with halogenated pyrimidines (Nayak and Murray, 1973), although this virus cannot be transmitted to cultured cells (Rhim et al., 1974). This virus remains poorly characterized and, although it is distinct by hybridization from other known viruses, it shows some similarities to B-type viruses serologically and morphologically (Dahlberg et al., 1980) and, like B-type viruses, has a divalent cation preference for Mg^{2+} for reverse transcriptase activity (Nayak and Murray, 1973). Sequences complementary to guinea pig viral RNA have been detected in guinea pig DNA by solution hybridization (Michalides et al., 1975; Dahlberg et al., 1980).

A retrovirus which more closely resembles C-type viruses (Sherwin et al., 1979) has also been isolated from agoutis. It is replication-competent and has an apparent xenotropic host range. This virus, termed DPC-1, is not related by hybridization to other known retroviruses, but its gag and pol gene products are immunologically related to primate, rat, mouse, pig, and cat retroviruses. Sequences related to this virus appear to be present in multiple copies in agoutis, but closely related genes are not present in other species.

III. BIOLOGICAL EFFECTS OF RETROVIRUSES

The effort to describe the number and variety of retroviruses and retrovirus-related genes in mice as well as other rodents was a direct result of the early observation that, in mice, chronic lifelong infection was clearly associated with a high incidence of neoplastic disease. While the incidence of naturally occurring retrovirus-associated disease in most natural populations or laboratory colonies of mice is low, certain strains of laboratory mice or certain wild mouse populations have a high incidence of spontaneous disease that is associated with the production and spread of high levels of retroviruses. These include the T-cell lymphomas in AKR mice (Fig. 5), lymphomas and hind limb paralysis in

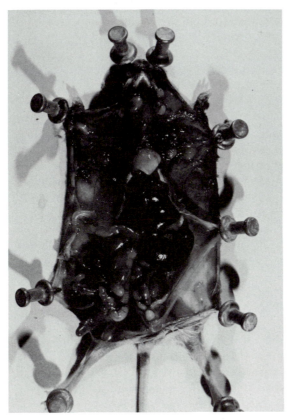

FIGURE 5. AKR/N mouse with thymic lymphoma and associated splenomegaly and enlarged nodes. [Provided by J. Silver.]

wild mice, and mammary adenocarcinomas in C3H mice. In addition to these naturally occurring viruses, a large number of pathogenic retroviruses have also been generated by animal passage of spontaneous tumor tissues, injection of nonpathogenic retroviruses, or irradiation of mice. These pathogenic retroviruses, although rare in nature, cause a high incidence of disease when injected back into mice and have provided the research community with valuable tools to aid in the understanding of the role of retroviruses in disease development. The majority of these viruses cause malignant diseases, but they are also associated with the development of neurological and autoimmune diseases, as well as an immune deficiency syndrome that has been proposed as a model for human AIDS. Finally, while the great majority of retrovirus-related sequences identified in the mouse genome either serve no known function as mouse genes or have no obvious deleterious effect on the host, the reinsertion of retroviral sequences into host chromosomes can disrupt or alter expression of nearby genes. A number of such integrations have now been described which cause developmental mutations or have lethal consequences.

A. Transforming Viruses

Naturally occurring mouse retroviruses are not directly responsible for the malignant diseases with which they are associated. Rather, they have been shown to recombine with endogenous cellular or viral sequences to generate transforming viruses or to activate cellular genes whose expression is responsible for the transformation. Transforming retroviruses can be divided into two groups: (1) those that cause disease after a short latency period and are thought to be directly responsible for the disease, and (2) those that cause disease after a long latency period and are thought to be only the first step in a series of events that are required for the generation of a transformed cell.

1. Acute Leukemia and Sarcoma Viruses

Many mouse retroviruses have been described which induce leukemias and sarcomas after a short latency period, and these are listed in Table IV. The pathogenicity of these viruses is due to the transduction by these viruses of particular proto-oncogenes or altered viral genes (pathogenes) from the host (Fig. 6). They are invariably defective for replication and require the presence of a helper virus, whose basic role is to allow the genome of the transforming virus to be integrated into a sufficient number of target cells. Due to the short latency period of disease induction, the fact that they can transform fibroblasts or hematopoietic cells in culture, and the polyclonal nature of the transformed cells, these viruses are thought to be directly responsible for the diseases that they induce. The exact mechanisms by which they induce disease are not known, but their biological activity is generally thought to be due to the uncoupling of growth factor pathways from their normal control.

The particular type of disease that each virus induces is determined primarily by the cellular gene which it has captured, but other factors, such as the site of injection of the virus and the particular regulatory sequences that the virus carries, can determine the cell types in which the virus is expressed and consequently influence disease specificity. Although the transforming protein encoded by each acute transforming virus is responsible for the neoplastic potential of these viruses, expression of these proteins may not always be sufficient for inducing all of the stages of transformation. In these cases, a secondary genetic event, such

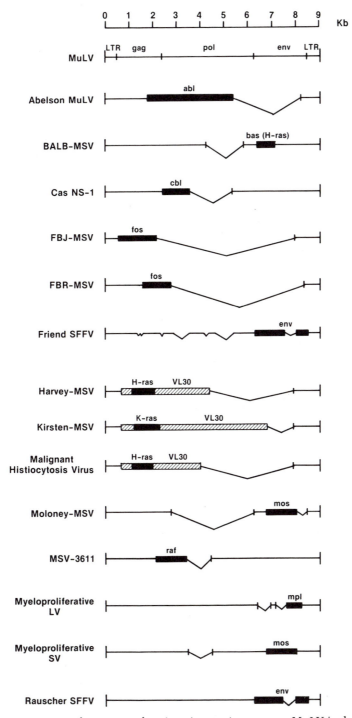

FIGURE 6. Genomes of acute transforming viruses. A non-acute MuLV is shown at the top for comparison. Sizes of genes and deletions should be considered approximate. Black rectangles are transforming sequences.

TABLE IV. Acute Transforming Viruses

Virus[a]	Derivation[b]	Transduced gene	Disease induced
Abelson MuLV	M-MuLV-infected mouse	abl	B-cell lymphoma
AKT-8 MuLV	Spontaneous AKR lymphoma	akt	T-cell lymphoma
BALB MSV	Spontaneous mouse tumor	bas (Ha-ras)	Hemangiosarcoma
Cas-NS-1	Cas-MCF-infected mouse	cbl	Pre-B-cell lymphoma
Cas-SFFV	Cas-BR-E-infected mouse	env	Erythroleukemia
C58 MSV-1	Cas-Br-E-infected mouse	ras	Sarcoma; erythroleukemia
FBJ MSV	Spontaneous mouse tumor	fos	Osteosarcoma
FBR MSV	Irradiated mouse	fos	Osteosarcoma
Friend SFFV	Tumor cells passed through mice	env	Erythroleukemia
Gazdar MSV	Spontaneous mouse tumor	mos	Sarcoma
Harvey MSV	M-MuLV-infected rat	Ha-ras	Sarcoma; erythroleukemia
Kirsten MSV	Ki-MuLV-infected mouse	Ki-ras	Sarcoma; erythroleukemia
Malignant histiocytosis virus	F-MuLV-infected mouse	Ha-ras	Histiocytosis
Moloney MSV	M-MuLV-infected mouse	mos	Sarcoma
MSV-3611	Transformed C3H cell line	raf	Sarcoma
Myeloproliferative LV	F-MuLV-infected mouse	mpl	Myeloproliferative disease
Myeloproliferative SV	M-MSV-infected mouse	mos	Myeloproliferative disease; sarcoma
Rat SV	Transformed rat cell line	Ha-ras	Sarcoma; erythroleukemia
Rauscher SFFV	Tumor cells passed through mice	env	Erythroleukemia

[a] MuLV, Murine leukemia virus; MSV, murine sarcoma virus; SV, sarcoma virus; SFFV, spleen focus-forming virus.
[b] M. Moloney; F, Friend; Ki, Kirsten.

as the integration of the acutely transforming virus or the helper virus into a particular site in the host DNA, may be required.

a. Harvey, Kirsten, and Related Murine Sarcoma Viruses

One of the first acutely transforming retroviruses isolated from a rodent was the Harvey sarcoma virus (Ha-MSV). As discussed earlier (Section II.B.1a), this virus was isolated from a rat injected with Moloney murine leukemia virus (Harvey, 1964). When filtered plasma from this rat was injected into newborn rats or BALB/c mice, it induced a rapid sarcoma, described as an undifferentiated mesenchymal tumor, within 2 weeks. In most cases, splenomegaly and erythroblastosis accompanied the development of the solid tumor. The sarcoma virus was shown to be a recombinant between the Moloney MuLV and the rat proto-oncogene Ha-ras (Scolnick and Parks, 1974). It encodes a 21-kDa phosphoprotein (Shih et al., 1979) that has guanine nucleotide binding activity (Scolnick et al., 1979b). A related virus, Kirsten murine sarcoma virus (Ki-MSV), was also isolated after passage of cell-free mouse virus extracts into rats (Kirsten and Mayer, 1967). The pathology of the disease associated with this virus, which is a recombinant between the mouse virus (Kirsten MuLV) and the rat proto-oncogene Ki-ras (Scolnick et al., 1976), is almost identical to that induced by Ha-MSV. Ki-MSV also encodes a 21-kDa phosphoprotein that has guanine nucleotide binding activity (Scolnick et al., 1979b).

Several other acutely transforming viruses which contain a transduced Ha-ras gene have also been isolated. The rat sarcoma virus was isolated by in vitro cocultivation of several chemically-transformed rat cells with Sprague-Dawley rat embryo cells releasing an endogenous ecotropic MuLV (Rasheed et al., 1978). This isolate transforms rat cells in tissue culture and induces fibrosarcomas in rats (Rasheed and Young, 1982), albeit at a low frequency. The BALB/c murine sarcoma virus (Ba-MSV) was isolated from a spontaneous chloroleukemia in an old BALB/c mouse (R. L. Peters et al., 1974). Unlike other ras-containing virus isolates, it transforms cells within the blood vessels and causes hemangiosarcomas in mice after a latency as short as 3 weeks. The virus appears to be the result of recombination between a B-tropic MuLV and the proto-oncogene bas, which is the mouse equivalent of the rat Ha-ras oncogene (P. R. Anderson et al., 1981). Another Ha-ras-containing virus, the malignant histiocytosis sarcoma virus (MHSV), has the ability to transform macrophages (Ostertag et al., 1980). MHSV was isolated by passage of Friend MuLV in newborn BALB/c mice and induces in either newborn or adult mice an acute histiocytosis associated with splenomegaly and severe anemia. The virus was shown to be a recombinant between Friend mink cell focus-inducing virus and Ha-MSV (Franz et al., 1985). The unique ability of this virus to transform cells of the mononuclear/phagocytic lineage appears to be due to sequences within the U3 region of the viral LTR (Friel et al., 1990). Finally, ras-containing viruses were recently isolated from NFS × NS.C58 F$_1$ hybrid mice that had been inoculated with a polytropic virus derived from a Cas-Br-E-infected NFS mouse (Fredrickson et al., 1987). NS.C58 MSV-1, which induces sarcomas and erythroblastosis in neonatal mice, was shown to contain sequences highly related to the bas oncogene present in Ba-MSV.

b. Moloney, Gazdar, and Myeloproliferative Sarcoma Viruses

Moloney murine sarcoma virus (M-MSV) was isolated by injecting a high dose of Moloney MuLV into BALB/c mice (Moloney, 1966). The virus transforms muscle tissue and induced rhabdomyosarcomas in 3–5 days when injected into neonatal mice. The tumors often regress and recur, although those mice that develop progressive tumors can die within weeks. Like mice infected with Harvey and Kirsten sarcoma viruses, the animals can also develop splenomegaly, but this is not due to erythroblastosis, but to a hyperplastic response to the tumor. M-MSV is a recombinant between Moloney MuLV and the mouse proto-oncogene *mos* (Scolnick *et al.*, 1975; Frankel and Fischinger, 1976; Hu *et al.*, 1977). This virus encodes a 37-kDa protein (Papkoff *et al.*, 1982) which is a serine/threonine kinase (Maxwell and Arlinghaus, 1985). Gazdar *et al.* (1972) isolated a similar sarcoma-inducing virus, designated Gazdar-MSV (Gz-MSV), from a spontaneous tumor in an NZW/NZB F_1 hybrid mouse. Its structure is highly related to that of M-MSV (Donoghue *et al.*, 1979).

After continuous passage of M-MSV-induced sarcomas in mice, followed by passage of cell-free extracts, Chirigos *et al.* (1968) isolated another *mos*-containing virus that causes a highly undifferentiated sarcoma in adult mice as well as extensive hematopoietic changes in the myeloid pathway (Ostertag *et al.*, 1980). This virus, designated myeloproliferative sarcoma virus (MPSV), is a recombinant between Moloney MuLV and endogenous *mos* sequences (Kollek *et al.*, 1984) and is predicted to encode a 34-kDa protein (Stacey *et al.*, 1984). The ability of this *mos*-containing virus, in contrast to M-MSV and G-MSV, to transform hematopoietic cells is thought to be due to unique sequences in the U3 region of the viral LTR (Stocking *et al.*, 1985).

c. FBJ and FBR Osteosarcoma Viruses

The FBJ virus was isolated from a spontaneous osteosarcoma in a CF-1 mouse (Finkel *et al.*, 1966). Serial passage of the tumor extract to newborn mice resulted in osteosarcomas with a latency as short as 3 weeks. The virus consists of a complex of N-tropic MuLV and a defective virus (Levy *et al.*, 1973) which is a recombinant between the MuLV and the proto-oncogene *fos* (Curran *et al.*, 1982). The transforming virus encodes a 55-kDa DNA-binding protein (Curran and Teich, 1982; Sambucetti and Curran, 1986).

The FBR osteosarcoma virus was isolated from a strontium-90-induced osteosarcoma in an X/Gf mouse (Finkel *et al.*, 1973, 1975). It is highly oncogenic in newborn X/Gf mice, with 100% of the mice developing osteosarcomas as early as 3 weeks. The isolate is a complex of

B-tropic MuLV and defective virus which, like the FBJ virus, is a recombinant between the MuLV and the proto-oncogene *fos* (van Beveren *et al.*, 1984). The transforming virus encodes a 75-kDa *gag*-fusion, DNA-binding protein (Curran and Verma, 1984).

d. MSV-3611

Another sarcoma-inducing retrovirus was isolated by IUdR induction of methylcholanthrene-transformed C3H/10 T 1/2 mouse cells (Rapp *et al.*, 1983a). This virus, designated murine sarcoma virus 3611 (MSV-3611), induces undifferentiated fibrosarcomas in newborn NFS mice within 4 weeks of injection. The defective sarcoma virus is a recombinant between an endogenous MuLV and the proto-oncogene *raf* and encodes a 75-kDa *gag–raf* protein that has serine kinase activity (Rapp *et al.*, 1983b; Moelling *et al.*, 1984).

e. Abelson Murine Leukemia Virus

Passage of Moloney MuLV into prednisolone-treated BALB/c mice resulted in the isolation of a new virus, Abelson MuLV (Ab-MuLV), that induces a pre-B-cell lymphosarcoma 3–5 weeks after intraperitoneal injection of newborn mice (Abelson and Rabstein, 1990a, b). Mice also show a characteristic bulging of the skull caused by tumor cell infiltration of the meninges. The transforming virus is a recombinant between Moloney MuLV and the proto-oncogene *abl* (Goff *et al.*, 1980). The most studied strain encodes a 120-kDa *gag–abl* fusion protein (Witte *et al.*, 1978; Reynolds *et al.*, 1978) which has tyrosine kinase activity (Witte *et al.*, 1980). The original strain encodes a highly-related 160-kDa protein (Goff *et al.*, 1981). Ab-MuLV can also cause a T-cell lymphoma if it is injected directly into the thymus (Cook, 1982). Induction of B-cell lymphomas, but not T-cell lymphomas, by Ab-MuLV is thought to require a second event induced by integration of the helper virus into host DNA (Green *et al.*, 1989; Poirier and Jolicoeur, 1989).

f. Cas-NS-1 Virus

Another virus that induces pre-B-cell lymphomas, Cas-NS-1 virus, was isolated from an NFS/N mouse injected at birth with Cas-Br-E virus (Langdon *et al.*, 1989). Injection of the virus into newborn NFS/N mice results in a high incidence of pre-B- and pro-B-cell lymphomas with a mean latency period of 4 months. The virus can also induce occasional myeloid tumors. Cas-NS-1 virus is the result of recombination between endogenous polytropic viral sequences and the *cbl* proto-oncogene. It encodes a 100-kDa *gag–cbl* DNA-binding protein that shares some homology with the yeast transcriptional activator GCN4.

g. Myeloproliferative Leukemia Virus

The myeloproliferative leukemia virus (MPLV) was isolated from a DBA/2 mouse inoculated as a newborn with Friend MuLV (Wendling *et al.*, 1986). It induces an acute myeloproliferative syndrome in adult mice 2–3 weeks after injection. The virus has no effect on lymphoid cells or fibroblasts. The transforming virus is a recombinant between Friend MuLV and cellular sequences designated *mpl* (Penciolelli *et al.*, 1987; Souyri *et al.*, 1990). The v-*mpl* sequences share strong structural homology with the hemopoietin receptor superfamily, suggesting that MPLV has transduced a truncated form of an as-yet-unidentified hematopoietic growth factor receptor (Souyri *et al.*, 1990).

h. Friend Spleen Focus-Forming Virus

Friend spleen focus-forming virus (F-SFFV) was isolated from a leukemic Swiss mouse that had been injected as a newborn with Erlich ascites cells (Friend, 1957). A cell-free extract from the leukemic mouse induced in adult mice a rapid erythroleukemia associated with hepatosplenomegaly and anemia (Fig. 7). Upon passage of the original isolate in other laboratories, a variant virus emerged which also caused a rapid erythroleukemia, but the mice became polycythemic rather than anemic (Mirand *et al.*, 1961; Sassa *et al.*, 1968). The two viruses have subsequently been designated FV-A, for the anemia-inducing strain, and FV-P, for the polycythemia-inducing strain. Both preparations consist of a complex of the defective spleen focus-forming virus (SFFV$_P$ or SFFV$_A$), which is responsible for the acute effects of the virus complex on erythroid cells, and Friend MuLV (Troxler *et al.*, 1977a, 1980). SFFV is a recombinant between Friend MuLV and endogenous envelope gene sequences related to those of polytropic viruses (Troxler *et al.*, 1977b) and encodes a unique 52/55-kDa envelope-related protein (Ruscetti *et al.*, 1979; Dresler *et al.*, 1979) that is responsible for the acute erythroleukemia (Wolff and Ruscetti, 1988). Recent studies suggest that the mechanism by which the envelope protein of SFFV$_P$ induces erythropoietin-independent erythroid hyperplasia is by interacting with and triggering the erythropoietin receptor (Ruscetti *et al.*, 1990; J. Li *et al.*, 1990; Hoatlin *et al.*, 1990; Casadevall *et al.*, 1991). If mice live long enough, a second, more malignant stage of disease becomes apparent (Tambourin *et al.*, 1979). This stage depends upon the integration of SFFV into a

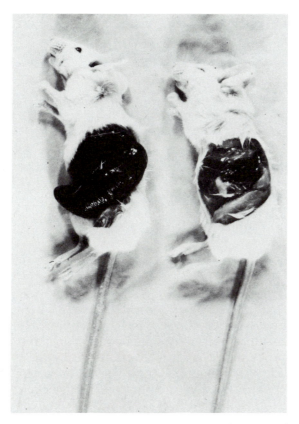

FIGURE 7. Splenomegaly induced by SFFV in an NIH Swiss mouse. The mouse on the left was injected 21 days earlier with F-MuLV/SFFV$_P$. For comparison, a control mouse is shown on the right.

particular site in the host DNA (Moreau-Gachelin *et al.*, 1988; Paul *et al.*, 1991).

Two additional viruses related to the anemia-inducing strain of Friend SFFV have also been isolated. Rauscher SFFV was isolated by the passage of lymphoma cells from a Swiss mouse through Swiss and BALB/c mice (Rauscher, 1962), and Cas-SFFV was isolated from a mouse that received a spleen homogenate from a lymphomatous mouse previously injected with Cas-Br-E virus (Langdon *et al.*, 1983). Both viruses cause a rapid erythroleukemia associated with anemia. Rauscher SFFV is a recombinant between Rauscher MuLV and endogenous polytropic virus-related envelope gene sequences that are highly related to those in Friend SFFV (Bestwick *et al.*, 1983, 1984). The genome of Cas-SFFV is not well characterized, although Cas-Br-E appears to have recombined with endogenous polytropic virus-related envelope gene sequences similar to those in Friend and Rauscher SFFVs.

i. AKT-8 Virus

A virus which can transform mink cells in culture was isolated from the thymus of a leukemic AKR mouse (Staal et al., 1977). The virus, designated AKT-8, induces thymic lymphomas in AKR or NFS × NS.C58v1 F₁ hybrid mice (Staal and Hartley, 1988). AKT-8 is a defective virus which is a recombinant between an AKR polytropic virus and endogenous cellular sequences designated akt (Staal, 1987). It encodes a 110-kDa gag–akt fusion protein (Sacks et al., 1978) that was recently shown to be a serine-threonine kinase containing an SH2-like region (Bellarosa et al., 1991).

2. Nonacute Transforming Viruses

A second group of transforming viruses (Table V) has been isolated which do not contain oncogenes or pathogens and which induce disease after a long latency period. These viruses are thought to cause disease due to an indirect mechanism, usually involving generation of recombinant viruses and/or the activation of cellular genes due to proviral integration. Since a long period occurs between virus injection and disease, a number of events initiated by the input virus may occur and be necessary for the resulting disease that develops. The fact that some studies have reported the detection of preleukemic cells in mice replicating nonacute transforming viruses is consistent with the idea that the diseases induced by this class of mouse retroviruses involve multiple steps (Haran-Ghera, 1980; Hays, 1982; Haas et al., 1984; Davis et al., 1987; Hays et al., 1989; Gokhman et al., 1990).

The mechanisms by which nonacute transforming retroviruses or their recombinant derivatives induce disease is not known. It has been suggested that the unique envelope proteins encoded by polytropic viruses may act as mitogens by binding to receptors on hematopoietic cells (McGrath and Weissman, 1979; Li and Baltimore, 1991). While this may induce hematopoietic hyperplasia, it is probably not sufficient to immortalize these cells. Rather, immortalization may be due to integration of the input virus or the polytropic viruses that they generate into specific sites in the mouse genome, leading to the activation of cellular genes involved in oncogenesis (Fig. 8). Viral insertion can activate the expression of cellular genes either by using the viral promoter to promote the expression of the gene or by putting the cellular promoter under the local influence of the viral enhancer. The cellular genes that are activated, like the cellular genes transduced by the acutely trans-

TABLE V. Nonacute Transforming Viruses

Virus	Origin	Primary disease	Polytropic viruses generated
AKR SL viruses	AKR leukemia lines	T-cell lymphoma	+
B-ecotropic MuLV	Healthy BALB/c mouse	T-cell lymphoma	+
Cas-Br-E MuLV	Diseased California wild mouse	Nonthymic lymphoma; other diseases	+
4070A, 1504A, 4886E	Healthy and diseased California wild mice	Nonthymic lymphoma	?
Friend MuLV	Tumor cells passaged through mice	Eythroleukemia	+
Gross passage A MuLV	Spontaneous AKR lymphoma	T-cell lymphoma	+
Ho MuLV	Healthy European wild mouse	T-cell lymphoma; erythroid, myeloid leukemia	+
Moloney MuLV	Spontaneous mouse tumor	T-cell lymphoma; myeloid leukemia	+
Mouse mammary tumor virus	Spontaneous mouse tumor	Mammary adenocarcinoma	−
Radiation LVs	Irradiated C57BL/6 mice	T- and B-cell lymphomas	+
Rauscher MuLV	Tumor cells passed through mice	Erythroleukemia	+
10A1 MuLV	1504A-infected mouse	Null-cell leukemia	+[a]

[a] 10A1 is itself a polytropic-related virus.

forming viruses, may either directly or indirectly have an effect on cell growth.

A number of different cellular genes that are altered by retroviral insertion have been identified in tumors induced by nonacute transforming viruses (Table VI). Many of these genes were previously identified as transduced genes in acute leukemia-inducing viruses (such as *Myc*, *Mos*, and *Myb*) or as cellular DNA segments capable of transforming transfected cells. The properties of some of these genes are known: *Fim-2*, which is identical to the *Fms* oncogene (Gisselbrecht *et al.*, 1987), encodes the CSF-1 receptor (Sherr *et al.*, 1985); *Lck* encodes a tyrosine protein kinase (Marth *et al.*, 1985); the products of *Myc*, *Myb*, and *Sfpi-1/PU.1* are DNA-binding proteins (Abrams *et al.*, 1982; Donner *et al.*, 1982; Moelling *et al.*, 1985; Goebl *et al.*, 1990; Paul *et al.*, 1991); and the *Il-3* gene encodes a hematopoietic growth factor (Ihle *et al.*, 1983). Retroviral integration does not always result in the activation of a gene. For example, integration into the p53 gene (*Trp53*) often results in its inactivation (Ben-David *et al.*, 1988; Munroe *et al.*, 1990). This has led to the

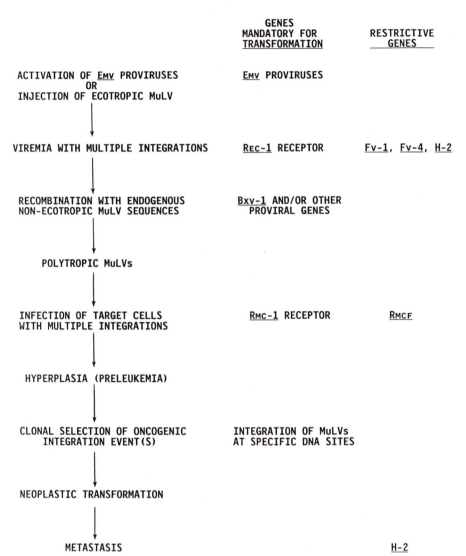

FIGURE 8. Multiple steps in the development of leukemia by nonacute transforming murine leukemia viruses. Leukemogenesis in the AKR mouse and in other strains of mice injected with ecotropic MuLVs involves multiple steps including the replication of ecotropic virus, the generation of polytropic viruses, and insertional mutagenesis due to integration of either ecotropic or polytropic MuLVs in specific sites in the DNA. In order for leukemogenesis to occur, the animal must have the necessary proviral genes (or be injected with virus), receptors for ecotropic and polytropic viruses, and viruses integrated into specific regions of the DNA that will lead to activation of oncogenes (see Section III.A.2). The host can carry other genes which restrict several of these events, including replication of ecotropic or polytropic viruses and the growth of virus-infected or transformed cells (see Section III.C).

TABLE VI. Common Sites of Retroviral Gene Insertion

Locus	Chromosome	Virus[a]	Disease	Reference
Ahi-1	10	M-MuLV	B-cell lymphoma	Poirier et al. (1988)
Dsi-1	4	M-MuLV[b]	Thymoma	Vijaya et al. (1987)
Evi-1	3	Endogenous MuLV	Myeloid leukemia	Mucenski et al. (1988)
Evi-2	11	Endogenous MuLV	Myeloid leukemia	Buchberg et al. (1988, 1990)
Fim-1	13	F-MuLV	Myeloid leukemia	Sola et al. (1986)
Fim-2 (Fms)	18	F-MuLV	Myeloid leukemia	Sola et al. (1986)
Fim-3	3	F-MuLV	Myeloid leukemia	Bordereaux et al. (1987)
Fis-1	7	F-MuLV	T-cell lymphoma; Myeloid leukemia	Silver and Kozak (1986)
Fli-1	9	F-MuLV	Erythroleukemia	Ben-David et al. (1990)
Gin-1	19	G-MuLV	T-cell lymphoma	Villermur et al. (1987)
Wnt-1 (Int-1)	15	MMTV	Mammary tumors	Nusse and Varmus (1982), Nusse et al. (1984)
Int-2	7	MMTV	Mammary tumors	G. Peters et al. (1983)
Int-3	17	MMTV	Mammary tumors	Gallahan et al. (1986)
Int-4	11	MMTV	Mammary tumors	Buchberg et al. (1989)
Int-5	9	MMTV	Mammary tumors	Morris et al. (1991)
Il-3	11	IAP	Myeloid leukemia	Ymer et al. (1985)
Lck	4	M-MuLV	T-cell lymphoma	Voronova and Sefton (1986)
Mlvi-2	15	M-MuLV[b]	T-cell lymphoma	Tsichlis et al. (1984)
Mlvi-3	15	M-MuLV[b]	T-cell lymphoma	Tsichlis et al. (1985)
Mlvi-4	15	M-MuLV[b]	T-cell lymphoma	Tsichlis et al. (1990)
Mos	4	IAP	Myeloma	Canaani et al. (1983)
Myb	10	M-MuLV, Cas-Br-E	Myeloid leukemia	Shen-Ong et al. (1984), Shen-Ong and Wolff (1987), Weinstein et al. (1986), Gonda et al. (1987)
Myc	15	M-MuLV, AKR-MCF	T-cell lymphoma	Corcoran et al. (1984), Selten et al. (1984), Y. Li et al. (1984), Van Lohuizen et al. (1989)
Pim-1	17	M-MuLV, M-MCF, AKR-MCF	T- and B-cell lymphoma	Cuypers et al. (1984), Hilkens et al. (1986), Nadeau and Phillips (1987)
Pim-2	17	M-MuLV	T-cell lymphoma	Breuer et al. (1989)
Pvt-1 (Mis-1, Mlvi-1)	15	M-MuLV	T-cell lymphoma	Tsichlis et al. (1983), Graham et al. (1985), Jolicoeur et al. (1985)
Sfpi-1[c]	2	F-SFFV	Erythroleukemia	Moreau-Gachelin et al. (1988)
Trp53-1	11	F-MuLV, F-SFFV	Erythroleukemia	Mowat et al. (1985)

[a] M, Moloney; F, Friend; G, Gross.
[b] Virus-infected rats.
[c] Formerly Spi-1.

idea that the p53 gene is a tumor-suppressor gene and its inactivation allows the outgrowth of cells transformed by another event.

Some virus-induced diseases are associated with integration at one common site, whereas others are associated with integration at more than one site. For full transformation of the cell, the virus may need to activate several cellular genes. This is more likely in virus-induced diseases associated with the generation of polytropic viruses, because both the input virus and the polytropic virus can infect and integrate into a single cell and activate different genes. Also, different diseases induced by the same MuLV have been shown to be associated with integration of the virus into different regions of the host DNA. For example, Moloney MuLV-induced myeloid leukemia is associated with integration of the virus into the *Myb* locus (Shen-Ong and Wolff, 1987), whereas Moloney MuLV-induced thymic lymphomas are associated with integration of the virus into several other regions of the host DNA. Also, Friend MuLV-induced erythroleukemia is associated with integration of Friend MuLV into the *Fli-1* locus (Ben-David *et al.*, 1990), whereas myeloid leukemia induced by Friend MuLV is associated with integration into *Fim-1, -2,* or *-3* (Sola *et al.*, 1986; Bordereaux *et al.*, 1987).

The induction of mammary tumors by MMTV is generally associated with integration of the virus into specific sites in the host DNA (*Int-1–Int-5*) (Table VI), resulting in activation of cellular genes, probably by enhancer insertion, that are directly or indirectly involved in transforming the cell. These integration sites differ from those identified in MuLV studies. Recent characterization of the MMTV locus *Int-1*, now termed *Wnt-1* (Nusse and Varmus, 1982; Nusse *et al.*, 1991), shows it to be related to the *Drosophila wingless* mutation (Rijsewijk *et al.*, 1987), a gene which functions in segmentation and which, in mammals, has an important role in brain development (McMahon and Bradley, 1990). The *Int-2* locus encodes a protein which is strikingly homologous to the basic fibroblast growth factor family (Dickson and Peters, 1987; Smith *et al.*, 1988).

The mouse leukemia viruses which are capable of inducing long-latency neoplastic disease have been isolated from a number of sources and induce a variety of diseases (Table V).

a. MuLVs from AKR Mice

Gross passage A virus was derived by serial passage in C3Hf/Bi mice of an extract from a spontaneous AKR lymphoma (Gross, 1957). It induces a high incidence of thymomas in adult or newborn C3Hf/Bi mice within 3 months. The Gross passage A preparation contained a mixture of ecotropic and polytropic viruses (Famulari *et al.*, 1982). The polytropic virus was found to be necessary and sufficient for induction of leukemias in C3Hf/Bi mice. The ecotropic virus also is leukemogenic, but the disease appears to involve the generation of a polytropic virus.

Cell lines developed from spontaneous thymic leukemias of AKR mice are another source of leukemogenic viruses. These N-ecotropic viruses, designated AKR SL viruses, induce T-cell leukemias in newborn AKR, C3Hf/Bi, and NFS mice (Pederson *et al.*, 1981). Thymomas induced in C3Hf/Bi and NFS mice by a molecular clone of one of these viruses, SL3-3, express polytropic viral envelope proteins (Famulari, 1983).

b. Ecotropic MuLVs

Ecotropic virus derived from normal BALB/c mice induces lymphosarcomas 6–15 months after injection into newborn BALB/c mice (Jolicoeur *et al.*, 1978). A similar virus induced from immunologically-triggered lymphoreticular tumors in young adult BALB/c × A F$_1$ hybrid mice causes lymphoreticular tumors when injected into newborn BALB/c mice (Armstrong *et al.*, 1980). Polytropic viruses can be isolated from these later tumors, but the role that they play in disease development is not known.

c. Moloney, Friend, and Rauscher MuLVs

Ecotropic Friend MuLV biologically cloned from the original virus stock of Friend (1957) induces erythroleukemia associated with severe anemia 4–6 weeks after injection into newborn BALB/c and NIH Swiss mice (Troxler and Scolnick, 1978). Polytropic viruses are generated in these mice (Troxler *et al.*, 1978; Ishimoto *et al.*, 1981) and appear to play an important role in the development of the early erythroleukemia (Ruscetti *et al.*, 1981, 1985). Animals that are resistant to the early disease can develop myeloid, lymphoid, or erythroid leukemia after a long latency (greater than 6 months) (Chesebro *et al.*, 1983; Silver and Fredrickson, 1983). In contrast to the development of the early erythroleukemia, these later diseases are not dependent upon the generation and replication of polytropic viruses (Chesebro *et al.*, 1983; Silver, 1984).

Ecotropic Moloney MuLV was originally isolated from a spontaneous tumor of BALB/c mice (Moloney, 1960). Injection of this MuLV into newborn NFS or BALB/c mice results in the development of thymic lymphomas beginning at about 3 months of age (Moloney, 1960; Gisselbrecht, 1978; Reddy *et al.*, 1980). Polytropic viruses are generated in these mice and they appear to play an important role in the generation of disease (Fischinger *et al.*, 1975; Vogt, 1979; Gisselbrecht *et al.*, 1981). Studies using recombinant viruses constructed between Friend and Moloney MuLV indicate that sequences in the LTRs of these viruses determine the target-cell specificity of disease (Chatis et al., 1983, 1984). Moloney MuLV can also cause myeloid leukemia when injected intravenously into adult BALB/c mice that have been pretreated with pristane to induce a chronic inflammatory response (Wolff et al., 1988). Unlike thymic lymphomas induced by Moloney MuLV, polytropic viruses are not detected and LTR sequences do not determine the myeloid cell specificity.

Another ecotropic MuLV, Rauscher MuLV, was isolated by passage of lymphoma cells from a Swiss mouse through Swiss and BALB/c mice (Rauscher, 1962). When injected into newborn NIH Swiss mice, it induces erythroblastosis in 9–25 weeks and thymic lymphoma in 12–29 weeks (Vogt, 1982). Generation of polytropic viruses was associated with both diseases, and a polytropic virus isolated from a thymic lymphoma was still able to induce erythroblastosis. Polytropic viruses can also be isolated from mice infected with the Rauscher MuLV/spleen focus-forming virus complex and these viruses cause a slow erythroleukemia in BALB/c mice (van Griensven and Vogt, 1980). Reddy et al. (1980) reported that cloned Rauscher MuLV induced B-cell lymphomas 10–11 weeks after injection into NIH Swiss mice.

d. MuLVs Isolated from Wild Mice

Both ecotropic (Cas-Br-E and 4996E) and amphotropic (4070A and 1504A) MuLVs isolated from diseased or normal tissues of wild mice trapped in southern California can cause B- or null-cell lymphomas 7–15 months after injection into neonatal NIH Swiss mice (Gardner, 1978). The incidence of disease is rather low (about 25%) when compared with diseases induced by retroviruses isolated from inbred mouse strains. In addition to inducing nonthymic lymphomas, Cas-Br-E MuLV has been shown to induce a number of other hematopoietic neoplasms, including B- and pre-B-cell lymphomas, erythroleukemias, and myelogenous and megakaryocytic leukemias (Fredrickson et al., 1984; Holmes et al., 1986). Polytropic viruses have been detected in the nonthymic lymphomas induced by Cas-Br-E MuLV (Hoffman et al., 1981), but it is not known whether they play a role in lymphoma development.

Passage of amphotropic virus 1504A, which induces a low incidence of lymphomas after a long latency, through newborn NIH Swiss mice resulted in the isolation of a more pathogenic virus (Rasheed et al., 1982). This virus, termed 10A1, causes a null-cell leukemia 1–2 months after injection into newborn NIH Swiss mice. Like polytropic viruses derived from ecotropic viruses, the 10A1 virus appears to be a recombinant between 1504A and env gene sequences present in mouse DNA (Lai et al., 1982; Ott et al., 1990). Another amphotropic MuLV, 4070A, which causes a low incidence of lymphomas after a long latency, was recently shown to induce a high incidence of myeloid leukemia 2–6 months after injection into adult DBA/2 mice that were undergoing a chronic inflammatory response induced by pristane (Wolff *et al.*, 1991).

An ecotropic MuLV isolated from a European mouse, HoMuLV, induces a high incidence of thymic lymphomas and a few erythroleukemias 5–10 months after injection into neonatal female NIH Swiss mice (Voytek and Kozak, 1988). Male mice develop a lower incidence of erythroid and myeloid leukemia in 8–16 months. Polytropic viruses could be detected in several thymic lymphomas examined.

e. Radiation Leukemia Viruses

Whole-body X-irradiation of C57BL/6 mice results in the induction of a number of different retroviruses and a high incidence of thymic lymphomas (Lieberman and Kaplan, 1959). Polytropic viruses isolated from these tumors have also been shown to induce T-cell lymphomas (Haas, 1980). Passage of cell-free splenic extracts from radiation-induced tumors gave rise to the Laterjet–Duplan strain of radiation leukemia virus (Laterjet and Duplan, 1962). This preparation, which contains ecotropic and polytropic MuLVs as well as a defective virus, induces an early lymphoproliferative disease analogous to human AIDS (Pattengale *et al.*, 1982; Mosier *et al.*, 1985) (see Section III.B.2) followed by B-cell lymphoma (Klinken *et al.*, 1988).

f. Mouse Mammary Tumor Virus

The mouse mammary tumor virus (MMTV), the only B-type retrovirus isolated from laboratory mice, induces mammary adenocarcinomas in mice after a latency of 4–9 months (Bittner, 1936; Hilgers and Bentvelzen, 1978). Unlike diseases induced by C-type retroviruses, tumors induced by MMTV are not associated with the generation of recombinant viruses. Rather, integration of MMTV DNA into specific regions of host chromosomes may be a critical event in oncogenesis by this virus.

3. Genetically-Engineered Transforming Viruses

In addition to those created in mice, a number of acutely transforming mouse retroviruses have also been generated in the laboratory by genetic engineering. For example, the oncogenic sequences present in several different avian retroviruses have been put into mouse retroviral vectors so that they could be studied in mice. These include MRSV, which contains the avian v-*src* gene and induces erythroleukemia in mice (S. M. Anderson and Scolnick, 1983); ME26, which contains the v-*myb* and v-*ets* oncogenes from the avian E26 virus and induces erythroid, myeloid, and lymphoid leukemias in mice (Yuan *et al.*, 1989; Ruscetti *et al.*, 1992; Aurigemma *et al.*, 1992); retroviruses which contain the avian v-*myc* oncogene and cause either lymphoid or myeloid tumors in mice (Brightman *et al.*, 1986; Baumbach *et al.*, 1986; Wolff *et al.*, 1988); a retrovirus containing v-*src* that causes rapidly fatal hemangiosarcomas in the brains of mice (Hevezi and Goff, 1991); and retroviruses containing both the v-*myc* and v-*raf* oncogenes (Rapp *et al.*, 1985) which cause a variety of tumors in mice, including lymphoblastic lymphomas and pancreatic or mammary adenocarcinomas, as well as plasmacytomas in pristane-primed mice (Morse *et al.*, 1986). Another laboratory-derived virus, RIM, which contains the c-*myc* gene under the control of the immunoglobulin heavy-chain-gene regulatory sequences, and the v-Ha-*ras* gene driven by a retroviral promoter (Clynes *et al.*, 1988), induces rapid plasmacytomas in pristane-treated mice. Finally, growth factor genes which may be activated by proviral insertion in some tumors have been put into retroviral vectors to test their pathogenicity. A vector containing the *Il-3* gene under control of retroviral regulatory sequences was shown to cause myeloid leukemia when injected into mice (P. M. C. Wong *et al.*, 1989; Chang *et al.*, 1989).

B. Retroviruses Associated with Nonmalignant Diseases and Mutations

1. Neurological Diseases

Wild mice from the Lake Casitas region in California express high levels of ecotropic MuLV and have a high incidence of hind-limb paralysis (Gardner *et al.*, 1973). The Cas-Br-E virus (also termed Cas-Br-M) isolated from these mice can cause the same disease when injected into various inbred strains of mice (Gardner, 1978). Although polytropic viruses are generated in the spleens of these mice and may be associated with the lymphomas that the mice often develop, these viruses cannot be detected in the brain and are not thought to play any role in the neurological disease (Hoffman *et al.*, 1981). The disease appears to be a direct consequence of replication of the ecotropic virus in the central nervous system (Fig. 9). If mice older than 10 days of age are injected with the virus, the animals do not become viremic (apparently due to an effective immune response against the virus) or develop hind-limb paralysis (Hoffman *et al.*, 1984). The ability of the Cas-Br-E virus to cause neurological disease is determined by the viral envelope gene (DesGrosseillers *et al.*, 1984; Paquette *et al.*, 1989), but LTR and *gag–pol* sequences appear to influence the tempo of the disease (DesGrosseillers *et*

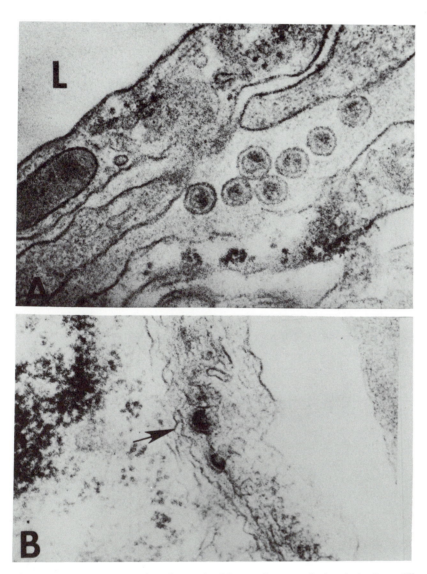

FIGURE 9. Electron micrographs of virus budding and accumulating in brain capillary endothelial cells from a mouse injected intracranially with Cas-Br-E MuLV: (A) Virions accumulating in the basal lamina. L, lumen. (B) Virus budding (arrow) from the abluminal surface. [Kindly provided by Dr. P. Hoffman.]

al., 1985; Jolicoeur and DesGrosseillers, 1985; Portis *et al.*, 1990). The mechanism by which the virus causes disease is not known. The damage to the nervous system is not an autoimmune response against the virus-infected cells. Rather, the virus appears to be causing direct damage to the central nervous system. Perhaps the envelope glycoprotein encoded by the Cas-Br-E virus preferentially binds to neuronal cells and blocks receptors for growth factor receptors. It is also possible that the viral envelope glycoprotein could be toxic to components of the central nervous system, or its accumulation there could somehow perturb membrane function. The fact that there can be a long latency (up to 12 months) between injection of the virus and the appearance of clinical signs of disease may suggest that a viral gene product is not directly responsible for the disease, or may just reflect the time that it takes for the central nervous system to show effects of the damage. Like mouse retroviruses that cause leukemia after a long latency, it is possible that the Cas-Br-E virus can activate by proviral insertion a cellular gene that is more directly responsible for the neurological damage.

In addition to viruses isolated from wild mice, variants of viruses isolated from laboratory strains of mice can also cause hind-limb paralysis. These include a temperature-sensitive mutant of Moloney MuLV (ts-M-MuLV) (McCarter *et al.*, 1977), as well as a nonleukemogenic variant of Friend MuLV which has been passaged through rats (Kai and Furuta, 1984; Masuda *et al.*, 1992). Like the Cas-Br-E virus, the neurovirulence associated with the ts-M-MuLV is attributed to its envelope gene (Szurek *et al.*, 1988), whose product accumulates in the cell due to defective processing (P. K. Y. Wong *et al.*, 1983). Also, mice injected with retroviruses that primarily cause malignant diseases, such as Harvey and Kirsten MSV, can develop neurological diseases, especially if the virus is injected directly into the brain (Pitts *et al.*, 1983). The pathological features of the diseases induced by the different viruses are not identical and may reflect different effects of each virus on the central nervous system. In some cases, hind-limb paralysis may be due to infiltration of leukemic cells into the central nervous system (Bedigian *et al.*, 1984).

2. Immune System Disorders

Certain strains of mice injected as adults with the Laterjet–Duplan strain of the radiation leukemia virus (also called LP-BM5) develop a severe immunodeficiency disease that is similar to human AIDS (Mosier *et al.*, 1985). These mice show polyclonal B-cell activation, progressive lymphadenopathy, splenomegaly, hypergammaglobulinemia, profound immunodeficiency (due to suppression of both humoral and cellular immunity), and, in the latter stages, B-cell lymphoma (Klinken *et al.*,

1988). The virus is a mixture of ecotropic MuLV, polytropic MuLV, and a defective virus. The defective virus, which encodes an unusual *gag*-related 60-kDa protein, appears to be responsible for the immunosuppression (Chattopadhyay *et al.*, 1989; Aziz *et al.*, 1989; Huang *et al.*, 1984; Huang and Jolicoeur, 1990). Whether or not the *gag*-related protein encoded by the defective virus is responsible and has a direct pathogenic effect on the immune system is not known; however, recent studies suggest that the *gag* protein of the defective virus shares "superantigenic" properties with some microbial antigens and minor lymphocyte stimulatory (*Mls*)-like antigens, resulting in selective expansion of T cells bearing Vβ5 (Hügin *et al.*, 1991). The temperature-sensitive mutant of Moloney MuLV that induces hind-limb paralysis can also induce a number of immunological abnormalities similar to those induced by LP-BM5 MuLV (P. K. Y. Wong *et al.*, 1989). These include the destruction of T cells, polyclonal activation of B cells, and hypergammaglobulinemia. The genetic determinants in this virus responsible for its effect on the immune system have not been identified.

A number of ecotropic MuLVs, such as Friend and Rauscher MuLV, have been shown to be immunosuppressive, and this effect could render mice infected with these viruses more susceptible to infectious diseases or cancer (Salamon and Wedderbrun, 1966; Bendinelli *et al.*, 1985). A retroviral envelope gene product, TMp15E, appears to be responsible for the immunosuppression (Cianciolo *et al.*, 1985; Mitani *et al.*, 1987).

In addition to suppressing the immune response, retroviruses may also be stimulatory to the immune system, leading to autoimmune disease. Since autoimmune-prone mice fail to develop immunological tolerance (Miller *et al.*, 1984), they make large amounts of autoantibodies to endogenous retroviral envelope proteins (Izui *et al.*, 1979). These antibodies appear to contribute to the pathogenesis of autoimmunity by forming complexes with retroviral envelope proteins (Izui *et al.*, 1981). This can result in glomerulonephritis and polyarteritis. While there is no evidence for an etiologic role for endogenous ecotropic or xenotropic retroviruses in autoimmune strains of mice, there may be a role for endogenous polytropic viral sequences in the disease. Support for this comes from the observation that autoimmune-prone mice express a novel polytropic MuLV-related RNA transcript that is present at much lower levels, or is undetectable, in control mouse strains (Krieg *et al.*, 1988; Krieg and Steinberg, 1990).

3. Other Biological Changes and Germline Mutations

In addition to causing a variety of diseases, replication of retroviruses can also result in a variety of nonpathogenic biological and genetic changes in the animal. For example, inoculation of newborn mice with certain ecotropic MuLVs results in deformed whiskers (Rowe, 1983). Also, a high level of expression of ecotropic and polytropic viruses early in life has been associated with progressive graying in C57BL mice (Morse et al., 1985). The mechanisms by which retroviruses cause these changes is not known.

Since the average mouse is estimated to contain more than 1000 copies of endogenous retroviral sequences, it is not surprising that their integration has caused heritable mutations. Chromosomal integration during the viral replicative cycle can occur in germline as well as somatic tissues (Jaenisch, 1976; Rowe and Kozak, 1980; Jenkins and Copeland, 1985) or by retrotransposition (Heidmann and Heidmann, 1991), and such novel integrations can either disrupt the integrity of cellular genes or alter expression of adjacent coding regions. Three germline mutations have been clearly shown to be caused by retroviral gene insertions: the dilute coat color mutation (d) (Jenkins et al., 1981), the lethal collagen I gene mutation (Mov-13) (Breindl et al., 1984; Harbers et al., 1984), and the hairless mutation (hr) (Stoye et al., 1988). Also, the regulatory element responsible for the androgen responsiveness of the mouse sex-limited protein (Slp) is within the 5' LTR of a proviral element (Stavenhagen and Robins, 1988), suggesting that expression of cellular genes can also be regulated by proviral sequences.

Because of the number of endogenous retroviral sequences and the age- and tissue-dependent patterns of expression often characteristic of these genes, it has been suggested that retroviruses may play some important role in normal differentiation and development (reviewed in Levy, 1978). It can be argued that retroviral genes are important for their hosts because they provide an important source of genetic diversity, particularly since it is clear that retroviral integration can alter normal gene function. Although germline copies tend to be relatively stable when examined over the decades of inbreeding which separate the inbred mouse strains (Buckler et al., 1982; Jenkins et al., 1982; Quint et al., 1982), the cumulative effects of integrations and excisions become significant when considered over the evolutionary life of the different rodent species. Furthermore, because different wild mouse populations contain different subtypes of retroviruses, it is possible that the presence of reiterated copies of these different genes may serve some role in evolution and even speciation.

Additional support for the idea that retroviral genes are important for the host has come from the observation that many MuLV-related sequences map near immune system genes (Blatt et al., 1983; Wejman et al., 1984; Rossomando and Meruelo, 1986). The physical proximity of these genes, however, is not surprising, given the large number of both types of genes, and no functional significance has been attributed to any of these associations. Studies on the chromosomal distribution of MMTVs have shown that these proviral genes can also map within or near immune system genes such as *Igk* (Yang and Dudley, 1991). In this case, however, there is now evidence to suggest that the products of various endogenous MMTVs function as superantigens to direct the clonal deletion of T cells expressing specific T-cell receptor V_β chains. Multiple *Mls* genes have been described and chromosomally mapped which are capable of effecting this deletion. Recent studies now demonstrate that many of these *Mls* loci map to the integration sites of specific MMTV proviruses and that the *Mls* allele governing V_β T-cell deletion is associated with the presence of the MMTV provirus in genetic crosses and among inbred strains (Woodland *et al.*, 1990, 1991; Frankel *et al.*, 1991; Dyson *et al.*, 1991). The conclusion that MMTVs encode *Mls* antigens is also supported by the observations that only T-cell lines which stimulate the proliferative response express the appropriate MMTV transcripts (Woodland *et al.*, 1991) and that one *Mls* superantigen is maternally transmitted in milk of mice which carry a milk-borne MMTV (Marrack *et al.*, 1991). Transfection experiments and examination of transgenic mice have now demonstrated that the viral superantigen is encoded by the MMTV LTR-associated *orf* gene (Choi *et al.*, 1991; Acha-Orbea *et al.*, 1991).

These observations suggest that the MMTV gene products can affect the T-cell repertoire, although the broader consequences of this process in the mouse immune system are not known. This process may function to eliminate potentially autoreactive T-cell clones or may exclude the V_β chains conferring reactivity to exogenous superantigens such as bacterial toxins. Although MMTVs are not common to all species of mice (Callahan *et al.*, 1982), virally encoded superantigens may be more broadly distributed, since a recent study has suggested that at least one other retrovirus, the MuLV which induces an AIDS-like disease in mice, MAIDs, also shows *Mls*-like properties in its ability to induce a T-cell proliferative response (Hügin *et al.*, 1991). On the other hand, arguing against a general role for retroviruses as endogenous superantigens in the mammalian immune system is the observation that other species, such as rats and humans, do not have *Mls*-type genes.

Other immune system responses to expression of endogenous viral sequences have been documented. Early characterization of G_{IX} antigen as a lymphocyte differentiation alloantigen also associated with the development of neoplastic disease (Stockert *et al.*, 1971) was amended when it was shown to be a MuLV-related SU*env* glycoprotein (Obata *et al.*, 1975; Tung *et al.*, 1975). Similarly, a drug-mediated tumor antigen has now been identified as a MuLV SU*env* (Grohmann *et al.*, 1990).

C. Host Genes Controlling Disease Susceptibility

A number of genes have been identified and chromosomally mapped which control susceptibility of mice to induction of diseases by retroviruses (Table VII, Fig. 8). These fall into two basic categories: those which affect viral replication, and those which affect the target cells for the virus. The replication of ecotropic viruses has been shown to be influenced by the *Fv-1* gene (Lilly, 1967), which apparently encodes a product that interacts with the virus to inhibit viral integration (Jolicoeur and Rassart, 1980; Yang *et al.*, 1980), and the *Fv-4* (*Akvr-1*) gene (S. Suzuki, 1975; Gardner *et al.*, 1980), which encodes an envelope glycoprotein that blocks the receptor for ecotropic viruses (Ikeda and Odaka, 1983). The replication of polytropic viruses is under control of the *Rmcf* gene (Hartley *et al.*, 1983), which encodes or controls the expression of an envelope glycoprotein that blocks the receptor for polytropic viruses (Ruscetti *et al.*, 1981; Bassin *et al.*, 1982). *Sxv* restricts replication of xenotropic MuLVs in laboratory mice and is thought to represent a polymorphism of the polytropic MuLV receptor locus (*Rmc-1*) (Kozak, 1985b); no comparable polymorphisms have been identified for the ecotropic and amphotropic receptors (*Rec-1* and *Ram-1*). The *Gv-1* gene product coordinately reduces transcription of nonecotropic MuLVs, but is not known to affect replication of exogenous viruses. Other genes, such as *Fv-2* (Odaka and Yamamoto, 1962; Lilly, 1970), *W* (Steeves *et al.*, 1968), and *S1* (Bennett *et al.*, 1968), have been shown to affect susceptibility to the spleen focus-forming virus by controlling the number or susceptibility of target cells for transformation. Mice can also be resistant to retrovirus-induced disease because of an effective immune response against either the virus or virus-infected cells (Chesebro and Wehrly, 1978, 1979). Cellular genes are also thought to be responsible for the resistance of adult mice of susceptible strains to leukemia (Ruscetti *et al.*, 1981) and neurological diseases induced by various MuLVs (Hoffman *et al.*, 1984) and to the B-cell lymphomas induced by Abelson MuLV (Risser *et al.*, 1978). Finally, animals resistant to certain virus-induced leukemias may still develop leukemias that are different from those induced in susceptible strains in terms of latency and target-cell specificity. For example, Friend MuLV induces an early erythroleu-

TABLE VII. Host Genes Controlling Susceptibility
to Murine Leukemia Viruses

Locus	Chromosome	Mechanism of resistance
Fv-1	4	Inhibits viral integration
Fv-2	9	Affects target cells for SFFV
Fv-4 (*Akvr-1*)	12	Blocks ecotropic viral receptors
Rmcf	5	Blocks polytropic viral receptors
S1	10	Affects target cells for SFFV
Sxv-1	1	Restricts replication of xenotropic viruses
W	5	Affects target cells for SFFV

kemia after 4–6 weeks in NFS and BALB/c mice, but the virus causes late lymphoid, myeloid, or erythroid disease in strains resistant to the early erythroleukemia or in adult mice of susceptible strains (Chesebro et al., 1983; Silver and Fredrickson, 1983). While the early erythroleukemia is associated with the replication of polytropic viruses (Ruscetti et al., 1981), the latter diseases are not (Chesebro et al., 1983; Silver, 1984) and are most likely due to proviral insertion of the input virus. The genetic factors responsible for these different responses have not been defined, although a single gene, termed *Fhe*, was shown to control the type of leukemia induced by Friend MuLV in C57BL/6 mice (Silver and Fredrickson, 1983).

D. Rodent Retroviruses as Models for Human Diseases

Rodents have provided us with a valuable tool for studying genes involved in the development of neoplastic and nonneoplastic diseases. Passage of rodent tumor cells and viruses through mice and rats has led to the isolation of a plethora of retroviruses that have captured potential pathogenic genes from the host or have activated such genes due to specific integration into the host DNA. Although most human diseases are not associated with retroviruses, it is likely that some of the genes captured or activated by rodent retroviruses, many of which have been shown to have counterparts in humans, may also play a role in some human diseases. Characterization of the activated rodent genes and their products, as well as determining how they interact with the host cell to cause a biological effect, will provide us with valuable information for understanding human diseases associated with activation of similar genes. Development of treatments to counteract the effects of the activated genes in the rodent may also lead to new therapeutic approaches to treat similar diseases in humans.

ACKNOWLEDGMENTS. We would like to thank Brenda Rae Marshall, Ann Ha, and Karen Cannon for preparing the manuscript. Photographs were kindly provided as follows: Drs. J. Hartley and T. Fredrickson (XC plaque and MCF focus), Dr. R. Bassin (S⁺L⁻ focus), Dr. J. Levy (MSV focus), Dr. J. Silver (AKR thymoma), Dr. M. Gonda (A-, B-, C-type particles), and Dr. P. Hoffman (Cas-Br-E-infected brain).

IV. REFERENCES

Aaronson, S. A., and Stephenson, J. R., 1973, Independent segregation of loci for activation of biologically distinguishable RNA C-type viruses in mouse cells, *Proc. Natl. Acad. Sci. USA* **70**:2055.

Aaronson, S. A., Scolnick, E. M., and Todaro, G. J., 1971, Induction of murine C-type viruses from clonal lines of virus free BALB/3T3 cells, *Science* **174**:157.

Abelson, H. T., and Rabstein, L. S., 1990a, Influence of prednisolone on Moloney leukemogenic virus in BALB/mice, *Cancer Res.* **30**:2208.

Abelson, H. T., and Rabstein, L. S., 1990b, Lymphosarcoma: Virus-induced thymic-independent disease in mice, *Cancer Res.* **30**:2213.

Abrams, H., Rohrschneider, L., and Eisenman, R., 1982, Nuclear location of the putative transforming protein of avian myelocytomatosis virus, *Cell* **29**:427.

Acha-Orbea, H., Shakhov, A. N., Scarpellino, L., Kolb, E., Muller, V., Vessaz-Shaw, A., Fuchs, R., Blochlinger, K., Rollini, P., Billotte, J., Sarafidou, M., MacDonald, H. R., and Diggelmann, H., 1991, Clonal deletion of Vβ14-bearing T cells in mice transgenic for mammary tumour virus, *Nature* **350**:207.

Adachi, A., Sakai, K., Kitamura, N., Nakanishi, S., Niwa, O., Matsuyama, M., and Ishimoto, A., 1984, Characterization of the *env* gene and long terminal repeat of molecularly cloned Friend mink cell focus-inducing virus DNA, *J. Virol.* **3**:813.

Albritton, L. M., Tseng, L., Scadden, D., and Cunningham, J. M., 1989, A putative murine ecotropic retrovirus receptor gene encodes a multiple membrane-spanning protein and confers susceptibility to virus infection, *Cell* **57**:659.

Anderson, K. P., Low, M.-A. L., Lie, Y. S., Lazar, R., Keller, G., and Dinowitz, M., 1990a, Defective endogenous retroviruslike particles of Chinese hamster ovary cells, in: *Proceedings Evergreen Society of Animal Cell Technology*, May 7–11, 1990, Avignon, France.

Anderson, K. P., Lie, Y. S., Low, M.-A. L., Williams, S. R., Fennie, E. H., Nguyen, T. P., and Wurm, F. M., 1990b, Presence and transcription of intracisternal A-particle-related sequences in CHO cells, *J. Virol.* **64**:2021.

Anderson, P. R., Devare, S. G., Tronick, S. R., Ellis, R. W., Aaronson, S. A., and Scolnick, E. M., 1981, Generation of BALB-MuSV and Ha-MSV by type C virus transduction of homologous transforming genes from different species, *Cell* **26**:129.

Anderson, S. M., and Scolnick, E. M., 1983, Construction and isolation of a transforming murine retrovirus containing the *src* gene of Rous sarcoma virus, *J. Virol.* **46**:594.

Armstrong, M. Y. K., Weininger, R. B., Binder, D., Himsel, C. A., and Richards, F. R., 1980, Role of endogenous murine leukemia virus in immunologically triggered lymphoreticular tumors II. Isolation of B-tropic mink cell focus-inducing (MCF) murine leukemia virus, *Virology* **104**:164.

Aurigemma, R. E., Blair, D. G., and Ruscetti, S. K., 1992, Transactivation of erythroid transcription factor GATA-1 by a *myb-ets*-containing retrovirus, *J. Virol.* **66**:3056.

Aziz, D. C., Hanna, Z., and Jolicoeur, P., 1989, Severe immunodeficiency disease induced by a defective murine leukaemia virus, *Nature* **338**:505.

Bassin, R. H., Simons, P. J., Chesterman, F. C., and Harvey, J. J., 1968, Murine sarcoma virus (Harvey): Characteristics of focus formation in mouse embryo cell cultures, and virus production by hamster tumor cells, *Int. J. Cancer* **3**:265.

Bassin, R. N., Tuttle, N., and Fischinger, P. J., 1970, Isolation of murine sarcoma virus-transformed mouse cells which are negative for leukemia virus from agar suspension cultures, *Int. J. Cancer* **6**:95.

Bassin, R. H., Ruscetti, S., Ali, I., Haapala, D. K., and Rein, A., 1982, Normal DBA/2 mouse cells synthesize a glycoprotein which interferes with MCF virus infection, *Virology* **123**:139.

Baumbach, W. R., Keath, E. J., and Cole, M. D., 1986, A mouse c-*myc* retrovirus transforms established fibroblast lines *in vitro* and induces monocyte-macrophage tumors *in vivo*, *J. Virol.* **59**:276.

Bedigian, H. G., Johnson, D. A., Jenkins, N. A., Copeland, N. G., and Evans, R., 1984, mice, *J. Virol.* **51**:586.

Bellacosa, A., Testa, J. R., Staal, S. P., and Tsichlis, P. N., 1991, A retroviral oncogene, *akt*, encoding a serine-threonine kinase containing an SH2-like region, *Science* **254**:274.

Ben-David, Y., Prideaux, V. R., Chow, V., Benchimol, S., and Bernstein, A., 1988, Inactivation of the p53 oncogene by internal deletion or retroviral integration in erythroleukemic cell lines induced by Friend leukemia virus, *Oncogene* **3**:179.

Ben-David, Y., Giddens, E. B., and Bernstein, A., 1990, Identification and mapping of a common proviral integration site *Fli-1* in erythroleukemia cells induced by Friend murine leukemia virus, *Proc. Natl. Acad. Sci. USA* **87**:1336.

Bendinelli, M. D., Matteucci, D., and Friedman, H., 1985, Retrovirus induced acquired immunodeficiencies, *Adv. Cancer Res.* **45**:125.

Bennett, M., Steeves, R. A., Cudkowicz, G., Mirand, E. A., and Russell, L. B., 1968, Mutant *S1* alleles of mice affect susceptibility to Friend spleen focus-forming virus, *Science* **162**:546.

Bentvelzen, P., and Daams, J. H., 1969, Hereditary infections with mammary tumor viruses in mice, *J. Natl. Cancer Inst.* **43**:1025.

Benveniste, R. E., and Todaro, G. J., 1973, Homology between type-C viruses of various species as determined by molecular hybridization, *Proc. Natl. Acad. Sci. USA* **70**:3316.

Benveniste, R. E., Heinemann, R., Wilson, G. L., Callahan, R., and Todaro, G. J., 1974, Detection of baboon type C viral sequences in various primate tissues by molecular hybridization, *J. Virol.* **14**:1332.

Benveniste, R., Callahan, R., Sherr, C., Chapman, V., and Todaro, G. J., 1977, Two distinct endogenous type C viruses isolated from the Asian rodent *Mus cervicolor:* Conservation of virogene sequences in related rodent species, *J. Virol.* **21**:849.

Bergs, V. V., Pearson, G., Chopra, H. C., and Turner, W., 1972, Spontaneous appearance of cytopathology and rat C-type virus (WF-1) in a rat embryo cell line, *Int. J. Cancer* **10**:165.

Besmer, P., Olshevsky, U., Baltimore, D., and Fan, H., 1979, Virus-like 30S RNA in mouse cells, *J. Virol.* **29**:1168.

Bestwick, R., Ruta, M., Kiessling, A., Faust, C., Linemeyer, D., Scolnick, E., and Kabat, D., 1983, Genetic structure of Rauscher spleen focus-forming virus, *J. Virol.* **3**:1217.

Bestwick, R. K., Boswell, B. A., and Kabat, D., 1984, Molecular cloning of biologically active Rauscher spleen focus-forming virus and the sequences of its *env* gene and long terminal repeat, *J. Virol.* **51**:695.

Bikel, I., 1976, Properties of two distinct classes of leukemia viruses induced from C3H mouse cells, *Virology* **69**:793.

Bittner, J. J., 1936, Some possible effects of nursing on the mammary gland tumor incidence in mice, *Science* **84**:162.

Blatt, C., Mileham, K., Haas, M., Nesbitt, N., Harper, M. E., and Simon, M. I., 1983, Chromosomal mapping of the mink cell focus-inducing and xenotropic *env* gene family in the mouse, *Proc. Natl. Acad. Sci. USA* **80**:6298.

Bordereaux, D., Fichelson, S., Sola, B., Tambourin, P. E., and Gisselbrecht, S., 1987, Frequent involvement of the *fim-3* region in Friend murine leukemia virus-induced mouse myeloblastic leukemias, *J. Virol.* **61**:4043.

Bosselman, R. A., Straaten, F., Van Beveren, C., Verma, I. M., and Vogt, M., 1982, Analysis of the *env* gene of a molecularly cloned and biologically active Moloney mink cell focus-forming proviral DNA, *J. Virol.* **44**:19.

Breindl, M., Harbers, K., and Jaenisch, R., 1984, Retrovirus-induced lethal mutation in collagen I gene of mice is associated with an altered chromatin structure, *Cell* **38**:9.

Breuer, M. L., Cuypers, H. T., and Berns, A., 1989, Evidence for the involvement of *pim-2*, a new common proviral insertion site, in progression of lymphomas, *EMBO J.* **8**:743.

Brightman, B. K., Pattengale, P. K., and Fan, H., 1986, Generation and characterization of a recombinant Moloney murine leukemia virus containing the *v-myc* oncogene of avian MC29 virus: *In vitro* transformation and in vivo pathogenesis, *J. Virol.* **60**:68.

Bryant, M. L., Roy-Burman, P., Gardner, M. B., and Pal, B. K., 1978, Genetic relationship of wild mouse amphotrophic virus to murine ecotropic and xenotropic viruses, *Virology* **88**:389.

Buchberg, A. M., Bedigian, H. G., Taylor, B. A., Brownell, E., Ihle, J. N., Nagata, S., Jenkins, N. A., and Copeland, N. G., 1988, Localization of *Evi-2* to chromosome 11: Linkage to other proto-oncogene and growth factor loci using interspecific backcross mice, *Oncogene Res.* **2**:149.

Buchberg, A. M., Brownell, E., Nagata, S., Jenkins, N. A., and Copeland, N. G., 1989, A comprehensive genetic map of murine chromosome 11 reveals extensive linkage conservation between mouse and human, *Genetics* **122**:153.

Buchberg, A. M., Bedigian, H. G., Jenkins, N. A., and Copeland, N. G., 1990, *Evi-2*, a common integration site involved in murine myeloid leukemogenesis, *Mol. Cell. Biol.* **10**:4658.

Buckler, C. E., Staal, S. P., Rowe, W. P., and Martin, M. A., 1982, Variation in the number of copies and in the genomic organization of ecotropic murine leukemia virus proviral sequences in sublines of AKR mice, *J. Virol.* **43**:629.

Calarco, P. G., and Szollosi, D., 1973, Intracisternal A particles in ova and preimplantation stages of the mouse, *Nature New Biol.* **243**:91.

Calarco, P. G., Callahan, R., Yasamura, T., and Hwang, T. T.-F., 1980, Pre-implantation mouse embryos of *Mus cervicolor* and *Mus pahari* express intracisternal-A particles, *J. Cell Biol.* **87**:A140.

Callahan, R., and Todaro, G. J., 1978, Four major endogenous retrovirus classes each genetically transmitted in various species of *Mus*, in: *Origins of Inbred Mice* (H. C. Morse III, ed.), pp. 689–713, Academic Press, New York.

Callahan, R., Lieber, M. N., and Todaro, G. J., 1975, Nucleic acid homology of murine xenotropic type C viruses, *J. Virol.* **15**:1378.

Callahan, R., Meade, C., and Todaro, G. J., 1979, Isolation of an endogenous type C virus related to the infectious primate type C viruses from the Asian rodent *Vandeleuria oleracea*, *J. Virol.* **30**:124.

Callahan, R., Kuff, E. L., Lueders, K. K., and Birkenmeier, E., 1981, Genetic relationship between the *Mus cervicolor* M432 retrovirus and the *Mus musculus* intracisternal type A particle, *J. Virol.* **40**:901.

Callahan, R., Drohan, W., Gallahan, D., D'Hoostelaere, L., and Potter, M., 1982, Novel class of mouse mammary tumor virus-related DNA sequences found in all species of *Mus*, including mice lacking the virus proviral genome, *Proc. Natl. Acad. Sci. USA* **79**:4113.

Canaani, E., Dreazen, O., Klar, A., Rechavi, G., Ram, D., Cohen, J. B., and Givol, D., 1983, Activation of the *c-mos* oncogene in a mouse plasmacytoma by insertion of an endogenous intracisternal A-particle genome, *Proc. Natl. Acad. Sci. USA* **80**:7118.

Casadevall, N., Lacombe, C., Gisselbrecht, S., and Mayeux, P., 1991, Multimeric structure of the membrane erythropoietin receptor of murine erythroleukemia cells (Friend cells). Cross-linking of erythropoietin with the spleen focus-forming virus envelope protein, *J. Biol. Chem.* **266**:16015.

Chan, H. W., Bryan, T., Moore, J. L., Staal, S. P., Rowe, W. P., and Martin, M. A., 1980, Identification of ecotropic proviral sequences in inbred mouse strains with a cloned subgenomic DNA fragment, *Proc. Natl. Acad. Sci. USA* **77**:5579.

Chang, J. M., Metcalf, D., Lang, R. A., Gonda, T. J., and Johnson, G. R., 1989, Nonneoplastic hematopoietic myeloproliferative syndrome induced by dysregulated multi-CSF (IL-3) expression, *Blood* **73**:1487.

Ch'ang, L.-Y., Yang, W. K., Myer, F. E., Koh, C. K., and Boone, L. R., 1989, Specific sequence deletions in two classes of murine leukemia virus-related proviruses in the mouse genome, *Virology* **168**:245.

Chatis, P. A., Holland, C. A., Hartley, J. W., and Rowe, W. P., 1983, Role for the 3' end of the genome in determining disease specificity of Friend and Moloney murine leukemia viruses, *Proc. Natl. Acad. Sci. USA* **80**:4408.

Chatis, P. A., Holland, C. A., Silver, J. E., Fredrickson, T. N., Hopkins, N., and Hartley, J. W., 1984, A 3' end fragment encompassing the transcriptional enhancers of nondefective Friend virus confers erythroleukemogenicity on Moloney leukemia virus, *J. Virol.* **52**:248.

Chattopadhyay, S. K., Rowe, W. P., Teich, N. M., and Lowy, D. R., 1975, Definitive evidence that the murine C-type virus inducing locus *Akv-1* is viral genetic material, *Proc. Natl. Acad. Sci. USA* **72**:906.

Chattopadhyay, S. K., Lander, M. R., Rands, E., and Lowy, D. R., 1980a, Structure of endogenous murine leukemia virus DNA in mouse genomes, *Proc. Natl. Acad. Sci. USA* **77**:5774.

Chattopadhyay, S. K., Lander, M. R., and Rowe, W. P., 1980b, Close similarity between endogenous ecotropic virus of *Mus musculus molossinus* and AKR virus, *J. Virol.* **36**:499.

Chattopadhyay, S. K., Oliff, A. I., Linemeyer, D. L., Lander, M. R., and Lowy, D. R., 1981, Genomes of murine leukemia viruses isolated from wild mice, *J. Virol.* **39**:777.

Chattopadhyay, S. K., Cloyd, M. W., Linemeyer, D. K., Lander, M. R., Rands, E., and Lowy, D. R., 1982a, Cellular origin and role of mink cell focus-forming viruses in murine thymic lymphomas, *Nature* **295**:25.

Chattopadhyay, S. K., Lander, M. R., Gupta, S., Rands, E., and Lowy, D. R., 1982b, Origin of mink cytopathic focus-forming (MCF) viruses: Comparison with ecotropic and xenotropic murine leukemia virus genomes, *Virology* **43**:416.

Chattopadhyay, S. K., Morse III, H. C., Makino, M., Ruscetti, S. K., and Hartley, J. W., 1989, Defective virus is associated with induction of murine retrovirus-induced immunodeficiency syndrome, *Proc. Natl. Acad. Sci. USA* **86**:3862.

Chesebro, B., and Wehrly, K., 1978, *Rfv-1* and *Rfv-2*, two *H-2*-associated genes that influence recovery from Friend leukemia virus-induced splenomegaly, *J. Immunol.* **120**:1081.

Chesebro, B., and Wehrly, K., 1979, Identification of a non-*H-2* gene (*Rfv-3*) influencing recovery from viremia and leukemia induced by Friend virus complex, *Proc. Natl. Acad. Sci. USA* **76**:425.

Chesebro, B., Portis, J. L., Wehrly, K., and Nishio, J., 1983, Effect of murine host genotype on MCF virus expression, latency, and leukemia cell type of leukemias induced by Friend murine leukemia helper virus, *Virology* **128**:221.

Chien, Y. H., Lai, M., Shih, T. Y., Verma, I. M., Scolnick, E. M., Roy-Burman, P., and Davidson, N., 1979, Heteroduplex analysis of the sequence relationships between the genomes of Kirsten and Harvey sarcoma viruses, their respective parental murine leukemia viruses and the rat endogenous 30S RNA, *J. Virol.* **31**:752.

Chirigos, M. A., Scott, D., Turner, W., and Perk, K., 1968, Biological, pathological and physical characterization of a possible variant of a murine sarcoma virus (Moloney), *Int. J. Cancer* **3**:223.

Chiu, I.-M., Callahan, R., Tronick, S. R., Schlom, J., and Aaronson, S. A., 1984, Major *pol* gene progenitors in the evolution of oncoviruses, *Science* **223**:364.

Choi, Y., Kappler, J. W., and Marrack, P., 1991, A superantigen encoded in the open reading frame of the 3' long terminal repeat of mouse mammary tumour virus, *Nature* **350**:203.

Chopra, H. C., Woodside, N. J., and Bogden, A. E., 1970, Virus particles in rat leukemias, *Cancer Res.* **30**:1544.

Christy, R. J., and Huang, R. C. C., 1988, Functional analysis of the long terminal repeats of intracisternal A-particle genes: Sequences within the U3 region determine both the efficiency and direction of promoter activity, *Mol. Cell. Biol.* **8**:1093.

Cianciolo, G. J., Copeland, T. D., Oroszlan, S., and Snyderman, R., 1985, Inhibition of lymphocyte proliferation by a synthetic peptide homologous to retroviral envelope proteins, *Science* **230**:453.

Cloyd, M. W., and Chattopadhyay, S. K., 1986, A new class of retrovirus present in many murine leukemia systems, *Virology* **151**:31.

Cloyd, M. W., Hartley, J. W., and Rowe, W. P., 1980, Lymphomagenicity of recombinant mink cell focus-inducing murine leukemia viruses, *J. Exp. Med.* **151**:542.

Cloyd, M. W., Thompson, M. M., and Hartley, J. W., 1985, Host range of mink cell focus-inducing viruses, *Virology* **140**:239.

Clynes, R., Wax, J., Stanton, L. W., Smith-Gill, S., Potter, M., and Marcu, K. B., 1988, Rapid induction of IgM-secreting murine plasmacytomas by pristane and immunoglobulin heavy-chain promoter/enhancer-driven c-*myc*/v-Ha-*ras* retrovirus, *Proc. Natl. Acad. Sci. USA* **85**:6067.

Cohen, J. C., and Varmus, H. E., 1979, Endogenous mammary tumour virus DNA varies among wild mice and segregates during inbreeding, *Nature* **278**:418.

Cole, M. D., Ono, M., and Huang, R. C. C., 1981, Terminally redundant sequences in cellular intracisternal A-particle genes, *J. Virol.* **38**:680.

Cook, W. D., 1982, Rapid thymomas induced by Abelson murine leukemia virus, *Proc. Natl. Acad. Sci. USA* **79**:2917.

Corcoran, L. M., Adams, J. M., Dunn, A. R., and Cory, S., 1984, Murine T lymphomas in which the cellular *myc* oncogene has been activated by retroviral insertion, *Cell* **37**:113.

Courtney, M. G., Elder, P. K., Steffen, D. L., and Getz, M. J., 1982, Evidence for an early evolutionary origin and locus polymorphism of mouse VL30 DNA sequences, *J. Virol.* **43**:511.

Curran, T., and Teich, N. M., 1982, Identification of a 39,000-dalton protein in cells transformed by the FBJ osteosarcoma virus, *Virology* **116**:221.

Curran, T., and Verma, I. M., 1984, FBR murine osteosarcoma virus: I. Molecular analysis and characterization of a 75000-Da *gag–fos* fusion product, *Virology* **135**:218.

Curran, T., Peters, G., van Beveren, C., Teich, N. M., and Verma, I. M., 1982, FBJ murine osteosarcoma virus: Identification and molecular cloning of biologically active proviral DNA, *J. Virol.* **44**:674.

Cuypers, H. T., Selten, G., Quint, W., Zijlstra, M., Maandag, E. R., Boelens, W., van Wezenbeek, P., Melief, C., and Berns, A., 1984, Murine leukemia virus-induced T-cell lymphomagenesis: Integration of proviruses in a distinct chromosomal region, *Cell* **37**:141.

Dahlberg, J. E., Tronick, S. R., and Aaronson, S. A., 1980, Immunological relationships of an endogenous guinea pig retrovirus with prototype mammalian type B and type D retroviruses, *J. Virol.* **33**:522.

Datta, S. K., and Schwartz, R. S., 1977, Mendelian segregation of loci controlling xenotropic virus production in NZB crosses, *Virology* **83**:449.

Datta, S. K., Tsichlis, P. N., Schwartz, R. S., Chattopadhyay, S. K., and Melief, C. J. M., 1978, Genetic differences unrelated to H-2 in H-2 congenic mice, *Immunogenetics* **7**:359.

Davis, B. R., Brightman, B. K., Chandy, K. G., and Fan, H., 1987, Characterization of a preleukemic state induced by Moloney murine leukemia virus: Evidence for two infection events during leukemogenesis, *Proc. Natl. Acad. Sci. USA* **84**:4875.

DesGroseillers, L., and Jolicoeur, P., 1983, Physical mapping of the *Fv-1* tropism host range determinant of BALB/c murine leukemia viruses, *J. Virol.* **48**:685.

DesGroseillers, L., and Jolicoeur, P., 1984, The tandem direct repeats within the long terminal repeat of murine leukemia viruses are the primary determinant of their leukemogenic potential, *J. Virol.* **52**:945.

DesGroseillers, L., Rassart, E., and Jolicoeur, P., 1983, Thymotropism of murine leukemia virus is conferred by its long terminal repeat, *Proc. Natl. Acad. Sci. USA* **80**:4203.

DesGroseillers, L., Barrette, M., and Jolicoeur, P., 1984, Physical mapping of the paralysis-inducing determinant of a wild mouse ecotropic neurotropic virus, *J. Virol.* **52**:356.

DesGroseillers, L., Rassart, E., Robitaille, Y., and Jolicoeur, P., 1985, Retrovirus-induced spongiform encephalopathy: The 3'-end long terminal repeat-containing viral sequences influence the incidence of disease and the specificity of the neurological syndrome, *Proc. Natl. Acad. Sci. USA* **82**:8818.

Dickson, C., and Peters, C., 1987, Potential oncogene product related to growth factor, *Nature* **326**:833.

Dickson, C., Smith, R., and Peters, G., 1981, *In vitro* synthesis of polypeptides encoded by the long terminal repeat region of mouse mammary tumour virus DNA, *Nature* **291**:511.

Donehower, L. A., Huang, A., and Hager, G. L., 1981, Regulatory and coding potential of the mouse mammary tumor virus long terminal redundancy, *J. Virol.* **37**:226.

Donner, P., Greiser-Wilke, I., and Moelling, K., 1982, Nuclear localization and DNA binding of the transforming gene product of avian myelocytomatosis virus, *Nature* **296**:262.

Donoghue, D. J., Sharp, P. A., and Weinberg, R. A., 1979, Comparative study of different isolates of murine sarcoma viruses, *J. Virol.* **32**:1015.

Dorner, A. J., Bonneville, F., Kriz, R., Kelleher, K., Bean, K., and Kaufman, R. J., 1991, Molecular cloning and characterization of a complete Chinese hamster provirus related to intracisternal A particle genomes, *J. Virol.* **65**:4713.

Dresler, S., Ruta, M., Murray, M. J., and Kabat, D., 1979, Glycoprotein encoded by the Friend spleen focus-forming virus, *J. Virol.* **30**:564.

Dunkel, V. C., Bast, R. C., Jr., Gerwin, B. I., Heine, U., Cottler-Fox, M., and Borsos, T., 1974, Presence of A-type and absence of C-type virus particles in a chemically induced guinea pig hepatoma, *J. Natl. Cancer Inst.* **53**:597.

Dyson, P. J., Knight, A. M., Fairchild, S., Simpson, E., and Tomonari, K., 1991, Genes encoding ligands for deletion of Vβ 11 T cells cosegregate with mammary tumour virus genomes, *Nature* **349**:531.

Eicher, E. M., and Lee, B. K., 1990, The NXSM recombinant inbred strains of mice: Genetic profile for 58 loci including the *Mtv* proviral loci, *Genetics* **125**:431.

Eicher, E. M., Hutchinson, K. W., Phillips, S. J., Tucker, P. K., and Lee, B. K., 1989, A repeated segment on the mouse Y chromosome is composed of retroviral-related, Y-enriched and Y-specific sequences, *Genetics* **122**:181.

Elder, J. H., Jensen, F. C., Bryant, M. L., and Lerner, R. A., 1977, Polymorphism of the major envelope glycoprotein (gp70) of murine C-type viruses: Virion associated and differentiation antigens encoded by a multi-gene family, *Nature* **267**:23.

Ellis, R. W., DeFeo, D., Maryak, J. M., Young, H. A., Shih, T. Y., Change, E. H., Lowy, D. R., and Scolnick, E. M., 1980, Dual evolutionary origin for the rat genomic sequences of Harvey murine sarcoma virus, *J. Virol.* **36**:408.

Evans, L. H., and Cloyd, M. W., 1985, Friend and Moloney murine leukemia viruses specifically recombine with different endogenous viruses to generate mink cell focus-forming viruses, *Proc. Natl. Acad. Sci. USA* **82**:459.

Famulari, N. G., 1983, Murine leukemia viruses with recombinant env genes: A discussion of their role in leukemogenesis, Curr. Top. Microbiol. Immunol. 103:76.

Famulari, N. G., Koehne, C. F., and O'Donnell, P. V., 1982, Leukemogenesis by Gross passage A murine leukemia virus: Expression of viruses with recombinant env genes in transformed cells, Proc. Natl. Acad. Sci. USA 79:3872.

Finkel, M. P., Biskis, B. O., and Jinkins, P. B., 1966, Virus induction of osteosarcomas in mice, Science 151:698.

Finkel, M. P., Reilly, Jr., C. A., Biskis, B. O., and Greco, I. L., 1973, Bone tumor viruses, in: Bone—Certain Aspects of Neoplasia (C. H. G. Price and F. G. M. Ross, eds.), pp. 353–366, Butterworths, London.

Finkel, M. P., Reilly, Jr., C. A., and Biskis, B. O., 1975, Viral etiology of bone cancer, Front. Radiat. Ther. Oncol. 10:28.

Fischinger, P. J., Nomura, S., and Bolognesi, D. P., 1975, A novel murine oncornavirus with dual eco- and xenotropic properties, Proc. Natl. Acad. Sci. USA 72:5150.

Frankel, A. E., and Fischinger, P. J., 1976, Nucleotide sequences in mouse DNA and RNA specific for Moloney sarcoma virus, Proc. Natl. Acad. Sci. USA 73:3705.

Frankel, W. N., Stoye, J. P., Taylor, B. A., and Coffin, J. M., 1989a, Genetic analysis of endogenous xenotropic murine leukemia viruses: Association with two common mouse mutations and the viral restriction locus Fv-1, J. Virol. 63:1763.

Frankel, W. N., Stoye, J. P., Taylor, B. A., and Coffin, J. M., 1989b, Genetic identification of endogenous polytropic proviruses by using recombinant inbred mice, J. Virol. 63:3810.

Frankel, W. N., Stoye, J. P., Taylor, B. A., and Coffin, J. M., 1990, A linkage map of endogenous murine leukemia proviruses, Genetics 124:221.

Frankel, W. N., Rudy, C., Coffin, J. M., and Huber, B. T., 1991, Mls genes are linked to endogenous mouse mammary tumor viruses of inbred mice, Nature 349:526.

Franz, T., Löhler, J., Fusco, A., Pragnell, I., Nobis, P., Padua, R., and Ostertag, W., 1985, Transformation of mononuclear phagocytes in vivo and malignant histiocytosis caused by a novel murine spleen focus-forming virus, Nature 315:149.

Fredrickson, T. N., Langdon, W. Y., Hoffman, P. M., Hartley, J. W., and Morse III, H. C., 1984, Histologic and cell surface antigen studies of hematopoietic tumors induced by Cas-Br-M murine leukemia virus, J. Natl. Cancer Inst. 72:447.

Fredrickson, T. N., O'Neill, R. R., Rutledge, R. A., Theodore, T. S., Martin, M. A., Ruscetti, S. K., Austin, J. B., and Hartley, J. W., 1987, Biologic and molecular characterization of two newly isolated ras-containing murine leukemia viruses, J. Virol. 61:2109.

Friel, J., Hughes, D., Pragnell, I., Stocking, C., Laker, C., Nowock, J., Ostertag, W., and Padua, R. A., 1990, The malignant histiocytosis sarcoma virus, a recombinant of Harvey murine sarcoma virus and Friend mink cell focus-forming virus, has acquired myeloid transformation specificity by alterations in the long terminal repeat, J. Virol. 64:369.

Friend, C., 1957, Cell-free transmission in adult Swiss mice of a disease having the character of a leukemia, J. Exp. Med. 105:307.

Gallahan, D., Kozak, C., and Callahan, R., 1986, A new common integration region (int-3) for the mouse mammary tumor virus on mouse chromosome 17, J. Virol. 61:218.

Gardner, M. B., 1978, Type C viruses of wild mice: Characterization and natural history of amphotropic, ecotropic, and xenotropic MuLV, Curr. Top. Microbiol. Immunol. 79:215.

Gardner, M. B., Henderson, B. E., Officer, J. E., Rongey, R. W., Parker, J. C., Oliver, C., Ester, J. D., and Huebner, R. J., 1973, A spontaneous lower motor neuron disease apparently caused by indigenous type-C RNA virus in wild mice, J. Natl. Cancer Inst. 51:1243.

Gardner, M. B., Chivi, A., Daugherty, M. F., Casagrande, J., and Estes, J. D., 1979, Congenital transmission of murine leukemia virus from wild mice prone to the development of lymphoma and paralysis, *J. Natl. Cancer Inst.* **62**:63.

Gardner, M. B., Rasheed, S., Pal, B. K., Estes, J. D., and O'Brien, S. J., 1980, *Akvr-1*, a dominant murine leukemia virus restriction gene, is polymorphic in leukemia-prone wild mice, *Proc. Natl. Acad. Sci. USA* **77**:531.

Gautsch, J. W., Elder, J. H., Schindler, J., Jansen, F. C., and Lerner, R. A., 1978, Structural markers on core protein p30 of murine leukemia virus: Functional correlation with *Fv-1* tropism, *Proc. Natl. Acad. Sci. USA* **75**:4170.

Gautsch, J. W., Elder, J. H., Jensen, F. C., and Lerner, R. A., 1980, *In vitro* construction of a B-tropic virus by recombination: B-tropism is a cryptic phenotype of xenotropic murine retroviruses, *Proc. Natl. Acad. Sci. USA* **77**:2989.

Gazdar, A. F., Chopra, H. C., and Sarma, P. S., 1972, Properties of a murine sarcoma virus isolated from a tumor arising in an NZW/NZB F_1 hybrid mouse. I. Isolation and pathology of tumors induced in rodents, *Int. J. Cancer* **9**:219.

Gazdar, A. F., Oie, H., Lalley, P. A., Moss, W., Minna, J. D., and Francke, U., 1977, Identification of mouse chromosomes required for murine leukemia virus replication, *Cell* **11**:949.

Gisselbrecht, S., Pozo, F., Debre, P., Hurot, M. A., Lacombe, M. J., and Levy, J. P., 1978, Genetic control of sensitivity to Moloney-virus-induced leukemias in mice, I. Demonstration of multigenic control, *Int. J. Cancer* **21**:626.

Gisselbrecht, S., Fischinger, P. J., Elder, J. H., and Levy, J. P., 1981, Isolation of two novel recombinant leukemogenic viruses from Moloney leukemia virus stocks, *Int. J. Cancer* **27**:531.

Gisselbrecht, S., Fichelson, S., Sola, B., Bordereaux, D., Hampe, A., Andre, C., Galibert, F., and Tambourin, P., 1987, Frequent c-*fms* activation by proviral insertion in mouse myeloblastic leukaemias, *Nature* **329**:259.

Goebl, M. G., Moreau-Gachelin, F., Ray, D., Tambourin, P., Tavitain, A., Klemsz, M. J., McKercher, S. R., Celada, A., Van Beveren, C., and Maki, R. A., 1990, The PU.1 transcription factor is the product of the putative oncogene *Spi-1*, *Cell* **61**:1165.

Goff, S. P., Gilboa, E., Witte, O. N., and Baltimore, D., 1980, Structure of the Abelson murine leukemia virus genome and the homologous cellular gene: Studies with cloned viral DNA, *Cell* **22**:777.

Goff, S. P., Witte, O. N., Gilboa, E., Rosenberg, N., and Baltimore, D., 1981, Genome structure of Abelson murine leukemia virus variants: Proviruses in fibroblasts and lymphoid cells, *J. Virol.* **38**:460.

Gokhman, I., Peled, A., and Haran-Ghera, N., 1990, Characteristics of potential lymphoma-inducing cells in mice sensitive or resistant to lymphomagenesis by radiation leukemia virus variants, *Cancer Res.* **9**:2554.

Gonda, T. J., Cory, S., Sobieszczuk, P., and Holtzman, D., 1987, Generation of altered transcripts by retroviral insertion within the c-*myb* gene in two murine monocytic leukemias, *J. Virol.* **61**:2754.

Graham, M., Adams, J. M., and Cory, S., 1985, Murine T lymphomas with retroviral inserts in the chromosomal 15 locus for plasmacytoma variant translocations, *Nature* **314**:740.

Green, P. L., Kaehler, D. A., Bennett, L. M., and Risser, R., 1989, Multiple steps are required for the induction of tumors by Abelson murine leukemia virus, *J. Virol.* **63**:1989.

Grohmann, U., Ullrich, S. J., Mage, M. G., Appella, E., Fioretti, M. C., Puccetti, P., and Romani, L., 1990, Identification and immunogenic properties of an 80-kDa surface antigen on a drug-treated tumor variant: Relationship to MuLV gp70, *Eur. J. Immunol.* **20**:629.

Gross, L., 1957, Development and serial cell-free passage of a highly potent strain of mouse leukemia virus, *Proc. Soc. Exp. Biol. Med.* **94**:767.

Haapala, D. K., Robey, W. G., Oroszlan, S. D., and Tsai, W. P., 1985, Isolation from cats of an endogenous type C virus with a novel envelope glycoprotein, *J. Virol.* **53**:827.

Haas, M., 1980, B-cell and T-cell malignant lymphomas in C57BL/6 mice induced by different recombinant retroviruses isolated from irradiated mice, in: *Abstracts of Papers Presented at the 7th Cold Spring Harbor Conference on Cell Proliferation: Viruses in Naturally Occurring Cancers*, p. 1073, Cold Spring Harbor Laboratory, Cold Spring Harbor, New York.

Haas, M., and Patch, V., 1980, Genomic masking and rescue of dual-tropic murine leukemia viruses: Role of pseudotype virions in viral lymphomagenesis, *J. Virol.* **35**:583.

Haas, M., Altman, A., Rothenberg, E., Bogart, M. H., and Jones, O. W., 1984, Mechanism of T-cell lymphomagenesis: Transformation of growth-factor-dependent T-lymphoblastoma cells to growth-factor-independent T-lymphoma cells, *Proc. Natl. Acad. Sci. USA* **81**:1742.

Haran-Ghera, N., 1980, Potential leukemic cells among bone marrow cells of young AKR/J mice, *Proc. Natl. Acad. Sci. USA* **77**:2923.

Harbers, K., Soriano, P., Muller, U., and Jaenisch, R., 1984, Insertion of retrovirus into the first intron of alpha 1(I) collagen gene leads to embyonic lethal mutation in mice, *Proc. Natl. Acad. Sci. USA* **81**:1504.

Hartley, J. W., and Rowe, W. P., 1975, Clonal cell lines from a feral mouse embryo which lack host-range restriction for murine leukemia viruses, *Virology* **65**:128.

Hartley, J. W., and Rowe, W. P., 1976, Naturally occurring murine leukemia viruses in wild mice. Characterization of a new "amphotropic" class, *J. Virol.* **19**:19.

Hartley, J. W., Rowe, W. P., and Huebner, R. J., 1970, Host-range restrictions of murine leukemia viruses in mouse embryo cell cultures, *J. Virol.* **5**:221.

Hartley, J. W., Wolford, N. K., Old, L. J., and Rowe, W. P., 1977, A new class of murine leukemia virus associated with development of spontaneous lymphomas, *Proc. Natl. Acad. Sci. USA* **74**:789.

Hartley, J. W., Yetter, R. A., and Morse III, H. C., 1983, A mouse gene on chromosome 5 that restricts infectivity of mink cell focus-forming recombinant murine leukemia viruses, *J. Exp. Med.* **158**:16.

Harvey, J. J., 1964, An unidentified virus which causes the rapid production of tumors in mice, *Nature* **204**:1104.

Hays, E. F., 1982, Bone marrow progenitor cells of AKR mice give rise to thymic lymphoma cells, *Leuk. Res.* **6**:429.

Hays, E. F., Bristol, G. O., Lugo, J. P., and Wang, X. F., 1989, Progression to development of lymphoma in the thymus of AKR mice treated neonatally with SL 3-3 virus, *Exp. Hematol.* **11**:1116.

Heidmann, O., and Heidmann, T., 1991, Retrotransposition of a mouse IAP sequence tagged with an indicator gene, *Cell* **64**:159.

Heine, U. O., Margulies, I., Demsey, A. E., and Suskind, R. G., 1979, Quantitative electron microscopy of intracytoplasmic type A particles at kinetochores of metaphase chromosomes isolated from Chinese hamster and murine cell lines, *J. Gen. Virol.* **45**:631.

Herr, W., 1984, Nucleotide sequence of AKV murine leukemia virus, *J. Virol.* **49**:471.

Hevezi, P., and Goff, S. P., 1991, Generation of recombinant murine retroviral genomes containing the v-src oncogene: Isolation of a virus inducing hemangiosarcomas in the brain, *J. Virol.* **65**:5333.

Hilgers, J., and Bentvelzen, P., 1978, Interaction between viral and genetic factors in murine mammary cancer, *Adv. Cancer Res.* **26**:143.

Hilkens, J., Cuypers, H. T., Selten, G., Kroezen, V., Hilgers, J., and Berns, A., 1986, Genetic mapping of *Pim-1* putative oncogene to mouse chromosome 17, *Somat. Cell Mol. Genet.* **12**:81.

Hoatlin, M. E., Kozak, S. L., Lilly, F., Chakraborti, A., Kozak, C. A., and Kabat, D., 1990, Activation of erythropoietin receptors by Friend viral gp55 and by erythropoietin and down-modulation by the murine Fv-2r resistance gene, *Proc. Natl. Acad. Sci. USA* **87**:9985.

Hoffman, P. M., Davidson, W. F., Ruscetti, S. K., Chused, T. M., and Morse III, H. C., 1981, Wild mouse ecotropic murine leukemia virus infection of inbred mice: Dual-tropic virus expression precedes the onset of paralysis and lymphoma, *J. Virol.* **39**:597.

Hoffman, P. M., Robbins, D. S., and Morse III, H. C., 1984, Role of immunity in age-related resistance to paralysis following murine leukemia virus infection, *J. Virol.* **52**:734.

Hoggan, M. D., Buckler, C. E., Sears, J. F., Rowe, W. P., and Martin, M. A., 1983, Organization and stability of endogenous xenotropic murine leukemia virus proviral DNA in mouse genomes, *J. Virol.* **45**:473.

Hoggan, M. D., O'Neill, R. R., and Kozak, C. A., 1986, Non-ecotropic murine leukemia viruses in BALB/c and NFS/N mice: Characterization of the BALB/c *Bxv-1* provirus and the single NFS endogenous xenotrope, *J. Virol.* **60**:980.

Hojman, F., Emanoil-Ravier, R., Lesser, J., and Périès, J., 1988, Biological and molecular characterization of an endogenous retrovirus present in CHO/HBs-A Chinese hamster cell line, *Dev. Biol. Stand.* **70**:195.

Holland, C. A., Wozhey, J., and Hopkins, N., 1983, The nucleotide sequence of the gp70 gene of murine retrovirus MCF 247, *J. Virol.* **47**:413.

Holmes, K. L., Langdon, W. Y., Fredrickson, T. N., Coffman, R. L., Hoffman, P. M., Hartley, J. W., and Morse III, H. C., 1986, Analysis of neoplasms induced by Cas-Br-M MuLV tumor extracts, *J. Immunol.* **137**:670.

Honacki, J. E., Kinman, K. E., and Koeppl, J. W., 1982, *Mammal Species of the World: A Taxonomic and Geographic Reference*, Allen Press and Association of Systematic Collections, Lawrence, Kansas.

Hopkins, N., and Jolicoeur, P., 1975, Variants of N-tropic leukemia virus derived from BALB/c mice, *J. Virol.* **16**:991.

Horowitz, J. M., and Risser, R., 1985, Molecular and biological characterization of the endogenous ecotropic provirus of BALB/c mice, *J. Virol.* **56**:798.

Howk, R. S., Troxler, D. H., Lowy, D., Duesberg, P. H., and Scolnick, E. M., 1978, Identification of a 30S RNA with properties of defective type C virus in murine cells, *J. Virol.* **25**:115.

Hu, S., Davidson, N., and Verma, I. M., 1977, A heteroduplex study of the sequence relationships between the RNAs of M-MSV and M-MLV, *Cell* **10**:469.

Huang, M., and Jolicoeur, P., 1990, Characterization of the *gag*/fusion protein encoded by the defective Duplan retrovirus inducing murine acquired immunodeficiency syndrome, *J. Virol.* **64**:5764.

Huang, M., Simard, C., and Jolicoeur, P., 1984, Immunodeficiency and clonal growth of target cells induced by helper-free defective retrovirus, *Science* **246**:1614.

Hügin, A. W., Vacchio, M. S., and Morse, H. C., 1991, A virus-encoded "superantigen" in a retrovirus-induced immunodeficiency syndrome of mice, *Science* **252**:424.

Ihle, J. N., Keller, J., Oroszlan, S., Henderson, L. E., Copeland, T. D., Fitch, F., Prystowsky, M. B., Goldwasser, E., Schrader, J. W., Palaszynski, E., Dy, M., and Lebel, B., 1983, Biologic properties of homogeneous interleukin 3, *J. Immunol.* **131**:282.

Ikeda, H., and Odaka, T., 1983, Cellular expression of murine leukemia virus gp70-related antigen on thymocytes of uninfected mice correlates with *Fv-4* gene-controlled resistance to Friend leukemia virus infection, *Virology* **128**:127.

Inaguma, Y., Miyashita, N., Moriwaki, K., Huai, W. C., Mei-Lei, J., Xinquiao, H., and Ikeda, H., 1991, Acquisition of two endogenous ecotropic murine leukemia viruses in distinct Asian wild mouse populations, *J. Virol.* **65**:1796.

Ishimoto, A., Adachi, A., Yorifuji, T., and Tsuruta, S., 1981, Rapid emergence of mink cell focus-forming (MCF) virus in various mice infected with NB-tropic Friend virus, *Virology* **113**:644.

Itin, A., and Keshet, E., 1983a, Nucleotide sequence analysis of the long terminal repeat of murine virus-like DNA (VL30) and its adjacent sequences: Resemblance to retrovirus proviruses, *J. Virol.* **47**:656.

Itin, A., and Keshet, E., 1983b, Apparent recombinants between virus-like (VL30) and murine leukemia virus-related sequences in mouse DNA, *J. Virol.* **47**:178.

Itin, A., and Keshet, E., 1986, A novel retroviruslike family in mouse DNA, *J. Virol.* **59**:301.

Itin, A., Rotman, G., and Keshet, E., 1983, Conservation patterns of mouse "virus-like" (VL30) DNA sequences, *Virology* **127**:374.

Izui, S., McConahey, P. J., Theofilopoulos, A. N., and Dixon, F. J., 1979, Association of circulating retroviral gp70–anti-gp70 immune complexes with murine systemic lupus erythematosus, *J. Exp. Med.* **149**:1099.

Izui, S., McConahey, P. J., Clark, J. P., Hang, L. M., Hara, I., and Dixon, F. J., 1981, Retroviral gp70 immune complexes in NZBxNZW F_2 mice with murine lupus nephritis, *J. Exp. Med.* **154**:517.

Jaenisch, R., 1976, Germ line integration and Mendelian transmission of exogenous Moloney leukemia virus, *Proc. Natl. Acad. Sci. USA* **73**:1260.

Jenkins, N. A., and Copeland, N. G., 1985, High frequency germline acquisition of ecotropic MuLV proviruses in SWR/J–RF/J hybrid mice, *Cell* **43**:811.

Jenkins, N. A., Copeland, N. G., Taylor, B. A., and Lee, B. K., 1981, Dilute (*d*) coat colour mutation of DBA/2J mice is associated with the site of integration of an ecotropic MuLV genome, *Nature* **293**:370.

Jenkins, N. A., Copeland, N. G., Taylor, B. A., and Lee, B. K., 1982, Organization, distribution, and stability of endogenous ecotropic murine leukemia virus DNA sequences in chromosomes of *Mus musculus*, *J. Virol.* **43**:26.

Jolicoeur, P., and Baltimore, D., 1976, Effect of *Fv-1* gene product on proviral DNA formation and integration in cells infected with murine leukemia viruses, *Proc. Natl. Acad. Sci. USA* **73**:2236.

Jolicoeur, P., and DesGroseillers, L., 1985, Neurotropic Cas-BR-E murine leukemia virus harbors several determinants of leukemogenicity mapping in different regions of the genome, *J. Virol.* **56**:639.

Jolicoeur, P., and Rassart, E., 1980, Effect of *Fv-1* gene product on synthesis of linear and supercoiled viral DNA in cells infected with murine leukemia virus, *J. Virol.* **33**:183.

Jolicoeur, P., Rosenberg, N., Cotellessa, A., and Baltimore, D., 1978, Leukemogenicity of clonal isolates of murine leukemia viruses, *J. Natl. Cancer Inst.* **60**:1473.

Jolicoeur, P., Rassart, E., Villeneuve, L., and Kozak, C., 1985, Mouse chromosomal mapping of a murine leukemia virus integration region (*Mis-1*) first identified in rat thymic leukemia, *J. Virol.* **56**:1045.

Kai, K., and Furuta, T., 1984, Isolation of paralysis-inducing murine leukemia viruses from Friend virus passaged in rats, *J. Virol.* **50**:970.

Kelloff, G., Huebner, R. J., Lee, Y. K., Toni, R., and Gilden, R., 1970, Hamster-tropic sarcomagenic and nonsarcomagenic viruses derived from hamster tumors induced by the Gross pseudotype of Moloney sarcoma virus, *Proc. Natl. Acad. Sci. USA* **65**:310.

Kelly, M., Holland, C. A., Lung, M. L., Chattopadhyay, S. K., Lowy, D. R., and Hopkins, N., 1983, Nucleotide sequence of the 3' end of MCF 247 murine leukemia virus, *J. Virol.* **45**:291.

Keshet, E., and Itin, A., 1982, Patterns of genomic distribution and sequence heterogeneity of a murine "retrovirus-like" multigene family, *J. Virol.* **43**:50.

Keshet, E., and Shaul, Y., 1981, Terminal direct repeats in a retrovirus-like repeated mouse gene family, *Nature* **289**:83.

Keshet, E., Shaul, Y., Kaminchik, J., and Aviv, H., 1980, Heterogeneity of virus-like genes encoding retrovirus-associated 30S RNA and their organization within the mouse genome, *Cell* **20**:431.

Khan, A. S., and Martin, M. A., 1983, Endogenous murine leukemia proviral long terminal repeats contain a unique 190-base-pair insert, *Proc. Natl. Acad. Sci. USA* **80**:2699.

Khan, A. S., Rowe, W. P., and Martin, M. A., 1982, Cloning of endogenous murine leukemia virus-related sequences from chromosomal DNA of BALB/c and AKR/J mice: Identification of an *env* progenitor of AKR-247 mink cell focus-forming proviral DNA, *J. Virol.* **44**:625.

Kim, J. W., Closs, E. I., Albritton, L. M., and Cunningham, J. M., 1991, Transport of cationic amino acids by the mouse ecotropic retrovirus receptor, *Nature* **353**:725.

King, S. R., Horowitz, J. M., and Risser, R., 1987, Nucleotide conservation of murine leukemia proviruses in inbred mice: Implication for viral origin and dispersal, *Virology* **157**:543.

Kirsten, W. H., and Mayer, L. A., 1967, Morphologic responses to a murine erythroblastosis virus, *J. Natl. Cancer Inst.* **39**:311.

Klement, V., Rowe, W. P., Hartley, J. W., and Pugh, W. E., 1969, Mixed culture cytopathogenicity: A new test for growth of murine leukemia viruses in tissue culture, *Proc. Natl. Acad. Sci. USA* **63**:753.

Klement, V., Nicolson, M. O., Nelson-Rees, W., and Gilden, R. V., 1973, Spontaneous production of a C-type RNA virus in rat tissue culture lines, *Int. J. Cancer* **12**:654.

Klinken, S. P., Fredrickson, T. N., Hartley, J. W., Yetter, R. A., and Morse III, H. C., 1988, Evolution of B cell lineage lymphomas in mice with a retrovirus-induced immunodeficiency syndrome, MAIDS, *J. Immunol.* **140**:1123.

Koch, W., Hunsmann, G., and Friedrich, R., 1983, Nucleotide sequence of the envelope gene of Friend murine leukemia virus, *J. Virol.* **45**:1.

Koch, W., Zimmerman, W., Oliff, A., and Friedrich, R., 1984, Molecular analysis of the envelope gene and long terminal repeat of Friend mink cell focus-inducing virus: Implications for the functions of these sequences, *J. Virol.* **49**:828.

Kollek, R., Stocking, C., Smadja-Joffe, F., and Ostertag, W., 1984, Molecular cloning and characterization of a leukemia-inducing myeloproliferative sarcoma virus and two of its temperature-sensitive mutants, *J. Virol.* **50**:717.

Kozak, C. A., 1983, Genetic mapping of a mouse chromosomal locus required for mink cell focus-forming virus replication, *J. Virol.* **48**:300.

Kozak, C. A., 1985a, Analysis of wild-derived mice for *Fv-1* and *Fv-2* murine leukemia virus restriction loci: A novel wild mouse *Fv-1* allele responsible for lack of host range restriction, *J. Virol.* **55**:281.

Kozak, C. A., 1985b, Susceptibility of wild mouse cells to exogenous infection with xenotropic leukemia viruses: Control by a single dominant locus on chromosome 1, *J. Virol.* **55**:690.

Kozak, C. A., and O'Neill, R. R., 1987, Diverse wild mouse origins of xenotropic, mink cell focus-forming, and two types of ecotropic proviral genes, *J. Virol.* **61**:3082.

Kozak, C. A., and Rowe, W. P., 1980, Genetic mapping of xenotropic murine leukemia virus-inducing loci in 5 mouse strains, *J. Exp. Med.* **152**:219.

Kozak, C. A., and Rowe, W. P., 1982, Genetic mapping of ecotropic murine leukemia inducing loci in six inbred strains, *J. Exp. Med.* **155**:524.

Kozak, C. A., Hartley, J. W., and Morse III, H. C., 1984, Laboratory and wild-derived mice with multiple loci for production of xenotropic murine leukemia virus, *J. Virol.* **51**:77.

Kozak, C., Peters, G., Pauley, R., Morris, V., Michalides, R., Dudley, J., Green, M., Davisson, M., Prakash, O., Vaidya, A., Hilgers, J., Verstraeten, A., Hynes, N., Diggelmann, H., Peterson, D., Cohen, J. C., Dickson, C., Sarkar, N., Nusse, R., Varmus, H., and Callahan, R., 1987, A standardized nomenclature for endogenous mouse mammary tumor viruses, *J. Virol.* **61**:1651.

Krieg, A. M., and Steinberg, A. D., 1990, Analysis of thymic endogenous retroviral expression in murine lupus, *J. Clin. Invest.* **86**:809.

Krieg, A. M., Steinberg, A. D., and Khan, A. S., 1988, Increased expression of novel full-length endogenous mink cell focus-forming-related transcripts in autoimmune mouse strains, *Virology* **162**:274.

Kuff, E. L., and Lueders, K. K., 1988, The intracisternal A-particle gene family: Structure and functional aspects, *Adv. Cancer Res.* **51**:183.

Kuff, E. L., Lueders, K. K., Ozer, H. L., and Wivel, N. A., 1972, Some structural and antigenic properties of intracisternal A particles occurring in mouse tumors, *Proc. Natl. Acad. Sci. USA* **69**:218.

Kuff, E. L., Lueders, K. K., and Scolnick, E. M., 1978, Nucleotide sequence relationship between intracisternal type A particles of *Mus musculus* and an endogenous retrovirus (M432) of *Mus. cervicolor*, *J. Virol.* **28**:66.

Kuff, E. L., Smith, L. A., and Lueders, K. K., 1981, Intracisternal A-particle genes in *Mus musculus:* A conserved family of retrovirus-like elements, *Mol. Cell. Biol.* **1**:216.

Lai, M. M. C., Rasheed, S., Shimizu, C. S., and Gardner, M. B., 1982, Genomic characterization of highly oncogenic *env* gene recombinant between amphotropic retrovirus of wild mouse and endogenous xenotropic virus of NIH Swiss mouse, *Virology* **117**:262.

Lamont, C., Culp, P., Talbott, R. L., Phillips, T. R., Trauger, R. J., Frankel, W. N., Wilson, M. C., Coffin, J. M., and Elder, J. H., 1991, Characterization of endogenous and recombinant proviral elements of a highly tumorigenic AKR cell line, *J. Virol.* **65**:4619.

Langdon, W. Y., Hoffman, P. M., Silver, J. E., Buckler, C. E., Hartley, J. W., Ruscetti, S. K., and Morse III, H. C., 1983, Identification of a spleen focus-forming virus in erythroleukemic mice infected with a wild-mouse ecotropic murine leukemia virus, *J. Virol.* **46**:230.

Langdon, W. Y., Theodore, T. S., Buckler, C. E., Stimpfling, J. H., Martin, M. A., and Morse III, H. C., 1984, Relationship between a retroviral germ line reintegration and a new mutation at the ashen locus in B10.F mice. Retroviral integration and an ashen mutation, *Virology* **133**:183.

Langdon, W. Y., Hartley, J. W., Klinken, S. P., Ruscetti, S. K., and Morse III, H. C., 1989, v-cbl, an oncogene from a dual-recombinant murine retrovirus that induces early B-lineage lymphomas, *Proc. Natl. Acad. Sci. USA* **86**:1168.

Laterjet, R., and Duplan, J. F., 1962, Experiments and discussion on leukemogenesis by cell-free extracts of radiation-induced leukemia in mice, *Int. J. Radiat. Biol.* **5**:339.

Lee, B. K., and Eicher, E. M., 1990, Segregation patterns of endogenous mouse mammary tumor viruses in five recombinant inbred strain sets, *J. Virol.* **64**:4568.

Lenz, J., and Haseltine, W. A., 1983, Localization of the leukemogenic determinants of SL3-3, an ecotropic, XC-positive murine leukemia virus of AKR mouse origin, *J. Virol.* **47**:317.

Levy, D. E., Lerner, R. A., and Wilson, M. C., 1985, The *Gv-1* locus coordinately regulates the expression of multiple endogenous murine retroviruses, *Cell* **41**:289.

Levy, J. A., 1973, Xenotropic viruses: Murine leukemia viruses associated with NIH Swiss, NZB, and other mouse strains, *Science* **182**:1151.

Levy, J. A., 1974, Autoimmunity and neoplasia. The possible role of C-type viruses, *Am. J. Clin. Pathol.* **62**:258.

Levy, J. A., 1975, Host range of murine xenotropic virus: Replication in avian cells, *Nature* **253**:140.

Levy, J. A., 1978, Xenotropic type C viruses, *Curr. Top. Microbiol. Immunol.* **79**:111.

Levy, J. A., and Pincus, T., 1970, Demonstration of biological activity of a murine leukemia virus of New Zealand Black mice, *Science* **170**:326.

Levy, J. A., Hartley, J. W., Rowe, W. P., and Huebner, R. J., 1973, Studies of FBJ osteosarcoma virus in tissue culture. I. Biologic characteristics of the "C"-type viruses, *J. Natl. Cancer Inst.* **51**:525.

Li, J., D'Andrea, A. D., Lodish, H. F., and Baltimore, D., 1990, Activation of cell growth by binding of Friend spleen focus-forming virus gp55 glycoprotein to the erythropoietin receptor, *Nature* **343**:762.

Li, J.-P., and Baltimore, D., 1991, Mechanism of leukemogenesis induced by mink cell focus-forming murine leukemia viruses, *J. Virol.* **65**:2408.

Li, Y., Holland, C. A., Hartley, J. W., and Hopkins, N., 1984, Viral integrations near c-*myc* in 10–20% of MCF 247-induced AKR lymphomas, *Proc. Natl. Acad. Sci. USA* **81**:6808.

Lieber, M. M., Benveniste, R. E., Livingston, D. M., and Todaro, G. J., 1973, Mammalian cells in culture frequently release type C viruses, *Science* **182**:56.

Lieber, M., Sherr, C., Potter, M., and Todaro, G., 1975, Isolation of type-C viruses from the Asian feral mouse *Mus musculus molossinus*, *Int. J. Cancer* **15**:211.

Lieberman, M., and Kaplan, H. S., 1959, Leukemogenic activity of filtrates from radiation-induced lymphoid tumor of mice, *Science* **130**:387.

Lilly, F., 1967, Susceptibility to two strains of Friend leukemia virus in mice, *Science* **155**:461.

Lilly, F., 1970, *Fv-2:* Identification and location of a second gene governing the spleen focus response to Friend leukemia virus in mice, *J. Natl. Cancer Inst.* **45**:163.

Lowy, D. R., Rowe, W. P., Teich, N., and Hartley, J. W., 1971, Murine leukemia virus: High frequency activation *in vitro* by 5-iododeoxyuridine and 5-bromodeoxyuridine, *Science* **174**:155.

Lueders, K. K., and Kuff, E. L., 1980, Intracisternal A-particle genes: Identification in the genome of *Mus musculus* and comparison of multiple isolates from a mouse gene library, *Proc. Natl. Acad. Sci. USA* **77**:3571.

Lueders, K. K., and Kuff, E. L., 1981, Sequences homologous to retrovirus-like genes of the mouse are present in multiple copies in the Syrian hamster genome, *Nucleic Acids Res.* **22**:5917.

Lueders, K. K., and Kuff, E. L., 1983, Comparison of the sequence organization of related retrovirus-like multigene families in three evolutionarily distant rodent genomes, *Nucleic Acids Res.* **13**:4391.

Lueders, K. K., and Meitz, J. A., 1986, Structural analysis of type II variants within the mouse intracisternal A-particle sequence family, *Nucleic Acids Res.* **14**:1495.

Lung, M. L., Hartley, J. W., Rowe, W. P., and Hopkins, N., 1983, Large RNase T_1-resistant oligonucleotides encoding p15E and the U3 region of the long terminal repeat distinguish two biological classes of mink cell focus-forming type C viruses of inbred mice, *J. Virol.* **45**:275.

Majors, J. E., and Varmus, H. E., 1981, Nucleotide sequences at host–proviral junctions for mouse mammary tumour virus, *Nature* **289**:253.

Manly, K. F., Givens, J. F., Taber, R. L., and Zeigel, R. F., 1978, Characterization of virus-like particles released from the hamster cell line CHO-K1 after treatment with 5-bromodeoxyuridine, *J. Gen. Virol.* **39**:505.

Marrack, P., Kushnir, E., and Kappler, J., 1991, A maternally inherited superantigen encoded by a mammary tumour virus, *Nature* **349**:524.

Marth, J. D., Peet, R., Krebs, E. G., and Perlmutter, R. M., 1985, A lymphocyte-specific protein kinase gene is rearranged and overexpressed in the murine T cell lymphoma LSTRA, *Cell* **43**:393.

Massey, A. C., Coppola, M. A., and Thomas, C. Y., 1990, Origin of pathogenic determinants of recombinant murine leukemia viruses: Analysis of *Bxv-1*-related xenotropic viruses from CWD mice, *J. Virol.* **64**:5491.

Masuda, M., Remington, M. P., Hoffman, P. M., and Ruscetti, S. K., 1992, Molecular characterization of a neuropathogenic and nonerythroleukemogenic variant of Friend murine leukemia virus PVC-211, *J. Virol.* **66**:2798.

Maxwell, S. A., and Arlinghaus, R. B., 1985, Serine kinase activity associated with Moloney murine sarcoma virus-124-encoded p37mos, *Virology* **143**:321.

McCarter, J. A., Ball, J. K., and Frei, J. V., 1977, Lower limb paralysis induced in mice by a temperature-sensitive mutant of Moloney leukemia virus, *J. Natl. Cancer Inst.* **59**:179.

McGrath, M. S., and Weissman, I. L., 1979, AKR leukemogenesis: Identification and biological significance of thymic lymphoma receptors for AKR retroviruses, *Cell* **17**:65.

McMahon, A. P., and Bradley, A., 1990, The *Wnt-1 (int-1)* proto-oncogene is required for development of a large region of the mouse brain, *Cell* **62**:1073.

Meitz, J. A., Grossman, Z., Lueders, K. K., and Kuff, E. L., 1987, Nucleotide sequence of a complete mouse intracisternal A-particle genome: Relationship to known aspects of particle assembly and function, *J. Virol.* **61**:3020.

Mercer, J. A., Lee, K. H., Nexo, B. A., Jenkins, N. A., and Copeland, N. G., 1990, Mechanism of chemical activation of expression of the endogenous ecotropic murine leukemia provirus *Emv-3*, *J. Virol.* **64**:2245.

Michalides, R., Schlom, J., Dahlberg, J., and Perk, K., 1975, Biochemical properties of the bromodeoxyuridine-induced guinea pig virus, *J. Virol.* **16**:1039.

Miller, M. L., Raveche, E. S., Laskin, C. A., Klinman, D. M., and Steinberg, A. D., 1984, Genetic studies in NZB mice. VI. Association of autoimmune traits in recombinant inbred lines, *J. Immunol.* **133**:1325.

Mirand, E. A., Prentice, T. C., and Hoffmann, J. G., 1961, Effect of Friend virus in Swiss and DBA/1 mice on ^{59}Fe uptake, *Proc. Soc. Exp. Biol. Med.* **106**:423.

Mitani, M., Cianciolo, G. J., Snyderman, R., Yasuda, M., Good, R. A., and Day, N. K., 1987, Suppressive effect on polyclonal B cell activation of a synthetic peptide homologous to a transmembrane component of oncogenic retroviruses, *Proc. Natl. Acad. Sci. USA* **84**:237.

Moelling, K., Heimann, B., Beimling, P., Rapp, U. R., and Sander, T., 1984, Serine- and threonine-specific protein kinase activities of purified *gag–mil* and *gag–raf* proteins, *Nature* **312**:558.

Moelling, K., Pfaff, E., Beug, H., Beimling, P., Bunte, T., Schaller, H. E., and Graf, T., 1985, DNA-binding activity is associated with purified Myb proteins from AMV and E26 viruses and is temperature-sensitive for E26 ts mutants, *Cell* **40**:983.

Moll, B., Hartley, J. W., and Rowe, W. P., 1979, Induction of B-tropic and N-tropic murine leukemia virus from B10.BR/SGLI mouse embryo cells by 5-iodo-2'-deoxyuridine, *J. Natl. Cancer Inst.* **63**:213.

Moloney, J. B., 1960, Biological studies on a lymphoid-leukemia virus extracted from sarcoma 37. I. Origin and introductory investigations, *J. Natl. Cancer Inst. Monogr.* **24**:933.

Moloney, J. B., 1966, A virus-induced rhabdomyosarcoma of mice, *Natl. Cancer Inst. Monogr.* **22**:139.

Moreau-Gachelin, F., Tavitian, A., and Tambourin, P., 1988, *Spi-1* is a putative oncogene in virally induced murine erythroleukemias, *Nature* **331**:277.

Morris, V. L., Rao, T. R., Kozak, C. A., Gray, D. A., Lee Chen, E. C. M., Cornell, T. J., Taylor, C. B., Jones, R. F., and McGrath, C. M., 1991, Characterization of *Int-5*, a locus associated with early events in mammary carcinogenesis, *Oncogene Res.* **6**:53.

Morse III, H. C., Chused, T. M., Hartley, J. W., Mathieson, B. J., Sharrow, S. O., and Taylor, B. A., 1979, Expression of xenotropic murine leukemia viruses as cell-surface gp70 in genetic crosses between strains DBA/2 and C57BL/6, *J. Exp. Med.* **149**:1183.

Morse III, H. C., Yetter, R. A., Stimpfling, J. H., Pitts, O. M., Fredrickson, T. N., and Hartley, J. W., 1985, Greying with age in mice: Relation to expression of murine leukemia viruses, *Cell* **4**:439.

Morse III, H. C., Hartley, J. W., Fredrickson, T. N., Yetter, R. A., Majumdar, C., Cleveland, J. L., and Rapp, U. R., 1986, Recombinant murine retroviruses containing avian v-*myc* induce a wide spectrum of neoplasms in newborn mice, *Proc. Natl. Acad. Sci. USA* **83**:6868.

Mosier, D. E., Yetter, R. A., and Morse III, H. C., 1985, Retroviral induction of acute lymphoproliferative disease and profound immunosuppression in adult C57BL/6 mice, *J. Exp. Med.* **161**:766.

Mowat, M., Cheng, A., Kimura, N., Bernstein, A., and Benchimol, S., 1985, Rearrangements of the cellular p53 gene in erythroleukemic cells transformed by Friend virus, *Nature* **214**:633.

Mucenski, M. L., Taylor, B. A., Ihle, J. N., Hartley, J. W., Morse III, H. C., Jenkins, N. A., and Copeland, N. G., 1988, Identification of a common ecotropic viral integration site, *Evi-1*, in the DNA of AKXD murine myeloid tumors, *Mol. Cell. Biol.* **8**:301.

Mühlbock, O., and Bentvelzen, P., 1968, The transmission of the mammary tumor viruses, *Perspect. Virol.* **6**:75.

Munroe, D. G., Peacock, J. W., and Benchimol, S., 1990, Inactivation of the cellular p53 gene is a common feature of Friend virus-induced erythroleukemia: Relationship of inactivation to dominant transforming alleles, *Mol. Cell. Biol.* **10**:3307.

Nadeau, H. J., and Phillips, S. J., 1987, The putative oncogene *Pim-1* in the mouse: Its linkage and variation among *t* haplotypes, *Genetics* **117**:533.

Nandi, S., and McGrath, C. M., 1973, Mammary neoplasia in mice, *Adv. Cancer Res.* **17**:353.

Nayak, D. P., and Murray, P. R., 1973, Induction of type C viruses in cultured guinea pig cells, *J. Virol.* **12**:177.

Nusse, R., and Varmus, H. E., 1982, Many tumors induced by the mouse mammary tumor virus contain a provirus integrated in the same region of the host genome, *Cell* **31**:99.

Nusse, R., de Moes, J., Hilkens, J., and van Nie, R., 1980, Localization of a gene for expression of mouse mammary tumor virus antigens in the GR/Mtv-2⁻ mouse strain, *J. Exp. Med.* **152**:712.

Nusse, R., Van Ooyen, A., Cox, D., Fung, Y. K. T., and Varmus, H., 1984, Mode of proviral activation of a putative mammary oncogene (*int-1*) on mouse chromosome 15, *Nature* **307**:131.

Nusse, R., Brown, A., Papkoff, J., Scambler, P., Shackleford, G., McMahon, A., Moon, R., and Varmus, H., 1991, A new nomenclature for *int-1* and related genes: The *Wnt* gene family, *Cell* **64**:231.

Obata, M. M., and Khan, A. S., 1988, Structure, distribution, and expression of an ancient murine endogenous retrovirus-like DNA family, *J. Virol.* **62**:4381.

Obata, Y., Ideka, H., Stockert, E., and Boyse, E. A., 1975, Relation of G_{IX} antigen of thymocytes to envelope glycoprotein of murine leukemia virus, *J. Exp. Med.* **141**:188.

Odaka, T., and Yamamoto, T., 1962, Inheritance of susceptibility to Friend mouse leukemia virus, *Jpn. J. Exp. Med.* **32**:405.

Okabe, H., Gilden, R. V., and Hatanaka, M., 1974, Specificity of the DNA product of RNA-dependent DNA polymerase in type C viruses: III. Analysis of viruses derived from Syrian hamsters, *Proc. Natl. Acad. Sci. USA* **71**:3278.

O'Neill, R. R., Buckler, C. E., Theodore, T. S., Martin, M. A., and Repaske, R., 1985, Envelope and long terminal repeat sequences of a cloned infectious NZB xenotropic murine leukemia virus, *J. Virol.* **53**:100.

O'Neill, R. R., Khan, A. S., Hoggan, M. D., Hartley, J. W., Martin, M. A., and Repaske, R., 1986, Specific hybridization probes demonstrate fewer xenotropic than mink cell focus-forming murine leukemia virus *env*-related sequences in DNAs from inbred laboratory mice, *J. Virol.* **58**:359.

O'Neill, R., Hartley, J., Repaske, R., and Kozak, C., 1987, Amphotropic proviral envelope sequences are absent from the *Mus* germ line, *J. Virol.* **61**:2225.

Ono, M., Cole, M. D., White, A. T., and Huang, R. C. C., 1980, Sequence organization of cloned intracisternal A particle genes, *Cell* **21**:465.

Ono, M., Toh, H., Miyata, T., and Awaya, T., 1985, Nucleotide sequence of the Syrian hamster intracisternal A-particle gene: Close evolutionary relationship of type A particle gene to types B and D oncovirus genes, *J. Virol.* **55**:387.

Opler, S. R., 1967, Observation of a new virus associated with guinea pig leukemia: Preliminary note, *J. Natl. Cancer Inst.* **38**:797.

Ostertag, W., Vehmeyer, K., Fagg, B., Pragnell, I. B., Paetz, W., Le Bousse, M. C., Smadja-Joffe, F., Klein, B., Jasmin, C., and Eisen, H., 1980, Myeloproliferative virus, a cloned murine sarcoma virus with spleen focus-forming properties in adult mice, *J. Virol.* **33**:573.

Ott, D., Friedrich, R., and Rein, A., 1990, Sequence analysis of amphotropic and 10A1 murine leukemia viruses: Close relationship to mink cell focus-inducing viruses, *J. Virol.* **64**:757.

Ou, C.-Y., Boone, L. R., Koh, C.-K., Tennant, R. W., and Yang, W. K., 1983a, Nucleotide sequences of *gag–pol* regions that determine the *Fv-1* host range property of BALB/c N-tropic and B-murine leukemia viruses, *J. Virol.* **48**:779.

Ou, C.-Y., Boone, L. R., and Yang, W. K., 1983b, A novel sequence segment and other nucleotide structural features in the long terminal repeat of a BALB/c mouse genomic leukemia virus-related DNA clone, *Nucleic Acids Res.* **11**:5603.

Papkoff, J., Verma, I. M., and Hunter, T., 1982, Detection of a transforming gene product in cells transformed by Moloney murine sarcoma virus, *Cell* **29**:417.

Paquette, Y., Hanna, Z., Savard, P., Brousseau, R., Robitaille, Y., and Jolicoeur, P., 1989, Retrovirus-induced murine motor neuron disease: Mapping the determinant of spongiform degeneration within the envelope gene, *Proc. Natl. Acad. Sci. USA* **86**:3896.

Paterson, B. M., Segal, S., Lueders, K. K., and Kuff, E. L., 1978, RNA associated with murine intracisternal type A particles codes for the main particle protein, *J. Virol.* **27**:118.

Pattengale, P. K., Taylor, C. R., Twomey, P., Hill, S., Jonasson, J., Beardsley, T., and Haas, M., 1982, Immunopathology of B-cell lymphomas induced in C57BL/6 mice by dual-tropic murine leukemia virus (MuLV), *Am. J. Pathol.* **107**:362.

Paul, R., Schuetze, S., Kozak, S. L., Kozak, C. A., and Kabat, D., 1991, The *Sfpi-1* proviral integration site of Friend erythroleukemia encodes the *ets*-related transcription factor Pu.1 and a small RNA, *J. Virol.* **65**:464.

Pedersen, F. S., Crowther, R. L., Tenney, D. Y., Reimold, A. M., and Haseltine, W. A., 1981, Novel leukemogenic retroviruses isolated from cell line derived from spontaneous AKR tumor, *Nature* **292**:167.

Peebles, P. T., 1975, An *in vitro* focus-induction assay for xenotropic murine leukemia virus, feline leukemia virus C, and the feline-primate viruses RD-114/CCC/M7, *Virology* **67**:288.

Penciolelli, J. F., Wendling, F., Robert-Lezenes, J., Barque, J. P., Tambourin, P., and Gissel-brecht, S., 1987, Genetic analysis of myeloproliferative leukemia virus, a novel acute leukemogenic replication-defective retrovirus, *J. Virol.* **61**:579.

Peters, G., Brookes, S., Smith, R., and Dickson, C., 1983, Tumorigenesis by mouse mammary tumor virus: Evidence for a common region for provirus integration in mammary tumors, *Cell* **33**:369.

Peters, R. L., Rabstein, L. S., VanVleck, R., Kelloff, G. J., and Huebner, R. J., 1974, Naturally occurring sarcoma virus of the BALB/cCR mouse, *J. Natl. Cancer Inst.* **53**:1725.

Phillips, S. J., Birkenmeier, E. H., Callahan, R., and Eicher, E. M., 1982, Male and female mouse DNAs can be discriminated using retroviral probes, *Nature* **297**:241.

Phillips, S. M., Stephenson, J. R., Greenberger, J. S., Lane, P. E., and Aaronson, S. A., 1976, Release of xenotropic type C RNA virus in response to lipopolysaccharide: Activity of lipid-A portion upon B lymphocytes, *J. Immunol.* **116**:1123.

Pikó, L., Hammons, M. D., and Taylor, K. D., 1984, Amounts, synthesis and some properties of intracisternal A particle-related RNA in early mouse embryos, *Proc. Natl. Acad. Sci. USA* **81**:488.

Pitts, O. M., Powers, J. M., and Hoffman, P. M., 1983, Vascular neoplasms induced in rodent central nervous system by murine sarcoma viruses, *Lab. Invest.* **49**:171.

Poirier, Y., and Jolicoeur, P., 1989, Distinct helper virus requirements for Abelson murine leukemia virus-induced pre-B- and T-cell lymphomas, *J. Virol.* **63**:2088.

Poirier, Y., Kozak, C., and Jolicoeur, P., 1988, Identification of a common helper provirus integration site in Abelson murine leukemia virus-induced lymphoma DNA, *J. Virol.* **62**:3985.

Policastro, P. F., Fredholm, M., and Wilson, M. C., 1989, Truncated *gag* products encoded by *Gv-1*-responsive endogenous retrovirus loci, *J. Virol.* **63**:4136.

Portis, J. L., Czub, S., Garon, C. F., and McAtee, F. J., 1990, Neurodegenerative disease induced by the wild mouse ecotropic retrovirus is markedly accelerated by long terminal repeat and *gag–pol* sequences from nondefective Friend murine leukemia virus, *J. Virol.* **64**:1648.

Quint, W., van der Putten, H., Janssen, F., and Berns, A., 1982, Mobility of endogenous ecotropic murine leukemia viral genomes within mouse chromosomal DNA and integration of a mink cell focus-forming virus-type recombinant provirus in the germ line, *J. Virol.* **41**:901.

Quint, W., Boelens, W., van Wezenbleek, P., Cuypers, T., Maandag, E. R., Selten, G., and Berns, A., 1984, Generation of AKR mink cell focus-forming viruses; a conserved single-copy xenotrope-like provirus provides recombinant long terminal repeat sequences, *J. Virol.* **50**:432.

Rapp, U. R., and Marshall, T. H., 1980, Cell surface receptors for endogenous mouse type C viral glycoproteins and epidermal growth factor: Tissue distribution *in vivo* and possible participation in specific cell–cell interaction, *J. Supramol. Struct.* **14**:343.

Rapp, U. R., and Nowinski, R. C., 1976, Endogenous ecotropic mouse type C viruses deficient in replication and production of XC plaques, *J. Virol.* **18**:411.

Rapp, U. R., Reynolds, F. H., and Stephenson, J. R., 1983a, New mammalian transforming retrovirus: Demonstration of a polyprotein gene product, *J. Virol.* **45**:914.

Rapp, U. R., Goldsborough, M. D., Mark, G. E., Bonner, T. I., Groffen, J., Reynolds, F. H., Jr., and Stephenson, J. R., 1983b, Structure and biological activity of v-*raf*, a unique oncogene transduced by a retrovirus, *Proc. Natl. Acad. Sci. USA* **80**:4218.

Rapp, U. R., Cleveland, J. L., Fredrickson, T. N., Holmes, K. L., Morse III, H. C., Jansen, H. W., Patschinsky, T., and Bister, K., 1985, Rapid induction of hemopoietic neoplasms in newborn mice by a *raf(mil)/myc* recombinant murine retrovirus, *J. Virol.* **55**:23.

Rasheed, S., and Young, H. A., 1982, Induction of fibrosarcoma by rat sarcoma virus, *Virology* **118**:219.

Rasheed, R., Bruszewski, J., Rongey, R. W., Roy-Burman, P., Charman, H. P., and Gardner, M. B., 1976a, Spontaneous release of endogenous ecotropic type C virus from rat embryo cultures, *J. Virol.* **18**:799.

Rasheed, S., Gardner, M. B., and Chan, E., 1976b, Amphotropic host range of naturally occurring wild mouse leukemia viruses, *J. Virol.* **19**:13.

Rasheed, S., Gardner, M. B., and Huebner, R. J., 1978, *In vitro* isolation of stable rat sarcoma viruses, *Proc. Natl. Acad. Sci. USA* **75**:2972.

Rasheed, S., Pal, B. K., and Gardner, M. B., 1982, Characterization of a highly oncogenic murine leukemia virus from wild mice, *Int. J. Cancer* **29**:245.

Rasheed, S., Gardner, M. B., and Lai, M. M. C., 1983, Isolation and characterization of new ecotropic murine leukemia viruses after passage of an amphotropic virus in NIH Swiss mice, *Virology* **130**:439.

Rassart, E., Nelback, L., and Jolicoeur, P., 1986, Cas-Br-E murine leukemia virus: Sequencing of the paralytogenic region of its genome and derivation of specific probes to study its origin and the structure of its recombinant genomes in leukemic tissues, *J. Virol.* **60**:910.

Rauscher, F. J., 1962, A virus-induced disease of mice characterized by erythrocytopoiesis and lymphoid leukemia, *J. Natl. Cancer Inst.* **29**:515.

Reddy, E. P., Dunn, C. Y., and Aaronson, S. A., 1980, Different lymphoid cell targets for transformation by replication-competent Moloney and Rauscher mouse leukemia viruses, *Cell* **19**:663.

Rein, A., 1982, Interference grouping of murine leukemia viruses: A distinct receptor for the MCF-recombinant viruses on mouse cells, *Virology* **120**:251.

Rein, A., and Schultz, A., 1984, Different recombinant murine leukemia viruses use different cell surface receptors, *Virology* **136**:144.

Reuss, F. U., and Schaller, H. C., 1991, cDNA sequence and genomic characterization of intracisternal A-particle-related retroviral elements containing an envelope gene, *J. Virol.* **65**:5702.

Reynolds, F. H., Sacks, T. L., Deobagkar, D. H., and Stephenson, J. R., 1978, Cells nonproductively transformed by Abelson murine leukemia virus express a high molecular weight polyprotein containing structural and nonstructural components, *Proc. Natl. Acad. Sci. USA* **75**:3974.

Rhim, J. S., Wuu, K. D., Ro, H. S., Vernon, M. L., and Huebner, R. J., 1974, Induction of guinea pig cells, *Proc. Soc. Exp. Biol. Med.* **147**:323.

Rijsewijk, F., Schuermann, M., Wagenaar, E., Parren, P., Weigel, D., and Nusse, R., 1987, The *Drosophila* homolog of the mouse mammary oncogene int-1 is identical to the segment polarity gene *wingless*, *Cell* **50**:649.

Ringold, G. M., 1983, Regulation of mouse mammary tumor virus gene regulation by glucocorticoid hormones, *Curr. Top. Microbiol. Immunol.* **106**:79.

Risser, R., Potter, M., and Rowe, W. P., 1978, Abelson virus-induced lymphomagenesis in mice, *J. Exp. Med.* **148**:714.

Rossomando, A., and Meruelo, D., 1986, Viral sequences are associated with many histocompatibility genes, *Immunogenetics* **23**:233.

Rowe, W. P., 1972, Studies of genetic transmission of murine leukemia virus by AKR mice, *J. Exp. Med.* **136**:1272.

Rowe, W. P., 1983, Deformed whiskers in mice infected with certain endogenous murine leukemia viruses, *Science* **221**:562.

Rowe, W. P., and Hartley, J. W., 1983, Genes affecting mink cell focus-inducing (MCF) murine leukemia virus infection and spontaneous lymphoma in F_1 hybrids, *J. Exp. Med.* **158**:353.

Rowe, W. P., and Kozak, C. A., 1980, Germ-line reinsertions of AKR murine leukemia virus genomes in *Akv-1* congenic mice, *Proc. Natl. Acad. Sci. USA* **77**:4871.

Rowe, W. P., Pugh, W. E., and Hartley, J. W., 1970, Plaque assay techniques for murine leukemia viruses, *Virology* **42**:1136.

Ruscetti, S., Linemeyer, D., Feild, J., Troxler, D., and Scolnick, E. M., 1979, Characterization of a protein found in cells infected with the spleen focus-forming virus that shares immunological cross-reactivity with the gp70 found in mink cell focus-inducing virus particles, *J. Virol.* **30**:787.

Ruscetti, S., Davis, L., Feild, J., and Oliff, A., 1981, Friend murine leukemia virus-induced leukemia is associated with the formation of mink cell focus-inducing viruses and is blocked in mice expressing endogenous mink cell focus-inducing xenotropic viral envelope genes, *J. Exp. Med.* **154**:907.

Ruscetti, S., Feild, J., Davis, L., and Oliff, A., 1982, Factors determining the susceptibility of NIH Swiss mice to erythroleukemia induced by Friend murine leukemia virus, *Virology* **117**:357.

Ruscetti, S., Matthai, R., and Potter, M., 1985, Susceptibility of BALB/c mice carrying various DBA/2 genes to development of Friend murine leukemia virus-induced erythroleukemia, *J. Exp. Med.* **162**:1579.

Ruscetti, S. K., Janesch, N. J., Chakraborti, A., Sawyer, S. T., and Hankins, W. D., 1990, Friend spleen focus-forming virus induces factor independence in an erythropoietin-dependent erythroleukemia cell line, *J. Virol.* **63**:1057.

Ruscetti, S., Aurigemma, R., Yuan, C.-C., Sawyer, S., and Blair, D. G., 1992, Induction of erythropoietin responsiveness in murine hematopoietic cells by the *gag-myb-ets*-containing ME26 virus, *J. Virol.* **66**:20.

Russell, P., Gregerson, D. S., Albert, D. M., and Reid, T. W., 1979, Characteristics of a retrovirus associated with a hamster melanoma, *J. Gen. Virol.* **43**:317.

Sacks, T. L., Reynolds, F. H., Deobagkar, D. N., and Stephenson, J. R., 1978, Murine leukemia virus (T-8)-transformed cells: Identification of a precursor polyprotein containing *gag* gene-coded proteins (p15 and p12) and a nonstructural component, *J. Virol.* **27**:809.

Salamon, M. H., and Wedderbrun, N., 1966, The immuno-depressive effects of Friend virus, *Immunology* **10**:445.

Salmons, B., Erfle, V., Brem, G., and Gunzburg, W. H., 1990, *Naf*, a *trans*-regulating negative-acting factor encoded within the mouse mammary tumor virus open reading frame region, *J. Virol.* **64**:6355.

Sambucetti, L. C., and Curran, T., 1986, The Fos protein complex is associated with DNA in isolated nuclei and binds to DNA cellulose, *Science* **234**:1417.

Sassa, S., Takaku, F., and Nakao, K., 1968, Regulation of erythropoiesis in the Friend leukemia mouse, *Blood* **31**:758.

Schlom, J., Hand, P. H., Teramoto, Y. A., Callahan, R., Todaro, G., and Schidlovsky, G., 1978, Characterization of a new virus from *Mus cervicolor* immunologically related to the mouse mammary tumor virus, *J. Natl. Cancer Inst.* **61**:1509.

Schmidt, M., Wirth, T., Kröger, B., and Horak, I., 1985, Structure and genomic organization of a new family of murine retrovirus-related DNA sequences (MuRRS), *Nucleic Acids Res.* **13**:3461.

Scolnick, E. M., and Parks, W. P., 1974, Harvey sarcoma virus: A second murine type-C sarcoma virus with rat genetic information, *J. Virol.* **13**:1211.

Scolnick, E. M., Howk, R. S., Anisowicz, A., Peebles, P. T., Scher, C. D., and Parks, W. P., 1975, Separation of sarcoma virus-specific and leukemia virus-specific genetic sequences of Moloney sarcoma virus, *Proc. Natl. Acad. Sci. USA* **72**:4650.

Scolnick, E. M., Goldberg, R. J., and Williams, D., 1976, Characterization of rat genetic sequences of Kirsten sarcoma virus: Distinct class of endogenous rat type C viral sequences, *J. Virol.* **18**:559.

Scolnick, E. M., Vass, W. C., Howk, R. S., and Duesberg, P. H., 1979a, Defective retrovirus-like 30S RNA species of rat and mouse cells are infectious if packaged by type C helper virus, *J. Virol.* **29**:964.

Scolnick, E. M., Papageorge, A. G., and Shih, T. Y., 1979b, Guanine nucleotide-binding activity as an assay for *src* protein of rat-derived murine sarcoma viruses, *Proc. Natl. Acad. Sci. USA* **76**:5355.

Selten, G., Cuypers, H. T., Zijlstra, M., Melief, C., and Berns, A., 1984, Involvement of c-*myc* in MuLV-induced T cell lymphomas in mice: Frequency and mechanisms of activation, *EMBO J.* **3**:3215.

Servenay, M., Kupiec, J. J., d'Auriol, L., Galibert, F., Périès, J., and Emanoil-Ravier, R., 1988, Nucleotide sequence of the Chinese hamster intracisternal A-particle genomic region corresponding to 5′ LTR/GAG, *Nucleic Acids Res.* **16**:7725.

Shen-Ong, G. L. C., and Cole, M. D., 1982, Differing populations of intracisternal A-particle genes in myeloma tumors and mouse subspecies, *J. Virol.* **42**:411.

Shen-Ong, G. L. C., and Cole, M. D., 1984, Amplification of a specific set of intracisternal A-particle genes in a mouse plasmacytoma, *J. Virol.* **49**:171.

Shen-Ong, G. L. C., and Wolff, L., 1987, Moloney murine leukemia virus-induced myeloid tumors in adult BALB/c mice: Requirement of c-*myb* activation but lack of v-*abl* involvement, *J. Virol.* **61**:3721.

Shen-Ong, G. L. C., Potter, M., Mushinski, J. F., Lavu, S., and Reddy, E. P., 1984, Activation of the c-*myb* locus by viral insertional mutagenesis in plasmacytoid lymphosarcomas, *Science* **226**:1077.

Sherr, C. J., Rettenmier, C. W., Sacca, R., Roussel, M. F., Look, A. T., and Stanley, E. R., 1985, The c-*fms* proto-oncogene product is related to the receptor for the mononuclear phagocyte growth factor, CSF-1, *Cell* **41**:665.

Sherwin, S. A., Rapp, U. R., Benveniste, R. E., Sen, A., and Todaro, G. I., 1978, Rescue of endogenous 30S retroviral sequences from mouse cells by baboon type c virus, *J. Virol.* **26**:257.

Sherwin, S. A., Bonner, T. I., Heine, U., and Todaro, G. J., 1979, The isolation of an endogenous retrovirus from the New World rodent *Dasyprocta punctata* (agouti), *Virology* **94**:409.

Shih, T. Y., Williams, D. R., Weeks, M. O., Maryak, J. M., Vass, W. C., and Scolnick, E. M., 1978, Comparison of the genomic organization of Kirsten and Harvey sarcoma viruses, *J. Virol.* **27**:45.

Shih, T. Y., Weeks, M. O., Young, H. A., and Scolnick, E. M., 1979, Identification of a sarcoma virus-coded phosphoprotein in nonproducer cells transformed by Kirsten or Harvey murine sarcoma virus, *Virology* **96**:64.

Shinnick, T. M., Lerner, R. A., and Sutcliffe, J. G., 1981, Nucleotide sequence of Moloney murine leukemia virus, *Nature* **293**:543.

Silver, J., 1984, Role of mink cell focus-inducing virus in leukemias induced by Friend ecotropic virus, *J. Virol.* **50**:872.

Silver, J. E., and Fredrickson, T. N., 1983, A new gene that controls the type of leukemia induced by Friend murine leukemia virus, *J. Exp. Med.* **158**:493.

Silver, J., and Kozak, C., 1986, Common proviral integration region on mouse chromosome 7 in lymphomas and myelogenous leukemias induced by Friend murine leukemia virus, *J. Virol.* **57**:526.

Siracusa, L. D., Jenkins, N. A., and Copeland, W. G., 1991, Identification and applications of repetitive probes for gene mapping in the mouse, *Genetics* **127**:169.

Smith, R., Peters, G., and Dickson, C., 1988, Multiple RNAs expressed from the *int-2* gene in mouse embryonal carcinoma cell lines encode a protein with homology to fibroblast growth factor, *EMBO J.* **7:**1013.

Sola, B., Fichelson, S., Bordereaux, D., Tambourin, P. E., and Gisselbrecht, S., 1986, *fim-1* and *fim-2:* Two new integration regions of Friend murine leukemia virus in myeloblastic leukemias, *J. Virol.* **60:**718.

Souyri, M., Vigon, I., Penciolelli, J.-F., Heard, J.-M., Tambourin, P., and Wendling, F., 1990, A putative truncated cytokine receptor gene transduced by the myeloproliferative leukemia virus immortalizes hematopoietic progenitors, *Cell* **63:**1137.

Staal, S. P., 1987, Molecular cloning of the akt oncogene and its human homologues AKT1 and AKT2: Amplication of AKT1 in a primary human gastric adenocarcinoma, *Proc. Natl. Acad. Sci. USA* **84:**5034.

Staal, S. P., and Hartley, J. W., 1988, Thymic lymphoma induction by the AKT8 murine retrovirus, *J. Exp. Med.* **167:**1259.

Staal, S. P., Hartley, J. W., and Rowe, W. P., 1977, Isolation of transforming murine leukemia viruses from mice with a high incidence of spontaneous lymphoma, *Proc. Natl. Acad. Sci. USA* **74:**3065.

Stacey, A., Arbuthnott, C., Kollek, R., Coggins, L., and Ostertag, W., 1984, Comparison of myeloproliferative sarcoma virus with Moloney murine sarcoma virus variants by nucleotide sequencing and heteroduplex analysis, *J. Virol.* **50:**725.

Stavenhagen, J. B., and Robins, D. M., 1988, An ancient provirus has imposed androgen regulation on the adjacent mouse sex-limited protein gene, *Cell* **55:**247.

Steeves, R. A., Bennett, M., Mirand, E. A., and Cudkowicz, G., 1968, Genetic control by the *W* locus of susceptibility to (Friend) spleen focus-forming virus, *Nature* **218:**372.

Steffen, D. L., Bird, S., and Weinberg, R. A., 1980, Evidence for the Asiatic origin of endogenous AKR-type murine leukemia proviruses, *J. Virol.* **35:**824.

Stockert, E., Old, L. J., and Boyse, E. A., 1971, The G_{IX} system. A cell surface allo-antigen associated with murine leukemia virus: Implications regarding chromosomal integration of the viral genome, *J. Exp. Med.* **133:**1334.

Stocking, C., Kollek, R., Bergholz, U., and Ostertag, W., 1985, Long terminal repeat sequences impart hematopoietic transformation properties to the myeloproliferative sarcoma virus, *Proc. Natl. Acad. Sci. USA* **82:**5746.

Stoye, J. P., and Coffin, J. M., 1988, Polymorphism of murine endogenous proviruses revealed by using virus class-specific oligonucleotide probes, *J. Virol.* **62:**168.

Stoye, J. P., Fenner, S., Greenoak, G. E., Moran, C., and Coffin, J. M., 1988, Role of endogenous proviruses as insertional mutagens: The hairless mutation of mice, *Cell* **54:**383.

Stoye, J. P., Moroni, C., and Coffin, J. M., 1991, Virological events leading to spontaneous AKR thymomas, *J. Virol.* **65:**1273.

Suzuki, A., Kitasato, H., Kawakami, M., and Ono, M., 1982, Molecular cloning of retrovirus-like env genes present in multiple copies in the Syrian hamster genome, *Nucleic Acids Res.* **10:**5733.

Suzuki, S., 1975, *Fv-4:* A new gene affecting the splenomegaly induction by Friend leukemia virus, *Jpn. J. Exp. Med.* **45:**473.

Sveda, M. M., and Soeiro, R., 1976, Host restriction of Friend leukemia virus: Synthesis and integration of the provirus, *Proc. Natl. Acad. Sci. USA* **73:**2356.

Szurek, P. F., Yuen, P. H., Jerzy, R., and Wong, P. K. Y., 1988, Identification of point mutations in the envelope gene of Moloney murine leukemia virus TB temperature-sensitive paralytogenic mutant *ts*1: Molecular determinants for neurovirulence, *J. Virol.* **62:**357.

Takayama, Y., O'Mara, M.-A., Spilsbury, K., Thwaite, R., Rowe, P. B., and Symonds, G., 1991, Stage-specific expression of intracisternal A-particle sequences in murine myelomonocytic leukemia cell lines and normal myelomonocytic differentiation, *J. Virol.* **65**:2149.

Tambourin, P. E., Wendling, F., Jasmin, C., and Smadja-Joffe, F., 1979, The physiopathology of Friend leukemia, *Leuk. Res.* **3**:117.

Taylor, B. A., and Rowe, L., 1989, A mouse linkage testing stock possessing multiple copies of the endogenous ecotropic murine leukemia virus genome, *Genomics* **5**:221.

Thomas, C. Y., and Coffin, J. M., 1982, Genetic alterations of RNA leukemia viruses associated with the development of spontaneous thymic leukemia in AKR/J mice, *J. Virol.* **43**:416.

Tihon, C., and Green, M., 1973, Cyclic AMP-amplified replication of RNA tumour virus-like particles in Chinese hamster ovary cells, *Nature New Biol.* **244**:227.

Troxler, D. H., and Scolnick, E. M., 1978, Rapid leukemia induced by cloned Friend strain of replicating murine type-C virus, *Virology* **85**:17.

Troxler, D. H., Parks, W. P., Vass, W. C., and Scolnick, E. M., 1977a, Isolation of a fibroblast nonproducer cell line containing the Friend strain of the spleen focus-forming virus, *Virology* **76**:606.

Troxler, D. H., Boyars, J. K., Parks, W. P., and Scolnick, E. M., 1977b, Friend strain of spleen focus-forming virus: A recombinant between mouse type C ecotropic viral sequences and sequences related to xenotropic virus, *J. Virol.* **22**:361.

Troxler, D. H., Yuan, E., Linemeyer, D., Ruscetti, S., and Scolnick, E. M., 1978, Helper-independent mink cell focus-inducing strains of Friend murine type-C virus: Potential relationship to the origin of replication-defective spleen focus-forming virus, *J. Exp. Med.* **148**:639.

Troxler, D. H., Ruscetti, S. K., Linemeyer, D. L., and Scolnick, E. M., 1980, Helper-independent and replication-defective erythroblastosis-inducing viruses contained within anemia-inducing Friend virus complex (FV-A), *Virology* **102**:28.

Tsichlis, P. N., Strauss, P. G., and Hu, L. F., 1983, A common region for proviral DNA integration of MoMuLV-induced rat thymic lymphomas, *Nature* **302**:445.

Tsichlis, P. N., Strauss, P. G., and Kozak, C. A., 1984, A cellular DNA region involved in the induction of thymic lymphomas (Mlvi-2) maps to mouse chromosome 15, *Mol. Cell. Biol.* **4**:997.

Tsichlis, P. N., Strauss, P. G., and Lohse, M. A., 1985, Concerted DNA rearrangements in Moloney murine leukemia virus-induced thymomas: A potential synergistic relationship in oncogenesis, *J. Virol.* **56**:258.

Tsichlis, P. N., Lee, J. S., Bear, S. E., Lazo, P. A., Patriotis, C., Gustafson, E., Shinton, S., Jenkins, N. A., Copeland, N. G., Huebner, K., Croce, C., Levan, G., and Hanson, C., 1990, Activation of multiple genes by provirus integration in the Mlvi-4 locus in T-cell lymphomas induced by Moloney murine leukemia virus, *J. Virol.* **64**:2236.

Tsuchida, N., Gilden, R. V., and Hatanaka, M., 1974, Sarcoma virus-related RNA sequences in normal rat cells, *Proc. Natl. Acad. Sci. USA* **71**:4503.

Tsuchida, N., Gilden, R. V., Hatanaka, M., Freeman, A. E., and Huebner, R. J., 1975, Type-C virus-specific nucleic acid sequences in cultured rat cells, *Int. J. Cancer* **15**:109.

Tumilowicz, J. J., and Cholon, J. J., 1971, Intracisternal type A particles and properties of a continuous cell line originating from a gerbil fibroma, *Proc. Soc. Exp. Biol. Med.* **136**:1107.

Tung, J. S., Vitetta, E. S., Fleissner, E., and Boyse, E. A., 1975, Biochemical evidence linking the G_{IX} thymocyte surface antigen to the gp69/71 envelope glycoprotein of murine leukemia virus, *J. Exp. Med.* **141**:198.

Van Beveren, C., Enami, S., Curran, T., and Verma, I. M., 1984, FBR murine osteosarcoma virus II. Nucleotide sequence of the provirus reveals that the genome contains sequences acquired from two cellular genes, *Virology* **135**:229.

van Griensven, L. J. L. D., and Vogt, M., 1980, Rauscher "mink cell focus-inducing" (MCF) virus causes erythroleukemia in mice: Its isolation and properties, *Virology* **101**:376.

Van Klaveren, P., and Bentvelzen, P., 1988, Transactivating potential of the 3' open reading frame of murine mammary tumor virus, *J. Virol.* **62**:4410.

Van Lohuizen, M., Breuer, M., and Berns, A., 1989, N-*myc* is frequently activated by proviral insertion in MuLV-induced T cell lymphomas, *EMBO J.* **8**:133.

van Nie, R., and Verstraeten, A. A., 1975, Studies of genetic transmission of mammary tumour virus by C3Hf mice, *Int. J. Cancer* **16**:922.

van Nie, R., Verstraeten, A., and DeMoes, J., 1977, Genetic transmission of mammary tumor virus by Gr mice, *Int. J. Cancer* **19**:383.

Van Pelt, F. G., Bentvelzen, P., Brinkhof, J., 'T Mannetje, A. H., and Zurcher, C., 1976, Immunofluorescence studies on the association between a C-type oncornavirus and renal glomerulopathy in *Praomys (Mastomys) natalensis*, *Clin. Immunol. Immunopathol.* **5**:105.

Varnier, O. E., Hoffman, A. D., Nexø, B. A., and Levy, J. A., 1984, Murine xenotropic type C viruses. V. Biologic and structural differences among three cloned retroviruses isolated from kidney cells from one NZB mouse, *Virology* **132**:79.

Verstraeten, A. A., and van Nie, R., 1978, Genetic transmission of mammary tumor virus in the DBAf mouse strain, *Int. J. Cancer* **21**:473.

Verwoerd, D. W., and Sarma, P. S., 1973, Induction of type C virus related functions in normal rat embryo fibroblasts by treatment with 5-iododeoxyuridine, *Int. J. Cancer* **12**:551.

Vijaya, S., Steffen, D. L., Kozak, C., and Robinson, H. L., 1987, Dsi-1, a region with frequent proviral insertions in Moloney murine leukemia virus-induced rat thymomas, *J. Virol.* **61**:1164.

Villar, C. J., Fredrickson, T. N., and Kozak, C. A., 1988, Effect of the *Gv-1* locus on Moloney ecotropic murine leukemia virus induced disease in inbred and wild mice, *Curr. Top. Microbiol. Immunol.* **137**:250.

Villermur, R., Monczak, Y., Rassart, E., and Jolicoeur, P., 1987, Identification of a new common provirus integration site in Gross passage A murine leukemia virus-induced mouse thymoma DNA, *Mol. Cell. Biol.* **7**:512.

Vogt, M., 1979, Properties of "mink cell focus-inducing" (MCF) virus isolated from spontaneous lymphoma lines of BALB/c mice carrying Moloney leukemia virus as an endogenous virus, *Virology* **93**:226.

Vogt, M., 1982, Virus cloned from the Rauscher virus complex induces erythroblastosis and thymic lymphoma, *Virology* **118**:225.

Voronova, A. F., and Sefton, B. M., 1986, Expression of a new tyrosine protein kinase is stimulated by retrovirus promoter insertion, *Nature* **319**:682.

Voytek, P., and Kozak, C., 1988, HoMuLV: A novel pathogenic ecotropic virus isolated from the European mouse, *Mus hortulanus*, *Virology* **165**:469.

Voytek, P., and Kozak, C. A., 1989, Nucleotide sequence and mode of transmission of the wild mouse ecotropic virus, HoMuLV, *Virology* **173**:58.

Wang, H., Kavanaugh, M. P., North, R. A., and Kabat, D., 1991, Cell-surface receptor for ecotropic murine retroviruses is a basic amino-acid transporter, *Nature* **352**:729.

Wang, H., Paul, R., Burgeson, R. E., Keene, D. R., and Kabat, D., 1991, Plasma membrane receptors for ecotropic murine retroviruses require a limiting accessory factor, *J. Virol.* **65**:6468.

Weinstein, Y., Ihle, J. N., Lavu, S., and Reddy, E. P., 1986, Truncation of the c-*myb* gene by a retroviral integration in an interleukin 3-dependent myeloid leukemia cell line, *Proc. Natl. Acad. Sci. USA* **83**:5010.

Wejman, J. C., Taylor, B. A., Jenkins, N. A., and Copeland, N. G., 1984, Endogenous xenotropic murine leukemia virus-related sequences map to chromosomal regions encoding mouse lymphocyte antigens, *J. Virol.* **50**:237.

Wendling, F., Varlet, P., Charon, M., and Tambourin, P., 1986, MPLV: A retrovirus complex inducing an acute myeloproliferative leukemic disorder in adult mice, *Virology* **149**:242.

Wirth, T., Gloggler, K., Baumruker, T., Schmidt, M., and Horak, I., 1983, Family of middle repetitive DNA sequences in the mouse genome with structural features of solitary retroviral long terminal repeats, *Proc. Natl. Acad. Sci. USA* **80**:3327.

Witte, O. N., Rosenberg, N., Paskind, M., Shields, A., and Baltimore, D., 1978, Identification of an Abelson murine leukemia virus-encoded protein present in transformed fibroblasts and lymphoid cells, *Proc. Natl. Acad. Sci. USA* **75**:2488.

Witte, O. N., Dasgupta, A., and Baltimore, D., 1980, Abelson murine leukemia virus protein is phosphorylated *in vitro* to form phosphotyrosine, *Nature* **283**:826.

Wivel, N. A., and Smith, G. H., 1971, Distribution of intracisternal A-particles in a variety of normal and neoplastic tissues, *Int. J. Cancer* **7**:167.

Wolff, L., and Ruscetti, S., 1988, The spleen focus-forming virus (SFFV) envelope gene, when introduced into mice in the absence of other SFFV genes, induces acute erythroleukemia, *J. Virol.* **62**:2158.

Wolff, L., Mushinski, J. F., Shen-Ong, G. L. C., and Morse III, H. C., 1988, A chronic inflammatory response: Its role in supporting the development of c-*myb* and c-*myc* related promonocytic and monocytic tumors in BALB/c mice, *J. Virol.* **141**:681.

Wolff, L., Koller, R., and Davidson, W., 1991, Acute myeloid leukemia induction by amphotropic murine retrovirus (4070A): Clonal integrations involve c-*myb* in some but not all leukemias, *J. Virol.* **65**:3607.

Wong, P. K. Y., Soong, M. M., MacLeod, R., Gallick, G. E., and Yuen, P. H., 1983, A group of temperature-sensitive mutants of Moloney leukemia virus which is defective in cleavage of *env* precursor polypeptide in infected cells also induces hind-limb paralysis in newborn CFW/D mice, *Virology* **125**:513.

Wong, P. K. Y., Prasad, G., Hansen, J., and Yuen, P. H., 1989, *ts*1, A mutant of Moloney murine leukemia virus-TB, causes both immunodeficiency and neurologic disorders in BALB/c mice, *Virology* **170**:450.

Wong, P. M. C., Chung, S.-W., Dunbar, C. E., Bodine, D. M., Ruscetti, S., and Nienhuis, A. W., 1989, Retrovirus-mediated transfer and expression of the interleukin-3 gene in mouse hematopoietic cells results in a myeloproliferative disorder, *Mol. Cell. Biol.* **9**:798.

Woodland, D., Happ, M. P., Bill, J., and Palmer, E., 1990, Requirement for cotolerogenic gene products in the clonal deletion of I-E reactive T cells, *Science* **247**:964.

Woodland, D. L., Happ, M. P., Gollub, K. J., and Palmer, E., 1991, An endogenous retrovirus mediating deletion of αβ T cells?, *Nature* **349**:529.

Yang, J.-N., and Dudley, J. P., 1991, The endogenous Mtv-8 provirus resides within the V_K locus, *J. Virol.* **65**:3911.

Yang, W. K., Kiggans, J. O., Yang, D. M., Ou, C. Y., Tennant, R. W., Brown, A., and Bassin, R. H., 1980, Synthesis and circulation of N- and B-tropic retroviral DNA in *Fv-1* permissive and restrictive mouse cells, *Proc. Natl. Acad. Sci. USA* **77**:2994.

Yelle, J., and Berthiaume, L., 1982, Expression of virus-like particles in normal and transformed gerbil cells, *Arch. Virol.* **74**:77.

Yetter, R. A., Hartley, J. W., and Morse III, H. C., 1983, *H-2*-linked regulation of xenotropic murine leukemia virus expression, *Proc. Natl. Acad. Sci. USA* **80**:505.

Yetter, R. A., Langdon, W. Y., and Morse III, H. C., 1985, Characterization of ecotropic murine leukemia viruses in SJL/J mice, *Virology* **141**:319.

Ymer, S., Tucker, W. R. J., Sanderson, C. J., Hapel, A. J., Campbell, H. D., and Young, I.

G., 1985, Constitutive synthesis of interleukin-3 by leukemia cell line WEHI-3B is due to retroviral insertion near the gene, *Nature* **317**:255.

Yotsuyanagi, Y., and Szöllösi, D., 1984, Virus-like particles and related expression in mammalian oocytes and preimplantation stage embryos, in: *Ultrastructure of Reproduction* (J. Van Blerkom and P. M. Motta, eds.), pp. 218–234, Martinus Nijhoff, Boston.

Young, H. A., Gonda, M. A., De Feo, D., Ellis, R. W., Nagashima, K., and Scolnick, E. M., 1980, Heteroduplex analysis of cloned rat endogenous replication-defective (30S) retrovirus and Harvey murine sarcoma virus, *Virology* **107**:89.

Yuan, C. C., Kan, N., Dunn, K. J., Papas, T. S., and Blair, D. G., 1989, Properties of a murine retroviral recombinant of avian acute leukemia virus E26: A murine fibroblast assay for v-*ets* function, *J. Virol.* **63**:205.

Index